Power
Electronic
Circuits

Power Electronic Circuits

Issa Batarseh

University of Central Florida

WILEY

John Wiley & Sons, Inc.

EXECUTIVE EDITOR   Bill Zobrist

MARKETING MANAGER   Katherine Hepburn

SENIOR PRODUCTION EDITOR   Valerie A. Vargas

SENIOR DESIGNER   Dawn Stanley

PRODUCTION MANAGEMENT SERVICES   Publication Services

This book was set in 10/12 Times Roman by Publication Services and printed and bound by Hamilton Printing. The cover was printed by Phoenix Color.

This book is printed on acid-free paper. ∞

Library of Congress Cataloging in Publication Data:
Author's name entry.
Brief title entry: Subtitle if any xxxxxxxxxxxxxxxxxxxxxxxxxxxxxxxxxxxxxxxxxxxxxxxxxxxxxxxxxxxx

(Series Statement)
Notes xxxxxxxxxxxxxxxxxxxxxxxxxxxxxxxxxxxxxxxx
L.C. Subject Heading   Joint Author Entry
Title and Series added entries

L.C. Call no.   Dewey Classification No.   L.C. Card No.
ISBN 0-471-12662-4
ISBN 0-471-45228-9 (WIE)

Printed in the United States of America

10 9 8 7 6 5 4 3 2 1

Dedicated to my late advisor

Professor Chu Quon Lee

University of Illinois at Chicago

Preface

In recent years the field of power electronics has witnessed unprecedented research and teaching growth worldwide, emerging as a specialization in electrical engineering. This growth is due to expanding market demand for power electronic circuits for the energy conversion process. The need for power electronics engineers equipped with knowledge of new energy conversion technologies has never been greater.

Power Electronic Circuits is intended as a textbook to teach the subject of modern power electronics to senior undergraduate and first-year graduate electrical engineering students. Because of the breadth of the field of power electronics, teaching this subject to undergraduate students is a challenge. This textbook is designed to introduce the basic concepts of power electronics to students and professionals interested in updating their knowledge of the subject. The objective of this textbook is to provide students with the ability to analyze and design power electronic circuits used in various industrial applications.

The prerequisites for this text are a first course in circuit analysis techniques and a basic background in electronic circuits. Chapter 3 gives an overview of diode switching circuits and basic analysis techniques that students will find useful in the remaining chapters.

Material Presentation

Since the text is intended to be used in a three-credit-hour course in power electronics, topics such as power semiconductor devices, machine drives, and utility applications are not included. Because of limited lecture times, one course at the undergraduate level cannot adequately cover such topics and still present all power electronic circuits used in energy conversion. This text contains sufficient material for a single-semester introductory power electronics course, while giving the instructor flexibility in topic treatment and course design.

The text is written in such a way as to equip students with the necessary background material in such topics as devices, switching circuit analysis techniques, converter types, and methods of conversion in the first three chapters. The presentation of the material is new and has been recommended by many power electronics faculty. The discussion begins by introducing high-frequency, nonisolated dc-to-dc converters in Chapter 4, followed by isolated dc-to-dc converters in Chapter 5. Resonant soft-switching converters are treated early on in Chapter 6. The traditional diode and SCR converters and dc-ac inverters are presented in the second part of the text, in Chapters 7, 8, and 9, respectively.

Examples, Exercises, and Problems

Unlike many existing texts, this text provides students with a large number of examples, exercises, and problems, with detailed discussion of resonant and softswitching dc-to-dc converters.

Examples are used to help students understand the material presented in the chapter. To drill students in applying the basic concepts and equations, and to help them understand basic circuit operations, several exercises are given within each chapter. The text has more than 250 problems at different levels of complexity and difficulty. These problems are intended not only to strengthen students' understanding of the materials presented, but also to introduce many new concepts and circuits. To help meet recent Accreditation Board for Engineering and Technology (ABET) requirements for design in the engineering curriculum, special emphasis is made on providing students with opportunities to apply design techniques. Such problems are designated with the letter "D" next to the problem number, such as D5.32. Students should be aware that such problems are open-ended without unique solutions.

A bibliography is included at the end of the text and a list of textbooks is given separately.

About the Text

Like the majority of textbooks, this book was developed from class notes the author prepared over the last eight years while teaching power electronics at the University of Central Florida. The author started teaching power electronics in 1991, when only a limited number of power electronic textbooks were available. Since then, a handful of additional textbooks have been published with very similar material coverage. Unlike many existing texts, *Power Electronic Circuits* targets mainly senior undergraduate students majoring in electrical engineering.

Web-Based Course Material

Ancillaries to this text are available on a dedicated Web site, www.batarseh.org, established for both faculty and students to provide them with access to

- A complete set of lecture notes
- Sample quizzes
- PSPICE- and Mathcad-based simulation examples
- A complete solutions manual
- Transparency masters
- Up-to-date text corrections and the opportunity to submit new corrections

Even though the author's Web site is very useful in providing students and faculty with teaching material, at the time of this writing it doesn't offer an interactive platform. The author believes that due to the nonlinear interdisciplinary nature of the field of power electronics, computer simulations and computer tools must play an important role in delivering effective power electronics education. This is why any useful Web-based education in the area of power electronics must include a platform that allows students to do design and simulation interactively online. To the author's knowledge, this has not been effectively achieved yet. However, one of the most useful interactive Web-based facilities in power electronics is the Interactive Power Electronics Seminar (iPES) developed by the Swiss Federal Institute of Technology, Zurich (ETH Zurich), http://www.ipes.ethz.ch/. The Java applets are designed to allow a degree of interactivity and animation to aid in learning the basics of power electronics typically taught in an introductory course. The Web site is designed as a list of seminar topics and made "available on the Internet for private use and non-commercial use in classroom only." Also, the site provides course translations in Japanese, Korean, Chinese, and Spanish!

Acknowledgments

Over the last five years, many of my power electronics students have had the opportunity to read my lecture notes and solve many problems. Their feedback and comments were instrumental in presenting, to the best of our knowledge, error-free examples, exercises, and problems. I am grateful for reviewers' constructive suggestions for the students' sake. I am highly indebted to Khalid Rustom, Chris Iannelo, and Osama Abdel-Ruhman for spending many hours reviewing the examples, exercises, and problem solutions. Without their tireless work, this text would suffer from numerous errors and omissions! I am very fortunate to have such dedicated people work with me. I was also fortunate to have Dr. Kazi Khairul Islam visit our power electronics laboratory at UCF. He helped greatly in providing insightful suggestions on the inverters chapter, as well as in developing the appendixes. His suggestions and comments are greatly appreciated.

Special thanks go to the young, bright students from Princess Sumaya University for Technology in Amman, Jordan, who helped in typing and solving text problems during their stay at the University of Central Florida (UCF). Their names are Osama Abdul Rahman, Reem Khair, Ruba Amarin, Amjad Awwad, Bashar Nuber, Abeer Sharawi, Rami Beshawi, and Mohhamad Abu-Sumra. Specifically, I wish to acknowledge the help of two of my students, Khalid Rustom and Abdelhalim Alsharqawi, in solving problems and providing helpful insights throughout their undergraduate and graduate study at UCF. I would also like to thank my graduate students Guangyong Zhu, Huai Wei, Joy Mazumdar, Songquan Deng, and Jia Luo for their help and suggestions. Finally, thanks are due to Shadi Harb, Maen Jauhary, Samanthi Nadeeka, and William Said for their help in typing the solutions manual and developing the text Web site. Indeed, it was a pleasure working with all my students who made writing this text a gratifying and enjoyable experience.

Also, I would like to express my sincere thanks to Bill Zobrist, John Wiley & Sons Executive Editor, for his patience and support during the writing of the text, and special thanks to Jan Fisher, Project Manager at Publication Services, for her understanding and for keeping me on schedule.

Finally, I will never forget my late advisor, Professor C. Q. Lee, for introducing me to the field of power electronics and for being my advisor, teacher, mentor, and friend. On behalf of all of his students, I am dedicating this text to him. Of course, without my parents' love, patience, and support, this text would have never been written.

Issa Batarseh
Orlando, Florida

Condensed Table of Contents

Contents

Chapter 1

Introduction

INTRODUCTION

This chapter is intended to give the reader an overview of the field of power electronics and its applications. Basic block diagrams will be provided for a power electronics system and its major functions. Different types of power electronic circuits used to achieve power conversion will be presented.

1.1   WHAT IS POWER ELECTRONICS?

To date, there is no widely accepted definition that clearly and specifically delimits the field of power electronics. In fact, many experts in the academic and industrial communities feel that the name itself does not do justice to the field, which is applications oriented and multidisciplinary in nature, and which also encompasses many subareas in electrical engineering.[1] Because of the multidisciplinary nature of the field of power electronics, experts must have a commanding knowledge of several electrical engineering subjects, such as electronic devices, electronic circuits, signal processing, magnetism, electrical machines, control, and power. In a very broad sense, power electronic circuits perform the task of processing one form of energy supplied by a source into a different form required at the load side. Hence, power electronics can be closely identified with the following subdiscipline areas of electrical engineering:

---

[1]Many schools today offer power electronics under either the "power" or the "electronics" area, while a limited number of schools present it separately.

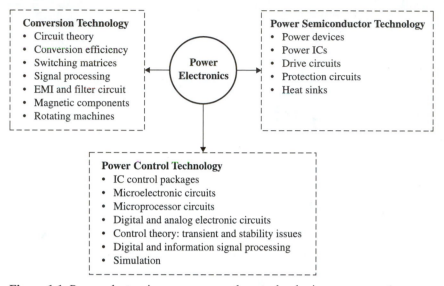

Figure 1.1 Power electronics encompasses three technologies: *power semiconductor, power conversion,* and *power control.*

electronics, power, and control. Here, *electronics* deals with the semiconductor devices and circuits used in signal processing to implement the control functions, *power* deals with both static and rotating equipment that uses electric power, and *control* deals with the steady-state stability of the closed-loop system during the power conversion process. Hence, the subject of power electronics deals specifically with the application of electronic semiconductor devices and circuits in the conversion and control of electric power. In summary, power electronics is a technology that brings together three fundamental technologies: power semiconductor technology, power conversion technology, and power control technology, as illustrated in Fig. 1.1.

A final observation is that in power electronic circuits there exist two types of switching devices: one type in the power stage that handles high power up to hundreds of gigawatts (which represents the muscle of the system) and another type in the feedback control circuit that handles low power up to hundreds of milliwatts, representing the brain or intelligence of the system. Hence, today's power electronic circuits are essentially digital electronic circuits whose switching elements manipulate power from milliwatts to gigawatts. As a result, one may conclude that the task of power electronics is to convert and control power using low-power switching devices that process power that is at much higher levels (a hundred times as great, or even greater). For example, a six-pulse SCR inverter with a power rating of a few kilowatts can process and control megawatts of power.

Recent Growth in Power Electronics

The field of power electronics has recently experienced unprecedented growth in terms of research and educational activities. Its applications have been steadily and rapidly expanded to cover many sectors of our society. This growth is due to several factors; paramount among them is the technological advancement by the semiconductor device industry, which has led to the introduction of very fast high-power capabilities and highly integrated power semiconductor devices. Other factors include (1) the revolutionary advances made in the microelectronics field that have led to the development of very efficient integrated circuits (ICs) used for the generation of control signals for processing and control purposes, (2) the ever-increasing demand for smaller-size and lighter-weight power electronic systems, and (3) the expanding market demand for new power electronic applications that require variable-speed motor drives, regulated power supplies, robotics, and uninterruptible power supplies. This increasing reliance on power electronic systems has made it mandatory that all such systems have radiated and conducted electromagnetic interference (EMI) limited within regulated ranges. The industry's interest in developing power systems with low harmonic content and with an improved power factor will continue to place the field of power electronics at the top of the research priority list.

1.2   THE HISTORY OF POWER ELECTRONICS

Before a review of the history of power electronics in the past century, it might be useful to cover the history of the development of ac and dc electricity in the last two decades of the nineteenth century. The inventions of the 1880s resulted in the present worldwide use of the ac electric power system, providing the energy form that must be processed for any power electronics application.

The History of dc and ac Electricity in the Late Nineteenth Century

It was decided in the middle of the nineteenth century that electrical energy was the most practical and economic form of energy for human use. Electricity was recognized as an excellent form of energy in terms of generation, transmission, and distribution. However, not long afterward, a heated debate began among scientists on whether the future of transmitting and distributing electricity to industries and homes would be based on alternating current (ac) or direct current (dc). George Westinghouse and Nikola Tesla (1856–1943) represented the ac camp, and Thomas Edison (1847–1931) represented the dc camp. After more than 15 years of intellectual debate, supported by new inventions and developmental and experimental studies, the ac advocates won; consequently, the entire world today uses an ac-based power distribution system.[2]

Thomas Edison was a self-educated inventor who was awarded 1033 patents over a 50-year period. He is best known for the invention of the phonograph and the incandescent lamp, which was invented in 1879 after many years of repeated experiments. In 1878, he formulated the concept of a centrally located power station from which power can be distributed to surrounding areas. On September 4, 1882, using dc generators (at that time called dynamos) driven by steam engines, he opened Pearl Street Station in New York City to supply electricity to 59 customers in a one-square-mile area. It was the first dc-based power station in the world, with a total power load of only 30 kW. It was the beginning of the electric utility industry, which grew at a remarkable rate. In 1884, Frank Sprague produced a practical dc motor for Edison's systems. This invention, coupled with the development of three-wire 220 VDC power, enabled Edison to distribute dc electrical power to larger areas and supply heavier loads and consequently more customers. Edison thereby prompted the adoption of

---

[2]Tesla and Edison worked together for a short time, and soon developed hatred for one another, resulting in Tesla opening his own business, believing in ac transmission systems. In 1912, both were nominated for the Nobel Prize in physics. Because of the feud between them, Tesla declared that he had nothing to do with Edison, and the prize was given to a third party!

dc-based power distribution systems. As transmission distances and load demands increased, Edison's dc systems ran into trouble. The dc distribution lines suffered very high power losses because of the high voltage and current that existed simultaneously. This severely limited the transmission distance and resulted in highly inefficient systems. In order to sustain the power level, Edison had to build a dc power station every 20 km! This was costly and very impractical. However, he didn't give up on the dc transmission idea and insisted that these problems could be overcome.

George Westinghouse and Nikola Tesla didn't hesitate to develop ac-based power distribution systems, despite Edison's plans to continue to construct dc transmission systems in New York. In 1885, Westinghouse took a major step in developing ac systems when he bought the American patents of L. Gaulard and J. D. Gibbs of Paris for ac systems. Westinghouse challenged the dc transmission system and went ahead with developing an ac system.

A major step in supporting ac systems occured in 1885, when William Stanley, an early associate of George Westinghouse, developed a commercially practical transformer, allowing the possible distribution of ac-based electricity. Using transformers, it was possible to transmit high-level voltages with a very low-level current, resulting in a very low voltage drop (low power dissipation) in the transmission line. In the winter of 1886, Stanley installed the first experimental ac distributed system in Great Barrington, Massachusetts, supplying power to 150 lamps in the covered area. In 1889, the first single-phase distributed power system was operational in the United States between Oregon City and Portland, covering a 21 km distance with 4 kVA of power.

The second major event that boosted the potential of ac systems took place on May 16, 1888, when Tesla presented a paper at the annual meeting of the American Institute of Electrical Engineers, discussing two-phase induction and synchronous motors. Basically, he showed that it is more practical and more efficient to use polyphase systems to distribute power. The first three-phase ac transmission power system was installed in Germany in 1891; it was rated at 12 kV and transmitted over a distance of 179 km. Two years later (1893), the first three-phase power transmission system in the United States was installed in California, rated at 2.3 kV and covering a distance of 12 km. Also in 1893, a two-phase distributed system was demonstrated at the Colombian Exposition in Chicago. The apparent advantages of ac, especially the three-phase systems, over the dc system led to the gradual replacement of dc by ac systems. Today, the transmission of electricity is done almost entirely by means of ac. However, dc transmission of electric power is used in some locations in Europe and is rarely used in the United States. Since the late nineteenth century, economic studies have shown that ac transmission is much more cost-effective, resulting in its worldwide acceptance.

The History of dc and ac Electricity in the Late Twentieth Century

Over the last 25 years, the technological advancement by the semiconductor device industry, the revolutionary advances made in the microelectronics field, and the ever-increasing demand for smaller and lighter-weight power systems for space, industrial, and residential applications has led to renewed interest in using dc transmission systems. Many experts believe that because of technological advances, it is now possible to develop dc transmission electric power systems economically and efficiently. Today's systems for conversion from ac to dc and back to ac can be produced using very fast, high-power, and highly integrated semiconductor devices. What we can achieve using today's technology was unimaginable only 10 years ago. This is why many power electronics researchers believe that the old debate between dc and ac camps is coming back under a new set of technological rules.

In the mid-1890s, ac was declared a winner over dc, and in the mid-1990s, the dc promoters, mostly power electronics experts, had another shot at the ac camp! History repeats itself! The twenty-first century might very well be friendlier to dc transmission system advocates! Who will win the next century is still to be seen!

The History of Modern Power Electronics

Many agree that the history of power electronics began in 1900, when the glass-bulb mercury-arc rectifiers were introduced, signaling the beginning of the age of vacuum tube electronics, also called glass tube–based industrial electronics. This period continued until 1947, when the germanium transistor was invented at Bell Telephone Laboratory by Bardeen, Bratain, and Shockley, signaling the end of the age of vacuum tubes and the beginning of the age of transistor electronics. During the 1930s and 1940s several new power electronic circuits (then known as industrial electronics) were introduced, including the metal-tank rectifier, the grid-controlled vacuum-tube rectifier, the thyratron motor, and gas/vapor tube switching devices such as hot-cathode thyratrons, ignatrons, and phanotrons. In the 1940s and early 1950s, solid-state magnetic amplifiers, using saturable reactors, were introduced.

The modern era of power electronics began in 1958, when the General Electric Company introduced a commercial thyristor, two years after it was invented by Bell Telephone Laboratory. Soon all industrial applications that were based on mercury-arc rectifiers and power magnetic amplifiers were replaced by silicon-controlled rectifiers (SCRs). In less than 20 years after commercial SCRs were introduced, significant improvements in semiconductor fabrication technology and physical operation were made, and many different types of power semiconductor devices appeared. The growth in power electronics was made possible with the microelectronic revolution of the 1970s and 1980s, in which the low-power IC control chips provided the brain and the intelligence to control the high-power semiconductor devices. Moreover, the introduction of microprocessors made it possible to apply modern control theory to power electronics. In the last 20 years, the growth in power electronics applications has been remarkable because of the introduction of very fast and high-power switching devices, coupled with the utilization of state-of-the-art control algorithms. Today power electronics is a mature technology. The future direction of the era of power electronics is hard to predict, but it is certain that as long as humans seek to improve the quality of life, produce a cleaner environment, and implement energy-saving measures, the demand for clean energy will continue to grow. This implies that power electronics must be used to address the tremendous changes in the way we generate, transmit, and distribute electricity as we cross the bridge into the new century. For a more detailed discussion of the modern history of power electronics, see the paper by D. Wyke.

1.3   THE NEED FOR POWER CONVERSION

With the invention of a practical transformer by Stanley in 1885 and of polyphase ac systems by Tesla in 1891, the advantages of low-frequency ac over dc were compelling to power engineers. The basis of utility power system generation, transmission, and distribution since the beginning of this century has been ac at a fixed frequency of either 50 or 60 Hz. The most outstanding advantage of ac over dc is the maintenance of high voltage over long transmission lines and the simplicity of designing distribution networks. However, the nature of the electricity being distributed is totally different from the nature of the energy required by the electrical load.

At the consumer end many applications may need dc or ac power at line, higher, lower, or variable frequencies. Therefore, it is necessary to convert the available ac systems to dc with precise control. Furthermore, in some cases the generated power is from dc sources such as batteries, fuel cells, or photovoltaic, and in other cases the available power is generated as variable-frequency ac from sources such as wind or gas turbines. The need for this power conversion, ac-to-dc, became more acute with the invention of vacuum tubes, transistors, ICs, and computers. Moreover, modern electric conversion goes beyond ac-to-dc conversion, as we shall shortly discuss.

In the late 1880s, power conversion from ac to dc was done by using ac motors along with dc generators in series (motor-generator set). The motor-generator arrangement was used in dc and with 50/60 Hz motors and generators. The difficulties of using the electromechanical conversion system include large weight and size, noisy operation, servicing and maintenance problems, short lifetime, low efficiency, limited range of conversion, and slow recovery time. To avoid the problems of electromechanical conversion systems, industrial engineers turned to linear electronics in the late 1960s, where power semiconductor devices are operated in their linear (active) region. To obtain electrical isolation, input line-frequency transformers were used, resulting in bulky, heavy power converter systems. Furthermore, with power devices operating in the linear region, the overall efficiency of the system is low. Compared with electromechanical systems and linear electronic systems, power electronics has many advantages, including (1) high energy conversion efficiency, (2) highly integrated power electronic systems, (3) reduced EMI and electronic pollution, (4) higher reliability, (5) use of environmentally clean voltage sources such as photovoltaic and fuel cells to generate electric power, (6) the integration of electrical and mechanical systems, and (7) maximum adaptability and controllability.

In short, all forms of electrical power conversion will be needed as long as consumers live in homes and use light, heat, electronic devices, and equipment and interface with industry.

1.4   POWER ELECTRONIC SYSTEMS

Most power electronic systems consist of two major modules: (1) The power stage (forward circuit) and (2) the control stage (feedback circuit). The power stage handles the power transfer from the input to the output, and the feedback circuit controls the amount of power transferred to the output.

A generalized block diagram of a power electronic system with n sources and m loads is given in Fig. 1.2 where,

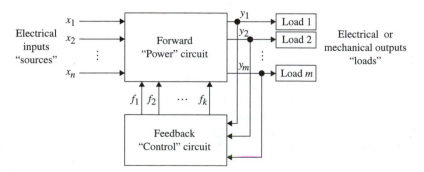

Figure 1.2 Simplified block diagram for a power electronic system.

$x_1, x_2, \ldots, x_n$ are input signals (voltage, current, or angular frequency).

$y_1, y_2, \ldots, y_m$ are output signals (voltages, currents, or angular frequency).

$p_{\text{in}}(t)$ is the total instantaneous input power in watts.

$p_{\text{out}}(t)$ is the total instantaneous output power in watts.

$f_1, f_2, \ldots, f_k$ are feedback signals: voltages or currents in an electrical system, or angular speed or angular position in a mechanical system.

Efficiency, η, is defined as follows:

$$\eta = \frac{p_{\text{out}}}{p_{\text{in}}} \times 100\%$$

Figure 1.3 shows a more detailed description of a block diagram for a power electronic input system with electrical and mechanical output loads. The main function of a power electronic circuit is to process energy from a given source to a required load. In many applications, the conversion process concludes with mechanical motion.

1.4.1   Classification of Power Converter Circuits

The function of the *power converter stage* is to perform the actual power conversion and processing of the energy from the input to the output by incorporating a matrix of power switching devices. The control of the output power is carried out through control signals applied to these switching devices. Broadly speaking, power conversion refers to the power electronic circuit that changes one of the following: voltage form (ac or dc), voltage level (magnitude), voltage frequency (line or otherwise), voltage waveshape (sinusoidal or nonsinusoidal, such as square, triangle, or sawtooth), or voltage phase (single- or three-phase).

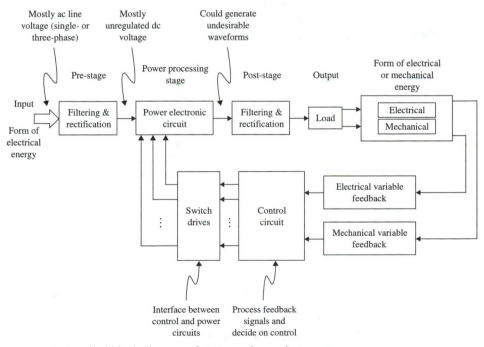

Figure 1.3 Detailed block diagram of a power electronic system.

Broadly speaking, there are four *conversion circuits* that are used in the majority of today's power electronic circuits:

a. ac-to-ac

b. ac-to-dc

c. dc-to-ac

d. dc-to-dc

In terms of the functional description, modern power electronic systems perform one or more of the following conversion functions:

1. Rectification (ac-to-dc)

2. Inversion (dc-to-ac)

3. Cycloconversion (ac-to-ac, different frequencies) or ac controllers (ac-to-ac, same frequency)

4. Conversion (dc-to-dc)

Rectification (ac-to-dc)

The term *rectification* refers to the power circuit whose function is to alter the ac characteristic of the line electric power to produce a "rectified" ac power at the load site that contains the dc value. Figure 1.4(*a*) and (*b*) shows the block diagram representation of an ac-to-dc converter and its typical input and output waveforms, respectively. To smooth out the output voltage by removing the unwanted ac component, an additional "filtering" circuit is added at the output side. Depending on the switch implementations, these converters are further divided into two types, *diode converter circuits* (*uncontrolled*) and *thyristor converters* (*phase-controlled*); each type can have either single-phase or three-phase input voltages. Both types are extensively used in various off-line applications. Rectification circuits will be discussed in Chapters 7 and 8. The topologies that perform the rectification function include half-wave, full-wave (full-bridge), semi-bridge, and transformer-coupled center-tapped. From the beginning of the industrial electronics era, the ac-to-dc line commutation converter class, utilizing thy-

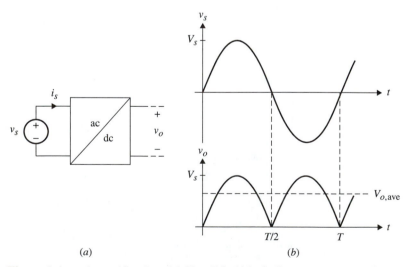

(*a*) (*b*)

Figure 1.4 ac-dc rectification. (*a*) Simplified block diagram representation. (*b*) Example of ac-dc conversion.

ristors, has been the most popular among power electronics converters because of their simplicity in design, efficiency, and higher current and voltage ratings.

Inversion (dc-to-ac)

The term *inversion* is used in power electronic circuits for the function that alters the dc source (e.g., a battery) with no ac components into an "inverted" ac power at the load that has no dc components, as shown in Fig. 1.5(*a*). Example input and output waveforms are shown in Fig. 1.5(*b*). The ac output can have an adjustable magnitude and frequency. Additional filtering is normally used to extract the fundamental component of $v_o(t)$ at $\omega = 2\pi/T$. Generally speaking, dc-to-ac inverters are classified as voltage-fed and current-fed inverter types as discussed in Chapter 9. In the last few years, resonant-link technology that has been successful in the design of PWM power supplies has been applied to the design of dc-to-ac inverters, producing ac outputs at variable voltages and variable frequencies. A resonant-link dc-to-ac inverter is a two-stage conversion circuit that takes the dc voltage and changes it to a high ac resonant voltage, which in turn is changed to a variable low-frequency output, as shown in Fig. 1.5(*c*). Generally speaking, since cascaded systems involve a two-stage conversion, power processing passes through more than one switching device and therefore increases conduction losses. However, in some cases, by using intermediate stages, it is possible to insert an electrical isolation transformer; in other cases, cascaded stages produce high-frequency resonant waveforms that could result in the soft switching of the power devices, which in turn could reduce the overall switching losses of the cascaded system. The ac-to-dc rectification and dc-to-ac inversion represent the broadest functions of power electronic circuits.

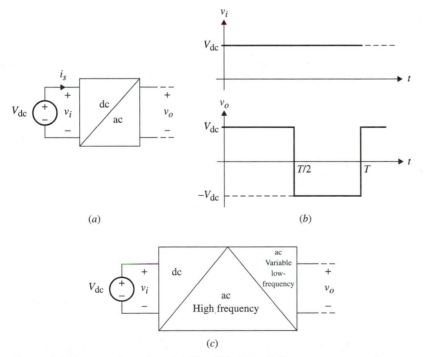

Figure 1.5 dc-to-ac inversion. (*a*) Simplified block diagram representation. (*b*) Example of dc-ac inversion. (*c*) Block diagram representation with high frequency inversion.

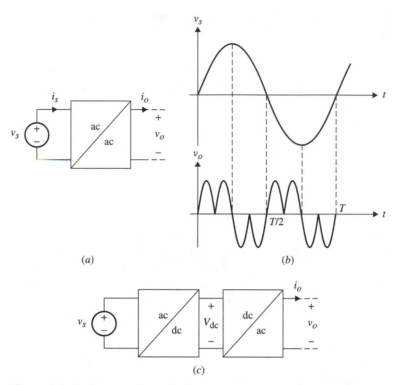

Figure 1.6 Cycloconversion. (*a*) One-stage ac-to-ac cycloconversion. (*b*) Example of ac-to-ac conversion waveforms. (*c*) Two-stage ac-to-ac cycloconversion.

Cycloconversion or Voltage Controllers (ac-to-ac)

The term *cycloconversion* is used for power electronic circuits that convert the ac input power at one frequency to an ac output power at a different frequency using one-stage conversion, as shown in Fig. 1.6. However, two-stage conversion is also possible, as shown in Fig. 1.6(*c*), i.e., ac-to-dc and then dc-to-ac, resulting in what is called a dc-link converter. An ac controller is a power electronic circuit that alters the rms ac input at the same frequency.

Conversion (dc-to-dc)

Dc-to-dc converters are used in power electronic circuits to convert an unregulated input dc voltage to a regulated or variable dc output voltage, as shown by the block diagram and its waveform of Fig. 1.7(*a*) and (*b*), respectively. These circuits dominate the power supply industry, such as in the switch-mode power supplies (SMPS). In high-power dc traction drive applications, dc-to-dc converters are known as *choppers*. High-frequency pulse-width-modulation (PWM) converters with and without output electrical isolation will be discussed in Chapters 4 and 5, respectively. The high-frequency resonant-type dc-to-dc converters, which use two-stage conversion, will be discussed in Chapter 6; here transformers can be used to provide electrical isolation and step-up/step-down features. Figure 1.7(*c*) shows such an implementation.

To achieve this, we need to design converters that operate in the hundreds of kilohertz and up to a few megahertz. In resonant converters, the switching devices are used in such a way that the turn-on or turn-off losses can be reduced or eliminated, depending on the converter operation. Such converters are known as "soft-switching" converters. These

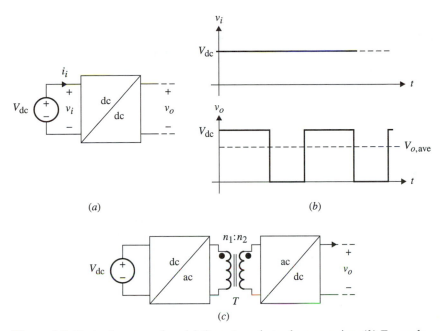

(a)   (b)

(c)

Figure 1.7 Dc-to-dc conversion. (a) One-stage dc-to-dc conversion. (b) Example of waveforms. (c) Two-stage dc-to-dc conversion.

converters are used in the design of high-power density dc power supplies for laptop computers, adapters, notebook computers, and aerospace and communication instrumentation.

Figure 1.8 shows a simplified block diagram of a power electronic conversion system representing four possible conversion functions. The literature is rich with power electronic circuit topologies for various applications. When selecting a topology for a given application, one has to consider several factors, including the basic conversion function required; the available switching devices and their characteristics, driving circuits, control and protection, and maximum switching losses; and finally, the cost, size, and weight.

Depending on the topologies used and the types of loads, power electronic circuits are capable of transferring power in only one direction (i.e., unidirectional power flow from the source to the load) in both directions (i.e., bidirectional power flow from the load to the source). The latter is said to operate in the regenerative mode. Since the polarities of

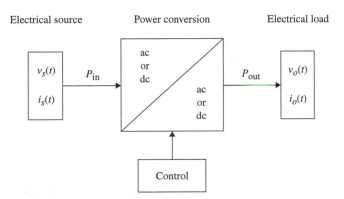

Figure 1.8 Simplified block diagram representation of the power electronic conversion function.

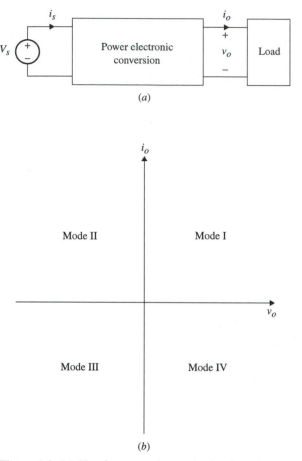

(a)

(b)

Figure 1.9 (a) Simple power electronic circuit with single voltage source and single load. (b) Possible converter modes of operation

the load current, i_o, and voltage, v_o, shown in Fig. 1.9(a) can be either positive or negative, there exist four modes of operation, as shown in Fig. 1.9(b).

In the first quadrant (mode I) the output voltage and current are always positive and the power flows unidirectionally to the load. Examples of such converters are the switch-mode dc-dc buck and boost converters. Similarly, in the third quadrant (mode III) the converter allows unidirectional power flow with both the output voltage and output current negative, as in the dc-dc buck-boost converter. The power electronic circuit with a dc motor can operate in modes I and II since the motor is capable of supporting the forward and reverse directions. Modes II and IV indicate that power is being transferred from the load to the source, as in the case of a dc generator. Converters that can operate in modes I and II or III and IV can support bidirectional power flow. Examples of converters operating in the different quadrants will be discussed in the following chapters.

1.4.2 Power Semiconductor Devices

Because of their high conversion efficiency, semiconductor switching devices are considered the heart of power electronic circuits, as we will see throughout the textbook. The control of power flow from the input to the output is done through a power-processing switching network made of switching devices and energy storage elements. Detailed discussion of the switching circuits will be given in the next chapter.

The need for new power electronic circuits to address the growing market demand for new applications has resulted in intensive research activities in the semiconductor industry to make available semiconductor devices with a wide range of power-handling capabilities and switching speeds.

The direction of power electronics has been, and will continue to be for many years to come, very much tied to the advancements made in power semiconductor technology. Due to the development of high-power, fast-switching, and high-efficiency thyristor-based unipolar and bipolar devices in the late 1950s and 1960s, the field of power electronics has emerged as a separate subarea in electrical engineering. Today, semiconductor switching devices have achieved unprecedented power-handling capabilities and switching speeds.

In the next chapter, we will study the i-v characteristics of five basic power devices widely used as switching elements in power electronic circuits. These devices include (1) power diodes, (2) bipolar junction transistors (BJTs), (3) metal oxide semiconductor field-effect transistors (MOSFETs), (4) thyristors or silicon-controlled rectifiers (SCRs), and (5) insulated gate bipolar transistors (IGBTs). Other available devices, such as triacs, diacs, gate-turn-off (GTO) thyristors, static induction transistors (SITs), static induction thyristors (SITHs), and MOS-controlled thyristors (MCTs) are, in one way or another, from the same families of these five basic devices. Many of these devices are rated in the hundreds of kilowatts! Even higher current and voltage ratings can be achieved by connecting devices in parallel and in series, respectively.

1.4.3   Converter Modeling and Control

In any power electronic circuit, the method used to control and stabilize power flow from the source to the load through the system is extremely important since the output must be regulated, whether it is voltage, current, or frequency. Over the last 30 years, tremendous efforts have gone into developing converter modeling methods that can be understood using well-known linear circuit analysis techniques. Moreover, the control of power electronic circuits has been greatly simplified by using microcomputer and commercial ICs that improve reliability, reduce the weight and size of hardware, and provide flexibility to the designer by changing the software algorithms. Also, the analysis and design of control circuits in power electronics have been made even easier with the availability of several simulation packages, such as PSPICE, MATLAB, and Saber. Converter modeling and analysis techniques are not included in the text since the emphasis is on circuits of the power stage.

1.5   APPLICATIONS OF POWER ELECTRONICS

Power electronics covers a wide range of residential, commercial, and industrial applications, including computers, transportation, aircraft/aerospace, information processing, telecommunication, and power utilities. Broadly speaking, these applications may be classified into three categories:

1. *Electrical applications.* Power electronics can be used to design ac and dc regulated power supplies for various electronic equipment, including consumer electronics, instrumentation devices, computers, aerospace, and uninterruptible power supply (UPS) applications. Power electronics is also used in the design of distributed power systems, electric heating and lighting control, power factor correction, and static var compensation.

2. *Electromechanical applications.* Electromechanical conversion systems are widely used in industrial, residential, and commercial applications. These applications include ac and dc machine tools, robotic drives, pumps, textile and paper mills, peripheral drives, rolling mill drives, and induction heating.

3. *Electrochemical applications.* Electrochemical applications include chemical processing, electroplating, welding, metal refining, production of chemical gases, and fluorescent lamp ballasts.

Table 1.1 gives a list of some power electronic application categories according to their conversion functions. We should mention that the examples given in Table 1.1 are not all-inclusive, but rather are given as an illustration of how wide the application spectrum of the field is. Finally, we note that the energy spectrum of the application of power electronics extends from a few watts, such as for a switching regulator, to a few megawatts, such as in high-voltage dc (HVDC) systems.

Table 1.1   Applications of Power Electronics by Conversion Functions (Partial List)

Conversion function	Applications
Uncontrolled ac-dc converters (diode circuits)	Front-end off-line regulated dc-to-ac power supplies and dc-ac inverters Battery chargers Welding dc motor drives
Phase-controlled converters (thyristor circuits)	Regulated dc power supplies ac and dc variable-speed motor control Battery chargers Flexible ac transmission system (FACTS) Utility interface of photovoltaic systems Regulated ac inverters Solid-state circuit breakers dc motor drives Induction heating Electromechanical processing (electroplating, anodizing, metal refining) HVDC systems Light dimmers Active power line conditioning (APLC) (var compensator, harmonic filters) Induction heating
dc-dc converters	High-frequency regulated dc power supplies using both isolated and nonisolated switch-mode and soft-switching resonant topologies Digital and analog electronics Solar energy conversion High-frequency quasi-resonant converters Electric vehicles and trams dc-fed forklifts Fuel cell conversion dc traction drives Distributed power systems Power factor correction Solid-state relays Capacitor chargers

(continued)

Table 1.1 (*continued*)

Conversion function	Applications
Linear-mode dc-dc converters	Low-power linear dc regulators
	Audio amplifiers
	RF amplifiers
Cycloconverters and ac controllers (ac-ac)	ac motor drives
	Rolling mill drives
	Static Scherbius drives
	Aircraft
	Frequency changers
	Solid-state power line conditioners
	Variable-speed constant-frequency (VSCF) systems
	Fluorescent lighting
	Light dimmers
	Induction heating
dc-to-ac inverters	Aircraft and space power supply systems
	ac variable-speed motor drives (lifts)
	Uninterruptible power supplies (UPS)
	Power factor correction
	Light dimmers
	Electric railroad systems
	Magnetically levitated (maglev) high-speed transportation systems
	Electric vehicles
Static switching	ac and dc circuit breakers
	Circuit protection
	Solid-state relays
Power ICs	Home and office automation
	Automobiles
	Telecommunications
	ac and dc drives
	dc power supplies

1.6 FUTURE TRENDS

It is hard to predict the direction of future research in the field of power electronics, or any field for that matter. However, based on today's research and teaching activities in power electronics, which are driven by market demands, energy conservation, and cost reduction, it is possible to identify some possible short-term future research activities in power electronics:

- Continued technological improvement of high-power and high-frequency semiconductor devices
- Continued development of power electronic converter topologies to attain further size and weight reduction with increased efficiency and performance
- Improvement in the design of driver circuits for switching devices
- Improvement in control techniques, including optimal and adaptive control
- Integration of power and control circuitry on "smart power" ICs and further development of application-specific modules
- Distributed power system (DPS) approach in applications such as VLSI mainframe computers, military VHSIC systems, and telecom switching equipment

- Power factor correction techniques and EMI reduction
- Additional applications of power electronics in flexible ac transmission systems (FACTS)

As power electronics becomes cheaper, it will extend into various new industrial, residential, aerospace, and telecommunications applications. Based on the growth of power electronics in recent years, future growth is projected to be even greater.

As long as we continue to seek improved standards of living, our quest for cheap and environmentally clean energy will continue. Power electronics will be used widely to address energy conservation and conversion efficiency. It has been reported that energy savings of more than 20% could be achieved with the help of power electronics. The role of power electronics will be greater as it becomes cheaper and more devices and systems become available. The challenge for power electronic circuit engineers is to keep developing new and optimal topologies to match market applications, and the challenge for power device engineers is to come up with new devices that can be used in these new topologies!

1.7 ABOUT THE TEXT AND ITS NOMENCLATURE

About the Text

This text is written in a way to benefit undergraduate students in electrical engineering the most. It is designed to be used in a one-semester power electronics course intended to be covered in an undergraduate introductory course in power electronics or first-year graduate power electronics course. Because of the interdisciplinary nature of power electronics, an expanded introduction (Chapters 1 to 3) has been provided to cover a brief orientation to the field, a review of power semiconductor devices, and general concepts that represent the cornerstone of power electronics.

Nomenclature

To avoid confusion and repetition, we establish some definitions, terminology, and notations to be used throughout the text.

1. *Time-dependent variables (pure ac).* All instant time-dependent variables, including current (i), voltage (v), and power (p), will be presented as lowercase letters with lowercase subscripts:

$$i_a(t), v_b(t), \text{ and } p_i(t)$$

   The "(t)" indicates "function of time." For simplicity, the time notation is dropped and the variables are given by i_a, v_b, and p_i.

2. *Average and constant variables.* All average or constant variables will be given in uppercase letters and lowercase subscripts, such as V_s, I_o, $V_{o,\text{rms}}$, and $I_{l,\text{min}}$. To distinguish peak, dc, and rms values and constants, additional subscripts will be added when necessary.

3. The ac voltage source and the dc voltage source will be given as follows:

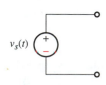

$$v_s(t) = V_s \sin \omega t \qquad \omega = 2 \pi f = 2 \pi / T$$

where

V_s = peak voltage

ω = angular frequency in radians/second

f = frequency in hertz

T = period in seconds

Rectified dc source:

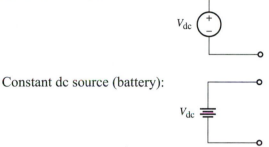

Constant dc source (battery):

4. *Time-dependent variables (ac and dc).* Like the conventional notation in electronics, a lowercase variable and an uppercase subscript indicate that the time variable has both ac and dc components:

$$i_L(t) = I_L + \hat{i}_L(t)$$

where

I_L = dc component

$\hat{i}_L(t)$ = ac component

The "$\frown$" notation indicates that the source of the ac component is perturbation around the dc value I_L. These notations are normally used when small-signal analysis and dynamic modeling of dc-dc converters are discussed.

5. The current and voltage sinusoidal harmonics of a periodical signal will be as follows:

$$i_s(t) = I_{dc} + I_{s,1} \sin \omega t + I_{s,2} \sin 2\omega t + \cdots + I_{s,n} \sin n\omega t$$

where

I_{dc} = average value of the signal

$I_{s,n}$ = peak value of the nth harmonic component of the signal, where $n = 1, 2, \ldots, \infty$

6. We will interchange the integration variables between time t and angular frequency ωt. For example,

$$\frac{1}{T} \int_0^{T/2} V_s \sin \omega t \, dt = \frac{1}{2\pi} \int_0^\pi V_s \sin \omega t \, d\omega t$$

7. *Three-phase circuit representation.* Balanced three-phase voltage sources consist of three equal sinusoidal voltages each shifted by 120° from the other in such a way that their phase sum is zero. Two well-known three-phase configurations will be used in this text:

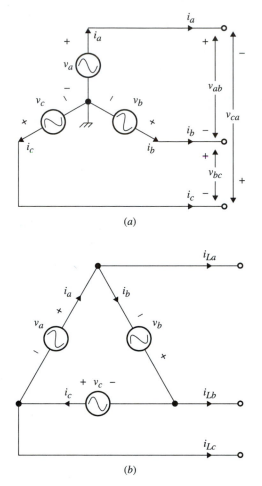

(a)

(b)

Figure 1.10 (a) Y-connected three-phase source. (b) Δ-connected three-phase source.

- Wye (Y) connected three-phase source as shown in Fig. 1.10(a):

Phase voltages: $\begin{cases} v_a = V_s \sin \omega t \\ v_b = V_s \sin(\omega t - 120°) \\ v_c = V_s \sin(\omega t - 240°) \end{cases}$

Line voltages: $\begin{cases} v_{ab} = \sqrt{3}\, V_s \sin(\omega t + 30°) \\ v_{bc} = \sqrt{3}\, V_s \sin(\omega t - 90°) \\ v_{ca} = \sqrt{3}\, V_s \sin(\omega t - 210°) \end{cases}$

Line (phase) currents: i_a, i_b, i_c

- Delta (Δ) connected three-phase source, as shown in Fig. 1.10(b):

Phase (line) voltages: $\begin{cases} v_a = V_s \sin \omega t \\ v_b = V_s \sin(\omega t - 120°) \\ v_c = V_s \sin(\omega t - 240°) \end{cases}$

Phase currents:
$$\begin{cases} i_a = I_s \sin \omega t \\ i_b = I_s \sin(\omega t - 120°) \\ i_c = I_s \sin(\omega t - 240°) \end{cases}$$

Line currents:
$$\begin{cases} i_{La} = \sqrt{3}I_s \sin(\omega t + 30°) \\ i_{Lb} = \sqrt{3}I_s \sin(\omega t - 90°) \\ i_{Lc} = \sqrt{3}I_s \sin(\omega t - 210°) \end{cases}$$

Chapter 2

Review of Switching Concepts and Power Semiconductor Devices

INTRODUCTION

In this chapter, an overview of power semiconductor switching devices will be given. Only devices that are available in the market and are currently used in power electronics applications will be considered. These devices include Unipolar and bipolar devices such as the power diode, Bipolar junction transistor (BJT), Metal Oxide semiconductor field-effect transistors (MOSFET), and insulated gate bipolar transistor (IGBT); and thyristor-based devices such as the silicon-controlled rectifier (SCR), gate turn-off (GTO) thyristor, triac, static induction transistor and thyristor, and MOS-controlled thyristor (MCT). Detailed discussion of the physical structure, fabrication, and physical behavior of these devices and their packaging is beyond the scope of this text. The emphasis here will be on the terminal i-v switching characteristics of the available devices and their current, voltage, and switching limits. Even though most of today's available semiconductor power devices are made of silicon or germanium, other materials, such as gallium arsenide, diamond, and silicon carbide, are currently being tested.

As stated in Chapter 1, one of the main contributions to the growth of the power electronics field has been the unprecedented advancement in semiconductor technology, especially with respect to switching speed and power-handling capabilities. The area of power electronics started with the introduction of the SCR in 1958. Since then, the field has grown in parallel with the growth of the power semiconductor device technology. In fact, the history of power electronics is very much connected to the development of switching devices, and it emerged as a separate discipline when high-power BJTs and MOSFET devices were introduced in the 1960s and 1970s. Since then, the introduction of new devices has been accompanied by dramatic improvements in power rating and switching performance.

In the 1980s, the development of power semiconductor devices took an important turn when new processing technology was developed that allowed the integration of MOS and BJT technologies on the same chip. Thus far, two devices using this new technology have been introduced: the IGBT and MCT. Many IC processing methods and types of equipment have been adopted for the development of power devices. However, unlike microelectronic ICs, which process information, power device ICs process power; hence, their packaging and processing techniques are quite different.

Since the development of thyristors in the late 1950s, power semiconductor device technology has been undergoing dynamic evolution, always following the evolution of microelectronics technology, and in the process, introducing many different kinds of devices with wide ranges of power ratings and frequency. Because of their functional importance, drive complexity, fragility, and cost, a power electronics design engineer must be equipped with a thorough understanding of the device's operation, limitations, and drawbacks, and related reliability and efficiency issues.

Power semiconductor devices represent the heart of modern power electronics, with two major desirable characteristics guiding their development:

1. Switching speed (turn-on and turn-off times)
2. Power-handling capabilities (voltage-blocking and current-carrying capabilities)

Improvements in semiconductor processing technology as well as in manufacturing and packaging techniques have allowed the development of power semiconductors for high voltage and high current ratings and fast turn-on and turn-off characteristics. The availability of different devices with different switching speeds, power-handling capabilities, sizes, costs, and other factors makes it possible to cover many power electronics applications, so that trade-offs must be made when it comes to selecting power devices.

2.1   THE NEED FOR SWITCHING IN POWER ELECTRONIC CIRCUITS

Do we have to use switches to perform electrical power conversion from the source to the load? The answer, of course, is no; there are many circuits that can perform energy conversion without switches, such as linear regulators and power amplifiers. However, the need for semiconductor devices to perform conversion functions is very much related to the converter efficiency. In power electronic circuits, the semiconductor devices are generally operated as switches—either in the *on* state or *off* state. This is unlike the case in power amplifiers and linear regulators, where semiconductor devices operate in the linear mode. As a result, a very large amount of energy is lost within the power circuit before the processed energy reaches the output. The need to use semiconductor switching devices in power electronic circuits is based on their ability to control and manipulate very large amounts of power from the input to the output with relatively very low power dissipation in the switching device, resulting in a very high-efficiency power electronic system.

Efficiency is an important figure of merit and has significant implications on the overall performance of the system. A low-efficiency power system means that large amounts of power are being dissipated in the form of heat, with one or more of the following implications:

1. The cost of energy increases due to increased consumption.

2. Additional design complications might be imposed, especially regarding the design of device heat sinks.

3. Additional components such as heat sinks increase the cost, size, and weight of the system, resulting in low power density.

4. High power dissipation forces the switch to operate at low switching frequencies, resulting in limited bandwidth and slow response, and most important, the size and weight of magnetic components (inductors and transformers) and capacitors remain large. Therefore, it is always desirable to operate switches at very high frequencies. But we will show later that as the switching frequency increases, the average switching power dissipation increases. Hence, a trade-off must be made between reduced size, weight, and cost of components versus reduced switching power dissipation, which means inexpensive low-switching-frequency devices.

5. Component and device reliability is reduced.

For more than 35 years, it has been shown that switching (mechanical or electrical) is the best possible way to achieve high efficiency. However, electronic switches are superior to mechanical switches because of their speed and power-handling capabilities as well as their reliability.

We should note that the advantages of using switches come at a cost. Because of the nature of switch currents and voltages (square waveforms), high-order harmonics are normally generated in the system. To reduce these harmonics, additional input and output filters are usually added to the system. Moreover, depending on the device type and power electronic circuit topology used, driver circuit control and circuit protection can significantly increase the complexity of the system and its cost.

EXAMPLE 2.1

The purpose of this example is to investigate the efficiency of four different power electronic circuits whose function is to take power from a 24 V dc source and deliver a 12 V dc output to a 6 Ω resistive load. In other words, the task of these circuits is to serve as dc *transformers* with a ratio of 2:1. The four circuits are shown in Fig. 2.1(*a*), (*b*), (*c*), and (*d*), representing a voltage divider circuit, zener regulator, transistor linear regulator, and switching circuit, respectively. The objective is to calculate the efficiency of these four power electronic circuits.

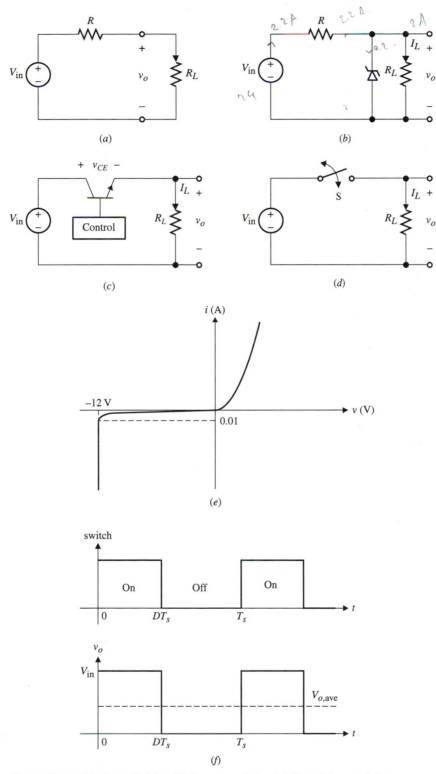

Figure 2.1 (a) Voltage divider. (b) Zener regulator. (c) Transistor regulator.
(d) Switching circuit. (e) Zener diode i-v switching characteristics. (f) Switching
waveforms for circuit (d).

SOLUTION

(a) *Voltage divider dc regulator.* The first circuit is the simplest, forming a voltage divider with $R = R_L = 6\ \Omega$ and $V_o = 12$ V. The efficiency, defined as the ratio of the average load power, P_L, to the average input power, P_{in}, is

$$\eta = \frac{P_L}{P_{in}}\%$$

$$= \frac{R_L}{R_L + R}\% = 50\%$$

In fact, the efficiency is simply $V_o/V_{in}\%$. As the output voltage becomes smaller, the efficiency decreases proportionally.

(b) *Zener dc regulator.* Since the desired output is 12 V, we select a zener diode with zener breakdown $V_Z = 12$ V. Assume the zener diode has the *i-v* characteristic shown in Fig. 2.1(*e*). Since $R_L = 6\ \Omega$, the load current, I_L, is 2 A. If we calculate R for $I_Z = 0.2$ A (10% of the load current), this results in $R = 5.45\ \Omega$. Since the input power is $P_{in} = 2.2$ A $\times 24$ V $= 52.8$ W and the output power is $P_{out} = 24$ W, the efficiency of the circuit is given by

$$\eta = \frac{24\ \text{W}}{52.8\ \text{W}}\%$$

$$= 45.5\%$$

(c) *Transistor dc regulator.* It is clear from Fig. 2.1(*c*) that for $V_o = 12$ V, the collector emitter voltage must be around 12 V. Hence, the control circuit must provide a base current, I_B, to put the transistor in the active mode with $V_{CE} \approx 12$ V. Since the load current is 2 A, the collector current is approximately 2 A (assume small I_B). The total power dissipated in the transistor can be approximated by the following equation:

$$P_{diss} = V_{CE}I_C + V_{BE}I_B$$
$$\approx V_{CE}I_C \approx 12 \times 2 = 24\ \text{W}$$

Therefore, the efficiency of the circuit is 50%.

(d) *Switching dc regulator.* Let us consider the switching circuit of Fig. 2.1(*d*) by assuming the switch is ideal and periodically turns on and off, as shown in Fig. 2.1(*f*). The output voltage waveform is also shown in Fig. 2.1(*f*). Even though the output voltage is not constant or pure dc, its average value is given by

$$V_{o,ave} = \frac{1}{T_s}\int_0^{T_s D} V_{in}\ dt = V_{in}D$$

where D is the duty ratio, which is the ratio of the *on*-time to the switching period, T_s. For $V_{o,ave} = 12$ V, we set $D = 0.5$, i.e., the switch has a duty cycle of 0.5 or 50%. In this case, the average output power is 48 W and the average input power is also 48 W, resulting in 100% efficiency! This is, of course, because we assumed the switch is ideal. However, let us assume that a BJT switch is used in the circuit with $V_{CE,sat} = 1$ V and I_B is small; then the average power loss in the switch is approximately 2 W, resulting in an overall efficiency of 96%. Of course, the switching circuit given in this example is oversimplified; the switch requires additional driving circuitry that is not shown, which also dissipates some power. Still, the example illustrates the high efficiency that can be achieved by a switching power electronic circuit compared to a linear power electronic circuit. Also, the difference between the linear circuit in Fig. 2.1(*b*) and (*c*) and the switched circuit of Fig. 2.1(*d*) is that the power delivered to the load in the latter case is pulsating between 0 and 96 W. If the application calls for constant power delivery with little output voltage ripple, then an *LC* filter must be added to smooth out the output voltage. This class of dc-dc converters will be studied in Chapters 4 and 5.

A final observation regards what is known as load regulation and line regulation. Line regulation is defined as the ratio between the change in the output voltage, ΔV_o, with respect to the change in the input voltage, ΔV_{in}. This is a very important design parameter in power electronics since the dc input voltage is obtained from a rectified line voltage that normally changes by ±20%. Therefore, any off-line power electronic circuit must have a limited or specified range of line regulation. If the input voltage in Fig. 2.1(*a*) and (*b*) is changed by 2 V, (i.e., $\Delta V_{in} = 2$ V) with R_L unchanged, the corresponding change in the output voltage, ΔV_o, is 1 V and 0.55 V, respectively. This is considered very poor line regulation. The circuits of Fig. 2.1(*c*) and (*d*) have much better line and load regulation since the closed-loop control compensates for the line and load variations.

2.2   SWITCHING CHARACTERISTICS

2.2.1   The Ideal Switch

It is always desirable to have power switches perform as near as possible to the ideal case. For a semiconductor device to operate as an ideal switch, it must possess the following features:

1. No limit on the amount of current (known as forward or reverse current) that the device can carry when in the conduction state (*on*-state)

2. No limit on the amount of device voltage (known as forward or reverse blocking voltage) when the device is in the nonconduction state (*off*-state)

3. Zero *on*-state voltage drop when in the conduction state

4. Infinite *off*-state resistance, i.e., zero leakage current when in the nonconduction state

5. No limit on the operating speed of the device when it changes state, i.e., zero rise and fall times

Typical switching waveforms for an ideal switch are shown in Fig. 2.2, where i_{sw} and v_{sw} are the current through and the voltage across the switch, respectively, and DT_s is the *on* time[1]. During the switching and conduction periods the power loss is zero, resulting in 100% efficiency. With no switching delays, an infinite operating frequency can be achieved. In short, an ideal switch has infinite speed, unlimited power-handling capabilities, and 100% efficiency. Semiconductor switching devices are available that can, for all practical purposes, perform as ideal switches for a number of applications.

2.2.2   The Practical Switch

The practical switch has the following switching and conduction characteristics:

1. Limited power-handling capabilities, i.e., limited conduction current when the switch is in the *on* state, and limited blocking voltage when the switch is in the *off* state

2. Limited switching speed, caused by the finite turn-on and turn-off times, which limits the maximum operating frequency of the device

3. Finite *on*-state and *off*-state resistances, i.e., the existence of forward voltage drop in the *on* state, and reverse current flow (leakage) in the *off* state

4. Because of characteristics 2 and 3, the practical switch experiences power losses in the *on* and *off* states (known as conduction loss) and during switching transitions (known as switching loss)

---

[1]Only the *on* time is shown since the emphasis is on the *on* and *off* times. The *off* time, $(1 - D)T_s$, has only conduction loss.

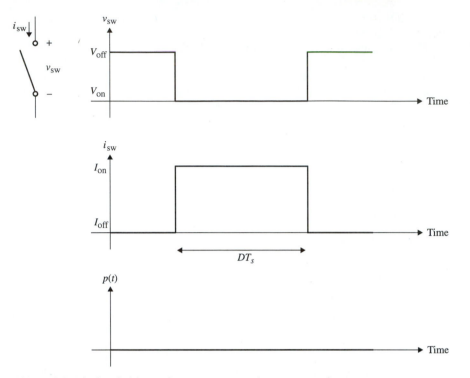

Figure 2.2 Ideal switching voltage, current, and power waveforms.

Typical switching waveforms for a practical switch are shown in Fig. 2.3(*a*). The average switching power and conduction power losses can be evaluated from these waveforms. The exact switching waveforms vary from one device to another, but Fig. 2.3(*a*) is a reasonably good representation. For simplicity, DT_s is shown to include the *off*-time and the *on*-time period of the switch. Moreover, other issues such as temperature dependence, power gain, surge capacity, and overvoltage capacity must be considered for specific devices in specific applications. A useful plot that illustrates how switching takes place from *on* to *off* and vice versa is called a *switching trajectory*, which is simply a plot of i_{sw} versus v_{sw}. Figure 2.3(*b*) shows several switching trajectories for the ideal and practical cases under resistive loads.

The average power dissipation, P_{ave}, over one switching cycle is given by

$$P_{ave} = \frac{1}{T_s}\int_0^{T_s} i_{sw} v_{sw}\, dt$$

$$= P_{ave,swit} + P_{ave,cond}$$

where $P_{ave,swit}$ is the average switching losses and $P_{ave,cond}$ is the average conduction losses, given by

$$P_{ave,swit} = \frac{1}{T_s}\left[\underbrace{\int_0^{t_{on}} i_{sw} v_{sw}\, dt}_{on \text{ switching loss}} + \underbrace{\int_{DT_s - t_{off}}^{DT_s} i_{sw} v_{sw}\, dt}_{off \text{ switching loss}}\right]$$

$$P_{ave,cond} = \frac{1}{T_s}\left[\int_{t_{on}}^{DT_s - t_{off}} I_{on} V_{off}\, dt + \int_{DT_s}^{T_s} I_{off} V_{on}\, dt\right]$$

$$= I_{on} V_{off}\left(D - \frac{(t_{on} + t_{off})}{T_s}\right) + I_{off} V_{on}(1 - D)$$

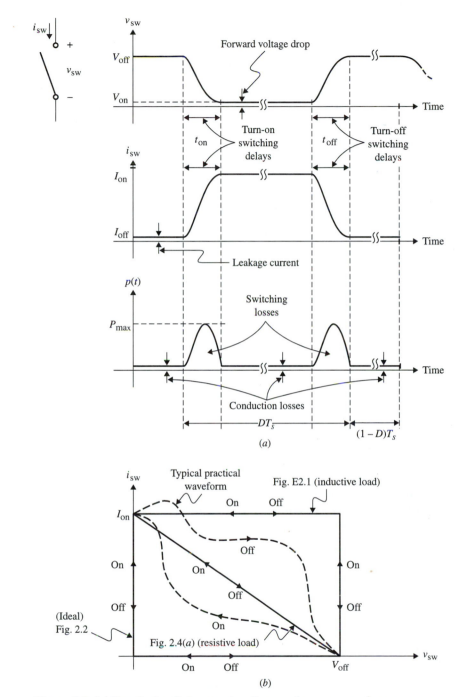

Figure 2.3 (a) Practical switch current, voltage, and power waveforms. (b) Switching trajectories under different load conditions.

If we assume the *on* and *off* times are small compared to T_s, then we have

$$P_{ave,cond} \approx \underbrace{I_{on} V_{off} D}_{on\ conduction\ loss} + \underbrace{I_{off} V_{on}(1-D)}_{off\ conduction\ loss}$$

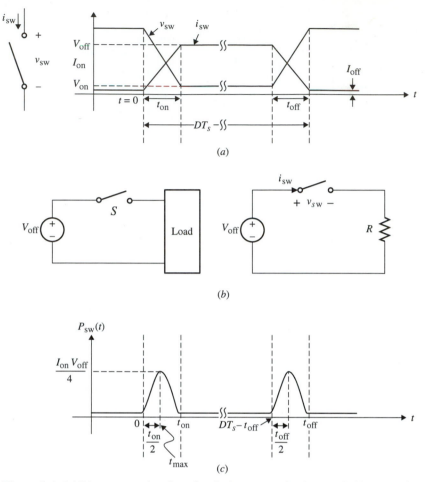

Figure 2.4 (*a*) Linear approximation of typical current and voltage switching waveforms. (*b*) Circuit implementation. (*c*) Instantaneous power waveform.

EXAMPLE 2.2

Consider a linear approximation of Fig. 2.3(*a*) as shown in Fig. 2.4(*a*) with $D = I$ (this assumes that T_s is the *on* time).

(a) Give a possible circuit implementation using a power switch whose switching waveforms are as shown in Fig. 2.4(*a*).

(b) Derive the expressions for the instantaneous switching and conduction power losses and sketch them.

(c) Determine the total average power dissipated in the circuit during one switching period.

(d) Find the maximum power.

SOLUTION

(a) First let us assume that the turn-on time, t_{on}, and turn-off time, t_{off}, the conduction voltage, V_{on}, and the leakage current, I_{off}, are part of the switching characteristics of the device and have nothing to do with the circuit topology.

When the switch is off, the blocking voltage across the switch is V_{off}, which can be represented as a dc voltage source of value V_{off} reflected somehow across the switch during the *off* state. When the switch is on, the current through the switch equals I_{on}; hence, a dc current is needed in series with the switch when it is in the *on* state. This suggests that when the switch turns off again, the current in series with the switch must be diverted somewhere else (this process is known as *commutation* and will be discussed later). As a result, a second switch is needed to carry the main current from the switch being investigated when it's switched off. However, since i_{sw} and v_{sw} are linearly related as shown in Fig. 2.4(*a*), a resistor will do the trick, and a second switch is not needed. Figure 2.4(*b*) shows a one-switch implementation, where S is the switch and R represents the switched load.

(b) The instantaneous current and voltage waveforms during the transition and conduction times are given as follows:

$$i_{sw}(t) = \begin{cases} \dfrac{t}{t_{on}}(I_{on} - I_{off}) + I_{off} & 0 \le t \le t_{on} \\ I_{on} & t_{on} \le t \le T_s - t_{off} \\ -\dfrac{t - T_s}{t_{off}}(I_{on} - I_{off}) + I_{off} & T_s - t_{off} \le t \le T_s \end{cases}$$

$$v_{sw}(t) = \begin{cases} -\dfrac{V_{off} - V_{on}}{t_{on}}(t - t_{on}) + V_{on} & 0 \le t \le t_{on} \\ V_{on} & t_{on} \le t \le T_s - t_{off} \\ \dfrac{V_{off} - V_{on}}{t_{off}}\left(t - (T_s - t_{off})\right) + V_{on} & T_s - t_{off} \le t \le T_s \end{cases}$$

It can be shown that if we assume $I_{on} \gg I_{off}$ and $V_{off} \gg V_{on}$, then the instantaneous power, $p(t) = i_{sw}v_{sw}$ can be given as follows:

$$p(t) = \begin{cases} -\dfrac{V_{off}I_{on}}{t_{on}^2}(t - t_{on})t & 0 \le t \le t_{on} \\ V_{on}I_{on} & t_{on} \le t \le T_s - t_{off} \\ -\dfrac{V_{off}I_{on}}{t_{off}^2}\left(t - (T_s - t_{off})\right)(t - T_s) & T_s - t_{off} \le t \le T_s \end{cases}$$

Figure 2.4(*c*) shows a plot of the instantaneous power where the maximum power during turn-on and turn-off is $V_{off}I_{on}/4$.

(c) The total average dissipated power is given by

$$P_{ave} = \frac{1}{T_s}\int_0^{T_s} p(t)\, dt = \frac{1}{T_s}\left[\int_0^{t_{on}} -\frac{V_{off}I_{on}}{t_{on}^2}(t - t_{on})t\, dt + \int_{t_{on}}^{T_s - t_{off}} V_{on}I_{on}\, dt \right.$$

$$\left. + \int_{T_s - t_{off}}^{T_s} -\frac{V_{off}I_{on}}{t_{off}^2}(t - (T_s - t_{off}))(t - T_s)\, dt\right]$$

Evaluation of this integral gives

$$P_{ave} = \frac{V_{off}I_{on}}{T_s}\left(\frac{t_{on} + t_{off}}{6}\right) + \frac{V_{on}I_{on}}{T_s}(T_s - t_{off} - t_{on})$$

The first expression represents the total switching loss, and the second expression represents the total conduction loss over one switching cycle. We notice that as the frequency increases, the average power increases linearly. Also, the power dissipation increases with an increase in the forward conduction current and the reverse blocking voltage.

(d) The maximum power occurs at the time when the first derivative of $p(t)$ during switching is set to zero, i.e.,

$$\left.\frac{dp(t)}{dt}\right|_{t=t_{max}} = 0$$

Solving this equation for t_{max}, we obtain the following values at turn-on and turn-off, respectively:

$$t_{max} = \frac{t_{on}}{2}$$

$$t_{max} = T_s - \frac{t_{off}}{2}$$

Solving for the maximum power, we obtain

$$P_{max} = \frac{V_{off}I_{on}}{4}$$

EXERCISE 2.1

Repeat Example 2.2 for the switching waveforms shown in Fig. E2.1.

ANSWER

(c)
$$P_{ave} = \frac{V_{off}I_{on}}{2T_s}(2t_d + t_{fall} + t_{rise})$$

(d)
$$P_{max} = V_{off}I_{on}$$

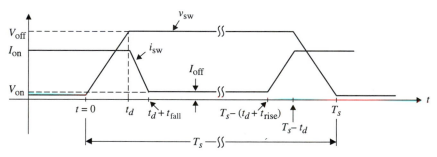

Figure E2.1 Waveforms for Exercise 2.1.

EXERCISE 2.2

Find the efficiency of the circuit in Fig. 2.1(d) assuming the switching characteristics for S are as shown in Fig. 2.4(a) with $t_{on} = 100$ ns, $t_{off} = 150$ ns, $T_s = 1$ μs, $I_{off} = 0$, $V_{on} = 0$, and $D = 1.0$.

ANSWER   80%

2.3   SWITCHING FUNCTIONS AND MATRIX REPRESENTATION

Since switches perform the duties of conversion, rectification, inversion, regulation, and so on, it is possible to use the block diagram of Fig. 2.5 as a useful representation of many power electronic circuits. This system has n inputs and m outputs that can be either voltages or currents. There are $n \times m$ switches, where each of the n input lines

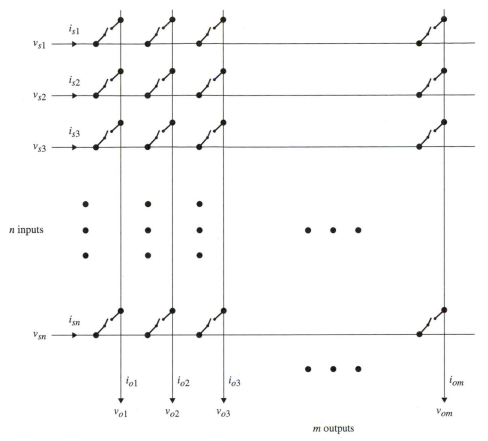

Figure 2.5 Switching matrix representation.

could be connected to any of the *m* outputs, resulting in what is known as a *switching matrix,* with the control of the switches described by a *switching function.* For illustration purposes, Fig. 2.6(*a*) and (*b*) shows the switching matrix representation for the single-phase full-bridge and the three-phase full-bridge, respectively. The switching function is a mathematical model for the switching matrix, describing the operation of the switches in the matrix. The literature is full of different techniques and technologies for the generation of switching functions. The switching function approach provides a compact matrix representation for the power converter and serves as a convenient tool for modeling all kinds of power conversion circuits. Due to the fact that no resistors are included in the structure, there is no power dissipation. In practical systems, additional energy storage elements are present to facilitate energy transfer; however, losses are associated with these components and the parts of the switching devices. If we assume ideal storage elements and ideal switches, then it is conceivable to achieve 100% efficiency in the switch matrix arrangement shown in Fig. 2.5. In order to process power bidirectionally, switches must be able to block voltages of either polarity and conduct current in either direction. One of the most challenging problems in designing and analyzing the system is the design and implementation of the switching network within the storage elements.

Theoretically, since we assume ideal switches and because there are no energy storage elements, the instantaneous input power must be equal to the instantaneous output power. Also, there are no restrictions on the form and frequency of the sources.

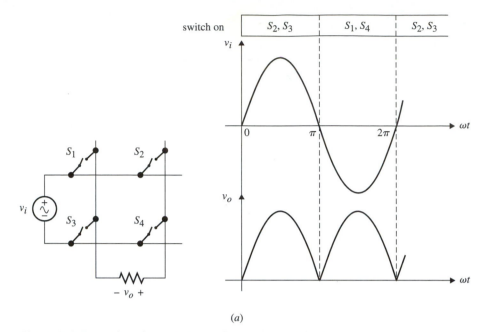

Figure 2.6 Examples of power electronic circuits. (*a*) Single-phase.

However, only one variable at the source terminal can be fixed (either voltage or current) and the corresponding terminal variable is determined by the switching function. For example, if $v_{s1}, v_{s2}, \ldots, v_{sn}$ represent fixed voltage sources, their corresponding currents $i_{s1}, i_{s2}, \ldots, i_{sn}$ are determined by the switching, as will be illustrated in Example 2.3. Similarly, if the output variables are represented by fixed output current sources $i_{o1}, i_{o2}, \ldots, i_{om}$, their corresponding terminal voltages are also determined by the switching functions. The reverse is also true for both the input and output terminals. Even though the switching matrix and its function determine the type of power conversion in a given power-processing circuit, the detailed implementation and terminal characteristics of the source and load sides are also a major part of the power conversion circuit. Normally the energy source is represented by an ideal voltage source that supplies a constant voltage over a wide range of currents. Similarly, the ideal current source can provide a constant current over a wide range of voltages. We will be using both types of energy sources throughout the book. As for the load side, the power conversion circuit must be designed to provide a stable and fixed output that can be represented by either a current source or a voltage source. If the output is to be a current source, the load is connected in series with an inductor; for a voltage source output, a capacitor is used.

Two important design issues need to be addressed in designing a power electronic switching circuit: (1) the "hardware," or physical implementation of the semiconductor switching matrix, and (2) the "software," or logical implementation that guarantees the operation of the switching matrix. The hardware implementation of the switching matrix is restricted by Kirchhoff's voltage law (KVL) and Kirchhoff's current law (KCL). These circuit laws must be observed at all times. KVL states that the algebraic sum of the voltage drops around a closed loop must be zero. Hence, there should not be a switching sequence that will allow two unequal voltage sources to be connected in parallel, or allow a short circuit across a voltage source (see the reference by Philip Krein). For example, to avoid establishing a short circuit across v_i in Fig. 2.6(*a*), S_1 and S_3 or S_2 and S_4 are not allowed to close simultaneously. Similarly, S_1 and S_3 of Fig. 2.6(*b*) cannot be closed simultaneously in order that two unequal voltage sources, v_a and v_b, are not connected in parallel.

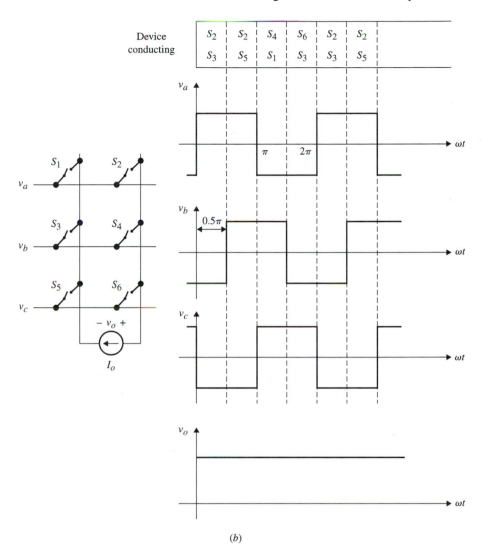

Device conducting	S_2	S_2	S_4	S_6	S_2	S_2
	S_3	S_5	S_1	S_3	S_3	S_5

(b)

Figure 2.6 (*continued*) Examples of power electronic circuits. (*b*) Three-phase.

Because KCL guarantees that the algebraic sum of the currents entering a node is zero, no switching sequence should allow two unequal current sources to be connected in series. For example, to avoid establishing an open circuit in series with a current source in Fig. 2.6(*b*), KCL dictates that neither S_1, S_3, and S_5 nor S_2, S_4, and S_6 be opened simultaneously. For more discussion about switching matrices and their associated switching functions, see the reference by Peter Wood.

EXAMPLE 2.3

Consider the single-switch, single-input power-processing circuit given in Fig. 2.7. Assume the source voltage, $v_s(t)$, is a triangular waveform with a peak voltage V_p and frequency $f = 1/T$, as shown in Fig. 2.7(*b*). Assume the switch is ideal and initially off, and its control works in such a way that it toggles every time $v_s(t)$ crosses zero. Use $V_p = 12$ V, R $= 10$ Ω and $T = 1$ ms.

(a) Sketch the waveforms for i_s and v_o.

(b) Calculate the average and rms values for the output voltage.

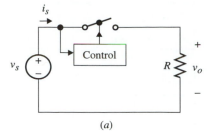

(a)

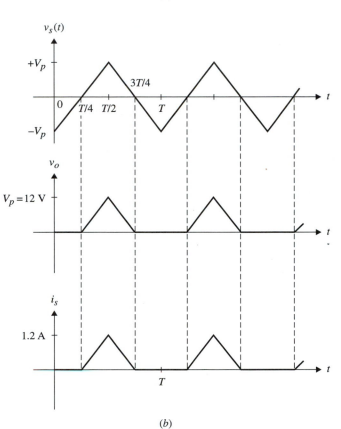

(b)

Figure 2.7 (a) Circuit and (b) waveforms for Example 2.3.

(c) Calculate the average input power, average output power, and efficiency.

(d) Repeat parts (a)–(c) by assuming $T = 1 \ \mu s$.

(e) Repeat parts (a)–(d) by assuming the switch has 1 V voltage drop when closed.

SOLUTION

(a) The output voltage and the source current waveform are shown in Fig. 2.7(b).

(b) The average output voltage is given by

$$V_o = \frac{1}{T}\int_0^T v_o(t)\,dt = \frac{1}{T}\left(\frac{1}{2}\frac{T}{2}V_p\right) = \frac{V_p}{4} = 3 \ V$$

The rms is given by

$$V_{o,\mathrm{rms}} = \sqrt{\frac{1}{T}\int_0^T v_o^2(t)\,dt} = \sqrt{\frac{1}{T}\left(\int_{T/4}^{T/2}\left(\frac{4V_p}{T}t - V_p\right)^2 dt + \int_{T/2}^{3T/4}\left(-\frac{4V_p}{T}t + 3V_p\right)^2 dt\right)}$$

$$= \frac{V_p}{\sqrt{6}} \approx 4.9 \text{ V}$$

(c) The average input power is calculated from

$$P_{\mathrm{in}} = \frac{1}{T}\int_0^T i_s(t)v_s(t)\,dt = \frac{1}{T}\left(\frac{1}{R}\int_{T/4}^{T/2}\left(\frac{4V_p}{T}t - V_p\right)^2 dt + \int_{T/2}^{3T/4}\frac{1}{R}\left(-\frac{4V_p}{T}t + 3V_p\right)^2 dt\right)$$

$$= \frac{1}{4RV_p}\left(\frac{2V_p^3}{3}\right) = \frac{V_p^2}{6R} = 2.4 \text{ W}$$

and the average output power,

$$P_{\mathrm{out}} = \frac{1}{T}\int_0^T i_o(t)v_o(t)\,dt = \frac{V_p^2}{6R} = 2.4 \text{ W}$$

The efficiency is

$$\eta = P_{\mathrm{out}}/P_{\mathrm{in}} = 100\%$$

(d) Same as above (because the results are independent of T).

(e) The average output voltage can be approximated by the following integration:

$$V_{o,\mathrm{ave}} = \frac{1}{T}\int_0^T v_o(t)\,dt = \frac{1}{T}\left(\int_{T/4}^{T/2}\left(\frac{4V_p}{T}t - V_p - 1\right)dt + \int_{T/2}^{3T/4}\left(\frac{-4V_p}{T}t + 3V_p - 1\right)dt\right)$$

$$= \frac{1}{4}(V_p - 2) = 2.5 \text{ V}$$

The rms voltage is

$$V_{o,\mathrm{rms}} = \sqrt{\frac{1}{T}\int_0^T v_o^2(t)\,dt} = \sqrt{\frac{1}{T}\left(\int_{T/4}^{T/2}\left(\frac{4V_p}{T}t - V_p - 1\right)^2 dt + \int_{T/2}^{3T/4}\left(\frac{-4V_p}{T}t + 3V_p - 1\right)^2 dt\right)}$$

$$= \sqrt{\frac{1}{6V_p}\left((V_p - 1)^3 + 1\right)} \approx 4.3 \text{ V}$$

It can be shown that the average input power is

$$P_{\mathrm{in}} = \frac{1}{T}\int_0^T i_s(t)v_s(t)\,dt$$

$$= \frac{1}{T}\left[\int_{T/4}^{T/2}\left(\frac{4V_p}{T}t - V_p\right)\frac{1}{R}\left(\frac{4V_p}{T}t - V_p - 1\right)dt + \int_{T/2}^{3T/4}\left(\frac{-4V_p}{T}t + 3V_p\right)\frac{1}{R}\left(\frac{-4V_p}{T}t + 3V_p - 1\right)dt\right]$$

$$= \frac{V_p(2V_p - 3)}{12R} = 2.1 \text{ W}$$

and the average output power,

$$P_{\mathrm{out}} = \frac{1}{T}\int_0^T i_o(t)v_o(t)\,dt$$

$$= \frac{1}{T}\left(\frac{1}{R}\int_{T/4}^{T/2}\left(\frac{4V_p}{T}t - V_p - 1\right)^2 dt + \int_{T/2}^{3T/4}\frac{1}{R}\left(\frac{-4V_p}{T}t + 3V_p - 1\right)^2 dt\right)$$

$$= \frac{1}{6RV_p}\left((V_p - 1)^3 + 1\right) \approx 1.85 \text{ W}$$

resulting in efficiency of $\eta = P_{\mathrm{out}}/P_{\mathrm{in}} \times 100\% = 1.85/2.1 \times 100\% \approx 88.2\%$.

EXERCISE 2.3

Repeat Example 2.3 for (a) $v_s = 12 \sin \omega t$ and (b) v_s a squarewave with peak-to-peak voltage equal to ± 12 V.

ANSWER

(a) –3.82 V, 6 V, 3.6 W, 3.6, 100%; same; –3.32 V, 5.37 V, 3.22 W, 2.89 W, 89.7%

(b) –6 V, 8.5 V, 7.2 W, 100%; same; –5.5 V, 7.8 V, 6.6 W, 6.05 W, 91.7%

2.4   TYPES OF SWITCHES

To implement a given switching function, the ideal switches in the switching matrix must be realized by practical power devices. Functionally speaking, any switch must have the ability to conduct current and/or the ability to block voltage by means of control signals. In the *on* state, the current conduction state is the task under consideration, and in the *off* state, the voltage blocking state is what we are considering. Practical switches have limitations in their conduction current and in their voltage blocking. Since the switch current can flow in the forward, reverse, or both directions and the voltage can be blocked in the forward, reverse, and both directions, there are nine different combinations of current carrying and voltage blocking directions. Four of these combinations are duplicates of another four, i.e., forward current carrying and reverse voltage blocking is the same as reverse current carrying and forward voltage blocking. As a result, switches are classified into five general types of restricted switches, as shown in Table 2.1 (for more details see the reference by Philip Krein).

1. *Forward current carrying and reverse voltage blocking.* This is an uncontrolled device with unidirectional current flow. Uncontrolled device means that turn-on and turn-off are not controlled by an external control signal but rather by the power circuit itself. An example of this type of switch is the diode since it carries the cur-

Table 2.1   Types of Semiconductor Switches, Their Controllability Features, and Their Possible Switch Implementations

Type	Current flow	Voltage blocking	Switch implementation
1	Forward	Reverse	Diode
2	Forward	Forward	Transistor
3	Forward	Bidirectional	SCR
4	Bidirectional	Forward	Transistor with flyback diode
5	Bidirectional	Bidirectional	Triac

rent only in the forward direction when the anode-cathode voltage is positive. The diode is an uncontrolled device since no external control signal can be applied to initiate *carrying* the forward current or to initiate *blocking* the reverse voltage. The controlling voltage is derived from either the source or the load or both.

2. *Forward current carrying and forward voltage blocking.* This is a controlled device with unidirectional current flow. Such a switch should be able to carry the current in the forward direction and block voltage in the forward direction. Of course, the diode is unable to block a forward voltage across it. Another type of switch is needed with an external control signal that allows the device to decide whether the forward current is to flow or not even when a forward voltage is applied. This is the same as implying that the switch performs a current conduction delay function. An example of such a switch is the transistor, which is able to block voltage in the forward direction when the base (gate) current is absent.

3. *Forward current carrying and bidirectional voltage blocking.* This switch can block current flow in both directions (i.e., supports forward and reverse voltage blocking) but carries the current only in the forward direction. An example is the silicon-controlled rectifier (SCR), to be studied later in this chapter.

4. *Bidirectional current carrying and forward voltage blocking.* This switch is similar to type 2, except the current can flow bidirectionally. Hence, the implementation is a transistor with a diode connected as shown in Table 2.1 to allow reverse current flow. The diode is known as a flyback or body diode since it picks the current in the reverse direction. This switch can carry the current in the forward direction through the transistor and in the reverse direction through the flyback diode. The base (gate) signal is used to allow the switch to determine whether to carry the current in the forward direction or to be in the voltage blocking state; because of the presence of the diode, the switch is unable to block reverse voltage.

5. *Bidirectional current carrying and bidirectional voltage blocking.* This switch is the most general power electronic switch and is similar to type 3 with additional characteristics that allow it to support current flow in both directions. Unlike type 4, where a flyback diode is added to allow reverse current flow, here the forward and reverse current flow must be controlled. An example of this type is the triac (*triode ac*), which is simply two SCRs connected in parallel and in opposite directions, as shown in Table 2.1.

2.5   AVAILABLE SEMICONDUCTOR SWITCHING DEVICES

In this section, the emphasis will be on the *i-v* switching characteristics of devices and their corresponding power ratings and possible applications. Selecting the most appropriate device for a given application is not an easy task, requiring knowledge about the device's characteristics and unique features, innovation, and engineering design experience. Unlike low-power (signal) devices, power devices are more complicated in structure, driver design, and operational *i-v* characteristics. This knowledge is very important for enabling power electronics engineers to design circuits that will make these devices close to ideal. In this section, we will briefly discuss two broad families of power devices:

Bipolar and Unipolar Devices
1. Power diodes
2. Bipolar junction transistors
3. Insulated gate bipolar transistors (IGBTs)
4. Metal oxide semiconductor field-effect transistors (MOSFETs)

Thyristor-Based Devices

1. Silicon-controlled rectifiers (SCRs)
2. Gate turn-off (GTO) thyristors
3. Triode ac switches (triacs)
4. Static induction transistors (SITs) and thyristors (SITHs)
5. MOS-controlled thyristors (MCTs)

2.5.1 Bipolar and Unipolar Devices

The Power Diode

The power diode is a two-terminal device composed of a *pn* junction and whose turn-on state cannot be controlled (uncontrolled switch). The diode turn-on and turn-off is decided by the external circuitry: A positive voltage imposed across it will turn it on and a negative current through it turns it off.

The symbol and the practical and ideal *i-v* characteristic curves of the power diode are shown in Fig. 2.8(*a*), (*b*), and (*c*), respectively. In the conduction state, the forward voltage drop, V_F, is typically 1 V or less. The diode current increases exponentially with the voltage across it; i.e., a small increase in V_F produces a large increase in I_F (see Problem 2.5). In the reverse-bias region, the device is in the *off* state and only a reverse saturation current, I_s, exists in the diode (also known as leakage current). The breakdown voltage, V_{BR}, is the maximum inverse voltage the diode is capable of blocking. V_{BR} is a diode-rated parameter with values up to a few kilovolts, and in normal operation the reverse voltage should not reach V_{BR}. Zener diodes are special diodes in which the breakdown voltage is approximately 6–12 V, controlled by the doping process.

In power circuits, power diodes have two important features:

1. Power-handling capabilities, including forward current carrying and reverse voltage blocking
2. Reverse recovery time (t_{rr}) at turn-off

The parameter t_{rr} is very significant because the speed of turning off the diode could be large enough to affect the operation of the circuit. At turn-on, the delay time is normally insignificant compared to the transient time in power electronic circuits.

Broadly speaking, two types of power diodes are available:

1. The bipolar diode, which is based on the *pn* semiconductor junction. Depending on the applications, a bipolar diode can be either the standard line-frequency

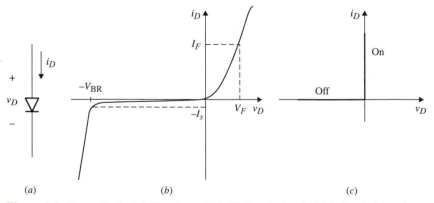

(*a*)   (*b*)   (*c*)

Figure 2.8 Power diode. (*a*) Circuit symbol. (*b*) Practical and (*c*) ideal switching characteristics.

type or the fast-recovery high-frequency type, with its t_{rr} varying between 50 ns and 50 μs. Typical voltage drop is 0.7–1.3 V, with reverse voltage blocking of 3 kV and forward current of 3.5 kA.

2. The Schottky diode, which is based on the metal-semiconductor junction. It has a lower forward voltage drop than the bipolar (about 0.5 V or less). Unlike the bipolar diode, whose current conduction depends on the minority and majority carriers, the Schottky diode current depends mainly on the majority carriers. Finally, the *i-v* characteristics of the Schottky diode are similar to the *i-v* characteristics of the bipolar diode. Unlike bipolar diodes, Schottky diodes generate fewer excess minority carriers than majority carriers; hence its current is primarily generated due to the drift of majority carriers. Due to its large-leakage circuit, it is normally used in low-voltage, high-current dc power supplies.

To study the reverse recovery characteristics of the diode, we consider the circuit of Fig. 2.9(*a*), which has a typical diode current waveform during turn-off as shown in Fig. 2.9(*b*). For simplicity we assume the switch is ideal. Such a circuit arrangement is normally encountered in switch-mode dc-to-dc converters with the switch replaced by either a BJT or a MOSFET. Initially we assume the diode is conducting with forward current I_0. At $t = t_0$, the switch is turned on, forcing the diode to turn off due to the dc input voltage V_{in}. The turn-on characteristics of the diode are simpler to deal with since turn-on only involves charging the diode depletion capacitor. The diode's forward conduction begins when its depletion capacitor has been charged. At turn-on, the diode voltage drop is larger than the normal forward drop during conduction. This transient voltage exists due to the large value of diode resistance at turn-on. This is why during the diode's turn-on time the power dissipation is much larger than when it is in the steady conduction state. The diode turn-off characteristics are more complex since significant stored charges exist in the body of the *pn* junction and at the junction.

As shown in Fig. 2.9(*b*), during turn-off, the diode current linearly decreases from its forward value, I_0, at $t = t_0$ to zero at $t = t_1$ and then continues to go negative until it reaches a negative peak value at $t = t_2$, known as the *reverse recovery current*, I_{RR}, at which the current starts to rise exponentially to zero at $t = t_3$. The time intervals can be broken down as follows: Between t_0 and t_1, the diode current is positive and the diode forward voltage is small. Hence, we assume the rate of change of diode current is constant and is determined by the total circuit inductance in series with the diode. In our example, since the switch is ideal, di_D/dt and I_{RR} are limited by the diode's and lead's parasitic inductances. At $t = t_1$, the current becomes zero and the diode should begin turning off by supporting reverse voltage. But because of the excess minority carriers in

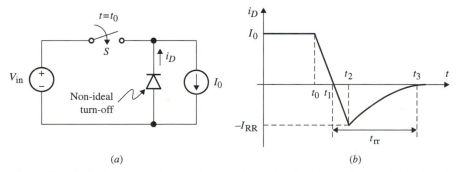

(*a*) (*b*)

Figure 2.9 Typical diode switching characteristics. (*a*) Switching circuit with *S* closed at $t = t_0$. (*b*) Diode current.

the *pn* junction that need to be removed before the diode's reverse voltage begins to rise, the diode remains in the conduction state for longer time, i.e., until $t = t_2$.

The delay from t_1 to t_2 is due to the minority carriers in the depletion region, whereas the delay from t_2 to t_3 is caused by the charge stored in the bulk of the semiconductor material. At $t = t_3$, all charge carriers are removed, causing the device to be fully switched off. The time it takes from the moment the diode current becomes negative until it becomes zero is the reverse recovery time, t_{rr}, as shown in Fig. 2.9(*b*).

Between t_2 and t_3, the junction behaves like a capacitor whose voltage goes from zero to the reverse voltage via a charging current in this interval. The total charge carriers that cause a negative diode current flow when it is turned off constitute the *reverse recovery charge*, Q_{rr}, and can be expressed in terms of I_{RR} and t_{rr}.

The time between t_2 and t_3 may be very short compared to t_{rr}, resulting in high *di / dt*. The ratio between $(t_3 - t_2)$ and t_{rr} is a parameter that defines what is known as *diode snappiness*. The smaller this ratio, the quicker the diode recovers its reverse blocking voltage, resulting in what is known as a *fast-recovery* or *hard-recovery* diode. Meanwhile, a diode with a high ratio of $(t_3 - t_2)$ to t_{rr} takes a relatively long time to bring its forward current to zero from its negative peak value. These diodes are known as *soft-recovery* diodes. The fast-recovery diodes have high *di / dt* and normally experience oscillation at turn-off. Standard or general-purpose diodes have soft recovery time and are used in low-speed applications where the frequency is less than a few kHz.

In general, an attempt to reduce either t_{rr} or I_{RR} will result in an increase in the other. The forward recovery voltage limits the efficiency because of device stresses and higher switching losses, and t_{rr} limits the frequency of operation. Power electronics engineers should keep in mind that the transient voltage at turn-on and the transient current at turn-off might affect the external circuitry and cause unwanted stresses. External snubber circuits are added to suppress these transient values.

EXAMPLE 2.4

Consider the switching circuit shown in Fig. 2.10 by modeling the circuit parasitic inductance as a lumped discrete value, L_s. Assume the switch was open for a long time before being turned on at $t = t_0$. Assume the same diode switching characteristics of Fig. 2.9(*b*), except that it is a fast-recovery diode with $t_3 - t_2 \approx 0$. Derive the expressions for I_{RR} and the peak switch current in terms of the diode reverse recovery time.

SOLUTION   While the diode is in the conduction state, its forward current is I_0. When the switch is closed at $t = t_0$, the diode voltage remains zero and its current is given by

$$i_D = I_0 - i_s$$

and i_s for $t \geq t_0$ is given by

$$i_s(t) = \frac{V_{in}}{L_s}(t - t_0) \quad t_0 \leq t \leq t_2$$

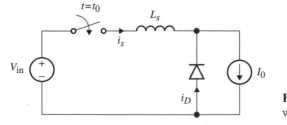

Figure 2.10 Diode switching circuit with parasitic inductor.

At $t = t_1$ the diode current becomes zero and i_s becomes I_0. Hence, the interval $t_1 - t_0$ is given by

$$t_1 - t_0 = \frac{I_0 L_s}{V_{in}}$$

Since $t_3 - t_2 \approx 0$, then $t_2 - t_1 \approx t_{rr}$ and I_{RR} is given by

$$I_{RR} = \frac{V_{in}}{L_s} t_{rr}$$

The peak switch current occurs at $t = t_2$ when $i_D = -I_{RR}$ and is given by

$$I_{s,peak} = \frac{V_{in}}{L_s} t_{rr} + I_0$$

At this point the diode is turned off and the peak inductor current is higher than the load current I_0. Since the load is highly inductive, its value cannot increase suddenly by the amount $V_{in} t_{rr} / L_s$ without creating high reverse voltage across the diode. As a result, a snubber circuit must be added across the diode to dissipate excess stored energy in the inductor.

Another important point is that when L_s becomes very small, a very large reverse recovery current occurs that could damage the diode and cause large switching losses.

The peak value of the reverse current, $-I_{RR}$, is a very important parameter and it can be less than, equal to, or larger than the forward current I_0, depending on the external circuitry connected to the diode, and the diode parasitic inductance. The fast-recovery diodes have low recovery time, normally less than 50 ns, and are used in applications such as high-frequency dc-dc converters, where the speed of recovery is critical.

The Bipolar Junction Transistors (BJT)

The schematic symbol and *i-v* characteristics for the bipolar junction transistors (BJT) are shown in Fig. 2.11(*a*), (*b*), and (*c*), respectively. It is a two-junction, three-terminal device with the minority carriers being the main conducting charges. The switching speed of the BJT is much faster than that of thyristor-type devices. A major drawback is the *second breakdown* problem.[2]

Unlike the SCR, the BJT is turned on by constantly applying a base signal. Power BJTs have two different properties from the low-power BJT and logic transistor: large blocking voltage in the *off* state and high forward current-carrying capabilities in the *on* state. BJT power ratings reach up to 1200 V and 500 A. These high rating values suggest that the power BJT's driving circuits are more complicated.

Because the BJT is a current-driven device, the larger the base current, the smaller β_{forced}[3] and the deeper the transistor is driven into saturation. In saturation, the collector-emitter voltage is almost constant and the collector current is determined largely by the external circuit to the switch. It is sometimes useful to define what is known as an *overdrive factor*, which gives a measure to how deep in saturation the transistor is. For example, if the transistor is at the edge of saturation with given base current I_B, then with an overdrive factor of 10 the base current becomes $10I_B$ and the transistor becomes deeper in saturation.

---

[2]Normally, the first breakdown voltage refers to the avalanche breakdown caused by the increase in the reverse bias voltage, which can be nondestructive. The second breakdown voltage is a destructive phenomenon caused by localized overheating spots in the device.

[3]β_{forced} is defined as the ratio I_C / I_B when the transistor is operating in the saturation mode.

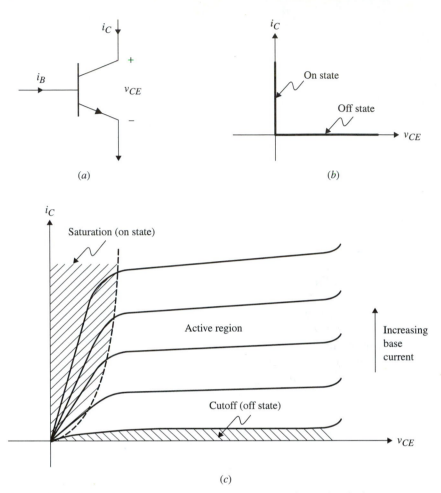

Figure 2.11 BJT switching characteristics. (*a*) *npn* transistor. (*b*) Ideal *i-v* characteristics. (*c*) Practical *i-v* characteristics.

Since the base thickness is inversely proportional to the current gain β, Darlington-connected BJT pairs have been developed in which the collectors of two devices are joined and the base of the first is connected to the emitter of the second, as shown in Fig. 2.12. This arrangement results in an overall gain that approximately equals the product of the individual β's of the two transistors. Transistor Q_1 serves as an auxiliary transistor, which provides the base current necessary to turn on Q_2. Because there is a high current gain, a smaller base current to Q_2 is needed to drive the power Darlington pair. Darlington power transistors are widely used in UPSs and various ac and dc motor drives up to hundreds of kilowatts and tens of kilohertz. A modern Darlington pair has ratings up to 1.2 kV with current up to 800 A and operating frequency up to several kilohertz.

Triple Darlingtons are also available, in which the current gain becomes proportional to the product of the three individual currents gains of the transistors. To turn off the Darlington switch, all base currents must become zero, resulting in slower switching speed compared to a single transistor. Also, the overall collector-emitter saturation voltage, $V_{CE,\text{sat}}$ is higher than for a single transistor, as will be illustrated in Exercise 2.5.

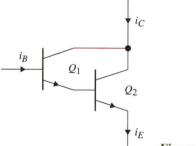

Figure 2.12 Darlington-connected BJT.

There are three regions of operation: saturation, active, and cutoff. As a power switch, the BJT must operate either in the saturation region (*on* state) or in the cutoff region (*off* state). The third state is when the transistor is in the linear region and is used as an analog amplifier.

To investigate the turn-on and turn-off processes, we consider a simple inverter circuit shown in Fig. 2.13(*a*) with its switching waveforms as shown in Fig. 2.13(*b*). The voltage v_I is the base driving voltage with positive polarity, V_1, to push positive current into the base, $I_{B1} = (V_I - V_{BE})/R_B$, and a negative polarity, V_2, to quickly discharge the base current, $I_{B2} = -(V_2 + V_{BE})/R_B$. At time $t = t_0$, V_1 is applied with positive dc voltage, $+V_1$. Because it takes time to charge the internal depletion capacitor to turn the junction on at $V_{BE} = 0.7$ V, a delay time, t_d, elapses before the collector current starts flowing. After the junction is turned on, the collector starts flowing exponentially through R_B and the emitter-base junction capacitor. During this period, the minority carriers are being stored in the transistor base region. The collector current increases until it reaches its maximum saturated value, I_{on}, determined by

$$I_{on} = \frac{V_{in} - V_{CE,\text{sat}}}{R}$$

The time it takes for the collector current to rise from 10% to 90% of its maximum value, I_{on}, is called the *rise time*. For simplicity, Fig. 2.13(*b*) shows the rise time from $I_C = 0$ to I_{on}. The total switching *on* time is given by $t_{on} = t_d + t_r$. To turn off the transistor, a negative (or zero) base voltage is normally applied, resulting in a base current I_{B2} being *pulled out* of the base as shown in Fig. 2.13(*b*). The collector current does not start decreasing until sometime later after the stored saturation charge in the base has been removed. This time is called the storage time, t_s; it is normally longer than the delay time, t_d, and usually determines the limiting range of the switching speed. If the base voltage is not negative (i.e., in the absence of I_{B2}), the entire base current must be removed through the process of recombination.

To turn on the BJT, a large current must be pushed to the base. This base current must be large enough to saturate the transistor. In the saturation region, both base-emitter and base-collector junctions are forward biased. This is why the BJT is known as a current-driven device. When it is operated in the saturation region, $I_B > I_C/\beta$, where β is the dc current gain. In this region a new dc current gain β is defined to indicate the depth of the transistor saturation. The saturation collector-emitter voltage is given as $V_{CE,\text{sat}}$, and β_{forced} is defined as

$$\beta_{\text{forced}} = \frac{I_C}{I_B}$$

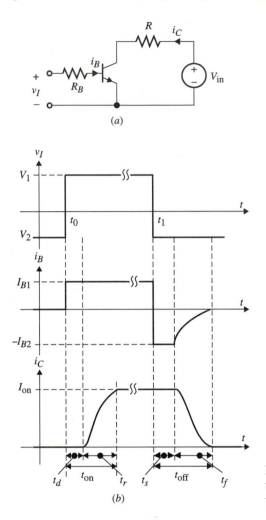

Figure 2.13 Switching characteristics for the BJT. (*a*) Circuit. (*b*) Switching waveforms.

where I_C and I_B are the collector and base currents in saturation, respectively, and $\beta_{\text{forced}} < \beta$. The smaller β_{forced}, the deeper the transistor is driven into saturation. Typically, β_{forced} can be as low as 1. Ideally, $V_{CE,\text{sat}} = 0$, but in practice this value varies between 0.1 and 0.6 V, depending on how deep in saturation the device is driven. The new ratio of collector to emitter current is much smaller than the case when the transistor is operated in the active mode. At the edge of saturation, $\beta_{\text{forced}} = \beta$.

The total power dissipation in the transistor is obtained by adding the input power supplied by the collector current and the input power supplied by the base current; hence, the total power dissipation is defined as follows:

$$P_{\text{diss}} = V_{CE}I_C + V_{BE}I_B$$

EXERCISE 2.4

Consider the transistor circuit shown in Fig. E2.4. Assume the transistor is operating in the saturation region with $V_{CE,\text{sat}} = 0.5$ V, $V_{BE} = 0.75$ V, $V_D = 0.7$ V, and $D = 0.5$. Sketch i_B, v_{CE}, and i_D. Determine the overall efficiency of the circuit (neglect the power supplied through the base).

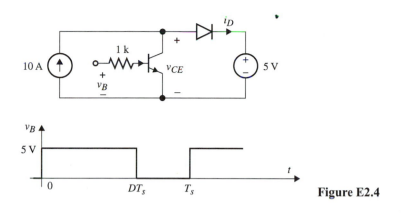

Figure E2.4

ANSWER $\eta = 80.6\%$

To determine the voltage, current, and power operational limits, normally a plot of *i-v* characteristics is given, as shown in Fig. 2.14. It gives the region in which the transistor can operate within its limits, the region is known as the safe operation area (SOA). It represents the permissible range of current, voltage, and power of the device in operation. The locus of switch voltage versus switch current during turn-on and turn-off must lie within the SOA.

EXERCISE 2.5

(a) Show that the current gain of the triple Darlington transistors shown in Fig. E2.5 is given by

$$\frac{i_C}{i_B} \approx \beta_1 \beta_2 \beta_3$$

(b) Assume transistor Q_1 has collector-emitter saturation voltage $V_{CE1,\text{sat}}$. Show that $V_{CE3,\text{sat}}$ for transistor Q_3 is given by

$$V_{CE3,\text{sat}} = V_{CE1,\text{sat}} + 2V_{BE}$$

Assume identical V_{BE} for the three transistors.

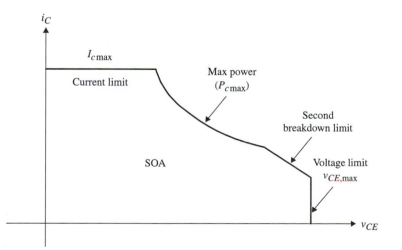

Figure 2.14 Safe Operation Area (SOA) for a BJT.

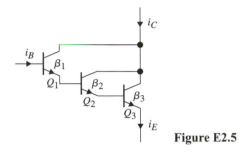

Figure E2.5

EXERCISE 2.6

Consider the simple BJT switch shown in Fig. E2.6. Determine β_{forced} for $R_B = 1 \text{ k}\Omega$, $R_B = 10 \text{ k}\Omega$, $R_B = 20 \text{ k}\Omega$. Use $V_{BE} = 0.7$ V, $V_{CE,\text{sat}} = 0.3$ V and assume an ideal diode.

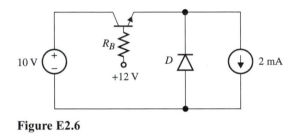

Figure E2.6

ANSWER 1.25, 12.5, 25

Initially, the BJT was developed to be used in linear audio output amplifiers. Soon BJT devices were used in switch-mode and high-frequency converters for aerospace applications to reduce the size and weight of magnetic components and filter capacitors. In applications where self-turn-off devices are needed, such as dc choppers and inverters, BJTs quickly replaced thyristors.

The Power MOSFET

In this section, an overview of power MOSFET semiconductor switching devices will be given. A detailed discussion of the physical structure, fabrication, and physical behavior of the device and its packaging is beyond the scope of this chapter. The emphasis here will be on the device's regions of operation and its terminal *i-v* switching characteristics.

Unlike the bipolar junction transistor, the metal oxide semiconductor field-effect transistor (MOSFET) device belongs to the *unipolar device* family, since it uses only the majority carriers in conduction. The development of metal oxide semiconductor technology for microelectronic circuits opened the way for the power MOSFET device in 1975. Selecting the most appropriate device for a given application is not an easy task, requiring knowledge about the device characteristics, and unique features, as well as innovation and engineering design experience. Unlike low-power (signal) devices, power devices are more complicated in structure, driver design, and operational *i-v* characteristics. This knowledge is very important in enabling a power electronics engineer to design circuits that will make these devices close to ideal.

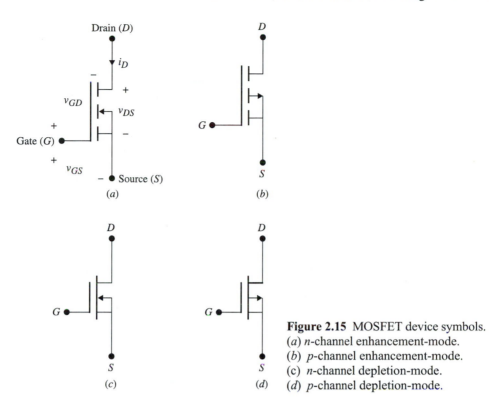

Figure 2.15 MOSFET device symbols.
(*a*) *n*-channel enhancement-mode.
(*b*) *p*-channel enhancement-mode.
(*c*) *n*-channel depletion-mode.
(*d*) *p*-channel depletion-mode.

The device symbols for *p*- and *n*-channel enhancement and depletion types are shown in Fig. 2.15. Figure 2.16 shows the *i-v* characteristics for the *n*-channel enhancement-type MOSFET. It is the fastest power switching device, with a switching frequency of more than 1 MHz, a voltage power rating up to 600 V, and a current rating as high as 40 A. MOSFET regions of operations will be studied shortly.

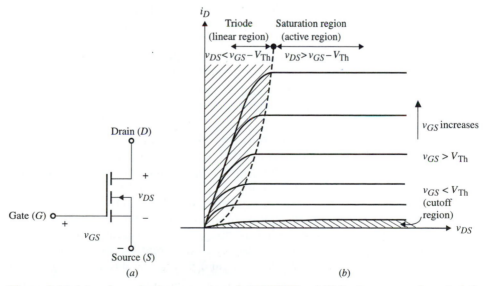

Figure 2.16 (*a*) *n*-channel enhancement-mode MOSFET and (*b*) its i_D vs. v_{DS} characteristics.

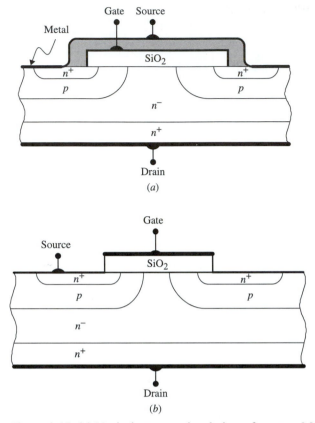

Figure 2.17 (*a*) Vertical cross-sectional view of a power MOSFET. (*b*) Simplified representation.

MOSFET Structure Unlike the lateral-channel MOSET devices used in many IC technologies, in which the gate, source, and drain terminals are located on the same surface of the silicon wafer, power MOSFETs use a vertical channel structure to increase the device's power rating. In the vertical channel structure, the source and drain are on opposite sides of the silicon wafer. Figure 2.17(*a*) shows a vertical cross-sectional view of a power MOSFET. Figure 2.17(*b*) shows a simplified representation. There are several discrete types of the vertical-structure power MOSFET available commercially today, such as the V-MOSFET, U-MOSFET, D-MOSFET, and S-MOSFET. The *pn* junction between the *p*-base region (also referred to as the body or bulk region) and the *n*-drift region provides the forward voltage blocking capabilities. The source metal contact is connected directly to the *p*-base region through a break in the n^+ source region to allow for a fixed potential to the *p*-base region during normal device operation. When the gate and source terminals are set at the same potential ($V_{GS} = 0$), no channel is established in the *p*-base region (i.e., the channel region remains unmodulated). The lower doping in the *n*-drift region is needed to achieve higher drain voltage blocking capabilities. For the drain-source current, i_D, to flow, a conductive path must be established between the n^+ and n^- regions through the *p*-base diffusion region.

On-State Resistance   When the MOSFET is in the *on* state (triode region), the channel of the device behaves like a constant resistance, $R_{DS(on)}$, that is linearly proportional to the change between v_{DS} and i_D, as given by the following relation:

$$R_{DS(on)} = \left. \frac{\partial v_{DS}}{\partial i_D} \right|_{V_{GS}=\text{constant}}$$

The total conduction (*on*-state) power loss for a given MOSFET with forward current I_D and *on*-resistance $R_{DS(on)}$ is given by

$$P_{\text{on,diss}} = I_D^2 R_{DS(on)}$$

The value of $R_{DS(on)}$ can be significant and varies between tens of milliohms and a few ohms for low-voltage and high-voltage MOSFET, respectively. The *on*-state resistance is an important data sheet parameter, since it determines the forward voltage drop across the device and its total power losses.

Unlike the current-controlled bipolar device, which requires base current to allow the current to flow in the collector, the power MOSFET is a voltage-controlled unipolar device and requires only a small amount of input (gate) current. As a result, it requires less drive power than the BJT. However, it is a nonlatching current like that of the BJT; i.e., a gate-source voltage must be maintained. Moreover, since only majority carriers contribute to the current flow, MOSFETs surpass all other devices in switching speed, with speeds exceeding a few megahertz. Comparing the BJT and the MOSFET, the BJT has higher power-handling capabilities and lower switching speed, while the MOSFET device has lower power-handling capabilities and relatively fast switching speed. The MOSFET device has a higher *on*-state resistance than the bipolar transistor. Another difference is that the BJT parameters are more sensitive to junction temperature compared to the MOSFET parameters. Unlike the BJT, MOSFET devices don't suffer from second breakdown voltages, and sharing current in parallel devices is possible.

Internal Body Diode   The modern power MOSFET has an internal diode called a *body diode* connected between the source and the drain, as shown in Fig. 2.18(*a*). This diode provides a reverse direction for the drain current, allowing a bidirectional switch implementation. Even though the MOSFET's body diode has adequate current and switching speed ratings, in some power electronic applications that require the use of ultra-fast diodes, an external fast-recovery diode is added in an anti-parallel fashion, with the body diode blocked by a slow-recovery diode, as shown in Fig. 2.18(*b*).

Internal Capacitors   Another important parameter that affects the MOSFET's switching behavior is the parasitic capacitances between the device's three terminals, namely, the gate-to-source (C_{gs}), gate-to-drain (C_{gd}), and drain-to-source (C_{ds}) capacitances, shown in Fig. 2.19(*a*). The values of these capacitances are nonlinear and a function of the device's structure, geometry, and bias voltages. During turn-on, capacitors C_{gd} and C_{gs} must be charged through the gate; hence, the design of the gate control circuit must take into consideration the variation in these capacitances. The largest variation occurs in the gate-to-drain capacitance as the drain-to-gate voltage varies. The MOSFET parasitic capacitances are given in terms of the device's data sheet parameters C_{iss}, C_{oss}, and C_{rss} as follows,

$$C_{gd} = C_{rss}$$
$$C_{gs} = C_{iss} - C_{rss}$$
$$C_{ds} = C_{oss} - C_{rss}$$

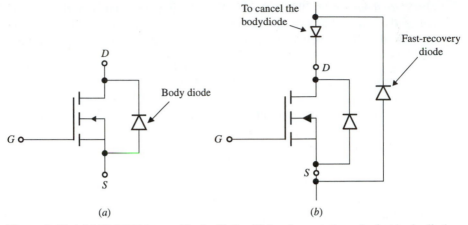

Figure 2.18 (a) MOSFET internal body diode. (b) Implementation of a fast body diode.

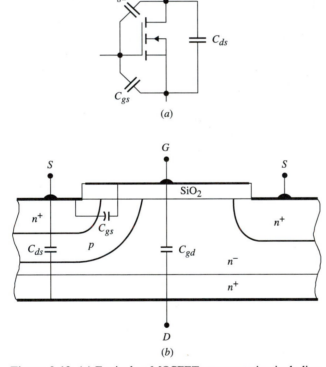

Figure 2.19 (a) Equivalent MOSFET representation including junction capacitances. (b) Representation of this physical location.

where

C_{rss} = small-signal reverse transfer capacitance

C_{iss} = small-signal input capacitance with the drain and source terminals shorted

C_{oss} = small-signal output capacitance with the gate and source terminals shorted.

The MOSFET capacitances C_{gs}, C_{gd}, and C_{ds} are nonlinear and are a function of the dc bias voltage. The variations in C_{oss} and C_{iss} are significant as the drain-to-source and gate-to-source voltages cross zero, respectively. The objective of the drive circuit is to charge and discharge the gate-to-source and gate-to-drain parasitic capacitances to turn the device on and off, respectively.

In power electronics, the aim is to use power switching devices to operate at higher and higher frequencies. Hence, the size and weight associated with the output transformer, inductors, and filter capacitors will decrease. As a result, MOSFETs are used extensively in power supply designs that require high switching frequencies, including switching and resonant-mode power supplies and brushless dc motor drives. Because of the device's large conduction losses, its power rating is limited to a few kilowatts. Because of its many advantages over BJT devices, modern MOSFET devices have received high market acceptance.

Regions of Operation   Most MOSFET devices used in power electronics applications are of the n-channel, enhancement type, like that shown in Fig. 2.16(a). For the MOSFET to carry drain current, a channel between the drain and the source must be created. This occurs when the gate-to-source voltage exceeds the device threshold voltage, V_{Th}. For $v_{GS} > V_{Th}$, the device can be either in the triode region, which is also called "constant resistance" region, or in the saturation region, depending on the value of v_{DS}. For a given v_{GS}, with a small v_{DS} ($v_{DS} < v_{GS} - V_{Th}$) the device operates in the triode region (saturation region in the BJT), and with a large v_{DS} ($v_{DS} > v_{GS} - V_{Th}$), the device enters the saturation region (active region in the BJT). For $v_{GS} < V_{Th}$, the device turns off, with the drain current almost equal to zero. Under both regions of operation, the gate current is almost zero. This is why the MOSFET is known as a voltage-driven device and, therefore, requires a simple gate control circuit.

The characteristic curves in Fig. 2.16(b) show that there are three distinct regions of operation, labeled as triode region, saturation region, and cutoff region. When used as a switching device, only the triode and cutoff regions are used; when it is used as an amplifier, the MOSFET must operate in the saturation region, which corresponds to the active region in the BJT.

The device operates in the cutoff region (*off* state) when $v_{GS} < v_{Th}$, resulting in no induced channel. In order to operate the MOSFET in either the triode or saturation region, a channel must first be induced. This can be accomplished by applying a gate-to-source voltage that exceeds v_{Th}, i.e.,

$$v_{GS} > V_{Th}$$

Once the channel is induced, the MOSFET can operate either in the triode region (when the channel is continuous with no pinch-off, resulting in the drain current being proportional to the channel resistance) or in the saturation region (the channel pinches off, resulting in constant I_D). The gate-to-drain bias voltage (v_{GD}) determines whether the induced channel undergoes pinch-off or not. This is subject to the following restrictions.

For the triode mode of operation, we have

$$v_{GD} > V_{Th}$$

and for the saturation region of operation, we have

$$v_{GD} < V_{Th}$$

Pinch-off occurs when $v_{GD} = V_{Th}$.

In terms of v_{DS}, the preceding inequalities may be expressed as follows:

1. For the triode region of operation,

$$v_{DS} < v_{GS} - V_{Th} \qquad v_{GS} > V_{Th}$$

2. For the saturation region of operation,

$$v_{DS} > v_{GS} - V_{Th} \qquad v_{GS} > V_{Th}$$

3. For the cutoff region of operation

$$v_{GS} < V_{Th}$$

It can be shown that the drain current, i_D, can be mathematically approximated as follows:

$$i_D = k[2(v_{GS} - V_{Th})v_{DS} - v_{DS}^2] \qquad \text{(triode region)}$$

$$i_D = k(v_{GS} - V_{Th})^2 \qquad \text{(saturation region)}$$

$$k = \frac{1}{2}\mu_n C_{ox}\left(\frac{W}{L}\right)$$

where

μ_n = electron mobility

C_{ox} = oxide capacitance per unit area

L = length of the channel

W = width of the channel

Typical values for these parameters are given in the PSPICE model discussed later. At the boundary between the saturation (active) and triode regions, we have

$$v_{DS} = v_{GS} - V_{Th}$$

resulting in the following equation for i_D:

$$i_D = kv_{DS}^2$$

The input transfer characteristic curve for i_D vs. v_{GS} when the device is operating in the saturation region is shown in Fig. 2.20.

The large-signal equivalent circuit model for an n-channel enhancement-type MOSFET operating in the saturation mode is shown in Fig. 2.21. The drain current is represented by a current source as a function of V_{Th} and v_{GS}.

If we assume the channel is pinched off, the drain-source current will no longer be constant but rather will depend on the value of v_{DS} as shown in Fig. 2.22. The increased

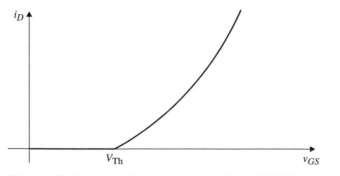

Figure 2.20 Input transfer characteristics for a MOSFET device operating in the saturation region.

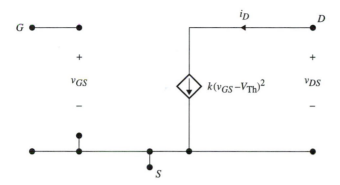

Figure 2.21 Large-signal equivalent circuit model.

value of v_{DS} results in reduced channel length, resulting in a phenomenon known as channel length modulation. If the v_{DS}-i_D lines are extended as shown in Fig. 2.22, they all intercept the v_{DS} axis at a single point labeled $-1/\lambda$, where λ is a positive constant MOSFET parameter. The term $(1 + \lambda v_{DS})$ is added to the i_D equation to account for the increase in i_D due to the channel length modulation. i_D is thus given by

$$i_D = k(v_{GS} - V_{Th})^2(1 + \lambda v_{DS}) \quad \text{(saturation region)}$$

From the definition of r_o given, it is easy to show that the MOSFET output resistance can be expressed as follows:

$$r_o = \frac{1}{\lambda k(v_{GS} - V_{Th})}$$

If we assume the MOSFET is operating under small-signal conditions, i.e., the variation in v_{GS} on the i_D vs. v_{GS} characteristic curve is in the neighborhood of the dc operating point Q at I_D and V_{GS}, as shown in Fig. 2.23, the i_D current source can be represented as the product of the slope g_m and v_{GS}, as shown in Fig. 2.24.

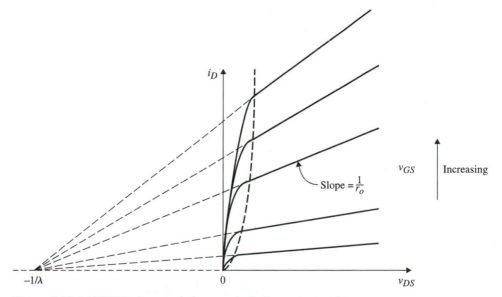

Figure 2.22 MOSFET characteristic curve including output resistance.

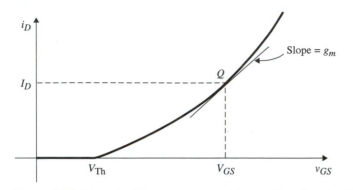

Figure 2.23 Linearized i_D vs. v_{GS} curve with operating dc point (Q).

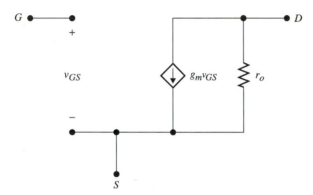

Figure 2.24 Small-signal equivalent circuit including MOSFET output resistance.

Input Capacitance   Because the MOSFET is a majority-carrier transport device, it is inherently capable of high-frequency operation. Still, the MOSFET has two limitations:

1. High input gate capacitances
2. Transient/delay due to carrier transport through the drift region

As stated earlier, the input capacitance consists of two components: the gate-to-source and gate-to-drain capacitances. The input capacitances can be expressed in terms of the device junction capacitances by applying the Miller theorem to Fig. 2.25(a). Using the Miller theorem, the total input capacitance, C_{in}, between the gate and source is given by

$$C_{in} = C_{gs} + (1 + g_m r_o)C_{gd}$$

The frequency response of the MOSFET circuit is limited by the charging and discharging times of C_{in}. Miller effect is inherent in any feedback transistor circuit with resistive load that exhibits a feedback capacitance from the input to the output. The objective is to reduce the feedback gate-to-drain resistance. The output capacitance between the drain and source, C_{ds}, does not affect the turn-on and turn-off MOSFET switching characteristics. Figure 2.26 shows how C_{gd} and C_{gs} vary under increased drain-source voltage, v_{DS}.

In power electronics applications, power MOSFETs are operated at high frequencies in order to reduce the size of the magnetic components. To reduce the switching

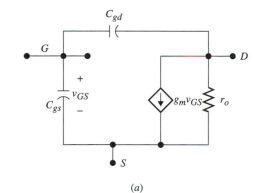

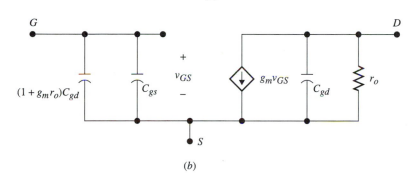

(b)

Figure 2.25 (*a*) Small-signal equivalent circuit including parasitic capacitances. (*b*) Applying the Miller theorem.

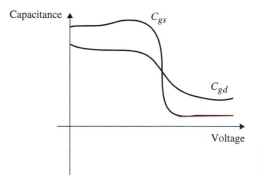

Figure 2.26 Variation of C_{gd} and C_{gs} as a function of v_{DS}.

losses, power MOSFETs are maintained in either the *on* state (conduction state) or *off* state (forward blocking state).

Safe Operation Area   The safe operation area (SOA) of a device provides the current and voltage limits the device must handle to avoid destructive failure. The typical SOA for a MOSFET device is shown in Fig. 2.27. The maximum current limit while the device is on is determined by the maximum power dissipation.

$$P_{\text{diss,on}} = I_{DS(\text{on})}^2 R_{DS(\text{on})}$$

As the drain-source voltage starts increasing, the device starts leaving the *on* state and enters the saturation (linear) region. During the transition time the device exhibits

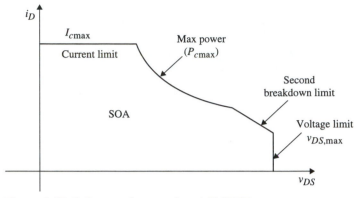

Figure 2.27 Safe operation area for a MOSFET.

large voltage and current simultaneously. At higher drain-source voltage values that approach the avalanche breakdown, it is observed that a power MOSFET suffers from a second breakdown phenomenon. The second breakdown occurs when the MOSFET is in the blocking state (off), and a further increase in v_{DS} will cause a sudden drop in the blocking voltage. The source of this phenomenon in MOSFETs is the presence of a parasitic n-type bipolar transistor, as shown in Fig. 2.28. The inherent presence of the body diode in the MOSFET structure makes the device attractive to applications in which bidirectional current flow is needed in the power switches.

Temperature Effect Today's commercial MOSFET devices have excellent response for high operating temperatures. The effect of temperature is more prominent on the *on*-state resistance, as shown in Fig. 2.29. As the *on*-state resistance increases, the conduc-

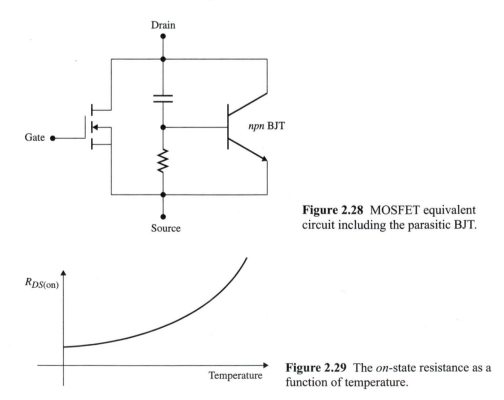

Figure 2.28 MOSFET equivalent circuit including the parasitic BJT.

Figure 2.29 The *on*-state resistance as a function of temperature.

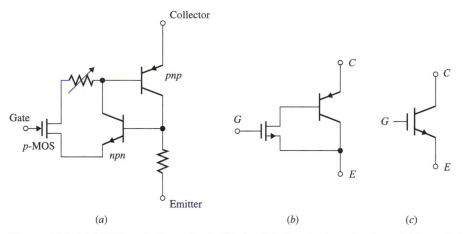

Figure 2.30 (*a*) IGBT equivalent circuit, (*b*) simplified equivalent circuit, and (*c*) symbol.

tion losses also increase. This large $v_{DS(\text{on})}$ limits the use of the MOSFET in high-voltage applications. The use of silicon carbide instead of silicon has reduced $v_{DS(\text{on})}$ manifold.

As the device technology keeps improving in terms of switch speeds and power-handling capabilities, it is expected that the MOSFET will continue to replace the BJT in all types of power electronic systems.

The Insulated Gate Bipolar Transistor (IGBT)

The detailed equivalent circuit model, the simplified two-transistor circuit model, and the schematic symbol for the insulated gate bipolar transistors (IGBT) are shown in Fig. 2.30(*a*), (*b*) and (*c*), respectively. Its *i-v* characteristic is similar to the MOSFET device and is not shown here. Since the IGBT architecture consists of a MOSFET and a BJT as shown in Fig. 2.30(*b*), it is clear that the IGBT has the high input impedance of the MOSFET along with the high current gain and small *on*-state conduction voltage of the BJT. The device was commercially introduced in 1983 and combines the advantages of MOSFETs, BJTs, and thyristor devices: the high current density allowed in BJT devices and the low-power gate drive needed in MOSFET devices.

The device is turned off by zero gate voltage, which removes the conducting channel. However, a negative base cannot turn off the *pnp* transistor current. As a result the turn-off time is higher in the IGBT than in the bipolar transistor. However, like the GTO (to be discussed shortly), the IGBT has a tail current at turn-off due to the recombination of carriers from the base region. At turn-on, a positive gate voltage is applied with respect to the emitter of the *npn* transistor, creating an *n*-channel in the MOS device that causes the *pnp* transistor to start conducting.

Its input capacitance is significantly smaller than that of the MOSFET device, and the device does not exhibit the second-breakdown phenomenon. It is faster than the BJT and can operate up to 20 kHz in medium-power applications. Currently, it is available at ratings as high as 1.2 kV and 400 A. The improvement in its fabrication is promising, and it is expected that it will replace the BJT in the majority of power electronics applications.

2.5.2   Thyristor-Based Devices

The generic term *thyristor* refers to the family of power semiconductor devices made of three *pn* junctions (four layers of *pnpn*) that can be latched into the *on* state through an external gate signal that causes a regeneration mechanism in the device. In this section, we will discuss four main members of the thyristor family that are currently used in power electronic circuits: The silicon-controlled rectifier (SCR), gate turn-off thyristor (GTO), triode ac switch (triac), static induction transistor (SIT), static induction thyristor (SITH), and MOS-controlled thyristor (MCT).

The Silicon-Controlled Rectifier

The silicon-controlled rectifier (SCR) is the oldest power controllable device utilized in power electronic circuits, introduced in 1958. Unlike the diode, the SCR can block voltages bidirectionally and carry current unidirectionally. Until the 1970s, when power transistors were presented, the conventional thyristor had been used extensively in various industrial applications. The SCR is a three-terminal device composed of a four-semiconductor *pn* junction. Unlike the diode, the SCR has a third terminal called the "gate" used for control purposes.

The symbol and *i-v* characteristics for the SCR are shown in Fig. 2.31(*a*) and (*b*), respectively. The ideal switching characteristic curves are shown in Fig. 2.31(*c*), where v_{AK} and i_A are the voltage across the anode-cathode terminals and the current through the anode, respectively.

The *latching current* is always less than the minimum trigger current specified in the device's data sheet. The *holding current* is the minimum forward current the SCR can carry in the absence of a gate drive. The *forward breakover voltage*, V_{BO}, is the voltage across the anode-cathode terminal that causes the SCR to turn on without the application of a gate current. *Reverse avalanche* (breakdown) occurs when v_{AK} is negatively large.

The normal operation of the SCR occurs when its gate is used to control the turn-on process by injecting a gate current i_G to allow the forward current to flow; v_{AK} is positive and can be turned off by applying a negative v_{AK} across it.

It must be noted that once the SCR is turned on, the gate signal can be removed. For this reason, this device is also known as a *latch device*. The gate current must be applied for a very short time and normally can go up to 100 mA. Once the SCR is turned on, it has a 0.5–2 V forward voltage.

The physical structure of the SCR consists of three *pn* junctions, as shown in Fig. 2.32. The different doping levels shown are used to help sustain a large block voltage and speed the breakdown process. Under no external bias voltage, the majority carriers diffuse across the junctions and recombine with the minority carriers, resulting in zero net current.

The Off State   Generally speaking, thyristor turn-off can be carried out by reversing the anode-cathode voltage (i.e., through natural ac commutation), or it can be turned off through forced commutation by switching a previously negatively charged capacitor across the SCR or by the insertion of a series impedance to reduce the forward current below the device's holding current.

In order to turn off the SCR, v_{AK} must be negative ($v_{AK} < 0$); of course, the triggering gate signal is immaterial (the presence of i_G only increases electron movement across the junction J_2). Under this condition, junctions J_1 and J_3 are reverse biased and

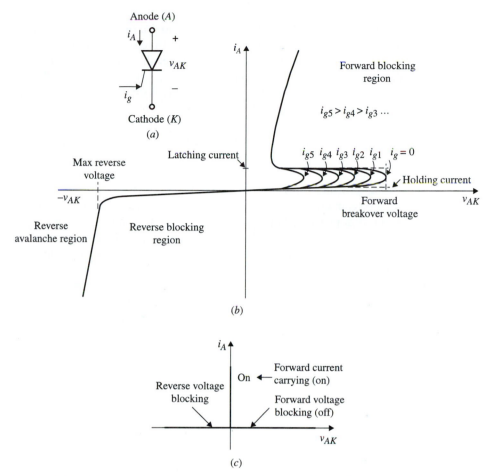

Figure 2.31 SCR switching characteristics. (*a*) Symbol. (*b*) *i-v* characteristics. (*c*) Ideal switching characteristics.

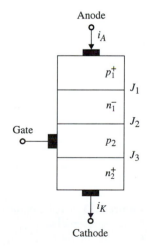

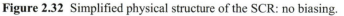

Figure 2.32 Simplified physical structure of the SCR: no biasing.

junction J_2 is forward biased. Like two reverse-bias diodes, the only currents that flow through the device are the leakage currents, I_{s1} and I_{s3}, of junctions J_1 and J_3, respectively. Notice that the larger the depletion region driven to the low doping level of n_1 compared to p_1, as shown in Fig. 2.33(a), the larger the reverse voltage and the wider the depletion regions of J_1 and J_3 become. Junction J_2 is forward biased with forward current, I_{F2}, equal to ($I_{s1} = I_{s3}$). The junction J_3 breaks down at lower voltages than J_1 due to different doping levels, and the large reverse blocking voltage is sustained by junction J_1.

Typical turn-off times for the SCR range from a few microseconds to 100 μs for low and high voltage ratings, respectively.

The On State   To turn on the SCR in the conduction state, a positive anode-cathode voltage must be applied ($v_{AK} > 0$) *and* a gate current must be injected to initiate the *on-state regeneration process*. For $v_{AK} > 0$ and $i_G = 0$, junctions J_1 and J_3 are in the forward-bias states and J_2 is in the reverse-bias state. Since n_1 is less doped than the p_2 region, the depletion region grows mostly into the n_1 region, as shown in Fig. 2.33(b).

Without a trigger current ($i_G = 0$), the junction J_2 is in the forward-blocking condition, and the only current that flows in the device is the small leakage current, I_{s2}, through J_2 since it is reverse biased. The forward current remains small until the critical forward breakover voltage, V_{BO}, is exceeded. At this point, the potential energy of the barrier at junction J_2 increases, accelerating the electron-hole pair generation until avalanche breakdown occurs, resulting in the thyristor being switched rapidly into the conduction state. This trigger mechanism should be avoided unless specified as safe by the manufacturer.

Gate triggering is achieved when a small pulse current, i_G, is injected at the gate, introducing an avalanche condition across J_2 and forward currents, I_{F1} and I_{F3}, that flow through J_1 and J_3, respectively. In order to explain the thyristor gate firing mechanism, normally the SCR is replaced by the model of two interconnected complementary *pnp* and *npn* transistors Q_1 and Q_2, respectively, as shown in Fig. 2.34.

We consider only the forward blocking and forward conducting states, since in the reverse blocking state the only current that flows is the leakage current of Q_1 and Q_2.

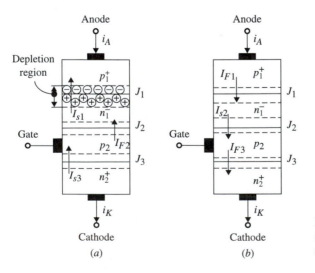

Figure 2.33   Depletion layer (a) under reverse bias ($v_{AK} < 0$) and (b) under forward bias ($v_{AK} > 0$).

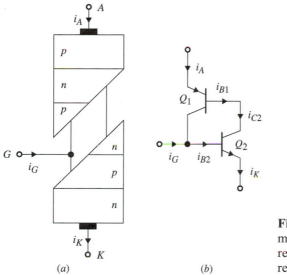

(a)   (b)

Figure 2.34 Two-transistor analogy model for the SCR. (a) *pn* junction representation. (b) Transistor circuit representation.

To derive the terminal current relation for the SCR, we turn to the large-signal model of the transistor, called the Ebers-Moll (EM) model, which is used for both transistor junctions, the *npn* and the *pnp*. Figure 2.35(a) and (b) shows the EM models for the *npn* and *pnp* transistors, respectively. Here α_{F1}[4] and α_{F2} denote the forward α, which is close to unity, and α_{R1} and α_{R2} denote the reverse α, which is normally very small (0.02). For simplicity we will assume α_{R1} and α_{R2} are zero. The currents $I_{D,BC}$ and $I_{D,CB}$ represent the diode leakage currents for the base-collector diode of Q_2 and the collector-base diode of Q_1 when both transistors are in the active mode. These currents are also known as the collector-to-base saturation current while the emitter is open-circuit, I_{CBO2}, for the *npn* transistor, and base-to-collector current, I_{BCO1}, for the *pnp* transistor. The direction of I_{CBO2} and I_{BCO1} are opposite to the direction of the leakage currents as shown in Fig. 2.35(a) and (b). Recall that, in the active region, the base-emitter junction is forward biased and the base-collector junction is reverse biased.

First, let us assume there's no gate triggering ($i_G = 0$). Replacing the equivalent EM models of Fig. 2.35(a) and (b) into Fig. 2.35(c), we obtain the two-transistor equivalent circuit model. With simple algebraic manipulation, we can show that

$$i_K = i_A = \frac{I_{CBO1} + I_{BCO2}}{1 - (\alpha_1 + \alpha_2)}$$

With $i_G = 0$, the only current that will flow is the leakage current (α_1 and α_2 are small). Normally $\alpha_1 + \alpha_2 \ll 1$ to keep it off. If $\alpha_1 + \alpha_2 = 1$, the SCR will enter a sustained breakdown, with the anode current limited only by the external circuitry.

To avoid entering the breakdown region, a gate signal is injected, resulting in the following forward current:

$$i_A = \frac{i_G \alpha_2 + I_{CBO1} + I_{BCO2}}{1 - (\alpha_1 + \alpha_2)} \tag{2.1}$$

---

[4]Forward α (α_F) is the same as the single transistor's α when operating in the active region, given by $\alpha = \beta/(1+\beta)$, where β is the transistor current gain, $\beta = I_C/I_B$.

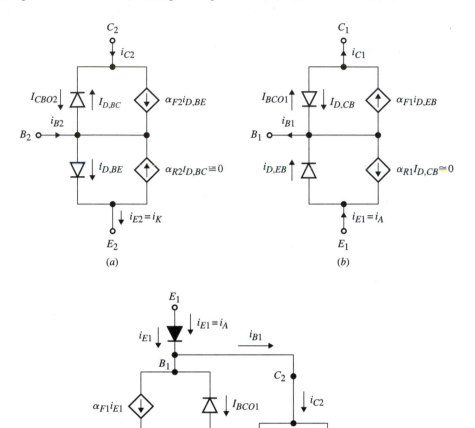

Figure 2.35 Complete Ebers-Moll model for (*a*) the *npn* transistor and (*b*) the *pnp* transistor. (*c*) Equivalent circuit model for Fig. 2.34(*b*).

$$i_K = \frac{(1 - \alpha_1)i_G + I_{CBO1} + I_{BCO2}}{1 - (\alpha_1 + \alpha_2)} \qquad (2.2)$$

For the generation process to start, we design the SCR for $\alpha_1 + \alpha_2 \approx 1$.

EXERCISE 2.7

Show that for the generation process to start, the following relation must be satisfied:

$$\beta_1 \beta_2 \leq 1$$

When the thyristor was invented, all the schemes for force-commutating mercury-arc rectifiers of the 1930s soon became thyristor-based circuits with expanded applications to include ac drives and UPSs. However, because of their cost and low efficiency, thyristor circuits did not penetrate the adjustable-speed-drive application area. Today's applications range from single phase-controlled rectifier circuits to static var compensation in utility systems. Because of its limited frequency of operation, the application of the thyristor has reached saturation.

The Gate Turn-off Thyristor (GTO)

The schematic symbol and the practical and ideal switching i-v characteristics for the gate turn-off thyristor (GTO) are shown in Fig. 2.36(a), (b), and (c), respectively. The device is as old as the SCR and was introduced commercially in 1962. Like the SCR, it can be turned on with a positive gate signal, but unlike the SCR, applying a negative gate signal, as shown in Fig. 2.36(b), turns off the GTO. Once the GTO is turned on or off, the gate signal can be removed. The device has a higher *on*-state voltage than the SCR at comparable currents. The GTO is normally an *off* device; it has a very poor turn-off current gain and it exhibits a second-breakdown problem at turn-off.

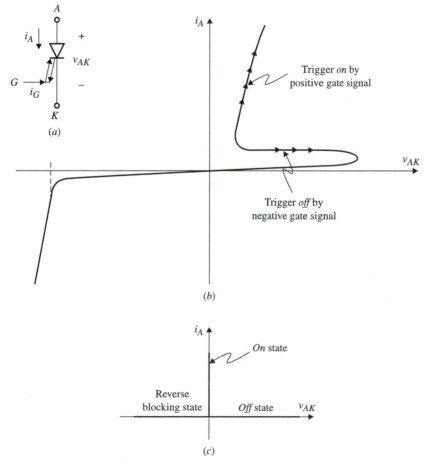

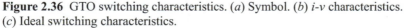

Figure 2.36 GTO switching characteristics. (*a*) Symbol. (*b*) *i*-*v* characteristics. (*c*) Ideal switching characteristics.

Because of its high switching power dissipation, the GTO's frequency of operation is limited to less than 1 kHz, and modern GTO devices are rated at 4.5 kV and at currents as high as 3 kA. The GTO is used in high current and voltage applications, such as voltage-fed inverters and induction heating resonant converters.

The Triode AC Switch (Triac)

Like the GTO, the *triode ac* (triac) switching device was introduced immediately after the SCR. In fact, the triac is nothing but a pair of SCRs connected in reverse-parallel on one integrated chip, as shown in Fig. 2.37. It is also known as a bidirectional SCR. The triac's equivalent circuit and the circuit schematic symbol are shown in Fig. 2.37(*a*) and (*b*), respectively. The device can be triggered in the positive and negative half-cycle of the ac voltage source by applying a positive or a negative gate signal, respectively. Today's triac ratings are up to 800 V at 40 A. The *i-v* characteristics and the ideal switching characteristics are shown in Fig. 2.37(*c*) and (*d*). Use of the triac is considerably limited due to its low rates of rise of voltage and current. Applications include light dimming, heating control, and various home appliances.

The Diac

Finally, we should mention another power device known as the diac, which is essentially a gateless triac constructed to break down at low forward and reverse voltages. The diac is mainly used as a triggering device for the triac.

Static Induction Transistors and Thyristors

In 1987 a device known as the static induction transistor (SIT) was introduced. One year later, the static induction thyristor (SITH) was introduced. The symbols for the SIT and SITH are shown in Fig. 2.38(*a*) and (*b*), respectively.

The SIT is a high-power and high-frequency device. The device is almost identical to the JFET, but with its special gate construction it has a lower channel resistance compared to the JFET.

The SIT and SITH are normally *on* devices and have no reverse voltage blocking capabilities. The SITH device turns off in the same way as the GTO, by applying a negative gate current, but it has a higher conduction drop than the GTO. Finally, both devices are majority-carrier devices with positive temperature coefficients, allowing device paralleling.

Among the SIT's major applications are audio and VHF/UHF amplifiers, microwaves, AM/FM transmitters, induction heating, and high-voltage, low-current power supplies. It has a large forward voltage drop compared to the MOSFET; hence, it is not normally used in power electronic converter applications. The applications of the SITH include static var compensators and induction heating.

The MOS-Controlled Thyristor

The simplified equivalent circuit model and the schematic symbol for a *p*-type MOS-controlled thyristor (MCT) are shown in Fig. 2.39(*a*) and (*b*), respectively. Its ideal *i-v* switching characteristic is similar to that of the GTO, as shown in Fig. 2.39(*c*). The device was commercially introduced in 1988. Like the GTO device, it has a high turn-off current gain.

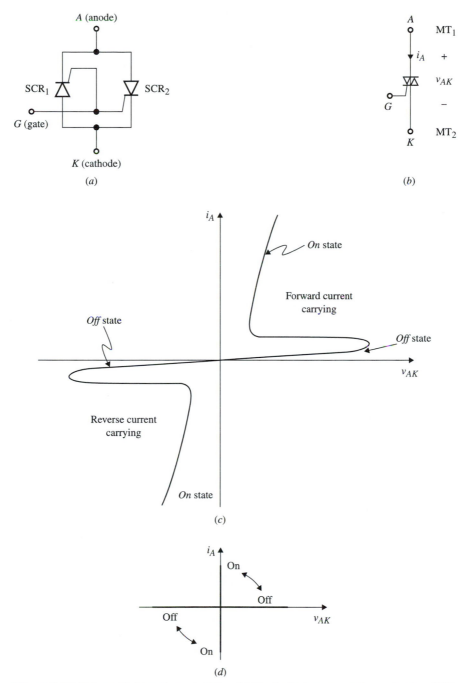

Figure 2.37 Triac switching characteristics. (*a*) Equivalent representation using two SCRs. (*b*) Symbol. (*c*) *i-v* characteristics. (*d*) Ideal switching characteristics.

The *p*-MCT is turned on by applying a negative gate voltage (less than –5 V) with respect to the cathode, turning on the *p*-FET and turning off the *n*-FET, initiating the regenerative mechanism in the SCR connected *npn* and *pnp* transistors. Similarly, applying a positive gate signal with respect to the cathode initiates the turn-off. The *n*-MCT has the same device structure, except that the *p*-FET and *n*-FET are

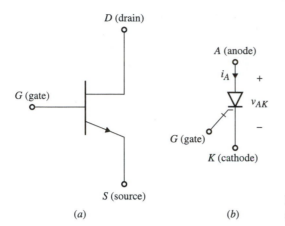

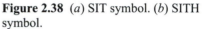

Figure 2.38 (*a*) SIT symbol. (*b*) SITH symbol.

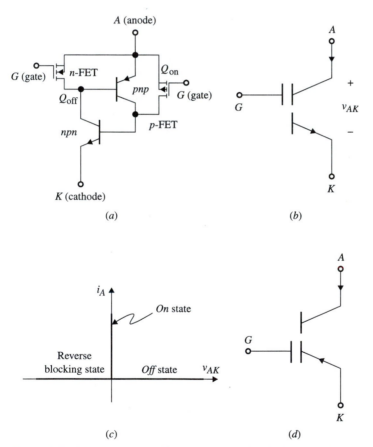

Figure 2.39 MCT switching characteristics. (*a*) Equivalent circuit.
(*b*) *p*-MCT symbol. (*c*) Ideal switching characteristics. (*d*) *n*-MCT symbol.

interchanged; hence, a positive and a negative gate signal turns the *n*-MCT on and off, respectively. The schematic symbol for the *n*-MCT is shown in Fig. 2.39(*d*).

The MCT's current and voltage ratings exceed 1 kV and 100 A and are continuously being improved. The device can be easily connected in series and in parallel combinations to boost power rating.

This device is serious competition for the IGBT. It has the same frequency of operation as the IGBT but with a smaller voltage drop and a higher operating temperature. Intensive efforts are under way to introduce a new improvement in the device, and it is expected to receive wider acceptance in medium- and high-power applications.

Other Power Devices

Other devices of the thyristor family include the reverse-conducting thyristor (RCT), which is nothing but a built-in anti-parallel body diode connected across the SCR to allow current to flow in the opposite direction, and the light-activated SCR (LASCR), which is used in high-voltage and high-current applications such as HVDC systems. Their power ratings go up to hundreds of kilovolts and hundreds of kiloamperes, and they provide complete electrical isolation between the power and control circuits.

2.6 COMPARISON OF POWER DEVICES

Depending on the applications, the power range processed in power electronics is very wide—from hundreds of milliwatts to hundreds of megawatts. Therefore, it is very difficult to find a single switching device type to cover all power electronics applications. Today's available power devices have tremendous power and frequency rating ranges as well as diversity. Their forward current ratings range from a few amperes to a few kiloamperes, their blocking voltage rating ranges from a few volts to a few kilovolts, and their switching frequency ranges from a few hundred hertz to a few megahertz, as illustrated in Table 2.2. This table gives only relative comparison between available power semiconductor devices because there is no straightforward technique that gives a ranking for these devices. Devices are still being developed very rapidly with higher current, voltage, and switching frequency ratings. Figure 2.40 shows a plot of frequency versus power, illustrating these rating ranges for various available power devices.

2.7 FUTURE TRENDS IN POWER DEVICES

It is expected that improvements in power-handling capabilities and increases in the frequency of operation of power devices will continue to drive the research and developments in semiconductor technology. From power MOSFETs to power MOS-IGBTs

Table 2.2 Comparison of Power Semiconductor Devices

Device type	Year made available	Rated voltage	Rated current	Rated frequency	Rated power	Forward voltage
Thyristor (SCR)	1957	6 kV	3.5 kA	500 Hz	100s MW	1.5–2.5 V
Triac	1958	1 kV	100 A	500 Hz	100s kW	1.5–2 V
GTO	1962	4.5 kV	3 kA	2 kHz	10s MW	3–4 V
BJT (Darlington)	1960s	1.2 kV	800 A	10 kHz	1 MW	1.5–3 V
MOSFET	1976	500 V	50 A	1 MHz	100 kW	3–4 V
IGBT	1983	1.2 kV	400 A	20 kHz	100s kW	3–4 V
SIT	1987	4 kV	600 A	100 kHz	10s kW	10–20 V
SITH	1975	4 kV	600 A	10 kHz	10s kW	2–4 V
MCT	1988	3 kV	2 kV	20–100 kHz	10s MW	1–2 V

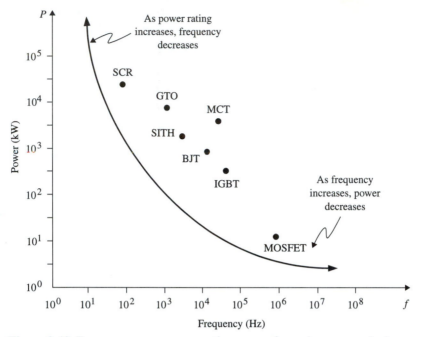

Figure 2.40 Frequency versus power rating ranges for various power devices.

to power MOS-controlled thyristors, the power rating has consistently increased by a factor of 5 from one type to another. Major research activities will focus on obtaining new device structures based on the MOS-BJT technology integration to rapidly increase power ratings. It is expected that the power MOS-BJT technology will capture more than 90% of the total power transistor market.

The continuing development of power semiconductor technology has resulted in power systems with driver circuits, logic and control, device protection, and switching devices designed and fabricated on a single chip. Such power IC modules are called "smart power" devices. For example, some of today's power supplies are available as ICs for use in low-power applications. There is no doubt that the development of smart power devices will continue in the near future, addressing more power electronics applications.

2.8   SNUBBER CIRCUITS

To relieve switches from overstress during switching, switching aid circuits, known as *snubber circuits*, are normally added to the power switching device. The objectives of snubber circuits may be summarized as (1) reducing the switching power losses in the main power device in the power electronic circuit, (2) avoiding second breakdowns, and (3) controlling the device's dv/dt or di/dt in order to avoid latching in *pnpn* devices. There are a wide range of turn-on and turn-off snubber circuits available in today's power electronic circuits. These include dissipative and nondissipative passive snubber circuits, and nondissipative active snubber circuits. In dissipative snubber circuits a capacitor is used to slow the device's voltage rise during turn-off, or an inductor to slow the device's current rise during turn-on. Figure 2.41(*a*) and (*b*) shows popular turn-off and turn-on snubber circuits, respectively. In Fig. 2.41(*a*), a capacitor is used to reduce the voltage rise dv_{sw}/dt across the switch during turn-off. In Fig. 2.41(*b*), a snubber inductor, L_s, is used to slow down the rise of the inductor current

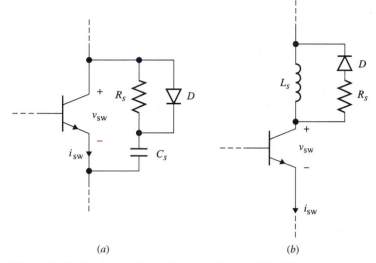

Figure 2.41 Passive snubber circuits: (*a*) turn-off and (*b*) turn-on snubber circuits.

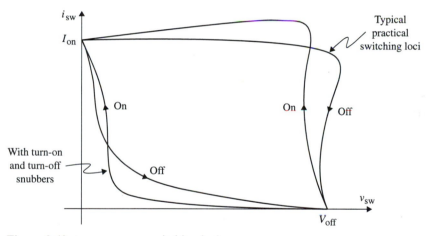

Figure 2.42 i_{sw} versus v_{sw} switching loci.

di_{sw}/dt (the inductor current equals the switch current, i_{sw}). Figure 2.42 shows the switching loci for a practical switch (transistor) with and without snubber circuits. For a detailed discussion on all types of snubber circuits and their design methods, refer to the references at the end of the textbook.

PROBLEMS

Ideal Switch Characteristics

2.1 Consider the switching circuit shown in Fig. P2.1 with a resistive load. Assume the switch is ideal and operating at a duty ratio of 40%.

(a) Sketch the waveforms for i_{sw} and v_{sw}.

(b) Determine the average output voltage.

(c) Determine the average output power delivered to the load.

(d) Determine the average output power supplied by the dc source.

(e) Determine the efficiency of the circuit.

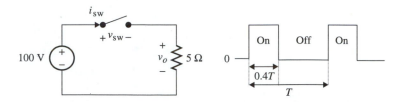

Figure P2.1

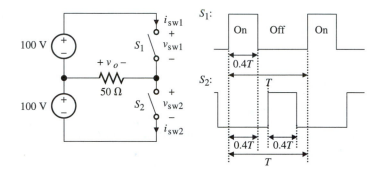

Figure P2.2

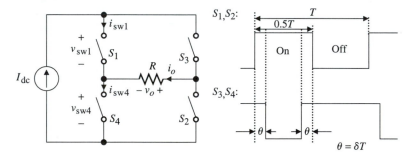

Figure P2.3

2.2 Consider the two-switch circuit given in Fig. P2.2. Assume ideal S_1 and S_2 with the shown switching sequence. Sketch i_{sw1}, v_{sw1}, i_{sw2}, and v_{sw2}. Determine (a) the average output voltage, (b) the average power delivered to the load, and (c) the efficiency.

2.3 The circuit shown in Fig. P2.3 is known as a current-driven full-bridge inverter. Assume S_1, S_2, S_3, and S_4 are ideal, with their switching waveforms as shown. Sketch the waveforms for i_o, v_o, v_{sw1}, i_{sw1}, v_{sw4}, and i_{sw4}, and derive the expression for the average output voltage.

Nonideal Switching Characteristics

2.4 Use Fig. E2.1 and by assuming that the switching waveform has the following parameters,

$I_{on} = 1$ A, $V_{off} = 150$ V, $I_{off} = 10$ μA, $V_{on} = 2$ V, $T_s = 10^{-4}$ s, $t_d = 90$ ns, $t_{fall} = 120$ ns, $t_{rise} = 100$ ns

determine the equation for the instantaneous power and find P_{ave}.

2.5 Repeat Problem 2.1 by assuming the switch has a 0.2 Ω *on*-state resistance and a 2 V forward drop during conduction.

2.6 Repeat Problem 2.3 by assuming each switch has a 0.2 Ω *on*-state resistance and a 2 V voltage drop during conduction.

2.7 Consider the circuit of Fig. P2.3 by assuming each switch has a forward voltage drop, V_F, and an *on*-state resistance r_{on}. Derive the expression for the circuit's efficiency in terms of I_{dc}, R, δ, V_F, and r_{on}. Determine δ for maximum efficiency.

2.8 Determine the conduction and switching average power dissipation for Example 2.2 by using $t_{on} = 5$ μs, $t_{off} = 8$ μs, $T_s = 150$ μs, $V_{off} = 150$ V, $V_{on} = 0.7$ V, and $I_{on} = 15$ A.

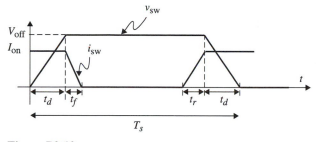

Figure P2.10

2.9 Repeat Problem 2.8 for Fig. E2.1 with $t_d = 4$ μs.

2.10 Consider a power switching device whose current and voltage waveforms are as shown in Fig. P2.10.

(a) Derive the expression for the instantaneous power and sketch it.

(b) Determine the average power dissipation.

(c) Calculate part (b) for $t_r = 120$ ns, $t_f = 180$ ns, $t_d = 90$ ns, $f_s = 100$ kHz, $V_{off} = 150$ V, and $I_{on} = 15$ A.

(d) Repeat parts (a)–(c) by assuming $V_{on} = 2$ V and $I_{off} = 10$ μA.

2.11 Use the linear approximated switching current and voltage given in Fig. 2.4 with zero forward voltage and zero leakage (reverse) current.

(a) Calculate the average power for $t_{on} = 5$ μs, $t_{off} = 8$ μs, $T_s = 150$ μs, $V_{off} = 150$ V, and $I_{on} = 15$ A.

(b) Repeat part (a) by assuming the t_{on} is measured from 10% to 90% of I_{on}, and t_{off} is 10% to 90% of V_{off}.

2.12 The switching current and voltage waveforms of Fig. 2.3 at turn-on are represented mathematically as follows:

$$i_{sw} = I_{on}(1 - e^{-t/\tau})$$

$$v_{sw} = V_{off}e^{-t/\tau}$$

where τ represents the time constant, which is a function of the on-state resistance of the device and its capacitance. Assume negligible I_{off} and V_{on}. Show that the switching power dissipation at turn-on is given by

$$P_{diss} = \frac{V_{off}I_{on}}{2T_s}\tau$$

Assume $t_{on} \gg \tau$ and $t_{off} \gg \tau$.

2.13 Assume the switching current and voltage of Fig. E2.1 at the turn-off interval ($0 \le t \le t_d + t_{off}$)

are represented as follows:

$$v_{sw}(t) = V_{off}$$

$$i_{sw}(t) = I_{on}e^{-t/\tau}$$

Derive the expression for the average switching power dissipation during the *off* time. Assume $t_d + t_f \approx 5\tau$.

2.14 The diode *i-v* characteristic curve in the forward region can be mathematically represented by

$$i_D = I_s e^{v_D/nV_T}$$

where

I_s = reverse saturation current

V_T = thermal voltage, equal to 25 mV at 20°C

n = empirical constant whose value depends on the semiconductor material and the physical construction of the device (normally between 1 and 2)

(a) Show that a decade change in the forward diode current results in a $2.3nV_T$ change in the forward voltage, mathematically expressed as follows:

$$V_{F2} - V_{F1} = 2.3nV_T\log\frac{I_{F2}}{I_{F1}}$$

(b) Consider the circuit of Fig. P2.14. Assume the diode has $I_F = 5$ A at $V_F = 1$ V with $n = 1.5$. Determine i_D and v_D for (i) $R = 10$ Ω and (ii) $R = 5$ Ω.

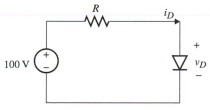

Figure P2.14

General Device Switching Problems

2.15 Consider Fig. P2.15, which shows an approximated reverse recovery turn-off characteristic for a power diode. Show that the following relation can express the total reverse recovery charge, Q_{rr}.

$$Q_{rr} = \frac{1}{2}t_{rr}t_{s1}\frac{di_1}{dt} = \frac{1}{2}t_{rr}t_{s2}\frac{di_2}{dt}$$

where di_1/dt and di_2/dt are the slopes of the diode current during t_{s1} and t_{s2}, respectively.

(a) Derive the expression for all the branch currents and v_D, and sketch them.

(b) Derive the expression for the capacitor C_s that is needed.

2.17 One way to speed the turn-off time of the Darlington connection is to include a diode between the bases of the two transistors as shown in Fig. P2.17. Discuss how the turn-off speed is improved by adding the diode D.

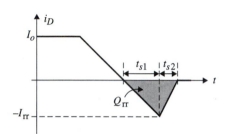

Figure P2.15

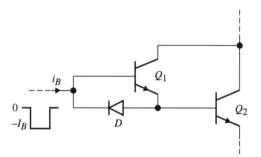

Figure P2.17

2.16 Consider the switching circuit with the series snubber R_s-C_s network connected across the diode as shown in Fig. P2.16. Assume the diode switching characteristic curve is given by Fig. P2.16(b) and the switch is ideal.

2.18 Consider the transistor switching circuit and its switching waveforms in Fig. P2.18(a) and (b), respectively.

(a) Sketch the waveforms for i_D and v_D.

(b) Calculate the average power dissipated in the transistor.

Use $f_s = 10$ kHz, $t_d = 150$ ns, $t_r = 100$ ns, $t_s = 100$ ns, and $t_f = 250$ ns.

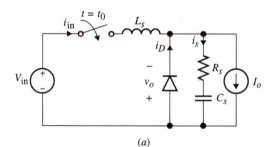

(a)

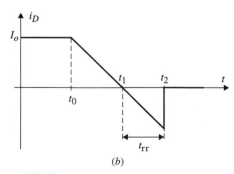

(b)

Figure P2.16

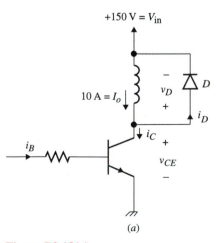

(a)

Figure P2.18(a)

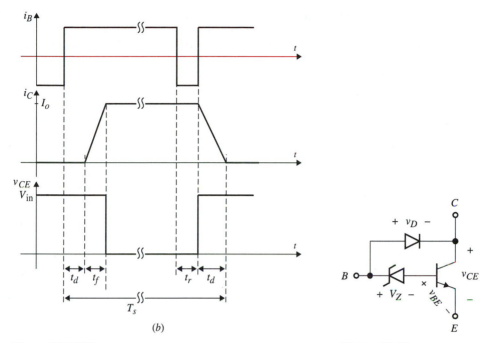

Figure P2.18(b)

Figure P2.19

2.19 Figure P2.19 shows a BJT switching circuit known as the Baker clamp, whose objective is to limit how deep in saturation the transistor is permitted to go. Show that if we assume $v_D = V_{BE}$, then the minimum V_{CE} is clamped to V_Z. If the collector is connected to a 15 V power supply through a collector resistor, R_C, and the base is connected to a 5 V power supply through a base resistor, R_B, design for V_Z, R_C,

and R_B so that the transistor is driven into saturation with $V_{CE,\text{sat}} = 1.8$ V and $\beta_{\text{forced}} = 10$.

2.20 Derive the SCR forward current relation given in Eqs. (2.1) and (2.2).

2.21 Consider the half-bridge inverter circuits with the unidirectional MOSFET switching a resistive load given in Fig. P2.21. Use $R = 10\ \Omega$.

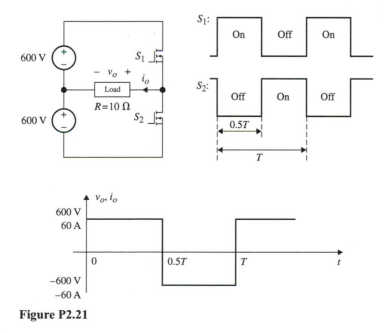

Figure P2.21

(a) Determine the average power delivered to the load when S_1 and S_2 are assumed ideal.

(b) Repeat part (a) for $R_{DS(on)} = 0.2\ \Omega$ for each MOSFET.

(c) Find the efficiency of the circuit in part (b).

2.22 Calculate β_{forced} in Fig. P2.22(a) and (b) for $R_B = 1\ k\Omega$ and $R_B = 10\ k\Omega$. Assume $V_{BE} = 0.75\ V$ and $V_{CE,sat} = 0.4\ V$.

2.23 Consider the Darlington transistor pair given in Fig. P2.23. Assume transistor Q_1 is driven into saturation with $V_{CE1,sat} = 0.3\ V$ and $\beta_{1,forced} = 0.4$. Calculate $\beta_{2,forced}$ and $V_{CE2,sat}$. Use $V_{BE1} = V_{BE2} = 0.7\ V$, and $V_B = +5\ V$.

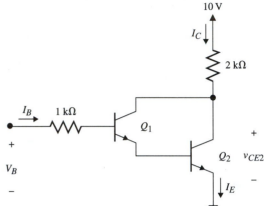

Figure P2.23

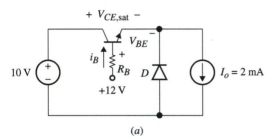

(a)

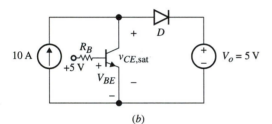

(b)

Figure P2.22

Chapter 3

Switching Circuits, Power Computations, and Component Concepts

INTRODUCTION

In this chapter, several general concepts will be discussed to provide the necessary background material in power electronics. The review material will cover switching diode circuits, power computation, harmonic analysis, and component concepts. Magnetic circuits, and inductor and transformer components will be covered in Appendix A.

3.1 SWITCHING DIODE CIRCUITS

Switching power electronic circuits process power more efficiently than linear power electronic circuits. This is why power electronic design engineers must understand the analysis of power switching circuits. In this section, we will discuss the analysis of switching circuits that include diodes, SCRs, and ideal switches under both dc and ac excitation.

Diode circuits are nonlinear and their analysis is normally not as straightforward as the analysis of linear circuits. By adding switches to diode circuits, additional nonlinearity is introduced, making the analysis even more complex. These circuits are encountered frequently in power electronics, such as in diode and SCR rectifier circuits, and in pulse-width-modulation (PWM) and resonant converters. To simplify the analyses, we will mostly assume that the diodes, SCRs, and switches are ideal.

3.1.1 Switching Diode Circuits under dc Excitation

We begin by analyzing diode resonant networks under dc excitation, which are often encountered in resonant-type dc-dc and dc-ac power electronic circuits. It will be shown later that resonance is used in power electronics to achieve several important functions, including filtering and soft switching. Let us first consider a series RLC circuit as shown in Fig. 3.1. Assume the switch is closed at $t = 0$ and $v_c(0^-)$ and $i_L(0^-)$ are the capacitor and inductor initial values, respectively, just before the switch is closed at $t = 0$.

The analysis of this circuit is straightforward and is given here as a review. Using KVL around the loop yields the following integral-differential equation in terms of the resonant current $i_L(t)$:

$$L\frac{di_L}{dt} + Ri_L + \frac{1}{C}\int_0^t i_L\, dt = V_{dc}$$

By taking the first derivative of this equation, we obtain a second-order differential equation:

$$\frac{d^2i_L}{dt^2} + \frac{R}{L}\frac{di_L}{dt} + \frac{1}{LC}i_L(t) = 0 \tag{3.1}$$

Since the excitation is a dc source, there exists only a *transient* (natural) solution,[1] whose roots are obtained from the *characteristic equation* given by

$$s^2 + \frac{R}{L}s + \frac{1}{LC} = 0 \tag{3.2a}$$

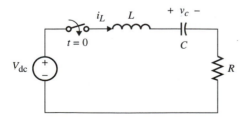

Figure 3.1 Series resonant RLC circuit with dc source.

---

[1] Also known as the *homogeneous solution*.

The roots of Eq. (3.2a) are given by

$$s_{1,2} = -\frac{R}{2L} \pm \sqrt{\left(\frac{R}{2L}\right)^2 - \frac{1}{LC}} \tag{3.2b}$$

Depending on the different values of these roots, there exist three different well-understood general solutions for the transient responses.

Case I (equal real roots). When the roots are real and equal, we obtain what is called a *critically damped* circuit, which occurs when the following circuit condition is met:

$$\left(\frac{R}{2L}\right)^2 = \frac{1}{LC}$$

The roots of Eq. (3.2b) then become

$$s_{1,2} = -\frac{R}{2L}$$

Under this condition, the general solution for the inductor current is given by

$$i_L(t) = (A_1 + A_2 t)e^{-(R/2L)t} \tag{3.3}$$

and the capacitor voltage is given by

$$v_c(t) = V_{dc} - L\frac{di_L}{dt} - Ri_L$$
$$= V_{dc} + \left[A_2\left(1 - \frac{R}{2L}t\right) - A_1\frac{R}{2L}\right]e^{-(R/2L)t} \tag{3.4}$$

The constants A_1 and A_2 are obtained from the given initial conditions.

Case II (unequal real roots). Under this condition, the circuit is known as *over-damped* and occurs when

$$\left(\frac{R}{2L}\right)^2 > \frac{1}{LC}$$

The general solution is given as follows:

$$i_L(t) = A_1 e^{s_1 t} + A_2 e^{s_2 t} \tag{3.5}$$

and the capacitor voltage is given by

$$v_c(t) = V_{dc} - (Ls_1 - R)A_1 e^{s_1 t} - (Ls_2 - R)A_2 e^{s_2 t} \tag{3.6}$$

where

$$s_1 = -\frac{R}{2L} + \sqrt{\left(\frac{R}{2L}\right)^2 - \frac{1}{LC}} \tag{3.7a}$$

$$s_2 = -\frac{R}{2L} - \sqrt{\left(\frac{R}{2L}\right)^2 - \frac{1}{LC}} \tag{3.7b}$$

Case III (complex pair of roots). Under this case, the circuit is known as *under-damped* and occurs when

$$\left(\frac{R}{2L}\right)^2 < \frac{1}{LC}$$

The responses are oscillatory with the general solution for $i_L(t)$ given by

$$i_L(t) = e^{-\alpha t}(A_1 \cos \omega_d t + A_2 \sin \omega_d t) \tag{3.8}$$

The voltage equation is given by

$$v_c(t) = (V_{\text{dc}} - Le^{-\alpha t})[(A_s\omega_d - A_1\alpha)\cos(\omega_d t) - (A_2\alpha + A_1\omega_d)\sin(\omega_d t)]$$
$$- Re^{-\alpha t}[A_2 \sin(\omega_d t) + A_1 \cos(\omega_d t)]$$

and the complex roots are given by

$$s_1 = -\alpha + j\omega_d$$
$$s_2 = -\alpha - j\omega_d$$

The parameters ω_d, α, and ω_0 are given by

$$\omega_d = \sqrt{\omega_0^2 - \alpha^2}$$

$$\alpha = \frac{R}{2L}$$

$$\omega_0 = \sqrt{\frac{1}{LC}}$$

where ω_0 is known as the *resonant frequency*, α is called the *damping factor*, and ω_d is known as the *damped resonant frequency*. The ratio α/ω_0 is defined as the *damping ratio*, δ:

$$\delta \equiv \frac{\alpha}{\omega_0} = \frac{R}{2\sqrt{L/C}}$$

It can be shown that the constants A_1 and A_2 for a given initial capacitor voltage, $v_c(0)$, and an initial inductor current, $i_L(0)$, for the series resonant RLC circuit of Fig. 3.1 can be found using Table 3.1.

For the critically damped case, we have $\delta = 1$, and for a purely capacitive-inductive circuit ($R = 0$), we have $\delta = 0$. In the latter case, the response is purely oscillatory. Such a response is encountered frequently in dc-dc soft-switching power electronic circuits. Another parameter that is normally given in the RLC circuit is the *quality factor*, Q_0, which is defined as

Table 3.1   Three Possible Cases and Their Response Constants for the Series Resonant RLC Circuit

Circuit type	A_1	A_2
Case I: critically damped	$i_L(0^-)$	$\dfrac{V_{\text{dc}} - v_c(0^-)}{L} + \dfrac{R}{2L}i_L(0^-)$
Case II: overdamped	$\dfrac{V_{\text{dc}} - v_c(0^-)}{L(s_1 - s_2)} + \dfrac{s_1 i_L(0^-)}{s_1 - s_2}$	$-\dfrac{V_{\text{dc}} - v_c(0^-)}{L(s_1 - s_2)} - \dfrac{s_2 i_L(0^-)}{s_1 - s_2}$
Case III: underdamped	$i_L(0^-)$	$\dfrac{V_{\text{dc}} - v_c(0^-)}{\omega_d L} - \dfrac{R}{2\omega_d L}i_L(0^-)$

$$s_1 = -\frac{R}{2L} + \sqrt{\left(\frac{R}{2L}\right)^2 - \frac{1}{LC}}, s_2 = -\frac{R}{2L} - \sqrt{\left(\frac{R}{2L}\right)^2 - \frac{1}{LC}}, \omega_d = \sqrt{\omega_0^2 - \alpha^2}, \alpha = \frac{R}{2L}, \omega_0 = \sqrt{\frac{1}{LC}}$$

$$Q_0 \equiv \frac{\omega_0 L}{R} = \frac{\sqrt{L/C}}{R}$$

$$= \frac{1}{2\delta}$$

(3.9)

The higher Q_0, the more oscillatory the current response becomes.

Another parameter of particular interest in power electronics is what is commonly referred to as the circuit *characteristic impedance*, Z_0, defined as

$$Z_0 \equiv \sqrt{\frac{L}{C}}$$

(3.10)

If R in the *RLC* circuit represents the load, then Q_0 is known as the *normalized load* and is given by

$$Q_0 = \frac{Z_0}{R}$$

(3.11a)

To give the same measure of oscillation in the series case for parallel resonant *RLC* circuits, the normalized load is defined as

$$Q_0 = \frac{R}{Z_0}$$

(3.11b)

The higher Q_0, the more oscillatory the voltage response becomes.

Normalized loads will be studied in dc-dc resonant converters in Chapter 6. Notice that Q_0 is defined the same as the quality factor of the resonant circuit. However, one should distinguish between the Q_0 of the resonant circuit in which R represents the losses in the resonant circuit that can be ignored, and the case when R represents the load, where Q_0 becomes a normalized load, i.e., a design parameter.

EXAMPLE 3.1

Consider the circuit of Fig. 3.2(a) with $R = 200$ Ω, $L = 2$ mH, $C = 0.01$ μF, and $V_{dc} = 20$ V. Derive the expressions for $i_L(t)$ and $v_c(t)$ for $t > 0$ after the switch is closed. Assume the initial inductor current and the initial capacitor voltage are zero, i.e., $i_L(0^-) = 0$ and $v_c(0^-) = 0$.

SOLUTION   Here we have $R/L = 100 \times 10^3$ rad/s, and $1/LC = 50 \times 10^9$ rad²/s². Since $(R/2L)^2 < 1/LC$, the circuit is *underdamped*; hence the roots of the characteristic equation are $s_1 = (-50 \times 10^3 + j218 \times 10^3)$ rad/s, and $s_2 = (-50 \times 10^3 - j218 \times 10^3)$ rad/s.

The general solution for the inductor current is given by

$$i_L(t) = e^{-50 \times 10^3 t}(A_1 \cos(218 \times 10^3 t) + A_2 \sin(218 \times 10^3 t)) \text{ A}$$

(3.12)

The constants A_1 and A_2 can be obtained from Table 3.1. However, for illustration purposes we will show how to find A_1 and A_2.

Since $i_L(0^+) = i_L(0^-) = 0$, then $A_1 = 0$. To solve for A_2, we use the capacitor initial condition.

Applying KVL to the circuit, we obtain

$$-V_{dc} + L\frac{di_L}{dt} + v_c + Ri_L = 0$$

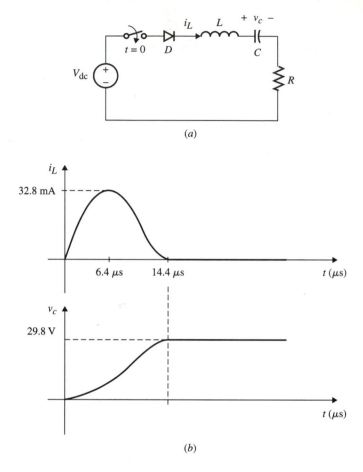

Figure 3.2 (a) Circuit for Example 3.1 and (b) its current and voltage waveforms.

Evaluating this equation at $t = 0^+$, we obtain the first derivative at $t = 0^+$:

$$\frac{di_L(0^+)}{dt} = \frac{V_{dc} - v_c(0^+)}{L} = 10 \text{ A/ms}$$

Taking the first derivative of Eq. (3.12), evaluating it at $t = 0^+$, and setting it to 10 A/ms, we obtain the following equation:

$$\left.\frac{di_L(t)}{dt}\right|_{t=0^+} = -50e^{-50t}A_2(\sin 218t) + 218e^{-50t}A_2(\cos 218t)\Big|_{t=0} = 10 \text{ A/ms}$$

Solving for A_2, we obtain $A_2 = 0.046$.

The inductor current and capacitor voltage expressions for $t > 0$ are given by

$$i_L(t) = 0.046e^{-50\times10^3 t}\sin(218 \times 10^3 t) \text{ A}$$

$$v_c(t) = V_{dc} - Ri_L - L\frac{di_L}{dt}$$

Substituting for $i_L(t)$, the capacitor voltage is given by

$$v_c(t) = 20 - e^{-50\times10^3 t}(4.6\sin(218 \times 10^3 t) - 20.056\cos(218 \times 10^3 t)) \text{ V}$$

The diode switches off when $i_L(t) = 0$, which occurs when $218 \times 10^3 t = \pi$, or $t = \pi/(218 \times 10^3) = 14.4 \ \mu\text{s}$, at which the capacitor voltage equals 29.76 V. The plots for i_L and v_c are given in Fig. 3.2(b).

EXAMPLE 3.2

Repeat Example 3.1 by replacing the diode across R as shown in Fig. 3.3(a). Again, assume $i_L(0^-) = 0$ and $v_c(0^-) = 0$.

SOLUTION   At $t = 0^+$ the diode begins conducting, because the inductor current just after $t > 0$ is positive (assuming an ideal diode). The equivalent circuit for $t \geq 0$ is shown in Fig. 3.3(b). This circuit has a damping ratio of zero, resulting in a purely sinusoidal response with $i_L(t)$ given by

$$i_L(t) = A_1 \cos \omega_0 t + A_2 \sin \omega_0 t$$

where $\omega_0 = 2.23 \times 10^5$ rad/s. Since the initial inductor current is zero, the constant $A_1 = 0$ and, from Table 3.1, $A_2 = 45 \times 10^{-3} = 0.045$; then the inductor current and capacitor voltage become

$$i_L(t) = 45 \sin \omega_0 t \text{ mA}$$
$$v_c(t) = 20(1 - \cos \omega_0 t) \text{ V}$$

The inductor and capacitor waveforms are shown in Fig. 3.3(c).

At $t = \pi/\omega_0$, the diode becomes reverse biased since i_L starts becoming negative. At this time the equivalent circuit model changes to a series RLC circuit with a new capacitor initial

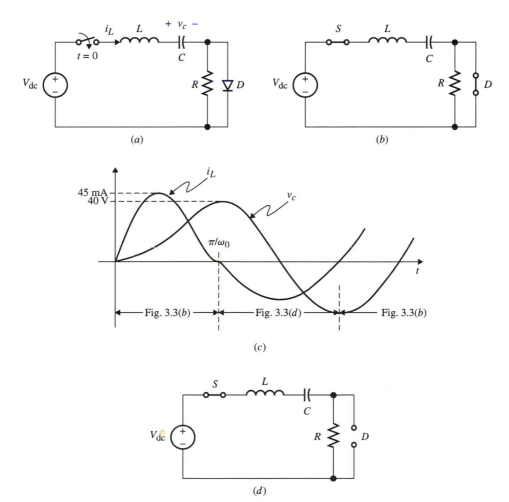

Figure 3.3 (a) RLC diode circuit for Example 3.2. (b) Equivalent circuit for $0 \leq t \leq \pi/\omega_0$. ($c$) The inductor and capacitor waveforms. (d) Equivalent circuit for $t \geq \pi/\omega_0$.

value, $v_c(\pi/\omega_0) = 40$ V, and $i_L(\pi/\omega_0) = 0$. Figure 3.3(d) shows the equivalent circuit for $t > \pi/\omega_0$. The solution for i_L and v_c is similar to that in Example 3.1 except that when i_L becomes zero again, the diode will conduct and the equivalent circuit mode of Fig. 3.3(b) again becomes valid. This process continues until the diode current decreases to zero.

EXERCISE 3.1

Repeat Example 3.2 with the diode direction reversed. Sketch the capacitor voltage and inductor current waveforms for $0 \le t < \pi/\omega_d$.

ANSWER   For $0 \le t < \pi/\omega_d$, $i_L(t) = 0.046e^{-50 \times 10^3 t} \sin(2.18 \times 10^5 t)$ A

$$v_c(t) = 20 - (4.6 \sin 2.18 \times 10^5 t + 20.1 \sin 2.18 \times 10^5 t)e^{-50 \times 10^3 t} \text{ V}$$

EXERCISE 3.2

Solve for $i_L(t)$ and $v_c(t)$ after the switch is closed at $t = 0$ in Fig. E3.2. This circuit is frequently encountered in dc-dc resonant converters. Assume the initial values are $v_c(0^-) = 0$ and $i_L(0^-) = 1.5I_g$. (The circuit that established this initial inductor current is not shown.)

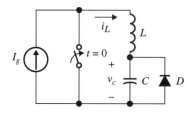

Figure E3.2 Circuit for Exercise 3.2.

ANSWER $i_L(t) = 1.5I_g \cos \omega_0 t$ A, $v_c(t) = 1.5I_g L \omega_0 \sin \omega_0 t$ V, $\omega_0 = \sqrt{1/LC}$ for $\omega_0 t \le \pi$

$$i_L(t) = -1.5I_g, \; v_c(t) = 0 \text{ for } \omega_0 t > \pi$$

EXERCISE 3.3

Repeat Exercise 3.2 by assuming the initial conditions are $v_c(0^-) = 0$ V and $i_L(0^-) = 0.5I_g$ A.

ANSWER $i_L(t) = 0.5I_g \cos \omega_0 t$ A, $v_c(t) = 0.5I_g L \omega_0 \sin \omega_0 t$ V, $\omega_0 = \sqrt{1/LC}$ for $\omega_0 t \le \pi$

$$i_L(t) = -0.5I_g, \; v_c(t) = 0 \text{ for } \omega_0 t > \pi$$

EXERCISE 3.4

Give switch implementations (unidirectional or bidirectional) for Exercises 3.2 and 3.3.

3.1.2   Switching Diode Circuits with an ac Source

Large classes of switching converter circuits use an ac source excitation rather than a dc source. The analysis of diode switching circuits with ac sources is carried out in two steps: first obtain the transient response (also known as the natural response) by setting the ac source to zero, and then obtain the steady-state response (also known as

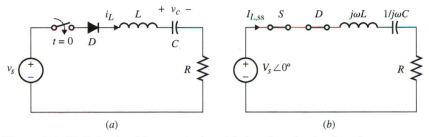

Figure 3.4 *RLC* circuit with ac excitation: (*a*) time domain circuit and (*b*) equivalent phasor domain circuit.

the *forced response*) by converting the circuit to the phasor domain. The final solution is the sum of the natural and the forced responses. Figure 3.4(*a*) shows an *RLC* circuit with an ac source, $v_s(t) = V_s \sin \omega t$ V. Assume the switch is turned on at $t = 0$, and we wish to solve for the inductor current.

For $t > 0$, KVL yields the following differential-integral equation in terms of $i_L(t)$:

$$-V_s \sin \omega t + L\frac{di_L}{dt} + \frac{1}{C}\int_{-\infty}^{t} i_L(t)\,dt + Ri_L(t) = 0$$

Taking the first derivative of this equation, we obtain

$$\frac{d^2 i_L}{dt^2} + \frac{R}{L}\frac{di_L}{dt} + \frac{1}{LC}i_L(t) = \frac{V_s \omega}{L}\cos \omega t \qquad (3.13)$$

The right-hand side of Eq. (3.13) represents the forced excitation. First we obtain the natural-response component of the complete solution by setting the forced function (source function v_s and its derivatives) to zero, to yield

$$\frac{d^2 i_L}{dt^2} + \frac{R}{L}\frac{di_L}{dt} + \frac{1}{LC}i_L(t) = 0 \qquad (3.14)$$

The natural response of i_L is the same as the response for the dc-source *RLC* circuit shown in Table 3.1 for the overdamped case; i.e., the transient response, $i_{L,\text{tran}}$, is given by

$$i_{L,\text{tran}}(t) = A_1 e^{s_1 t} + A_2 e^{s_2 t} \qquad (3.15)$$

The steady-state response component is obtained by transferring the circuit to the phasor domain as shown in Fig. 3.4(*b*). The steady-state response, $I_{L,\text{ss}}$, is easily obtained:

$$I_{L,\text{ss}} = \frac{V_s \angle 0}{R + j\left(\omega L - \dfrac{1}{\omega C}\right)} = \frac{V_s \angle -\theta}{|Z|} \qquad (3.16)$$

where

$$|Z| = \sqrt{R^2 + \left(\omega L - \frac{1}{\omega C}\right)^2} \quad \text{and} \quad \theta = \tan^{-1}\left(\frac{\omega L - \dfrac{1}{\omega C}}{R}\right)$$

In the time domain, $i_{L,\text{ss}}$ is given by

$$i_{L,\text{ss}}(t) = \frac{V_s}{|Z|}\sin(\omega t - \theta) \qquad (3.17)$$

The total response is obtained by adding Eqs. (3.15) and (3.17), to yield

$$i_L(t) = i_{L,\text{tran}}(t) + i_{L,\text{ss}}(t)$$

$$= A_1 e^{s_1 t} + A_2 e^{s_2 t} + \frac{V_s}{|Z|} \sin(\omega t - \theta) \tag{3.18}$$

The constants A_1 and A_2 are obtained from the initial condition of the inductor current and capacitor voltage at $t = 0^+$.

EXAMPLE 3.3

Consider the ac diode circuit shown in Fig. 3.5(a) with a dc source in the load side representing either a charged battery or a back electromotive force (emf) to excite the armature circuit in a dc motor. Sketch the waveforms for i_o, v_D, and v_o. Assume $v_s = 100 \sin 377t$. What is the average value of v_o?

SOLUTION The diode will turn on when $v_D > 0$, which occurs at $t = t_1$ when $v_s(t_1) = 60$ V, i.e.,

$$100 \sin 377t_1 = 60 \text{ V}$$

resulting in $t_1 = 1.7$ ms. For $0 \le t \le t_1$ the diode is off, and for the interval $t_1 \le t < t_2$ the diode is on, where $t_2 = T/2 - t_1 = 6.63$ ms.

During $t_1 \le t \le t_2$, the output voltage, v_o, equals v_s and the output current is given by

$$i_o(t) = \frac{v_s - V_{dc}}{R} = 12.5 \sin 377t - 7.5 \text{ A}$$

For all other times, namely $0 \le t < t_1$ and $t_2 \le t < T$, $i_o = 0$ and $v_o = V_{dc} = 60$ V.
The average output voltage is obtained from the following equation:

$$V_{o,\text{ave}} = \frac{1}{t}\left[\int_{t_1}^{t_2} V_s \sin \omega t \, dt + V_{dc}(T/2 + 2t_1)\right]$$

$$= \frac{1}{16.67 \text{ ms}}\left[\int_{1.7 \text{ ms}}^{6.63 \text{ ms}} 100 \sin 377t \, dt + 60(8.33 \text{ ms} + 3.4 \text{ ms})\right] = 67.79 \text{ V}$$

The voltage and current waveforms are shown in Fig. 3.5(b).

EXERCISE 3.5

Repeat Example 3.3 by including a freewheeling diode across the load as shown in Fig. E3.5.

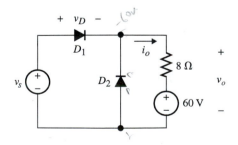

Figure E3.5 Circuit for Exercise 3.5.

ANSWER   67.79 V

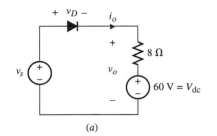

(a)

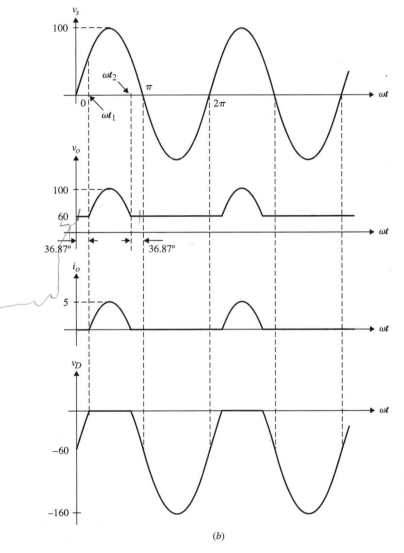

(b)

Figure 3.5
(a) ac diode circuit for Example 3.3.
(b) The voltage and current waveforms.

EXERCISE 3.6

Repeat Exercise 3.5 by reversing the polarity of the 60 V dc source.

ANSWER   31.83 V

EXAMPLE 3.4

Consider the circuit shown in Fig. 3.6(a) with a dc source in the load side. Assume ideal diodes, zero initial inductor current, and $V_{dc} < V_s$. Find the expressions for i_{D1}, i_{D2}, i_o, and v_o for $0 < t < 2T$, where T is the period of v_s given by $T = 2\pi/\omega$. Assume $v_s = V_s \sin \omega t$ with $V_s = 100$ V, $\omega = 377$ rad/s, $T = 16.67$ ms.

SOLUTION   At $t = 0$ the switch closes and v_s is switched into the circuit. However, when $v_s \leq V_{dc}$, D_1 remains off and so does D_2.

At $t = t_1$, $v_s = V_{dc}$, which forces D_1 to turn on and D_2 to turn off, resulting in the equivalent circuit of mode 1 for $t > t_1$ shown in Fig. 3.6(b). t_1 can be determined by the following equation:

$$t_1 = \frac{1}{\omega}\sin^{-1}\left(\frac{V_{dc}}{V_s}\right) = 0.67 \text{ ms}$$

The differential equation for mode 1 is given by

$$\frac{di_L}{dt} + \frac{R}{L}i_L = \frac{v_s - V_{dc}}{L}$$

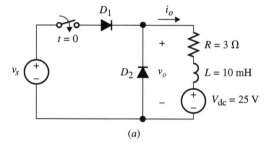

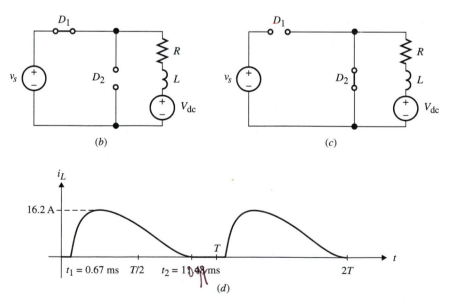

Figure 3.6 (a) Circuit for Example 3.4. (b) Mode 1: $0 \leq t < T/2$. (c) Mode 2: $T/2 < t < T$. (d) Plot of $i_L(t)$ for $0 < t < 2T$.

The transient response of $i_L(t)$ is

$$i_{L,\text{tran}}(t) = I_1 e^{-(t-t_1)/\tau}$$

where $\tau = L/R = 3.33$ ms, and the steady-state response is given by

$$i_{L,\text{ss}}(t) = \frac{V_s}{|Z|}\sin(\omega t - \theta) - \frac{V_{dc}}{R} = 20.75\sin(377t - 51.5°) - 8.33 \text{ A}$$

where $|Z| = \sqrt{R^2 + (\omega L)^2} = 4.82\ \Omega$ and $\theta = \tan^{-1}(\omega L/R) = 51.5°$.
   The overall response is given by

$$i_L(t) = I_1 e^{-(t-t_1)/\tau} + \frac{V_s}{|Z|}\sin(\omega t - \theta) - \frac{V_{dc}}{R} \tag{3.19}$$

The constant I_1 is obtained by setting $i_L = 0$ at $t = t_1$ in Eq. (3.19), which yields

$$I_1 = -\frac{V_s}{|Z|}\sin(\omega t_1 - \theta) + \frac{V_{dc}}{R}$$
$$= 20.82 \text{ A}$$

The general solution for $i_L(t)$ is given by

$$i_L(t) = 24.57 e^{-(t-t_1)/\tau} + 20.75\sin(377t - 51.5°) - 8.33 \qquad 0 \le t < T/2 \tag{3.20}$$

where

$$i_L(T/2) = 10.03 \text{ A} \tag{3.21}$$

   At $t = T/2$ the source voltage becomes negative, forcing D_2 to turn on and D_1 to turn off. The equivalent circuit for mode 2 is shown in Fig. 3.6(c). The inductor current at $t = T/2$ becomes the initial inductor current for the next cycle at $t = T/2$.
   This mode remains until the inductor current becomes zero again, and the transient response of $i_L(t)$ at mode 2 is given by

$$i_{L,\text{tran}}(t) = I_2 e^{-(t-T/2)/\tau}$$

and the steady-state response is given by

$$i_{L,\text{ss}}(t) = -\frac{V_{dc}}{R} = -8.33 \text{ A}$$

resulting in the following expression for $i_L(t)$:

$$i_L(t) = I_2 e^{-(t-T/2)/\tau} - 8.33 \text{ A}$$

The constant I_2 is obtained by setting $i_L = 10.03$ at $t = T/2$, which yields

$$I_2 = 10.03 + 8.33 = 18.36 \text{ A}$$

and the overall response for mode 2 can be given by

$$i_L(t) = 18.36 e^{-(t-T/2)/\tau} - 8.33 \text{ A} \qquad T/2 \le t < t_2$$

where t_2 is the time when $i_L(t)$ decreases to zero. It can be found by setting $i_L = 0$ at $t = t_2$, which yields

$$t_2 = -\tau \ln\frac{8.33}{18.36} + T/2 = 10.97 \text{ ms}$$

After that $i_L(t)$ will remain zero until $t = T + t_1$. The plot for $i_L(t)$ for $0 < t < 2T$ is shown in Fig. 3.6(d).

3.2 SWITCHING SCR CIRCUITS

Consider the SCR circuit shown in Fig. 3.7(a) driven by an ac voltage source $v_s = V_s \sin \omega t$. Assume the switch is turned on at $t = 0$ and the SCR is triggered by applying a gate current, i_g, at $t = t_1$.

At $t = 0$, when the switch is closed, it is still an open circuit and $i_o = 0$. At $t = t_1$, the SCR is triggered and since $v_s(t_1) = V_s \sin \omega t_1 > 0$, the SCR turns on, resulting in the equivalent circuit shown in Fig. 3.7(b).

The circuit is equivalent to the case of the ac diode circuit for $t \geq 0$. The differential equation representing this circuit is given by

$$\frac{d^2 i_o}{dt^2} + \frac{R}{L}\frac{di_o}{dt} + \frac{1}{LC}i_o = \frac{V_s \omega}{L}\cos \omega t \qquad t > t_1$$

The general solution for $i_o(t)$ for $t > t_1$ is given by

$$i_o(t) = \frac{V_s}{|Z|}\sin[\omega t - \theta] + A_1 e^{s_1(t-t_1)} + A_2 e^{s_2(t-t_1)} \qquad (3.22)$$

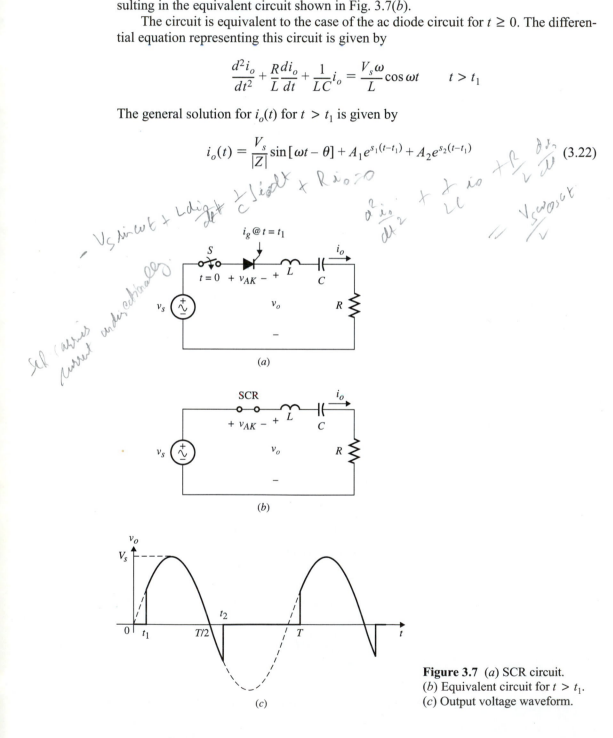

(a)

(b)

t_2

(c)

Figure 3.7 (a) SCR circuit. (b) Equivalent circuit for $t > t_1$. (c) Output voltage waveform.

where

$$|Z| = \sqrt{R^2 + \left(\omega L - \frac{1}{\omega C}\right)^2}$$

$$\theta = \tan^{-1}\frac{\omega L - \dfrac{1}{\omega C}}{R}$$

The constants A_1 and A_2 are obtained by applying the initial conditions of the inductor and the capacitor.

The output voltage, v_o, for $t > 0$ is given by

$$v_o = \begin{cases} 0 & 0 \le t < t_1 \\ V_s \sin \omega t & t_1 \le t < t_2 \\ 0 & t_2 \le t < T \end{cases}$$

where t_1 is the time at which the SCR is first turned on and t_2 is when the SCR is turned off, because $i_o(t)$ is zero for $t_2 > T/2$. If the SCR gate current is applied at $t = t_1$ into the new cycle, v_o becomes a periodical waveform, as shown in Fig. 3.7(c), whose average value is given by

$$V_{o,\text{ave}} = \frac{1}{T}\int_{t_1}^{t_2} V_s \sin \omega t \, dt = \frac{V_s}{2\pi}(\cos\alpha - \cos\beta) \tag{3.23}$$

where $\alpha = \omega t_1$ and $\beta = \omega t_2$. The angle α is known as the *firing angle*, and $(\beta - \alpha)$ is known as the SCR *conduction angle*. Notice that by varying α, we can vary the average output voltage. Such circuits will be studied in detail in Chapter 8.

EXERCISE 3.7

Assume the SCR circuit in Fig. 3.7(a) is triggered at $\alpha = 30°$ (i.e., $t = 1.39$ ms) after the switch is closed at $t = 0$. Derive the expression for $i_L(t)$ and $v_c(t)$ for $0 < \omega t < 2\pi$, assuming zero initial condition. Use $R = 200\ \Omega$, $L = 2$ mH, $C = 0.01\ \mu$F, and $v_s(t) = 20\sin 377t$.

ANSWER

$$i_L(t) = e^{-\alpha(t-t_1)}[A_1\cos\omega_d(t-t_1) + A_2\sin\omega_d(t-t_1)] + 75.4 \times 10^{-6}\sin(\omega t + 89.96°) \quad t_1 < t < t_2$$

where

$$\alpha = 50 \times 10^3,\ \omega_d = 217{,}945,\ t_1 = 1.39\ \text{ms},\ t_2 = 1.404\ \text{ms},\ A_1 = -65.35 \times 10^{-6},\ A_2 = 22.92 \times 10^{-3}$$

EXERCISE 3.8

Repeat Exercise 3.7 by placing an ideal diode as shown in Fig. E3.8.

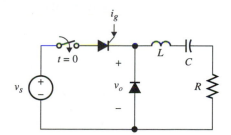

Figure E3.8 Circuit for Exercise 3.8.

ANSWER $i_L(t) = e^{-\alpha t_1}[A_1 \cos(\omega_d t) + A_2 \sin(\omega_d t)] + 75.4 \times 10^{-6} \sin(\omega t + 89.96°)$ A $t_1 < t < t_2$

where

$$\alpha = 50 \times 10^3,\ \omega_d = 217{,}945,\ t_1 = 4.17 \text{ ms},\ A_1 = -75.4 \times 10^{-6},\ A_2 = -17.3 \times 10^{-3}$$

3.3   BASIC POWER AND HARMONIC CONCEPTS

In this section we will review some basic power concepts applied to sinusoidal and nonsinusoidal current waveforms that are of particular importance in power electronic circuits.

3.3.1   Average, Reactive, and Apparent Powers

Power Flow

As stated earlier, the function of a power electronic circuit is to process power by performing some conversion function through a set of switching actions dictated by some control circuit. The direction of power flow in a power electronic circuit is an important concept since it relates to identifying the input and output ports in the circuit. Normally, the situation exists as shown in Fig. 3.8(*a*), where the power flow is from the input side to the output side. Here P_{in} and P_{out} represent the average input power from the source side (input ac or dc) and the average output power at the load side (output ac or dc), respectively. One may conclude from Fig. 3.8(*a*) that the average power flow is from the input terminal (source side) to the output terminal (load side); hence, the direction of power flow becomes the basis for defining the power circuit port. However, since some power electronic circuits have a dc source in the load side, it is possible to have the power flow in the opposite direction; i.e., the circuit is capable of bidirectional power flow, as shown in Fig. 3.8(*b*). As a result, one has to be careful in identifying the source and load sides. A good discussion on this issue, supported by several examples, is given in the reference by Kassakian et al.

The efficiency of the power-processing circuit is an extremely important parameter since it has a direct impact on the cost, performance, size, and weight of the system, as discussed in Chapter 1. For Fig. 3.8(*a*), the efficiency, η, is defined by

$$\eta = \frac{P_{out}}{P_{in}} \times 100\%$$

$$= \frac{P_{out}}{P_{out} + P_{loss}} \times 100\%$$

If the power circuit consists of ideal switching devices that operate in either the *on* or the *off* state, and lossless energy storage elements such as capacitors, inductors, and transformers, then the overall efficiency of the power processing circuit is 100%.

Average Values and rms

For a given periodical voltage signal, $v(t)$, with period T, its average value is defined by

$$V_{ave} = \frac{1}{T} \int_0^T v(t)\ dt \tag{3.24}$$

and the *root-mean-square* (rms) value is given by

$$V_{rms} = \sqrt{\frac{1}{T} \int_0^T v^2(t)\ dt} \tag{3.25}$$

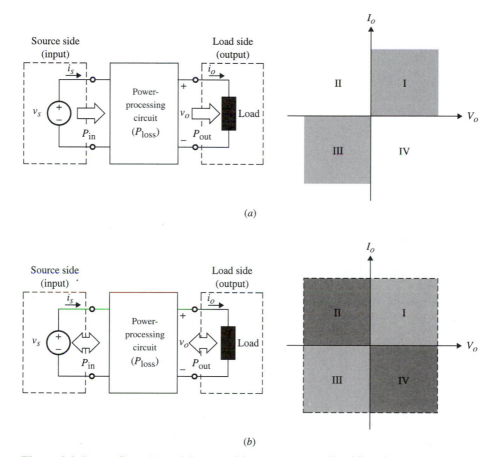

Figure 3.8 Power flow. (*a*) Unidirectional input-to-output. (*b*) Bidirectional.

Instantaneous Power

The instantaneous power, $p(t)$ delivered to a load element that has v(t) across it and $i(t)$ through it, as shown in Fig. 3.9(*a*), is given by

$$p(t) = v(t)i(t) \tag{3.26}$$

The voltage and current waveforms can be sinusoidal, periodical, or constant as shown by the arbitrary waveforms in Fig. 3.9(*c*). Notice that the instantaneous power in this figure can be either positive, zero, or negative. For example, for $0 \leq t \leq t_1$, the circuit element absorbs or dissipates power (positive), for $t_1 \leq t \leq t_2$, the power transfer to the element is zero, and for $t_2 \leq t \leq t_3$, the load generates power (positive) or returns power (negative) to the source.

Average Power

If the voltage and current repeat periodically, then we can define the total *average power* either generated or dissipated by the circuit element, which is given by

$$P_{\text{ave}} = \frac{1}{T}\int_0^T p(t)\, dt$$

$$= \frac{1}{T}\int_0^T i(t)v(t)\, dt$$

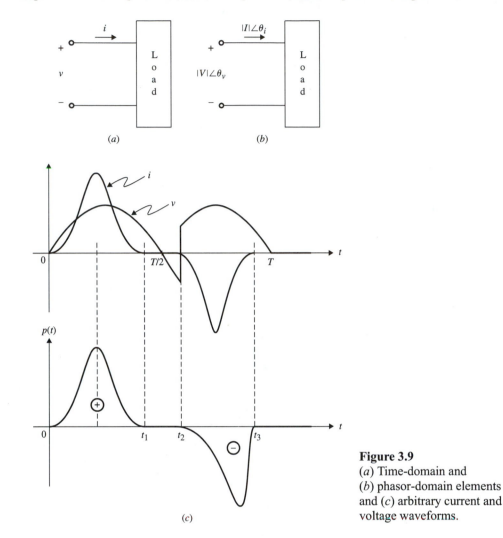

Figure 3.9
(*a*) Time-domain and
(*b*) phasor-domain elements
and (*c*) arbitrary current and
voltage waveforms.

The average power is also known as *real power,* which comes from the complex-number representation. The linear time-domain circuit element of Fig. 3.9(*a*) is redrawn as a phasor-domain circuit element in Fig. 3.9(*b*).

Apparent Power

The phasor magnitudes $|I_s|$ and $|V_s|$ represent the peaks of the source current and voltage, respectively, and θ_i, θ_v represent the current and voltage phase shifts, respectively. If we let the total impedance of the element be given by

$$Z = R + jX = |Z|e^{j\theta}$$

where X is the total reactance of a capacitive circuit, $-1/\omega C$, or a total reactance of an inductive circuit, ωL, then the total input complex power may be given by

$$P_T = \frac{VI^*}{2} = \frac{|V||I|}{2}e^{j(\theta_v - \theta_i)}$$

$$= V_{\mathrm{rms}}I_{\mathrm{rms}}e^{j\theta}$$

$$= Se^{j\theta}$$

where I^* is the complex conjugate of I. The parameter S is known as the *apparent power*, and θ represents the total phase shift between $i(t)$ and $v(t)$ and is known as the *power factor angle*.

Normally, P_T is expressed in terms of real and imaginary complex numbers as follows:

$$P_T = P + jQ \qquad (3.27)$$

The real part, P, is the *average power*, which is given by

$$\begin{aligned} P &= S\cos\theta \\ &= V_{rms}I_{rms}\cos\theta \end{aligned} \qquad (3.28)$$

and Q is the *reactive power*, which is given by

$$\begin{aligned} Q &= S\sin\theta \\ &= V_{rms}I_{rms}\sin\theta \end{aligned} \qquad (3.29)$$

From these expressions, S can be expressed mathematically as

$$S = \sqrt{P^2 + Q^2} \qquad (3.30)$$

The units of P are watts, representing the power being dissipated. Units for Q^2 are volt-amperes reactive (var), representing the reactive power being stored in the inductor and/or capacitor, and the units for S are volt-amperes (VA), representing the rms product of the voltage and current values. The reactive power is not a useful parameter and it is normally desired to make Q equal to zero, which means the total power is equal to the real or average power. Knowing the Q of the circuit helps the designer to compensate so that the load always draws real power. This case corresponds to a unity power factor, to be discussed shortly.

EXERCISE 3.9

Consider the one-port network of Fig. E3.9 with $i(t)$ a triangular waveform. Determine the average and rms current values, and the average power absorbed by the network in steady state under the following cases: (i) purely resistive, with $R = 0.5\ \Omega$; (ii) purely inductive, $L = 1$ mH; (iii) purely capacitive, $C = 1\ \mu F$; and (iv) resistive-inductive, with $R = 0.5\ \Omega$ and $L = 1$ mH.

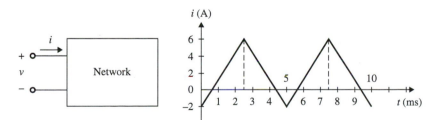

Figure E3.9 A network and its input current waveform.

ANSWER   2 A, 3.1 A, 4.66 W, 0 W, 0 W, 4.66 W

---

[2]This Q has nothing to do with the resonant circuit parameter Q discussed earlier, and the one to be discussed in Chapter 6.

3.3.2   Sinusoidal Waveforms

3.3.2.1   Instantaneous and Average Powers

First let us consider the case for the linear, one-port network shown in Fig. 3.10(*a*). Since the network consists of linear components, its instantaneous source current and voltage expressions may be represented as follows:

$$i_s(t) = I_s \sin(\omega t - \theta_i) \tag{3.31a}$$

$$v_s(t) = V_s \sin(\omega t - \theta_v) \tag{3.31b}$$

where θ_i and θ_v are the port current and voltage phase shifts, respectively.

The *instantaneous power*, $p(t)$, is given by

$$\begin{aligned} p(t) &= i_s v_s \\ &= I_s V_s \sin(\omega t - \theta_i)\sin(\omega t - \theta_v) \end{aligned} \tag{3.32}$$

The waveforms for $i_s(t)$, $v_s(t)$, and $p(t)$ are shown in Fig. 3.10(*b*).

The input average power can be calculated from the following integral:

$$\begin{aligned} P_{\text{ave}} &= \frac{1}{T}\int_0^T p(t)\,dt \\ &= \frac{1}{T}\int_0^T i_s(t)v_s(t)\,dt \end{aligned} \tag{3.33}$$

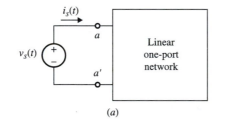

(*a*)

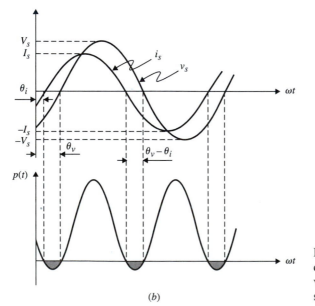

(*b*)

Figure 3.10 (*a*) Linear one-port circuit. (*b*) Typical current and voltage waveforms with phase shift.

Substituting for i_s and v_s from Eqs. (3.31) and using the trigonometric identities

$$\cos(\theta_1 \pm \theta_2) = \cos\theta_1 \cos\theta_2 \mp \sin\theta_1 \sin\theta_2$$
$$\sin(\theta_1 \pm \theta_2) = \sin\theta_1 \cos\theta_2 \pm \cos\theta_1 \sin\theta_2$$

the instantaneous power may be expressed as follows:

$$p(t) = \frac{I_s V_s}{2}[\cos(\theta_v - \theta_i) - \cos(2\omega t + \theta_v + \theta_i)] \tag{3.34}$$

Substitute Eq. (3.34) in the integral of Eq. (3.33), and the average power becomes

$$P_{\text{avc}} = \frac{I_s V_s}{2}[\cos(\theta_v - \theta_i)] \tag{3.35}$$

In terms of the rms parameters, the average power is given by

$$P_{\text{ave}} = I_{s,\text{rms}} V_{s,\text{rms}} \cos(\theta_v - \theta_i) \tag{3.36}$$

3.3.2.2   Power Factor

The *power factor* is a very important parameter in power electronics because it gives a measure of the effectiveness of the real power utilization in the system. It also represents a measure of the distortion of line voltage and line current and the phase shift between them. Let us consider Fig. 3.10(a) in providing the basic definition of the power factor.

The power factor (pf) is defined as the ratio of the average power measured at the terminals a-a' of Fig. 3.10(a) and the rms product of v_s and i_s, as given in Eq. (3.37).

$$\text{Power factor} = \frac{\text{Real power (average)}}{\text{Apparent power}} \tag{3.37}$$

For purely sinusoidal current and voltage waveforms, the average power is given in Eq. (3.36) and the apparent power is given by $I_{s,\text{rms}} V_{s,\text{rms}}$. As a result, Eq. (3.37) yields

$$\text{Power factor} = \frac{I_{s,\text{rms}} V_{s,\text{rms}} \cos\theta}{I_{s,\text{rms}} V_{s,\text{rms}}} \tag{3.38}$$
$$= \cos\theta$$

Hence in linear power systems, when the line voltage and line currents are purely sinusoidal, the power factor is equal to the cosine of the phase angle between the current and voltage. However, in power electronic circuits, due to the switching of active power devices, the phase angle representation alone is not valid. This is why we will shortly define power factor for terminals whose currents and/or voltages are nonsinusoidal (distorted).

The angle θ is known as the *power factor angle*; therefore, the power factor varies between 0 and 1, depending on the type of network. For $\theta > 0$, the current lags the voltage, representing an inductive-resistive load as shown in Fig. 3.11(b). The network load is said to be having a lagging power factor. Similarly, for $\theta < 0$, the current leads the voltage, representing a capacitive-resistive load with a leading power factor, as shown in Fig. 3.11(a).

Let us calculate the power factor for resistive, inductive, and capacitive two-terminal networks.

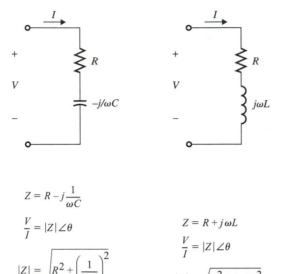

$$Z = R - j\frac{1}{\omega C}$$

$$\frac{V}{I} = |Z|\angle\theta$$

$$|Z| = \sqrt{R^2 + \left(\frac{1}{\omega C}\right)^2}$$

$$\theta = \tan^{-1}\frac{-1}{\omega RC} < 0°$$

(*a*)

$$Z = R + j\omega L$$

$$\frac{V}{I} = |Z|\angle\theta$$

$$|Z| = \sqrt{R^2 + (\omega L)^2}$$

$$\theta = \tan^{-1}\frac{\omega L}{R} > 0°$$

(*b*)

Figure 3.11 (*a*) Leading power factor. (*b*) Lagging power factor.

Resistive Network   The voltage and current relation is given by

$$v_s = i_s R$$

and the power factor angle is $\theta = \theta_v - \theta_i = 0$, resulting in a power factor equal to 1.

Capacitive Network   The capacitor current and voltage relation is given by

$$i_s = C\frac{dv_s}{dt}$$

In the phasor domain, we have

$$\frac{|V_s|\angle\theta_v}{|I_s|\angle\theta_i} = -j\frac{1}{\omega C}$$

$$\frac{|V_s|}{|I_s|}\angle\theta_v - \theta_i = \frac{1}{\omega C}\angle-90°$$

Therefore, the power factor angle is $\theta_v - \theta_i = \theta = -90°$, resulting in a zero power factor. This means the purely capacitive circuit has no average power delivered (as expected from an ideal capacitor). This is a leading power factor because current leads voltage by 90°.

Inductive Network
In the phasor domain, we have

$$\frac{V_s\angle\theta_v}{I_s\angle\theta_i} = +j\omega L$$

$$\frac{V_s}{I_s}\angle\theta_v - \theta_i = \omega L\angle90°$$

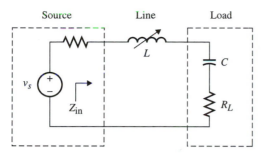

Figure 3.12 Circuit for Example 3.5.

The power factor angle is $\theta = +90°$, resulting in a lagging power factor because current lags voltage by 90°.

EXAMPLE 3.5

Determine L in the circuit of Fig. 3.12 so that the power factor becomes unity.

SOLUTION   The total impedance seen by the source is given by

$$Z_{in} = j\omega L - j\frac{1}{\omega C} + R_L$$

For unity power factor, the phase angle of Z_{in} must be zero, i.e., Z_{in} is a real number. Setting the imaginary part to zero yields

$$\omega L - \frac{1}{\omega C} = 0$$

and solving for the inductor value, we obtain

$$L = \frac{1}{\omega^2 C}$$

EXERCISE 3.10

Consider the circuit of Fig. E3.10 with a 10 kVA load and $v_s = 100 \sin \omega t$.

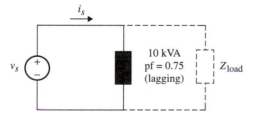

Figure E3.10   Circuit for Exercise 3.10.

It is required to compensate the load to improve the power factor. Determine the type and value of the load impedance, Z_{load}, that must be added in parallel to achieve pf = 0.97, and repeat the process to achieve pf = 1.0.

ANSWER   For pf = 0.97, $Z_{Load} = -j1.056\Omega(C = 2.5$ mF for $\omega = 60$ Hz)
and for pf = 1.0, $Z_{Load} = -j0.756\Omega(C = 3.5$ mF for $\omega = 60$ Hz)

3.3.3   Nonsinusoidal Waveforms

Harmonics cause serious problems of interference with sensitivity measurements and communication systems. To reduce or eliminate line voltage and current harmonics, it is necessary to add filters in the ac input side. Ac and dc filter design is a very specialized topic, and the literature is rich in the analysis, design, and implementation of such filters. For this reason the topic will not be addressed in this textbook. The voltage source of an ideal electrical power system supplies energy at a constant and single frequency with constant voltage amplitude under all load conditions. However, in practical power systems, voltage sources with a single, constant frequency and fixed amplitude are not available.

The importance of studying current and voltage harmonics has grown recently because of the widespread use of power switching devices in various power electronics applications. The topic of a system's harmonics is specialized and cannot be fully addressed in a textbook like this. Nevertheless, some important harmonics issues will be addressed here. The reader is encouraged to consult the references listed at the end of the book.

Because of the switching nature of the majority of power electronic circuits, the line (source) current is highly distorted while the line voltage remains nearly sinusoidal. Only a limited number of circuits have almost sinusoidal line voltage and current. In steady state, the nonsinusoidal nature of the line current produces unwanted oscillatory components at different frequencies. Such signals are called harmonics or harmonic components. Under some load conditions, these harmonics have high amplitudes that result in highly undesirable effects. These harmonics must be removed or at least significantly reduced. As a result, it becomes necessary to study the harmonics and power factor values in nonsinusoidal current waveforms. We will assume that the distorted waveforms are in steady state with a given fundamental frequency. The existence of these harmonics affects the overall efficiency, performance, and cost of the power electronic system. It will be shown that because of this distortion, the apparent power rating (volt-ampere) of the source must be higher than the real power needed by the load.

The best tool available to study harmonics is the Fourier analysis method. The basis for harmonic calculations is the Fourier theorem, introduced by the French mathematician Jean Baptiste Joseph Fourier in 1822. First we review the Fourier analysis technique and harmonic components.

3.3.3.1   Fourier Analysis

The Fourier theorem states that, in steady state, any given periodic function $f(t)$ can be represented by the sum of a constant F_0 and infinite sine and cosine functions $(f_1, f_2, \ldots)$ defined by the following formula:

$$f(t) = F_0 + f_1(t) + f_2(t) + \cdots + f_n(t)$$

$$= F_0 + \sum_{n=1}^{\infty} (a_n \cos n\omega t + b_n \sin n\omega t) \tag{3.39}$$

The constant coefficient can be obtained by taking the integral of both sides of Eq. (3.39) from 0 to T to give the following expression for F_0:

$$F_0 = \frac{1}{T} \int_0^T f(t)\, dt \tag{3.40}$$

From Eq. (3.40) we see that F_0 represents the average (dc) value of $f(t)$.

Also, it can be shown that the coefficients a_n and b_n are evaluated from the following integrals:

$$a_n = \frac{2}{T} \int_0^T f(t) \cos n\omega t\, dt \qquad n = 1, 2, 3, \ldots, \infty \qquad (3.41a)$$

$$b_n = \frac{2}{T} \int_0^T f(t) \sin n\omega t\, dt \qquad n = 1, 2, 3, \ldots, \infty \qquad (3.41b)$$

These equations constitute a frequency domain representation of $f(t)$[3] and suggest that any nonsinusoidal waveform with frequency ω (the waveform repeated every time period $T = 2\pi/\omega$) can be expressed as the sum of a constant and infinite sinusoidal waveforms at multiples ($n\omega$, with $n = 1, 2, \ldots$) of the original frequency. The frequency of the original waveform (at $n = 1$) is known as the *fundamental frequency*. The values of $f(t)$ at $n = 1, 2, 3, \ldots, \infty$ are known as the Fourier components of the waveform $f(t)$, or the harmonic components of $f(t)$. At $n = 1$, the component is usually called the *fundamental component* or fundamental harmonic, denoted by the subscript 1.

By defining the triangle shown in Fig. 3.13, it is also possible to represent $f(t)$ in the following form:

$$f(t) = F_0 + \sum_{n=1}^{\infty} F_n \sin(n\omega t + \theta_n) \qquad (3.42)$$

where F_n is the peak value of the nth harmonic and θ_n is the phase shift, which are given by

$$F_n = \sqrt{a_n^2 + b_n^2}$$

$$\theta_n = \tan^{-1}\left(\frac{a_n}{b_n}\right)$$

If the waveform has any of the forms of symmetry shown in Table 3.2, calculation of the integrals for a_n and b_n can be significantly reduced. In power electronic circuits, odd symmetry waveforms are more frequently encountered than even symmetry waveforms.

Another useful representation of $f(t)$ is in terms of its frequency spectrum as shown in Fig. 3.14. It is a plot of the magnitude of the harmonic component against its frequency.

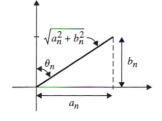

Figure 3.13 Phase representation for Eq. (3.42).

---

[3]It is also common to represent a_n and b_n in terms of frequency as follows:

$$a_n = \frac{1}{\pi} \int_0^{2\pi} f(\omega t) \sin n\omega t\, d(\omega t) \quad b_n = \frac{1}{\pi} \int_0^{2\pi} f(\omega t) \cos n\omega t\, d(\omega t) \quad \text{and} \quad F_0 = \frac{1}{2\pi} \int_0^{2\pi} f(\omega t)\, d(\omega t)$$

Table 3.2 Simplification of Fourier Coefficients Due to Function Symmetry

Odd symmetry $f(t) = -f(-t)$	$a_n = 0$ (for all n)	$b_n = \frac{4}{T}\int_0^{T/2} f(t)\sin n\omega t\, dt$
Even symmetry $f(t) = f(-t)$	$a_n = \frac{4}{T}\int_0^{T/2} f(t)\cos n\omega t\, dt$	$b_n = 0$ (for all n)
Half-wave symmetry $f(t) = -f(t + T/2)$	$a_n = \begin{cases} \frac{4}{T}\int_0^{T/2} f(t)\cos n\omega t\, dt & n \text{ odd} \\ 0 & n \text{ even} \end{cases}$	$b_n = \begin{cases} \frac{4}{T}\int_0^{T/2} f(t)\sin n\omega t\, dt & n \text{ odd} \\ 0 & n \text{ even} \end{cases}$
Odd and half-wave symmetry $f(t) = -f(t + T/2)$ $f(t) = -f(-t)$	$a_n = 0$ (for all n)	$b_n = \begin{cases} \frac{8}{T}\int_0^{T/4} f(t)\sin n\omega t\, dt & n \text{ odd} \\ 0 & n \text{ even} \end{cases}$
Even and half-wave symmetry $f(t) = -f(t + T/2)$ $f(t) = f(-t)$	$a_n = \begin{cases} \frac{8}{T}\int_0^{T/4} f(t)\cos n\omega t\, dt & n \text{ odd} \\ 0 & n \text{ even} \end{cases}$	$b_n = 0$ (for all n)

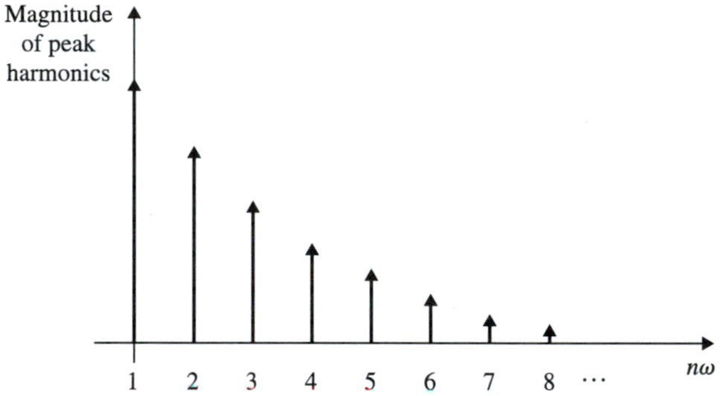

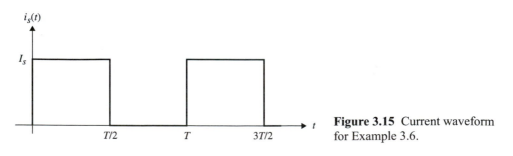

Figure 3.14 Frequency spectrum representation of $f(t)$.

As n increases the peak of the higher harmonic components decreases. The peak value at F_1, i.e., the fundamental component, is normally referred to as the "wanted" harmonic or component, whereas the higher components represent the "unwanted" components.

EXAMPLE 3.6

Consider a half-wave power electronic circuit whose line current $i_s(t)$ waveform is shown in Fig. 3.15. Calculate the harmonics of $i_s(t)$.

Figure 3.15 Current waveform for Example 3.6.

SOLUTION   The average value is clearly $I_s/2$, so we have

$$F_0 = \frac{I_s}{2}$$

The coefficients a_n are given by

$$a_n = \frac{1}{\pi}\int_0^{2\pi} f(\omega t)\cos(n\omega t)\,d(\omega t)$$

$$= \frac{1}{\pi}\int_0^{\pi} I_s \cos(n\omega t)\,d(\omega t) = \frac{I_s}{n\pi}\sin(n\omega t)\Big|_0^{\pi} = 0 \quad n = 1, 2, \ldots$$

The coefficients b_n are given by

$$b_n = \frac{1}{\pi}\int_0^{2\pi} f(\omega t)\sin(n\omega t)\,d(\omega t)$$

$$= \frac{1}{\pi}\int_0^{\pi} I_s \sin(n\omega t)\,d(\omega t) = \frac{-I_s}{n\pi}\cos(n\omega t)\Big|_0^{\pi} = \frac{2I_s}{n\pi} \quad n = 1, 3, 5, \ldots$$

EXERCISE 3.11

Determine the average and rms values for the output voltage waveform for the half-wave rectifier circuit shown in Fig. E3.11.

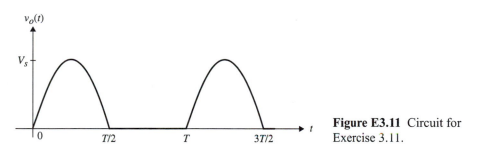

Figure E3.11 Circuit for Exercise 3.11.

ANSWER $V_s/\pi, V_s/2$

3.3.3.2   Line Current Harmonics

If we apply the Fourier equations to the line current i_s and the line voltage v_s, which are periodic nonsinusoidal waveforms with period T and zero dc components, then $i_s(t)$ and $v_s(t)$ are given as follows:

$$i_s(t) = I_{dc} + \sum_{n=1}^{\infty} I_{sn}\sin(n\omega t + \theta_{ni})$$

$$= I_{dc} + I_{s1}\sin(\omega t + \theta_{1i}) + \sum_{n=2}^{\infty} I_{sn}\sin(n\omega t + \theta_{ni})$$

(3.43)

$$v_s(t) = V_{dc} + \sum_{n=1}^{\infty} V_{sn} \sin(n\omega t + \theta_{nv})$$

$$= V_{dc} + V_{s1} \sin(\omega t + \theta_{1v}) + \sum_{n=2}^{\infty} V_{sn} \sin(n\omega t + \theta_{nv})$$

(3.44)

The fundamental current and voltage components are

$$i_{s1}(t) = I_{s1} \sin(\omega t + \theta_{1i}) \tag{3.45a}$$

$$v_{s1}(t) = V_{s1} \sin(\omega t + \theta_{1v}) \tag{3.45b}$$

where I_{s1} and V_{s1} are the peak values of the current and voltage fundamental components, respectively.

The rms values of $i_s(t)$ and $v_s(t)$ can be easily obtained using Eqs. (3.43) and (3.44):

$$I_{s,rms}^2 = \frac{1}{T} \int_0^T i_s^2(t)\, dt$$

$$I_{s,rms}^2 = I_{dc}^2 + \left(\frac{I_{s1}}{\sqrt{2}}\right)^2 + \left(\frac{I_{s2}}{\sqrt{2}}\right)^2 + \cdots + \left(\frac{I_{sn}}{\sqrt{2}}\right)^2$$

(3.46)

$$= I_{dc}^2 + I_{s1,rms}^2 + I_{s2,rms}^2 + \cdots + I_{sn,rms}^2 \qquad n = 1, 2, \ldots, \infty$$

Similarly,

$$V_{s,rms}^2 = V_{dc}^2 + \left(\frac{V_{s1}}{\sqrt{2}}\right)^2 + \left(\frac{V_{s2}}{\sqrt{2}}\right)^2 + \cdots + \left(\frac{V_{sn}}{\sqrt{2}}\right)^2$$

(3.47)

$$= V_{dc}^2 + V_{s1,rms}^2 + V_{s2,rms}^2 + \cdots + V_{sn,rms}^2 \qquad n = 1, 2, \ldots, \infty$$

These equations are obtained since the integration of the product of two different frequency components over T is zero.

The instantaneous power is given by

$$p(t) = i_s(t)v_s(t)$$

$$= I_{dc}V_{dc} + I_{dc}\sum_{n=1}^{\infty} V_{sn} \sin(n\omega t - \theta_{nv}) + V_{dc}\sum_{n=1}^{\infty} I_{sn} \sin(n\omega t - \theta_{ni})$$

$$+ \left(\sum_{n=1}^{\infty} I_{sn} \sin(n\omega t - \theta_{ni})\right)\left(\sum_{n=1}^{\infty} V_{sn} \sin(n\omega t - \theta_{nv})\right)$$

(3.48)

Evaluating the average value of $p(t)$ in Eq. (3.48) over the fundamental frequency, ω, shows that the second and third terms in Eq. (3.48) are zeros and the fourth term is simply the sum of expressions similar to Eq. (3.46) but evaluated at each harmonic component as shown below:

$$P_{ave} = I_{dc}V_{dc} + I_{s1,rms}V_{s1,rms}\cos(\theta_{v1} - \theta_{i1}) + I_{s2,rms}V_{s2,rms}\cos(\theta_{v2} - \theta_{i2}) + \cdots$$

$$= I_{dc}V_{dc} + \sum_{n=1}^{\infty} I_{sn,rms}V_{sn,rms}\cos\theta_n$$

(3.49)

where $\theta_n = \theta_{vn} - \theta_{in}$ $(n = 1, 2, \ldots, \infty)$, which represents the phase shift between the nth voltage and current harmonics.

It is clear from the above equation that true dc power can be obtained only if both the line current and voltage have dc components. The second term represents the average power at the source terminal obtained from the rms value of the harmonic components.

3.3.3.3   Total Harmonic Distortion

Since the wanted portion of the distorted waveform $i_s(t)$ is the fundamental component, the difference between the "desired" rms value of $i_s(t)$ and the "wanted" value is appropriately called the *distorted* portion of $i_s(t)$, defined as

$$i_{s,\text{dist}} = i_s(t) - i_{s1}(t) = \sum_{n=2}^{\infty} i_{sn}(t) \tag{3.50a}$$

$$v_{s,\text{dist}} = v_s(t) - v_{s1}(t) = \sum_{n=2}^{\infty} v_{sn}(t) \tag{3.50b}$$

The relative measure of the distortion is defined through an index called the *total harmonic distortion* (THD), which is the ratio of the rms value of the distorted waveform and the rms value of the fundamental component. The THD expression for the current and voltage are given in Eqs. (3.51a) and (3.51b), respectively, assuming no dc components.

$$\begin{aligned}
\text{THD}_i &= \frac{I_{\text{dist,rms}}}{I_{s1,\text{rms}}} = \frac{\sqrt{I_{s2,\text{rms}}^2 + I_{s3,\text{rms}}^2 + I_{s4,\text{rms}}^2 + \cdots}}{I_{s1,\text{rms}}} \\
&= \sqrt{\left(\frac{I_{s2,\text{rms}}}{I_{s1,\text{rms}}}\right)^2 + \left(\frac{I_{s3,\text{rms}}}{I_{s1,\text{rms}}}\right)^2 + \cdots}
\end{aligned} \tag{3.51a}$$

$$\begin{aligned}
\text{THD}_v &= \frac{V_{\text{dist,rms}}}{V_{s1,\text{rms}}} = \frac{\sqrt{V_{s2,\text{rms}}^2 + V_{s3,\text{rms}}^2 + V_{s4,\text{rms}}^2 + \cdots}}{V_{s1,\text{rms}}} \\
&= \sqrt{\left(\frac{V_{s2,\text{rms}}}{V_{s1,\text{rms}}}\right)^2 + \left(\frac{V_{s3,\text{rms}}}{V_{s1,\text{rms}}}\right)^2 + \cdots}
\end{aligned} \tag{3.51b}$$

In terms of the rms of the original waveform, Eqs. (3.51a) and (3.51b) may be rewritten as

$$\text{THD}_i = \sqrt{\left(\frac{I_{s,\text{rms}}}{I_{s1,\text{rms}}}\right)^2 - 1} \tag{3.52a}$$

$$\text{THD}_v = \sqrt{\left(\frac{V_{s,\text{rms}}}{V_{s1,\text{rms}}}\right)^2 - 1} \tag{3.52b}$$

It is also common to refer to THD as a percentage.

EXERCISE 3.12

Calculate THD$_i$ for $i_s(t)$ given in Example 3.6 and THD$_v$ for v_s of Exercise 3.11.

ANSWER   121%, 100%

3.3.3.4 Power Factor

The equations to calculate the power factor for distorted waveforms are more complex compared to the sinusoidal case discussed earlier. Applying the definition of the power factor given in Eq. (3.37) to the distorted current and voltage waveforms of Eqs. (3.43) and (3.44) and the average power given in Eq. (3.49) (with zero dc components), pf may be expressed as

$$\text{pf} = \frac{\sum_{n=1}^{\infty} I_{sn,\text{rms}} V_{sn,\text{rms}} \cos \theta_n}{I_{s,\text{rms}} V_{s,\text{rms}}} = \frac{\sum_{n=1}^{\infty} I_{sn,\text{rms}} V_{sn,\text{rms}} \cos \theta_n}{\sqrt{\sum_{n=1}^{\infty} I_{sn,\text{rms}}^2} \sqrt{\sum_{n=1}^{\infty} V_{sn,\text{rms}}^2}} \tag{3.53}$$

This expression for pf can be significantly simplified if we assume the line voltage is purely sinusoidal and distortion is limited to $i_s(t)$; thus it can be shown that pf can be expressed as

$$\text{pf} = \frac{I_{s1,\text{rms}}}{I_{s,\text{rms}}} \cos \theta_1 \tag{3.54}$$

where θ_1 is the phase angle between the voltage $v_s(t)$ and the fundamental component of $i_s(t)$. This assumption is valid in many power electronics applications. The line voltage is normally undistorted, and the line current is what gets distorted, i.e.,

$$v_s(t) = V_s \sin \omega t \tag{3.55a}$$

$$i_s(t) = \text{distorted (nonsinusoidal)} \tag{3.55b}$$

The current is expressed in terms of the Fourier series as follows:

$$i_s(t) = I_1 \sin(\omega t + \theta_1) + I_2 \sin(2\omega t + \theta_2) + \cdots + I_n \sin(n\omega t + \theta_n) \tag{3.56}$$

resulting in the average power given by

$$P_{\text{ave}} = \frac{1}{T} \int_0^T v_s i_s \, dt$$

$$= \frac{1}{T}\left[\int_0^T (V_s \sin \omega t)(I_1 \sin(\omega t + \theta_1) + I_2 \sin(2\omega t + \theta_2) + \cdots + I_n \sin(n\omega t + \theta_n)) \, dt \right]$$

$$= \frac{1}{T}\left[\int_0^T V_s I_1 \sin \omega t \sin(\omega t + \theta_1) \, dt \right]$$

$$= \frac{V_s I_1}{2} \cos \theta_1$$

$$P_{\text{ave}} = V_{s,\text{rms}} I_{s1,\text{rms}} \cos \theta_1$$

Hence, the power factor is given by

$$\text{Power factor} = \frac{I_{s1,\text{rms}} V_{s,\text{rms}} \cos \theta_1}{I_{s,\text{rms}} V_{s,\text{rms}}} = \frac{I_{s1,\text{rms}}}{I_{s,\text{rms}}} \cos \theta_1 \tag{3.57}$$

The expression $I_{s1,\text{rms}}/I_{s,\text{rms}}$ is caused by the distortion of the line current and is appropriately called the *distortion power factor*, k_{dist}. The term $\cos \theta_1$ is caused by the displacement angle between the line voltage and the fundamental current component and is commonly known as the *displacement power factor*, k_{disp}. Hence, the power factor in power electronics is more useful if it is represented as a product of k_{dist} and k_{disp}:

$$\text{pf} = k_{\text{dist}} k_{\text{disp}} \tag{3.58}$$

where

$$k_{\text{disp}} = \cos \theta_1$$
$$k_{\text{dist}} = I_{s1,\text{rms}}/I_{s,\text{rms}}$$

In terms of k_{dist}, it can be shown that the current THD can be expressed as

$$\text{THD}_i = \sqrt{\frac{1}{k_{\text{dist}}^2} - 1} \tag{3.59}$$

EXAMPLE 3.7

Calculate k_{dist}, k_{disp}, THD_i, and THD_v for the waveforms shown in Fig. 3.16.

SOLUTION   Since the voltage is purely sinusoidal, $\text{THD}_v = 0\%$.
To obtain THD_i, first we obtain the rms value of $i_s(t)$ and its fundamental component.

$$I_{s,\text{rms}}^2 = \frac{1}{2\pi} \int_{-\theta}^{2\pi-\theta} i_s^2 \, d\omega t = \frac{1}{2\pi} \left[\int_{-\theta}^{\pi-\theta} I_o^2 \, d\omega t + \int_{\pi-\theta}^{2\pi-\theta} (-I_o^2) \, d\omega t \right]$$

$$= I_o^2$$

The fundamental component of $i_s(t)$ is given by

$$I_{s1} = I_{s1}' \sin \omega t + I_{s1}'' \cos \omega t$$

where

$$I_{s1}' = \frac{1}{\pi} \left[\int_{-\theta}^{\pi-\theta} I_o \sin \omega t \, d\omega t + \int_{\pi-\theta}^{2\pi-\theta} -I_o \sin \omega t \, d\omega t \right]$$

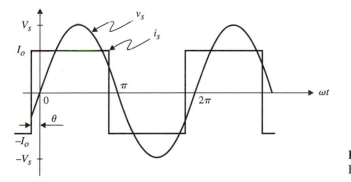

Figure 3.16 Waveforms for Example 3.7.

If we let $\omega\tau = \omega t + \theta$, the integral becomes

$$I'_{s1} = \frac{1}{\pi}\left[\int_0^\pi I_o \sin(\omega\tau - \theta)\,d\omega t - \int_\pi^{2\pi} I_o \sin(\omega\tau - \theta)\,d\omega t\right]$$

$$= \frac{I_o}{\pi}[-\cos(\pi - \theta) + \cos(-\theta) + \cos(2\pi - \theta) + \cos(-\theta)]$$

$$= \frac{4I_o}{\pi}\cos\theta$$

Similarly, $I''_{s1} = (4I_o/\pi)\sin\theta$. Therefore, the peak fundamental component is given by

$$I_{s1} = \sqrt{(I'_{s1})^2 + (I''_{s2})^2} = \frac{4I_o}{\pi}$$

Hence, the fundamental component of $i_s(t)$ is given by

$$i_{s1}(t) = I_{s1}\sin(\omega t + \theta)$$

and the rms of I_{s1} is given by

$$I_{s1,\text{rms}} = \frac{4}{\sqrt{2}}\frac{I_o}{\pi} = \frac{2\sqrt{2}}{\pi}I_o$$

k_{dist}, k_{disp}, and THD are given by the following expressions:

$$k_{\text{dist}} = \frac{2\sqrt{2}}{\pi}$$

$$k_{\text{disp}} = \cos\theta$$

$$\text{THD} = \left(\sqrt{\frac{\pi^2}{8} - 1}\right) \times 100\%$$

$$= 48.34\%$$

EXERCISE 3.13

You have to give the voltage waveform for the calculation of pf

Calculate THD$_i$ and pf for the current waveform shown in Fig. 3.15, by using up to the fifth harmonic.

ANSWER   117%, 0.64

EXERCISE 3.14

here too.

Calculate THD$_i$ and pf for the current waveform shown in Fig. E3.14.

ANSWER   136%, 0.51

3.4 CAPACITOR AND INDUCTOR RESPONSES

The transient and steady-state values of the capacitor and inductor voltage and current are well understood by undergraduate electrical engineering students. However, a brief review of such responses might be useful to some readers at this point.

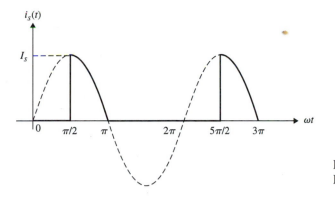

Figure E3.14 Waveform for Exercise 3.14.

3.4.1   Capacitor Transient Response

Consider the *RC* circuit of Fig. 3.17(*a*) with a dc excitation and an ideal switch. Assume the switch is open for $t < t_0$ and at $t = t_0$, the switch is closed. The capacitor voltage for $t < t_0$ is equal to the dc source, V_{dc}.

For $t > t_0$ the time-domain capacitor current is given in terms of its voltage,

$$i_c = C\frac{dv_c}{dt} \tag{3.60}$$

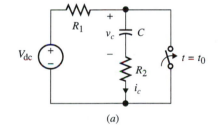

(*a*)

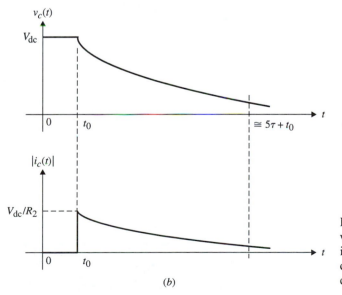

(*b*)

Figure 3.17 (*a*) *RC* circuit with dc excitation and an ideal switch. (*b*) The current change through the capacitor.

Substituting for $i_c = (0 - v_c(t))/R$ and solving for $v_c(t)$, we obtain the following general solution:

$$v_c(t) = v_{c,f} + (v_{c,i} - v_{c,f})e^{-(t-t_0)/\tau} \tag{3.61}$$

where

$v_{c,f}$ = final value at $t = \infty$

$v_{c,i}$ = initial value at $t = t_{0^+}$

τ = circuit time constant, with $\tau = (R_1 \| R_2)C$ R_2C

Since the capacitor voltage does not change instantaneously, we have

$$v_c(t_{0^-}) = v_c(t_{0^+})$$

From the circuit diagram, at $t = \infty$ the capacitor becomes fully discharged through R_2 with time constant R_2C. The capacitor's final value is given by

$$v_{c,f} = v_c(\infty) = 0$$

The final expressions for $v_c(t)$ and $i_c(t)$ for $t > t_0$ are

$$v_c(t) = V_{dc}e^{-(t-t_0)/\tau} \tag{3.62a}$$

$$i_c(t) = \frac{-V_{dc}}{R_2}e^{-(t-t_0)/\tau} \tag{3.62b}$$

The plot is shown in Fig. 3.17(*b*). The steady state is reached at approximately 5τ.

At $t = t_{0^-}$ the capacitor current is zero, but at $t = t_{0^+}$, it suddenly becomes $-V_{dc}/R_2$. This brings us to another important statement about the capacitor current: *The current through the capacitor can change instantaneously in switching circuits, as shown in Fig. 3.17(b).*

EXAMPLE 3.8

Consider the circuit of Fig. 3.18(*a*), where the switch has been closed for $t < 0$. At $t = 0$, the switch is opened. Sketch the capacitor voltage and current waveforms for $t > 0$. Assume the capacitor is initially uncharged.

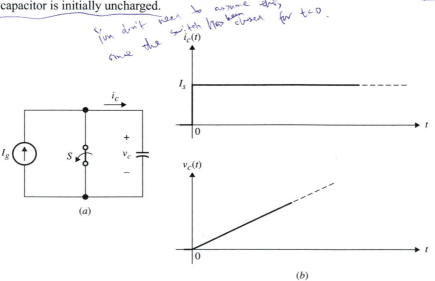

(a)

(b)

Figure 3.18 (*a*) Circuit for Example 3.8. (*b*) Capacitor current and voltage waveforms.

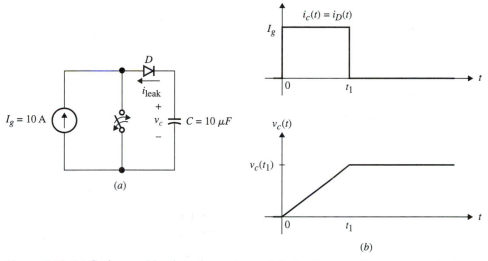

Figure 3.19 (a) Series combination of capacitor and diode. (b) Capacitor current and voltage waveforms.

SOLUTION   The capacitor voltage equation is given by

$$C\frac{dv_c}{dt} = i_c$$

$$= I_g$$

Therefore, $v_c(t)$ is

$$v_c(t) = \frac{I_s}{C}t$$

The capacitor current and voltage waveforms are shown in Fig. 3.18(b).

Since the capacitor voltage does not change instantaneously, the switch in this example is not allowed to close again unless a way is found by which the switch voltage is prevented from appearing across the capacitor (i.e., voltage diversion). This can be accomplished by adding a diode in series with the capacitor as shown in Fig. 3.19(a). As long as the switch is closed, the capacitor voltage remains constant.

At $t = t_1$, the switch is closed again, forcing D to turn off, since its anode voltage is pulled to the ground. The capacitor voltage at $t = t_1$ is given by

$$v_c(t_1) = \frac{I_g}{C}t_1 V$$

In practice, the energy stored in the capacitor dissipates through the capacitor's equivalent resistance and the diode's leakage current, as illustrated in Exercise 3.15.

EXERCISE 3.15

Consider the circuit of Fig. 3.19(a) with the switch waveform shown in Fig. 3.20. Assume an ideal switch and ideal diode, except it has a 100 μA leakage current. What are the capacitor voltages at $t_1 = 5$ μs, 10 μs, 0.5 s? At what time does the capacitor voltage become zero again?

ANSWER   5 V, 10 V, 5 V, 1 s

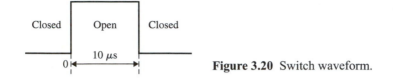

Figure 3.20 Switch waveform.

3.4.2 Capacitor Steady-State Response

Now let us consider the circuit of Fig. 3.17(a) with the switch turned on and off repeatedly according to the waveform shown in Fig. 3.21(a). After a few switching cycles, the capacitor voltage reaches steady state, at which its value at the beginning of each switching cycle is the same. Mathematically, this can be shown as follows:

$$v_c(nT + t_0) = v((n+1)T + t_0) \tag{3.63}$$

Figure 3.21(b) and (c) shows the steady-state capacitor voltage and current waveforms, respectively. Now we come to another important property of the capacitor: In steady state, the average capacitor current is zero.

$$I_{c,ave} = \frac{1}{T}\int_{nT}^{(n+1)T} i_c\,dt = \frac{1}{T}\int_{nT}^{(n+1)T} C\frac{dv_c}{dt}\,dt = \frac{C}{T}[v_c((n+1)T) - v_c(nT)] \\ = 0 \tag{3.64}$$

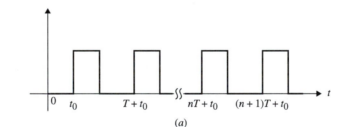

(a)

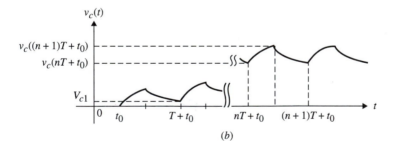

(b)

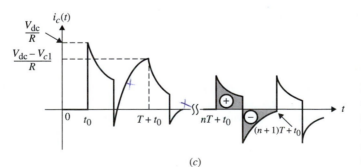
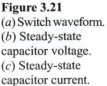

(c)

Figure 3.21
(a) Switch waveform.
(b) Steady-state capacitor voltage.
(c) Steady-state capacitor current.

This is illustrated by the equal negative and positive shaded areas for $i_c(t)$ shown in Fig. 3.21(c).

3.4.3   Inductor Transient Response

Similarly, let us look at the transient response for the inductor current by considering the dual circuit of Fig. 3.17(a), which is shown in Fig. 3.22. Assume the switch is open for a long time before it is closed at $t = t_0$. Assume that initially the inductor current is zero.

For $t > t_0$, the time-domain inductor voltage is given by

$$v_L = L\frac{di_L}{dt} \tag{3.65}$$

Substituting for $v_L = R(I_{dc} - i_L(t))$, where $R = R_1 \parallel R_2$, and solving for $i_L(t)$, we obtain the following general solution for $i_L(t)$:

$$i_L(t) = i_{L,f} + (i_{L,i} - i_{L,f})e^{-(t-t_0)/\tau} \tag{3.66}$$

where

$i_{L,f}$ = final value at $t = \infty$

$i_{L,i}$ = initial value at $t = t_{0^+}$

τ = circuit time constant, with $\tau = L/R$

Since the inductor current does not change instantaneously, we have

$$I_L(t_{0-}) = I_L(t_{0+}) \tag{3.67}$$

From the circuit, it is clear that at $t = \infty$, $i_L(t = \infty) = i_{L,f} = I_{dc}$ and its voltage becomes zero.

The final expression for $i_L(t)$ is given by

$$i_L(t) = I_{dc}(1 - e^{-(t-t_0)/\tau})$$

and the inductor voltage is given by

$$v_L(t) = RI_{dc}e^{-(t-t_0)/\tau}$$

Again at $t = t_{0-}$, $v_L(t_{0-}) = 0$, and at $t = t_{0+}$, the inductor voltage is

$$v_L(t_{0+}) = RI_{dc}$$

We conclude that *the voltage across the inductor can change instantaneously under switching action,* as shown by the inductor current and voltage waveforms in Fig. 3.23.

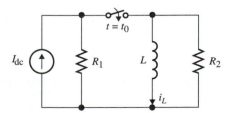

Figure 3.22 Inductor switching circuit.

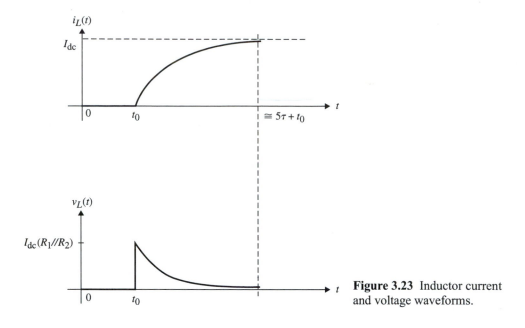

Figure 3.23 Inductor current and voltage waveforms.

EXERCISE 3.16

Consider the switching circuit given in Fig. E3.16 with an ideal diode. Assume the initial inductor current is zero and the switch is turned on at $t = 0$ and turned off at $t = 10 \ \mu s$. Sketch the inductor current waveform for $t > 0$. If the diode has a 5 Ω forward resistance, what is i_L at $t = 0.5 \ \mu s$, 10 μs, and 100 μs. At what time does the inductor becomes discharged?

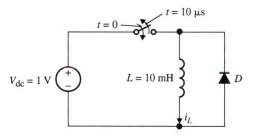

Figure E3.16 Switching circuit for Exercise 3.16.

ANSWER   50 μA, 1 mA, 0.95 mA, 10 ms.

Notice in the preceding exercise that when the switch is opened again at $t = 10$ ms, the inductor current gets trapped and the diode turns on and diverts the current from the switch branch to the diode branch. This is known as *current commutation* and will be covered extensively in Chapters 5 and 6. In practice, trapped energy in the inductor would dissipate through the inductor and diode resistances. To improve the converter's efficiency, the trapped energy is normally allowed to be reconnected by returning it to the source through a feedback circuit along with the use of transformers, to be illustrated in Chapter 5.

3.4.4   Inductor Steady-State Response

Let us reconsider the circuit of Fig. 3.22 except that the switch turns on and off repeatedly according to the waveform shown in Fig. 3.24(a).

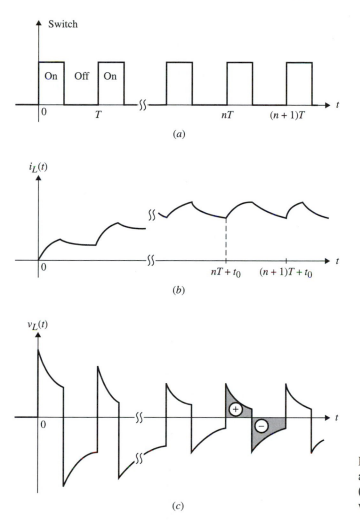

Figure 3.24 (*a*) Switch action and inductor (*b*) current and (*c*) voltage waveforms for Fig. 3.22.

After a few switching cycles, the capacitor reaches steady state, at which its value at the beginning of each switching cycle is the same. Mathematically, we express the steady-state condition as

$$i_L(nT + t_0) = i_L((n+1)T + t_0) \tag{3.68}$$

This condition makes it possible to show that the average value of the steady-state inductor voltage is zero:

$$V_{L,\text{ave}} = \frac{1}{T}\int_{nT}^{(n+1)T} v_L \, dt = \frac{1}{T}\int_{nT}^{(n+1)T} L\frac{di_L}{dt}dt = \frac{L}{T}[i_L((n+1)T) - i_L(nT)] \tag{3.69}$$
$$= 0$$

The steady-state current and voltage waveforms are shown in Fig. 3.24(*b*) and (*c*), respectively. The zero average inductor voltage is illustrated by the equal negative and positive shaded areas for $v_L(t)$ shown in Fig. 3.24(*c*).

PROBLEMS

In all problems assume ideal diodes unless stated otherwise.

3.1 Derive the expressions for $i_L(t)$ and $v_c(t)$ in the switching circuits shown in Fig. P3.1. Assume all initial conditions are zero. Use $V_{dc} = 20$ V, $I_{dc} = 2$ A, $R = 200$ Ω, $C = 0.01$ μF, and $L = 2$ mH, and sketch i_L and v_c.

3.2 Assuming the switch in Fig. P3.2 is opened at $t = 0^+$, derive the expressions for i_L, v_c, and i_{sw} and sketch them for $t > 0$. Assume $i_L(0) = I_g$ and $v_c(0) = V_o$.

3.3 Consider the switching diode circuits of Fig. P3.3 for $t > 0$ when the switch is turned on. Derive expressions for $i_L(t)$, $i_{sw}(t)$, and $v_c(t)$. Assume $i_L(0) = 1.0I_g$, and $v_c(0) = v_o$. Suggest possible switch implementations (unidirectional or bi-directional).

3.4 Consider the transistor switching circuits of Fig. P3.4. At $t = 0$ the transistor is turned on by a signal to the base. Derive and sketch the waveforms for i_Q, i_L, v_c, and v_o. Assume $v_c(0) = -20$ V and $i_L(0) = 0$. Compare the two circuits.

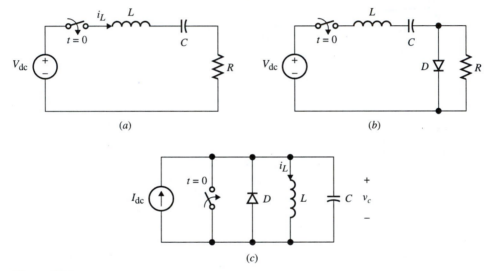

(a) (b)

(c)

Figure P3.1

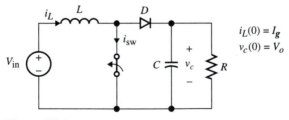

Figure P3.2

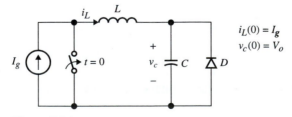

Figure P3.3

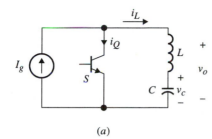

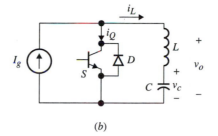

(a)   (b)

Figure P3.4 .

3.5 Consider the diode circuit of Fig. P3.5 with switch S_1 opened and S_2 closed at $t = 0$. Assume the initial conditions are given as $v_{c1}(0) = 0$ and $i_L(0) = -0.5I_g$. Derive the expressions for v_{c2} and i_L and sketch them. Use $I_g = 2$ A, $C_1 = C_2 = 0.01$ μF, and $L = 2$ mH.

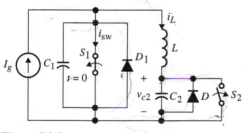

Figure P3.5

3.6 Assume the switch in Fig. P3.6 is opened at $t = 0$. Derive the expressions and sketch the waveforms for v_{c1}, v_{c2}, i_L, and i_D. Assume $v_{c1}(0) = v_{c2}(0) = 0$ and $i_L(0) = 0$.

3.7 Consider the capacitor current shown in Fig. P3.7 for a given switch-mode power supply with $C = 1$ μF, and assume $v_c(0) = 100$ V.
(a) Sketch v_c showing the peak values and times.
(b) Determine the ripple voltage across the capacitor.

3.8 Consider the SCR-diode switching circuit shown in Fig. P3.8 with the switch closing at $t = 0$.

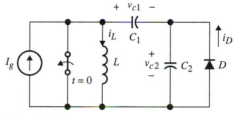

Figure P3.6

This circuit is a voltage commutation circuit known as an impulse-commuted chopper used to force the turn-off of the SCR by an additional switch S. Assume an ideal diode and that the capacitor is initially charged to $-V_o$, and the SCR is on for $t < 0$.

(a) Sketch the waveforms of i_{SCR}, v_c, i_D, and v_o.
(b) Show that the time, Δt, it takes for i_c to reach zero again after $t > 0$ is given by

$$\Delta t = \frac{(V_{dc} + V_o)C}{I_o}$$

Discuss the drawback of such an arrangement.

3.9 For the SCR to turn off in the circuit of Fig. P3.8, the capacitor voltage must first be charged to a large negative value. This is done through an external circuit consisting of a diode and an inductance as shown in Fig. P3.9. The purpose of D_1

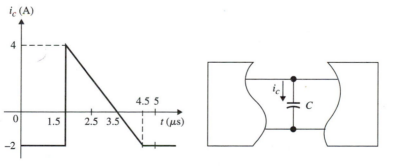

Figure P3.7

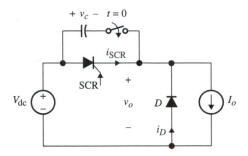

Figure P3.8

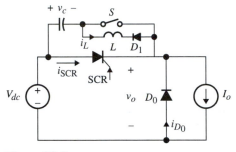

Figure P3.9

and L is that when the switch is turned off, the capacitor voltage returns to its original negative value, $-V_0$. Assume S and SCR are open for a long time with a capacitor initial value equal to $+V_0$. At $t = 0$, the SCR is triggered. Sketch the waveforms for i_{D1}, i_{D0}, i_{SCR}, v_c, and v_o. Compare this circuit with the circuit given in Problem 3.8.

3.10 Consider the SCR circuit shown in Fig. P3.10. Because the SCR turns off naturally due to the fact that its current becomes zero, the circuit is known as a self-commutated circuit. At $t = 0$, the SCR is turned on by applying i_g. Derive the expressions for $v_c(t)$, $i_L(t)$, and the SCR commutation time. Let $L = 0.1$ mH, $C = 47$ μF, and $V_{dc} = 120$ V. Assume the initial capacitor voltage is (i) zero and (ii) $-V_i$ (where $|V_i| < V_{dc}$).

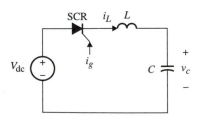

Figure P3.10

3.11 Design for L and C in Problem 3.10 so that the SCR commutation time is 100 μs and the peak capacitor voltage does not exceed 200 V.

3.12 Consider a source terminal connected to a power electronic circuit whose current and voltage waveforms are shown in Fig. P3.12, with $t_1 = 4$ μs and $T = 10$ μs.

(a) Calculate the average and rms source current and source voltage values.

(b) Calculate the average input power.

3.13 Consider the circuit of Fig. P3.13, where v_s is a train of pulses as shown. Assume $RC \gg T$ so that V_o is assumed constant and equal to 75 V. Sketch the steady-state inductor current waveform, i_L.

3.14 The switch in Fig. P3.14 is closed at $t = 0$. Assume ideal diodes and all inductor initial conditions are zero. Obtain the expressions for the inductor current and sketch them for $0 \leq t \leq 2T$. Assume $v_s(t)$ is sinusoidal with $\omega = 377$ rad/s and $V_s = 100$ V. Let $L = 1$ mH, $R = 1$ Ω, and $V_{dc} = 20$ V.

3.15 Consider a power source with its terminal voltage and current given as

$$v_s(t) = 100 + 80\sin(\omega t - 100°)$$
$$+ 70\cos(2\omega t + 120°) + 25\sin 3\omega t \text{ V}$$

$$i_s(t) = 12 + 10\sin(\omega t + 25°)$$
$$+ 5\sin(2\omega t - 30°) + 2\cos 3\omega t \text{ A}$$

where ω is the fundamental angular frequency. Calculate:

(a) The rms value of $i_s(t)$ and $v_s(t)$

(b) The average input power supplied by the source

(c) The rms values of the fundamental components of $i_s(t)$ and $v_s(t)$

(d) THD_i and THD_v

(e) k_{dist}, k_{disp}, and pf

3.16 Consider the phasor circuit of Fig. P3.16 with $Z_L = 20\angle{-36°}\Omega$, $V_S = 80\angle 0°$ V.

(a) Determine the circuit's real, apparent, and reactive powers and the input power factor.

(b) Determine the type and value of the load needed to be connected between a and a' to achieve a unity power factor.

3.17 Prove the following integrals:

$$F_n = \int_0^T \sin(n\omega t)\sin(m\omega t)dt$$

$$= \int_0^T \cos(n\omega t)\cos(m\omega t)dt$$

$$= \begin{cases} 0 & n \neq m \\ \dfrac{T}{2} & n = m \end{cases}$$

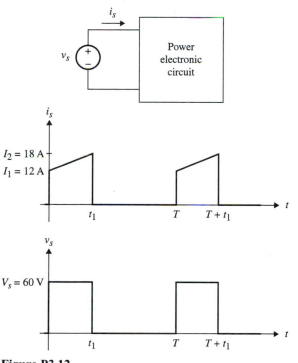

Figure P3.12

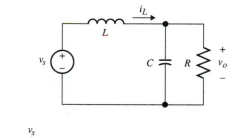

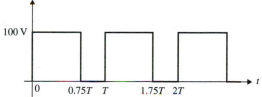

Figure P3.13

and

$$F_n = \int_0^T \sin(n\omega t)\cos(m\omega t)\,dt$$

$$= \begin{cases} 0 & n \neq m \\ -\dfrac{T}{2} & n = m \end{cases}$$

where $\omega = 2\pi/T$ and $n = 1, 2, \ldots, \infty, m = 1, 2, \ldots, \infty$.

3.18 Sketch the steady-state i_L and v_c waveforms by assuming the switch in Fig. P3.18 is repeatedly opened and closed as shown. Use $T = 10$ ms.

3.19 Show that the Fourier series for the half-wave rectifier output voltage of Fig. P3.19 is given by

$$v_o(t) = \frac{V_s}{\pi} + \frac{2V_s}{\pi}\sin \omega t$$

$$- \frac{2V_s}{\pi}\sum_{n=2,4,6,\ldots}^{\infty}\frac{1}{n^2 - 1}\cos n\omega t$$

3.20 Show that the Fourier series for the output voltage waveform of the full-wave rectifier of Fig. P3.20 is given by

$$v_o(t) = \frac{2V_s}{\pi} - \frac{2V_s}{\pi}\sum_{n=2,4,6,\ldots}^{\infty}\frac{2}{n^2 - 1}\cos n\omega t$$

3.21 By considering only the first four terms of the Fourier series given in Problem 3.19, verify the answers given in Exercise 3.11.

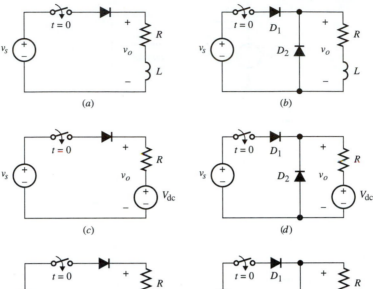

(a)

(b)

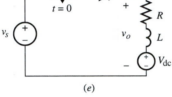

(c)

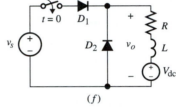

(d)

(e)

(f)

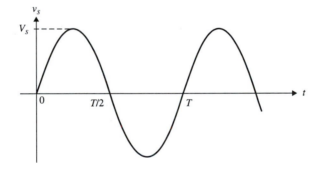

Figure P3.14

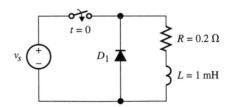

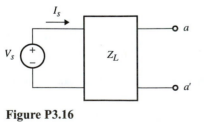

Figure P3.16

Switch waveform

On Off

 DT

 T

t

Figure P3.18

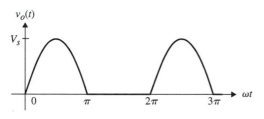

Figure P3.19

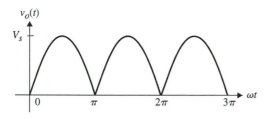

Figure P3.20

3.22 Derive the Fourier coefficient equations a_n and b_n given in Eqs. (3.41).

3.23 The phase current waveform in a three-phase full-wave rectifier SCR circuit under highly inductive load is shown in Fig. P3.23. Show that the harmonics of $i(t)$ are given by the following expression:

$$i(t) = \frac{2\sqrt{3}I_o}{\pi}\left[\cos\omega t - \frac{\cos 5\omega t}{5} + \frac{\cos 7\omega t}{7}\right.$$
$$\left. - \frac{\cos 11\omega t}{11} + \frac{\cos 13\omega t}{13} - \cdots\right]$$

with no triple harmonics in $i(t)$ and $n = 6k \pm 1$ ($k = 0, 1, 2, \ldots$). Show that the rms magnitude of the nth harmonic is $\sqrt{6}I_o/\pi n$.

Figure P3.23

3.24 Show that the Fourier series for the phase current of the six-pulse SCR converter shown in Fig. P3.24 is given by

$$i(t) = \frac{2\sqrt{3}I_o}{\pi}\left[\cos\omega t + \frac{\cos 5\omega t}{5} - \frac{\cos 7\omega t}{7}\right.$$
$$- \frac{\cos 11\omega t}{11} + \frac{\cos 13\omega t}{13}$$
$$\left. + \frac{\cos 17\omega t}{17} + \cdots + \frac{\cos n\omega t}{n}\right]$$

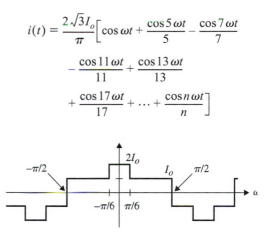

Figure P3.24

3.25 **(a)** Figure P3.25(a) shows a typical ac line current waveform in a single-phase, full-wave SCR-controlled rectifier circuit under resistance load and sinusoidal ac source. Show that the fundamental line current component is given by

$$i_{s1}(t) = I_{s1}\sin(\omega t + \theta)$$

where

$$\omega = \frac{2\pi}{T}$$

$$\theta = \tan^{-1}\frac{\cos 2\alpha - 1}{2\pi - 2\alpha + \sin 2\alpha}$$

$$I_{s1} = \sqrt{1 + 2(\pi - \alpha)^2 + 2(\pi - \alpha)\sin 2\alpha - \cos 2\alpha}$$
$$\times \frac{I_s}{\sqrt{2}\pi}$$

(b) Find $i_{s1}(t)$ for the half-wave SCR-controlled line current shown in Fig. P3.25(b).

3.26 Show that the high-harmonics Fourier coefficients for Fig. P3.26 are given by,

$$a_n = 0$$

$$b_n = \frac{1}{\pi}(2\alpha + \sin 2\alpha) + \sum_{n=3,5,7\ldots}^{\infty}\left[\frac{2}{\pi(1-n)}\sin\alpha(1-n)\right.$$
$$\left. - \frac{2}{\pi(1+n)}\sin\alpha(1+n) + \frac{4}{\pi(1-n)}\sin\alpha\cos\alpha\right]$$

What value of h will achieve 95% input power factor assuming the waveform represents a line current with a sinusoidal line voltage.

3.27 Figure P3.27 shows the ac line current for a full-wave, single-phase SCR-controlled converter under highly inductive load. Show that the a_n coefficients are given by

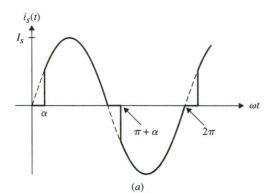

(a)

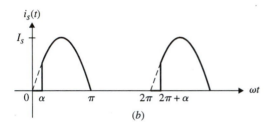

(b)

Figure P3.25

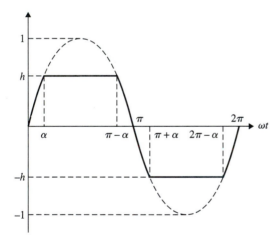

Figure P3.26

$$a_n = \frac{4I_o}{n\pi}\sin\frac{n(\pi-\alpha)}{2}$$

for $n = 1, 3, 5, 7, \ldots$.

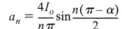

Figure P3.27

3.28 Show that the Fourier components of $i_s(t)$ of Fig. P3.28 are given by

$$i_s(t) = \frac{4I_o}{n\pi}\sum_{n=1,3,5,\ldots}^{\infty}\cos n\alpha\sin n\omega t$$

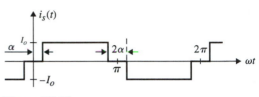

Figure P3.28

3.29 Determine the Fourier components for the waveforms given in Fig. P3.29.

General Problems

3.30 In the circuit of Fig. P3.30, the switch is opened at $t = 0$ after it has been closed for a long time. Derive the expressions for $i_L(t)$, $i_D(t)$, and $v_c(t)$ for $t > 0$, and plot them.

3.31 Show that the load current after the switch is closed at $t = 0$ in the diode circuit in Fig. P3.31 is given by

$$i_o(t) = 0 \qquad\qquad 0 \leq t < t_1$$

$$i_o(t) = \frac{V_s}{|Z|}\sin(\omega t_1 - \theta) - \frac{V_{dc}}{R} + Ae^{-t/\tau} \quad t_1 \leq t < t_2$$

$$i_o(t) = 0 \qquad\qquad t_2 \leq t < T$$

where

$$\tau = L/R$$

$$\theta = \tan^{-1}\frac{\omega L}{R}$$

$$A = \frac{V_{dc}}{R} - \frac{V_s}{|Z|}\sin(\omega t_1 - \theta)$$

Assume the inductor is not initially charged, and here t_1 is the time when the diode turns on and is given by $\omega t_1 = \sin^{-1}(V_{dc}/V_s)$, and t_2 is the time at which the diode turns off before the next cycle starts at $t = T$. Find the expression for t_2.

3.32 The circuit of Fig. P3.32(a) is known as a single-phase bridge inverter whose purpose is to convert the dc input voltage, V_{dc}, to an ac output voltage, v_o. If the switching sequence of S_1, S_2, S_3, and S_4 is done in such a way that the output voltage is shown in Fig. P3.32(b):

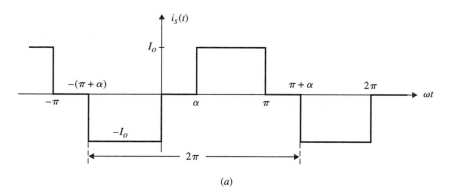

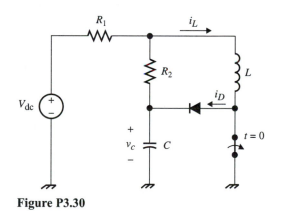

Figure P3.29

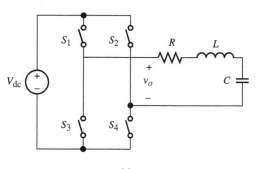

Figure P3.30

where

$$V_n = \frac{4V_{in}}{n\pi}\sin\frac{n\alpha}{2} \quad n = 1, 3, 5, 7, 9, \ldots$$

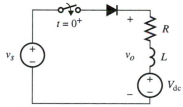

Figure P3.31

(a) Show that the Fourier series for v_o is given by

$$v_o = V_1 \sin\omega t + V_3 \sin 3\omega t + V_5 \sin 5\omega t + \cdots$$
$$+ V_n \sin n\omega t$$

(a)

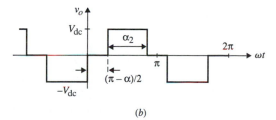

(b)

Figure P3.32

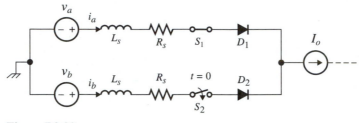

Figure P3.33

(b) Use the above results to find the Fourier series for $i_o(t)$.

3.33 Consider the two-phase switching commutation circuit shown in Fig. P3.33. Assume S_1 has been on for a long time prior to $\omega t = 150°$, when S_2 is turned on. (a) Sketch the waveforms for $i_a(t)$ and $i_b(t)$, and (b) determine the time during which both D_1 and D_2 were on. Assume $v_a(t) = 100 \sin 377t$ and $v_b(t) = 100 \sin(377t - 120°)$ V. Use $L_s = 1$ mH and $R_s = 2\ \Omega$.

3.34 The circuit shown in Fig. P3.34 is known as a forced commutation circuit whose voltage wave-

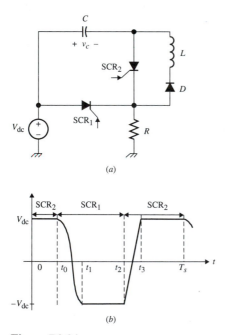

(a)

(b)

Figure P3.34

form over one switching period, T_s, is shown for $t > 0$. Derive the expression for $v_c(t)$ over one switching cycle. The conduction states of SCR_1 and SCR_2 are shown on the waveform.

3.35 Figure P3.35 shows a diac-triac switching circuit used in heater controller, motor speed variation, and light dimmer applications. Design for R

and R_L so that the triac triggers at 30° and 210° during the positive and negative cycles, respectively. Assume $v_s(t) = 110 \sin 2\pi 60t$ V and the diac breakover voltage is 24 V.

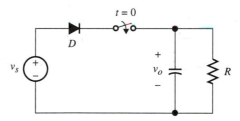

Figure P3.35

3.36 Assume the switch is turned on at $t = 0$ and the capacitor is initially discharged in the circuit of Fig. P3.36. Derive the expression for $v_c(t)$ and sketch it for $0 < t < 40$ ms, where $v_s(t) = 110 \sin 2\pi 50t$, $R = 10\ k\Omega$, and $C = 1\ \mu F$.

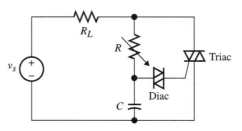

Figure P3.36

3.37 Figure P3.37 shows a self-oscillating LC circuit that allows the SCR to turn off naturally without using an additional auxiliary SCR. This circuit is known as a series resonant turn-off circuit. (a) Derive the expression for $v_c(t)$ and $i_L(t)$, and (b) determine the power rating of the SCR. Assume the SCR is first triggered at $t = 0$ and then repeatedly every $T = 2$ ms. Assume $v_c(0^-) = 30$ V and $i_L(0^-) = 0$.

3.38 The circuit given in Fig. P3.38 represents one possible implementation of a family of dc-dc converters known as *soft-switching converters*.

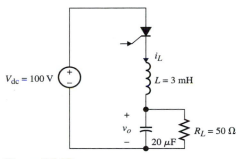

Figure P3.37

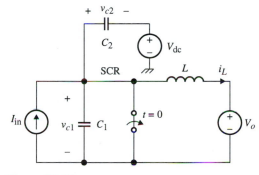

Figure P3.38

(a) If we assume the switch is turned on at $t = 0$, show that the capacitor voltage for $t > 0$ is given by

$$v_{c1} = V_o + (V_{dc} - V_o)\cos\omega_o t$$

where $\omega_o = \sqrt{C_{eq}L}$ and $C_{eq} = C_1 + C_2$.

(b) In order to turn S back on while the capacitor voltage across C_1 is zero, the voltage across it must be allowed to reach zero again during the *off* time of the switch. Show that for zero-voltage switching to occur, the following condition must be met:

$$V_o < \frac{V_{dc}}{2}$$

3.39 Figure P3.39 shows output and input waveforms in a cycloconverter-type power electronic circuit, where the output frequency is one-half of the input voltage frequency. Find the Fourier series representation for such waveforms.

3.40 Figure P3.40 shows a typical inductor current waveform in a switched-mode power supply when operating at the boundary of continuous- and discontinuous-mode operations.

(a) Determine the Fourier components for $i(t)$.

(b) Calculate its average and rms values.

(c) Calculate THD$_i$.

3.41 Repeat Problem 3.40 for the inductor current shown in Fig. P3.41, which represents a discontinuous conduction mode of operation in a switched-mode power supply.

3.42 Figure P3.42 shows a half-wave rectifier waveform with a sine-squared pulse represented mathematically for $0 < t < T/2$ by the following equation:

$$i_s(t) = I_o\sin^2\omega t$$

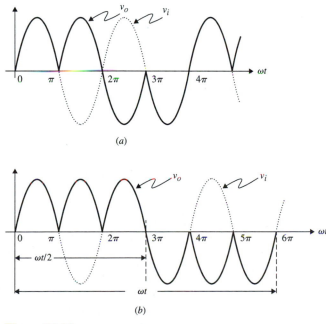

(a)

(b)

Figure P3.39

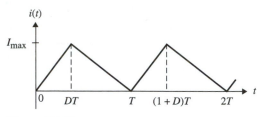

Figure P3.40

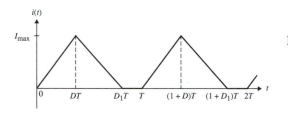

Figure P3.41

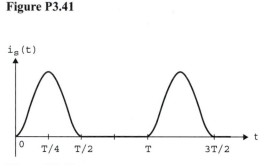

Figure P3.42

(a) Determine the Fourier coefficients.

(b) Calculate the rms and THD_i values.

3.43 Consider a sinusoidal line current as shown in Fig. P3.43 that can be represented mathematically for $0 < t < T$ by the following equation:

$$i_s(t) = I_o \sin^n \omega t$$

where n is an odd integer. Assuming that the line voltage is given by $v_s(t) = V_s \sin \omega t$ V, derive the expression for the power factor and THD_i.

3.44 The simplified equivalent circuit of a dc-to-dc converter known as a buck converter is shown in Fig. P3.44(a). Assume the switch is turned on and off according to the waveform shown in Fig. P3.44(b). Sketch the steady-state waveforms for v_{sw}, i_{sw}, and i_o and derive the expression for the average output power.

3.45 Figure P3.45 shows eight different switched-mode topologies, of which only one is valid. Identify this topology and state which circuit law (KVL or KCL) is violated for each of the other topologies.

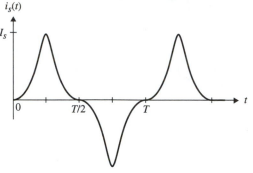

Figure P3.43

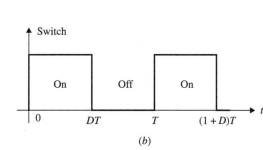

(a)

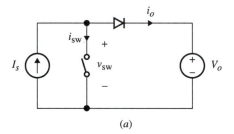

(b)

Figure P3.44

3.46 Sketch the waveforms for i_{D1}, i_{D2}, and v_o and find the average output voltage for the circuit of Fig. P3.46(a), where $v_{s1}(t)$ and $v_{s2}(t)$ are as shown in Fig. P3.46(b).

3.47 Repeat Problem 3.46 by replacing D_1 and D_2 with two switches S_1 and S_2 as shown in Fig. P3.47. Assume that S_1 is conducting only when $|v_{s1}(t)| > V_s/3$ and S_2 is the complement of S_1.

3.48 Consider the full-bridge rectifier of Fig. P3.48 with a current source $i_s(t) = I_s \sin \omega t$ and a constant output voltage, V_o. Sketch v_s and i_o and find the average output power.

3.49 Consider the full-bridge rectifier of Fig. P3.49 with a voltage source $v_s(t) = V_s \sin \omega t$ and a constant output current, I_o. Sketch i_s and v_o and find the average output power.

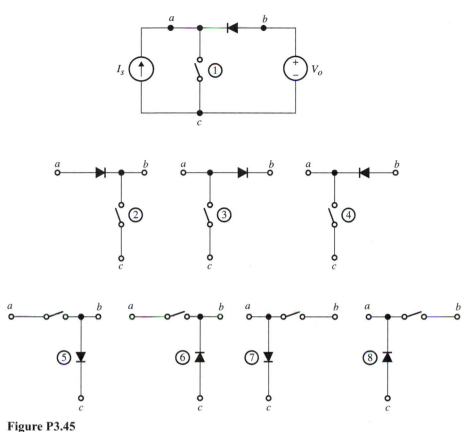

Figure P3.45

3.50 Figure P3.50 shows four different switched-mode topologies, of which only one is valid. Identify this topology and state the circuit law (KCL or KVL) that is violated for each of the other topologies.

3.51 Figure P3.51(a) shows an equivalent circuit for a switched-mode converter—Cuk (pronounced *chook*). Show that the ratio between the average output voltage, V_o, and the average input voltage, V_{in}, is given by the following relation:

$$\frac{V_o}{V_{in}} = \frac{D}{1-D}$$

The switch waveform is shown in Fig. P3.51(b). (*Hint:* Use the average voltage V_x as given by $V_{in} - V_o$.)

3.52 Figure P3.52(a) shows a switched-mode power electronic converter known as a buck-boost converter. This circuit is used to convert a dc input voltage to another dc level at the load. If we as-

sume that the inductor, L, is large and the average output current is represented by a constant current source, then the equivalent circuit becomes as shown in Fig. P3.52(b). Assume the switch is turned on and off as shown in Fig. P3.52(c).

(a) Show that the capacitor current is given by Fig. P3.52(d).

(b) Sketch the steady-state average capacitor voltage.

3.53 Determine the power factor for the circuits given in Problems 3.48 and 3.49.

3.54 Consider a transformer with a turn ratio $N_1/N_2 = 200$, and rated at 240 kVA (44 kV/220 V) used to step down a 60 Hz voltage in a distributed system.

(a) Determine the rated primary and secondary currents.

(b) Determine the load impedance seen between the primary terminals when the circuit is fully loaded.

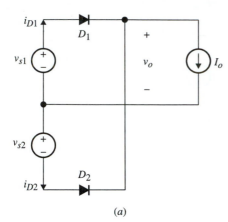

(a)

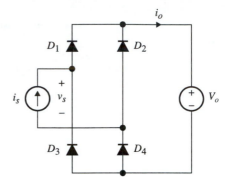

Figure P3.48

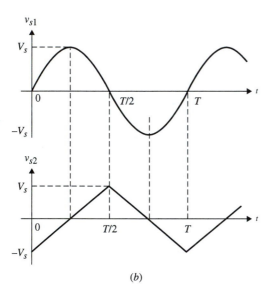

(b)

Figure P3.46

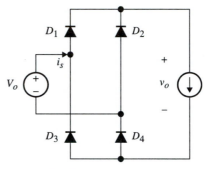

Figure P3.49

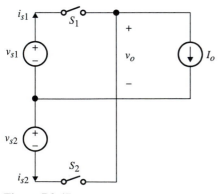

Figure P3.47

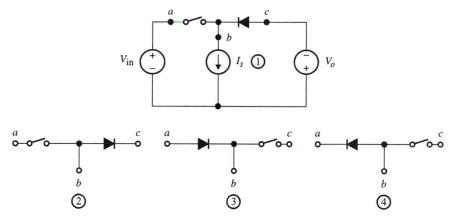

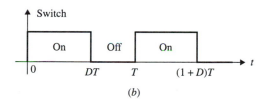

Figure P3.50

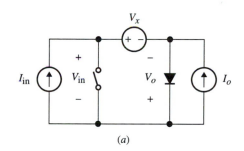

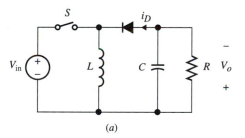

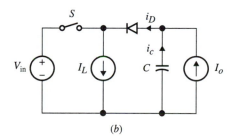

Figure P3.51

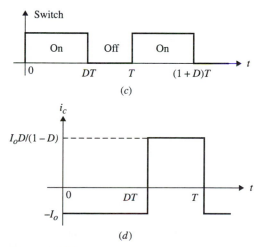

Figure P3.52

Chapter 4

Nonisolated Switch-Mode dc-dc Converters

INTRODUCTION

In this chapter we discuss converter circuits that are used in power electronic systems to change the system voltages from one dc level to another dc level. Once again, switching devices will be used to process energy from the input to the output. Since the input here is dc, which comes from a post-filtering stage, these devices are normally operated at much higher frequencies than the line frequency, reaching as high as a few hundred kilo-hertz. This is why such converter circuits are known as high-frequency dc-to-dc switching converters or regulators. The term *regulator* is used since the circuit's main commercial application is in systems that require a stable and *regulated* dc output voltage. Depending on whether or not an output transformer is used, high-frequency dc-to-dc switching converters are classified as isolated or nonisolated. In this chapter and in the next, the emphasis will be on the steady-state analysis and design of several well-known second- and fourth-order dc-to-dc converters, each having its own features and applications. We will consider those topologies that do not use high-frequency isolation transformers as part of their power stage. Moreover, a large number of applications require output electrical isolation and multiple outputs that cannot be achieved using the basic topologies discussed in this chapter. The isolated and magnetically coupled topologies will be discussed in Chapter 5. Such topologies are the most popular in the power supply industry and are used in various types of electronic equipment whose design requires outputs with electrical isolation and multi-outputs.

4.1 POWER SUPPLY APPLICATIONS

4.1.1 Linear Regulators

A typical block diagram of a linear regulator power supply is shown in Fig. 4.1. The front end of the linear regulator is a 60 Hz transformer, T_1, used to provide input electrical isolation and to step up or step down the line voltage, and this is followed by a full-wave bridge rectifier to convert the ac input to a dc input by adding a large filtering capacitor at the input of the linear regulator. This input to the linear regulator, V_{in}, is unregulated dc and cannot be used to drive the load directly. Using a linear circuit that provides a stable dc output regulates the dc voltage at the output, V_o.

For many years, most power supplies available in the market were of the linear regulator type, in which a series-pass active element is used to regulate the output voltage. In general, the active semiconductor element is used as a variable resistance to dissipate unwanted or excess voltage. Such an arrangement results in large amounts of power being dissipated in the active element, which can cause the efficiency to drop to as low as 40%. Because of this low efficiency, linear regulators have not been used

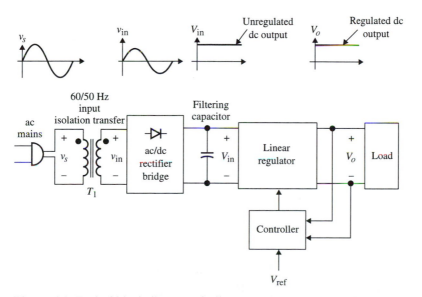

Figure 4.1 Typical block diagram of a linear regulator power supply.

for medium- and high-power applications since the early 1970s, when switched-mode dc-dc converters entered the marketplace. Despite the fact that linear regulators are simple to use and provide tight control, good output voltage ripples, and a low component count, their disadvantages are so numerous that their practical use is limited. Because of their high power losses, they suffer from high thermal dissipation, resulting in low power density and low efficiency.

Figure 4.2 illustrates a simplified circuit to show how the series active element is connected in a linear regulator. V_{in} represents the unregulated input voltage, V_{Excess} is the excess voltage to be dissipated across the linear device, and V_o is the output voltage. V_{Excess}, which is the difference between V_{in} and V_o, must be large enough to keep the transistor in the linear active mode, acting as a variable resistor used only to absorb the difference in the voltage. This is why V_{Excess} is considered one of the key design parameters of linear regulation. To illustrate how the linear resistor works, we present a simplified topology showing the series element regulator in Fig. 4.3, where V_{ref} is generated by a zener diode, and R_1 and R_2 are used as a voltage divider. The magnitude of i_B determines how deeply the transistor is driven in the saturation region. The comparator is used to compare the output voltage with a fixed reference voltage. As the output voltage increases, the base current, i_B, decreases and the excess voltage, v_{CE}, increases, hence reducing the output voltage. Similarly, if the output voltage decreases, the result is a reduced v_{CE} and an increased output.

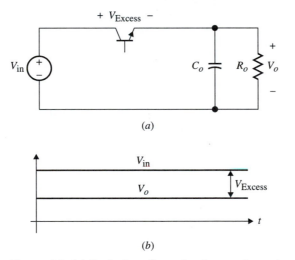

(a)

(b)

Figure 4.2 (a) Typical configuration for a series active element used in a linear regulator. (b) Average voltage waveforms.

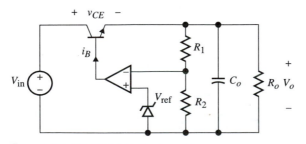

Figure 4.3 Simplified representation of a simple series element regulator.

4.1.2   Switched-Mode Power Supplies

The development of the power semiconductor switch made it possible for power electronics engineers to design power supplies with much higher efficiencies compared with linear regulators. Since transistors are used as switching devices, such power supplies are known as switched-mode power supplies or simply switching converters. In recent years, switching converters have become very popular due to recent advances in semiconductor technology. Today switching devices are available with very high switching speeds and very high power-handling capabilities. It is possible to design switched-mode power supplies with efficiency greater than 90% with low cost and relatively small size and light weight.

Unlike linear regulators, switching converters use power semiconductor devices to operate in either the on-state (saturation or conduction) or the off-state (cutoff or nonconduction). Since either state will lead to low switching voltage or low switching current, it is possible to convert dc to dc with higher efficiency using a switching regulator. Figure 4.4 shows a simplified block diagram for a switched mode ac-to-dc power converter with multi-output application. Compared with the block diagram of Fig. 4.2, a switching network and high-frequency output electrical isolation transformer T_2 are added. The objective is to control the *on* time of the power devices to regulate the dc output voltage. The post-filtering is used to reduce the output voltage ripple. This chapter will discuss the detailed power stage operation of the dc-to-dc block shown in Fig. 4.4 without including the high-frequency isolation transformer. Because of the regulation method used, these converters are known as pulse-width-modulation (PWM) converters.

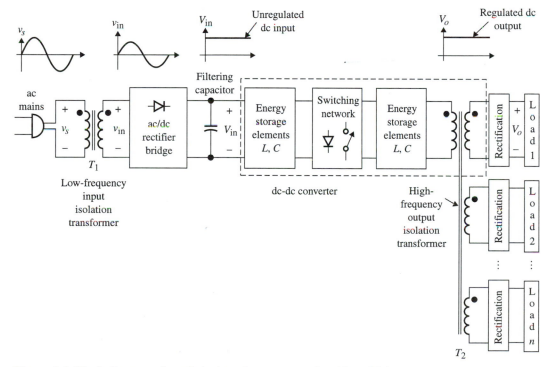

Figure 4.4 Block diagram of a switched-mode power supply with multiple outputs.

4.2   CONTINUOUS CONDUCTION MODE

Steady-state analyses of the basic direct-connected second-order converters such as the buck (step-down), boost (step-up), and buck-boost (step-up/down), and fourth-order converters such as Cuk and SEPIC converters will be presented in this chapter. Both the continuous and discontinuous conduction modes of operation, as well as some nonideal effects will be included in the analysis. The steady-state analysis of isolated or magnetically coupled converters (which are derived from these basic converters) such as the flyback, forward, push-pull, half- and full-bridge, and Weinberg converters will be discussed in Chapter 5.

First, we introduce the basic concept of a switched-mode circuit and the conversion technique in switching converters. Consider the simplest switching voltage converter, shown in Fig. 4.5. We assume that the switch is ideal and it is turned on at $t = t_0$ and turned off at t_1 alternately as shown in Fig. 4.6(a), where $f = 1/T$ is the switching frequency.

The waveforms for the output voltage, v_o, and the output current, i_o, are shown in Fig. 4.6(b) and (c), respectively, where V_{in} is the dc input voltage. The average output voltage is given by

$$
\begin{aligned}
V_o &= \frac{1}{T} \int_{t_0}^{T+t_0} v_o(t) \, dt \\
&= \frac{1}{T} \int_{t_0}^{t_1} V_{in} \, dt = \frac{t_1 - t_0}{T} V_{in}
\end{aligned}
\tag{4.1}
$$

If we let D be defined as the duty ratio or duty cycle,

$$
D = \frac{\text{On time}}{\text{Switching period}} = \frac{t_1 - t_0}{T}
\tag{4.2}
$$

then the average output voltage, V_o, is given by

$$
V_o = D V_{in}
\tag{4.3}
$$

EXERCISE 4.1

If the output ripple factor $K_{o,\text{ripple}}$ is defined by the relation

$$
K_{o,\text{ripple}} = \frac{\sqrt{V_{o,\text{rms}}^2 - V_o^2}}{V_o}
$$

where $V_{o,\text{rms}}$ and V_o are the rms and average values of $v_o(t)$, respectively, determine the output ripple factor for Fig. 4.5.

ANSWER $K_{o,\text{ripple}} = \sqrt{\dfrac{1}{D} - 1}$

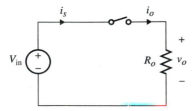

Figure 4.5 A simple switching circuit.

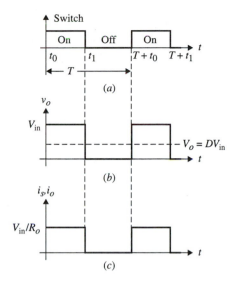

Figure 4.6 Switching waveform for Fig. 4.5.

It is clear from Eq. (4.3) that the average output voltage is less than the applied dc voltage. Assuming an ideal switch, theoretically speaking, the efficiency of this converter is 100%. The drawback of this simple switching circuit is that the output voltage is not constant, but a chopped dc with high ripple voltage. This results in high harmonics being generated at the load. This will not be acceptable for an electronics load when the output must be regulated to a fixed dc level with little ripple. However, in some application where a precise output voltage is not required, such as heating, light dimming, electroplating, and mechanical applications, this simple arrangement might be used, especially if the frequency is very high.

One approach to smooth the output voltage is to use a low-pass filter at the output of the circuit in order to filter the high-switching-frequency components of the output voltage. The filter may consist of a simple capacitor and inductor. The capacitor is used to hold a dc value across the output resistor, and the inductor is used within the circuit to serve as a nondissipative storage element needed to store energy from the input source and deliver it to the load. It can be argued that one single-pole, single-throw switch will not be sufficient to perform energy processing from the input to the output; rather, two switches or a single-pole, double-throw switch is required. This can be easily justified since the inductor current cannot be instantaneously interrupted. When one switch is switched, resulting in a sudden change in the inductor current, a second switch must be switched so that the continuity of the inductor current is maintained. Hence, a practical representation for a switched-mode converter must include either two switches, as shown in Fig. 4.7(a), or a single switch, as shown in Fig. 4.7(b). The inductor and capacitor elements are used as energy storage components to allow energy transfer from the input to the output. The output capacitor forms a low-pass filter to produce dc output voltage with little ripple. Most topologies, either isolated or nonisolated, will consist of one inductor and one output capacitor—hence the name *second-order* voltage converter. Fourth-order voltage converters consist of two inductors and two capacitors, which are considered series or combinations of second-order converters.

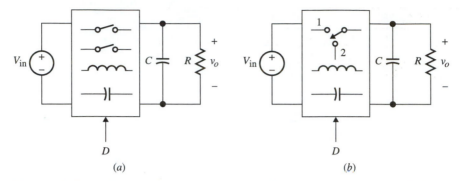

(a)   *(b)*

Figure 4.7 Typical block diagram representations for switching converters with (*a*) two independent switches and (*b*) one single-pole, double-throw switch.

In second-order converters, depending on the arrangements of the switches and *L*, three possible topologies can be obtained, namely buck, boost, and buck-boost, as shown in Fig. 4.8. Since these switches cannot be turned on and off simultaneously, single-pole, double-throw switches are used, which can be either on or off. In the following sections these converters will be thoroughly analyzed by deriving their steady-state characteristics. Since all switched-mode converters have pulsating currents and voltages, the dc output voltage can be obtained by adding a low-pass filter. It will be shown that in most of these converters an *LC*-type low-pass filter section is present. In fourth-order circuits like the Cuk and SEPIC, multiple *LC* low-pass filter sections are added.

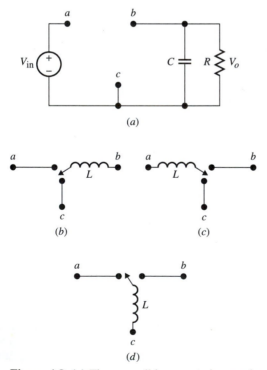

Figure 4.8 (*a*) Three possible ways to insert a low-pass *LC* filter between a dc source and a capacitive load: (*b*) buck converter, (*c*) boost converter, and (*d*) buck-boost converter.

EXAMPLE 4.1

Consider the simple PWM converter of Fig. 4.5 with $V_{in} = 28$ V and $f_s = 50$ kHz. Assume an ideal switch.

(a) Design for D and R so that the power delivered to the load is 25 W at an average output current of 1.5 A.

(b) Find the output ripple factor.

(c) Find the converter efficiency.

SOLUTION

(a) For 25 W average output power and average output current of 1.5 A, the average output voltage is given by

$$V_o = \frac{P_o}{I_o} = \frac{25}{1.5} = 16.7 \text{ V}$$

The load resistance is

$$R = \frac{16.7}{1.5} = 11.13 \ \Omega$$

For $V_o = 16.7$ V and $V_{in} = 28$ V, the duty cycle is

$$D = \frac{16.7}{28} = 0.6$$

(b) The rms value for the output voltage is

$$V_{o,\text{rms}} = \sqrt{D} V_{in} = 21.69 \text{ V}$$

The ripple factor is given by

$$K_{o,\text{ripple}} = \sqrt{\frac{21.69^2 - 16.7^2}{16.7^2}} = 0.83$$

This circuit results in 83% output voltage ripple, which is by no means acceptable in many dc power supply applications.

(c) Since the switch is ideal, the converter efficiency is 100%, which can be illustrated as follows. The average output power is given by

$$P_o = \frac{1}{T_s} \int_0^{T_s} v_o i_o \ dt$$

$$= \frac{1}{T_s} \int_0^{T_s} \frac{v_o^2}{R} \ dt = \frac{1}{RT_s} \int_0^{DT_s} V_{in}^2 dt$$

$$= \frac{D V_{in}^2}{R}$$

and the average input power is given by

$$P_{in} = \frac{1}{T_s} \int_0^{T_s} V_{in} i_{in} \ dt = \frac{V_{in}}{T_s} \int_0^{DT_s} i_{in} \ dt$$

$$= \frac{V_{in}}{T_s} \int_0^{DT_s} i_o \ dt = \frac{V_{in}}{T_s R} \int_0^{DT_s} v_o \ dt$$

$$= \frac{D V_{in}^2}{R}$$

As expected, the average input and output powers are equal.

EXERCISE 4.2

Repeat Example 4.1 by assuming the switch has a 1.8 V voltage drop across it when conducting.

ANSWER   0.64, 0.75, 93.6%

4.2.1   The Buck Converter

Topology and Basic Operation

Figure 4.9(*a*) and (*b*) shows the circuit configuration for a buck converter with a single- and two-switch implementation. Figure 4.9(*c*) shows the transistor-diode implementation. This topology is known as a buck converter because it steps down the average output voltage below the input voltage.

Throughout this chapter to obtain the steady-state characteristic equations, we will assume that power switching devices and the converter components are lossless. Moreover, the exact steady-state analysis of these converters requires solving second-order nonlinear systems. Such analysis is complex and because of the nature of the output voltage, it is not necessary. Since these converters' function is to produce dc output, the

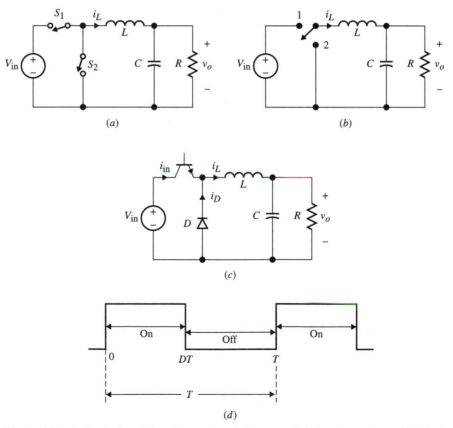

(a)   *(b)*

(c)

(d)

Figure 4.9 The buck (step-down) converter. (*a*) Two-switch implementation. (*b*) Single-pole, double-throw switch implementation. (*c*) Transistor-diode implementation. (*d*) Switching waveform for the power switch.

output voltage $v_o(t)$ consists of the desired dc and the undesired ac components. Practically, the output ripple due to switching is very small (less than 1%) compared to the level of the dc output voltage. As a result, we will assume the output ripple voltage is small and can be neglected when evaluating converter voltage gains, i.e., $v_o = V_o$. In other words, the ripple-free output voltage assumption is made since the output time constant for the filter capacitor and the output resistor, RC, is very large. Moreover, the analysis will be based on the converter operating in the steady-state condition, i.e., the converter currents and voltages have reached their steady-state values. These assumptions can be summarized and represented mathematically as follows:

1. Since we assume lossless components and ideal switching devices, the average input power, P_{in}, and the average output power, P_o, are equal:

$$P_{in} = P_o \tag{4.4}$$

2. Since we assume steady-state operation, the inductor current and the capacitor voltage are periodic over one switching cycle, i.e.,

$$i_L(t_0) = i_L(t_0 + T) \tag{4.5a}$$

$$v_c(t_0) = v_c(t_0 + T) \tag{4.5b}$$

where t_0 is the initial switching time and T is the switching period.

3. Since we assume ideal capacitors and inductors, the average inductor voltage and the average capacitor current are zero:

$$I_c = \frac{1}{T}\int_{t_0}^{T+t_0} i_c(t)\, dt = 0 \tag{4.6}$$

$$V_L = \frac{1}{T}\int_{t_0}^{T+t_0} v_L(t)\, dt = 0 \tag{4.7}$$

In fact, Eq. (4.7) is a representation of Faraday's law, which states that voltage time during charging equals voltage time during discharging. This is also known as the *volt-second principle*. These two relations suggest that the total energy stored in the capacitor or the inductor over one switching cycle is zero. Finally, throughout the analysis in this chapter, the typical switching waveform for the power devices given in Fig 4.9(*d*) will be used to represent the switching action of the power switch. For simplicity we set the initial switching time to zero, $t_0 = 0$.

Again, D is known as the duty ratio or duty cycle, defined in Eq. (4.2). The power transistor is turned on for a period of DT and turned off for the remaining time $(1 - D)T$. Depending on whether the switch is turned on or off, the inductor current will be either charging through V_{in} or discharging through the diode, respectively. As a result, there are two modes of operation. We first consider *mode 1*, when the switch is on, shown in Fig. 4.10(*a*).

As shown in the figure, when the switch is on, the input voltage, V_{in}, forces the diode into the reverse bias region. To determine the voltage conversion ratio, the average input and output currents, and the output voltage, we use the inductor current as a state variable in the following equation:

$$\begin{aligned} V_{in} &= v_L + V_o \\ &= L\frac{di_L}{dt} + V_o \end{aligned} \tag{4.8}$$

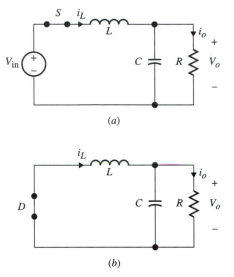

(a)

(b)

Figure 4.10 Equivalent circuit modes for the buck converter. (a) Mode 1: The power switch is on. (b) Mode 2: The power switch is off.

Equation (4.8) can be rearranged as follows:

$$\frac{di_L}{dt} = \frac{1}{L}(V_{in} - V_o) \tag{4.9}$$

Integrating Eq. (4.9) from $t = 0$ to t with $I_L(0)$ as the initial condition, we obtain

$$i_L(t) = \frac{1}{L}(V_{in} - V_o)t + I_L(0) \tag{4.10}$$

Equation (4.10) suggests that the inductor current charges linearly with a slope of $(V_{in} - V_o)/L$, where $I_L(0)$ is the initial inductor current value at $t = 0$, when the switch is first turned on. This equation applies as long as the switch is on. However, the equivalent circuit model changes when the power switch is turned off at $t = DT$, resulting in the equivalent circuit of mode 2 shown in Fig. 4.10(b), during which the diode is conducting.

As shown in Fig. 4.10(b), in order for the inductor current to maintain its continuity, the diode is forced to conduct by becoming forward biased so that the diode "picks up" the current in the direction shown. The diode is known as *flyback* or *free-wheeling* because of the manner in which it is forced to turn on. The resultant equation that describes mode 2 operation is

$$\frac{di_L}{dt} = -\frac{1}{L}V_o \tag{4.11}$$

Integrating both sides of Eq. (4.11) for $t \geq DT$ with $i_L(DT)$ as an initial condition, we obtain

$$i_L(t) = -\frac{V_o}{L}(t - DT) + I_L(DT) \tag{4.12}$$

where $I_L(DT)$ is the initial inductor current when the switch is first turned off.

Equation (4.12) suggests that the inductor current starts discharging at $t = DT$ with the slope of $-V_o/L$, as shown in Fig. 4.11(a). In steady-state operation we have

$$I_L(0) = I_L(T) \tag{4.13}$$

Evaluating Eq. (4.10) at $t = DT$ and Eq. (4.12) at $t = T$ and using Eq. (4.13), we obtain the following two relations for $I_L(0)$ and $I_L(DT)$:

$$I_L(DT) = \frac{1}{L}(V_{in} - V_o)DT + I_L(0) \tag{4.14a}$$

$$I_L(0) = -\frac{V_o}{L}(1 - D)T + I_L(DT) \tag{4.14b}$$

The steady-state current and voltage waveforms are shown in Fig. 4.11. $I_{L\max}$ and $I_{L\min}$ are the inductor current values at the instants the switch is turned off and on, respectively.

Voltage Conversion

Next we use the preceding relations to derive expressions for the voltage conversion, and average input and output currents. From Eqs. (4.14) we obtain

$$\frac{V_o}{V_{in}} = D \tag{4.15}$$

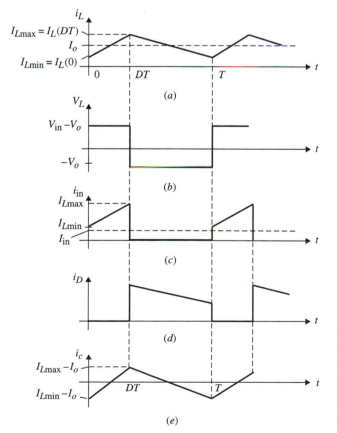

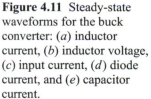

Figure 4.11 Steady-state waveforms for the buck converter: (a) inductor current, (b) inductor voltage, (c) input current, (d) diode current, and (e) capacitor current.

Hence, the maximum output voltage gain is 1. We should point out that Eq. (4.15) can be obtained easily by using the volt-second principle across the inductor, which is given as follows:

Mode 1   Mode 2

(interval DT)   (interval $(1-D)T$)

$\overbrace{\text{(Inductor voltage)(time)}} + \overbrace{\text{(Inductor voltage)(time)}} = 0$

$$v_L(t)DT + v_L(t)(1-D)T = 0 \tag{4.16}$$

where v_L equals $(V_{in} - V_o)$ and $-V_o$ during time intervals DT and $(1-D)T$, respectively.

We can make two observations on the buck voltage gain equation $V_o = DV_{in}$. First, since all the converter components (L, C, D, Q) are ideal, they don't dissipate any power, resulting in 100% voltage efficiency. Second, the average input and output voltage ratio has a linear control characteristic curve, as shown in Fig. 4.12. By varying the value of the duty cycle, D, we can control the average output voltage to the desired level.

Average Input and Output Currents

The input current, i_{in}, as illustrated in Fig. 4.11(c), with an average value of I_{in}, is given by

$$I_{in} = \frac{1}{T}\int_0^T i_{in}(t)\,dt \tag{4.17}$$

Since $i_{in} = i_L$ in mode 1, we substitute for $i_L(t)$ from Eq. (4.10), and by evaluating the integral between $t=0$ and $t=DT$, we obtain

$$I_{in} = \frac{1}{2L}(V_{in} - V_o)D^2T + I_L(0)D \tag{4.18}$$

Using Eq. (4.14b), we obtain

$$I_{in} = \frac{1}{2}(I_{L\max} + I_{L\min})D \tag{4.19}$$

where $I_{L\max}$ and $I_{L\min}$ represent $I_L(DT)$ and $I_L(0)$, respectively.

Similarly, by inspection, the average output current is given by

$$I_o = I_L = \frac{I_{L\min} + I_{L\max}}{2} = \frac{V_o}{R} \tag{4.20}$$

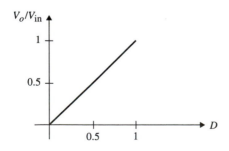

Figure 4.12 Ideal output control characteristic curve for the buck converter.

From Eqs. (4.14) and (4.20) we can solve for the maximum and minimum inductor currents, to obtain

$$I_{L\max} = DV_{in}\left(\frac{1}{R} + \frac{(1-D)T}{2L}\right)$$ (4.21)

$$I_{L\min} = DV_{in}\left(\frac{1}{R} - \frac{(1-D)T}{2L}\right)$$ (4.22)

Substituting these equations in Eqs. (4.19) and (4.20), we obtain

$$I_o = \frac{DV_{in}}{R}$$

$$I_{in} = \frac{D^2 V_{in}}{R}$$

Hence, the current gain is given by

$$\frac{I_o}{I_{in}} = \frac{1}{D}$$ (4.23)

This relation can be obtained by equating the average input and output power, to yield

$$I_{in}V_{in} = I_o V_o$$

$$\frac{I_{in}}{I_o} = \frac{V_o}{V_{in}} = D$$

From Eqs. (4.15) and (4.23), it is clear that the current and voltage relations for the converter are equivalent to a dc transformer model with a ratio of D, as shown in Fig. 4.13. The sinusoidal curve and straight line drawn across the transformer windings indicate that the transformer is capable of transferring ac and dc, respectively.

Critical Inductor Value

It is clear that for $I_{L\min} \neq 0$, the converter will operate in the *continuous conduction mode (ccm)*. To find the minimum inductor value that is needed to keep the converter in the ccm, we set $I_{L\min}$ to zero and solve for L:

$$I_{L\min} = DV_{in}\left(\frac{1}{R} - \frac{(1-D)T}{2L}\right) = 0$$

$$L_{crit} = \left(\frac{1-D}{2}\right)TR$$ (4.24)

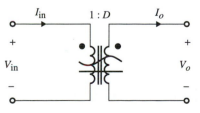

Figure 4.13 Equivalent circuit representation for the buck converter, referred to as a dc-dc transformer.

where L_{crit} is the critical inductance minimum value for a given D, T, and R before the converter enters the discontinuous conduction mode (dcm) of operation.

Output Voltage Ripple

Since we have assumed that the output voltage has no ripple, the entire ac output current from the inductor passes through the parallel capacitor, and only dc current is delivered to the load resistor. In practice, the value of the output capacitor is an important design parameter since it influences the overall size of the dc-to-dc converter and how much of the switching frequency ripple is being removed. Having said that, it is design practice to choose a larger output capacitor in order to limit the ac ripple across V_o. Theoretically speaking, if $C \longrightarrow \infty$, the capacitor acts like a short circuit to the ac ripple, resulting in zero output voltage ripple. If we assume C is finite, then there exists a voltage ripple superimposed on the average output voltage. In order to derive an expression for the capacitor ripple voltage, we first obtain an expression for the capacitor current, which is given by the following relation:

$$i_c(t) = i_L(t) - I_o$$

As a result, the initial capacitor current at $t = 0$ is given by

$$I_c(0) = I_L(0) - I_o$$

$$= -\frac{I_{L\max} - I_{L\min}}{2}$$

and at $t = DT$,

$$I_c(DT) = I_L(DT) - I_o$$

$$= +\left(\frac{I_{L\max} - I_{L\min}}{2}\right)$$

The resultant capacitor current and voltage are shown in Fig. 4.14.

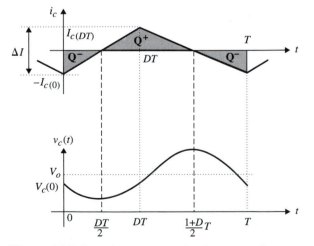

Figure 4.14 Capacitor current and voltage waveforms.

The instantaneous capacitor current can be expressed in terms of ΔI fr〈 Eqs. (4.21) and (4.22), as shown in the following equations:

$$i_c(t) = \frac{I_{L\max} - I_{L\min}}{DT}t - \frac{I_{L\max} - I_{L\min}}{2} = \frac{\Delta I}{DT}t - \frac{\Delta I}{2} \qquad 0 \le t < DT \quad (4.25a)$$

$$i_c(t) = -\frac{I_{L\max} - I_{L\min}}{(1-D)T}(t - DT) + \frac{I_{L\max} - I_{L\min}}{2}$$

$$= \frac{-\Delta I}{(1-D)T}(t - DT) + \frac{\Delta I}{2} \qquad 0 \le t < DT \qquad (4.25b)$$

where $\Delta I = (V_{in}(1-D)TD)/L$.

From the capacitor voltage-current relation, $i_c = C(dv_c/dt)$, the capacitor voltage, $v_{c1}(t)$, can be expressed by the following integral for $t \ge 0$:

$$v_{c1}(t) = \frac{1}{C}\int_0^t i_c \, dt + V_c(0)$$

where $V_c(0)$ is the initial capacitor voltage at $t = 0$. Substituting for $i_c(t)$ from Eq. (4.25a), we obtain the following equation:

$$v_{c1}(t) = \frac{1}{C}\int_0^t \left(\frac{\Delta I}{DT}t - \frac{\Delta I}{2} \right) dt + V_c(0)$$

Evaluating this integral yields

$$v_{c1}(t) = \frac{1}{C}\frac{\Delta I}{DT}\frac{t^2}{2} - \frac{\Delta I}{2C}t + V_c(0) \qquad 0 \le t < DT \qquad (4.26)$$

Similarly, for $t \ge DT$, the capacitor voltage is given by

$$v_{c2}(t) = \frac{1}{C}\int_{DT}^t i_c \, dt + V_c(DT)$$

where $V_c(DT)$ is the initial capacitor voltage when the switch is turned off at $t = DT$. From Eq. (4.27) $v_{c2}(t)$ is given by

$$v_{c2}(t) = \frac{-\Delta I}{C(1-D)T}\frac{(t-DT)^2}{2} + \frac{\Delta I}{2C}(t - DT) + V_c(DT) \qquad DT \le t < T \qquad (4.27)$$

Since the capacitor voltage is in the steady state, we have $v_{c1}(t = DT) = v_{c2}(t = DT)$ and $v_{c1}(0) = v_{c2}(T)$, resulting in the following boundary conditions for the capacitor voltage:

$$V_c(T) = V_c(DT) = V_c(0)$$

Since the average capacitor voltage is V_o, then we have in general

$$V_o = \frac{1}{T}\left[\int_0^{DT} v_{c1}(t) \, dt + \int_{DT}^T v_{c2}(t) \, dt \right]$$

Substituting for v_{c1} and v_{c2} from Eqs. (4.26) and (4.27), we obtain

$$V_o = \frac{\Delta I}{12C}(1 - 2D)T + V_c(0) \qquad (4.28)$$

Substitute for $\Delta I = (DV_{in}(1 - D)T)/L$ in Eq. (4.28) to yield

$$V_c(0) = DV_{in}\left[1 - \frac{(1-D)(1-2D)}{12CL}T^2\right] \tag{4.29}$$

Hence, the capacitor initial values at $t = 0$ and $t = DT$ are equal, as expected since the capacitor current is symmetrical. Since the peak capacitor voltage occurs when the inductor current is zero, we have the capacitor minimum voltage occurring at $t = DT/2$, which is obtained from Eq. (4.26):

$$
\begin{aligned}
V_{c,\,min} &= \frac{1}{C}\frac{\Delta I}{2DT}\left(\frac{DT}{2}\right)^2 - \frac{\Delta I}{2C}\left(\frac{DT}{2}\right) + V_c(0) \\
&= -\frac{\Delta I}{8C}DT + V_c(0)
\end{aligned}
\tag{4.30a}
$$

and the maximum capacitor voltage occurring at $t = (1 + D)T/2$ as obtained from Eq. (4.27):

$$
\begin{aligned}
V_{c,\,max} &= -\frac{\Delta I}{2C(1-D)T}\left(\frac{(1+D)}{2}T - DT\right)^2 + \frac{\Delta I}{2C}\left(\frac{(1+D)}{2}T - DT\right) + V_c(DT) \\
&= \frac{\Delta I}{8C}(1-D) + V_c(DT)
\end{aligned}
\tag{4.30b}
$$

Substituting for $V_c(0)$ from Eq. (4.29) and using $\Delta I = (DV_{in}(1 - D)T)/L$, it can be shown that $V_{c,\,min}$ and $V_{c,\,max}$ are expressed as follows:

$$V_{c,\,min} = V_o\left[1 - \frac{(1-D)(2-D)}{24CL}T^2\right] \tag{4.31a}$$

$$V_{c,\,max} = V_o\left[1 + \frac{(1-D^2)}{24CL}T^2\right] \tag{4.31b}$$

Hence, the variation in the capacitor peak voltage is given by

$$\Delta V_c = V_{c,\,max} - V_{c,\,min}$$

and from Eqs. (4.31) we obtain

$$\Delta V_c = \frac{V_o}{8LCf^2}(1-D)$$

Sometimes it is useful to express the ratio of the ripple to the output voltage,

$$\frac{\Delta V_c}{V_0} = \frac{1-D}{8LCf^2} \tag{4.32}$$

This term is known as the output voltage ripple and represents the regulation. As expected, when the filtering capacitor and the frequency increase, the voltage ripple decreases.

Using Capacitor Charge to Evaluate ΔV_c

Another useful way to evaluate the expression for ΔV_c without having to obtain the exact expression for $v_c(t)$ is to use the total charge, Q, deposited on the capacitor current interval. Figure 4.14 shows the waveform for i_c and v_c with areas of positive charge (+) and negative charge (−). Because of waveform symmetry, $t = DT/2$ and $t = (1+D)T/2$ represent the i_c zero crossing times when the capacitor voltage is minimum, $V_{c,\,min}$, and maximum, $V_{c,\,max}$, respectively. Hence, the capacitor voltage ripple is ΔV_c. The total charge stored in the capacitor between $t = DT/2$ and $t = (1+D)T/2$ is obtained from the following equation:

$$\frac{dQ}{dt} = C\frac{dv_c}{dt}$$

So the total charge Q between capacitor current i_c zero crossings $(DT/2 \le t < (1+D)T/2)$ is given by

$$\Delta Q = C\Delta V_c \tag{4.33}$$

However, since the total charge is related to the current according to the relation

$$i = \frac{dQ}{dt}$$

then we have

$$\Delta Q = \frac{1}{T/2}\int_{DT/2}^{(1+D)T/2} i\,dt = \text{Area under the curve}$$

$$= \frac{1}{2}\left(\frac{1+D}{2}T - \frac{D}{2}T\right)\frac{1}{2}\Delta I \tag{4.34}$$

$$= \frac{1}{2}\frac{T}{2}\frac{1}{2}\Delta I$$

From Eqs. (4.33) and (4.34), we obtain

$$\Delta V_c = \frac{T}{8C}\Delta I$$

Substituting for $\Delta I = (DV_{in}(1-D)T)/L$, we obtain

$$\frac{\Delta V_c}{V_o} = \frac{1-D}{8LC}T^2$$

$$= \frac{1-D}{8LCf^2} \tag{4.35}$$

EXAMPLE 4.2

Consider a buck converter with the following circuit parameters: $V_{in} = 20$ V, $V_o = 15$ V, and $I_o = 5$ A, for $f = 50$ kHz. Determine: (a) D, (b) L_{crit}, (c) maximum and minimum inductor currents for $L = 100L_{crit}$, (d) average input and output power, and (e) capacitor voltage ripple for $C = 0.47\ \mu F$.

SOLUTION

(a) $D = 0.75$

(b) Using $R = 3\ \Omega$ and $T = 20\ \mu s$, the critical inductor value is given by

$$L_{crit} = \left(\frac{1-D}{2}\right)TR = 7.5\ \mu H$$

(c) For $L = 100L_{crit} = 750 \ \mu H = 0.75 \ mH$, we have

$$I_{Lmin} = DV_{in}\left(\frac{1}{R} - \frac{(1-D)T}{2L}\right)$$

$$= (0.75)(20)\left(\frac{1}{3} - 3.33 \times 10^{-3}\right)$$

$$I_{Lmin} = 4.95 \ A$$

$$I_{Lmax} = (0.75)(20)\left(\frac{1}{3} + 3.33 \times 10^{-3}\right)$$

$$I_{Lmax} = 5.05 \ A$$

(d) Since it is an ideal converter, the average output and input powers are given by

$$P_{in} = P_o = V_o I_o = (15)(5) = 75 \ W$$

(e) The capacitor voltage ripple is given by

$$\frac{\Delta V_o}{V_o} = \frac{1-D}{8LCf^2}$$

$$= \frac{(1-0.75)}{8(0.75 \ mH)(0.47 \ \mu F)(50 \times 10^3)^2}$$

$$\frac{\Delta V_o}{V_o} = 0.035 = 3.5\%$$

EXAMPLE 4.3

Design a buck converter with the following specifications: $\Delta V_0/V_o = 0.5\%$, $V_{in} = 20 \ V$, $P_o = 12 \ W$, $f = 30 \ kHz$, and $D = 0.4$.

SOLUTION   In order to design this converter, we need to calculate the values for L, C, and R.
The output voltage is given by

$$V_o = DV_{in} = 8 \ V$$

Hence, the output current is

$$I_o = \frac{P_o}{V_o} = 1.5 \ A$$

The output resistance is

$$R = \frac{8}{1.5} = 5.33 \ \Omega$$

The critical inductance for ccm is given by

$$L_{crit} = \frac{1-D}{2}TR$$

$$= \left(\frac{1-0.4}{2}\right)\left(\frac{1}{30 \times 10^3}\right)5.33$$

$$= 53.3 \ \mu H$$

Let us select $L = 600$ μH. Based on this value, the maximum and minimum inductor currents are given by

$$I_{L\max} = DV_{in}\left(\frac{1}{R} + \frac{(1-D)T}{2L}\right)$$

$$= (0.4)(20)\left(\frac{1}{5.33} + 0.0167\right)$$

$$= 1.63 \text{ A}$$

$$I_{L\min} = (0.4)(20)\left(\frac{1}{5.33} - 0.0167\right)$$

$$= 1.37 \text{ A}$$

The ripple voltage is given by

$$\frac{\Delta V_o}{V_o} = \frac{1-D}{8LCf^2} = 0.005$$

Solving for C,

$$C = \frac{1-D}{(8Lf^2)0.005}$$

$$C = 27.78 \ \mu\text{F}$$

EXERCISE 4.3

Redesign Example 4.2 to achieve an output ripple voltage not to exceed 1% and an inductor current ripple not to exceed 10% at the average load current.

ANSWER   0.15 mH, 8.33 μF, 3 Ω

EXERCISE 4.4

Determine the diode and transistor average and rms current values for Exercise 4.3.

ANSWER   1.25 A, 3.75 A, 2.5 A, 4.33 A

EXERCISE 4.5

Show that the expression for the peak capacitor voltage at $t = (1+D)T/2$ is as given by Eq. (4.31b).

4.2.2   The Boost Converter

Basic Topology and Voltage Gain

Other possible switch and transistor-diode arrangements are shown in Fig. 4.15(*a*) and (*b*), respectively. This topology is known as a boost converter since the output voltage is higher than the input, as will be shown in this section.

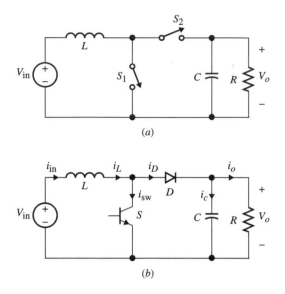

Figure 4.15 Boost converter. (*a*) Two-switch implementation. (*b*) Transistor-diode implementation.

Similar to the case for the buck converter, we assume all the converter components are ideal and the transistor switching waveform is as shown in Fig. 4.9(*d*). When the switch is turned on, the equivalent circuit of mode 1 is shown in Fig. 4.16(*a*). This is a charging interval, and the voltage across the inductor is V_{in}, and $i_L(t)$ is given by

$$i_L(t) = \frac{1}{L}V_{in}t + I_L(0) \qquad 0 \leq t < DT \tag{4.36}$$

where $I_L(0)$ is the initial inductor current value at $t = 0$. When the switch is turned off at $t = DT$, the resultant equivalent mode 2 circuit is shown in Fig. 4.16(*b*).

The inductor voltage is $V_{in} - V_o$, and $i_L(t)$ is given by

$$i_L(t) = \frac{1}{L}(V_{in} - V_o)(t - DT) + I_L(DT) \qquad DT \leq t < T \tag{4.37}$$

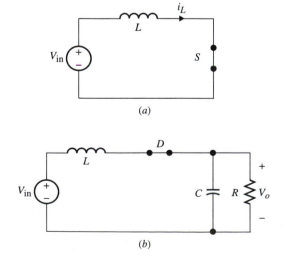

Figure 4.16 Equivalent circuit modes for the boost converter. (*a*) Mode 1: The switch is on. (*b*) Mode 2: The switch is off.

Evaluating Eqs. (4.36) and (4.37) at $t = DT$ and $t = T$, respectively, and using the fact that $I_L(T) = I_L(0)$, we obtain

$$I_L(DT) - I_L(0) = \frac{1}{L}V_{in}(DT) \qquad (4.38a)$$

$$I_L(DT) - I_L(0) = -\frac{1}{L}(V_{in} - V_o)(1 - D)T \qquad (4.38b)$$

From Eqs. (4.38a) and (4.38b), the resulting voltage conversion is given by

$$\frac{V_o}{V_{in}} = \frac{1}{1 - D} \qquad (4.39)$$

Hence, the voltage gain is always greater than 1. Also from Eqs. (4.38), the inductor ripple current is given by

$$
\begin{aligned}
\Delta I &= I_L(DT) - I_L(0) \\
&= I_{Lmax} - I_{Lmin} \\
&= \frac{1}{L}V_{in}DT
\end{aligned} \qquad (4.40a)
$$

Substituting for V_{in} from Eq. (4.39), we obtain

$$\Delta I = \frac{1}{L}V_o D(1 - D)T \qquad (4.40b)$$

Key current and voltage waveforms are given in Fig. 4.17.

Average Input and Output Currents

The input current is the same as the inductor current as shown in Fig. 4.17(a). Hence, the average input current by inspection is given by

$$I_{in} = \frac{I_{Lmax} + I_{Lmin}}{2} \qquad (4.41)$$

The average output current is the same as the average diode current and is given by

$$I_o = \left(\frac{I_{Lmax} + I_{Lmin}}{2}\right)(1 - D) = \frac{V_o}{R} \qquad (4.42)$$

Since we assume an ideal converter, the average input and output powers must be equal. Using Eqs. (4.41) and (4.42), we get

$$V_{in}I_{in} = V_o I_o$$

resulting in

$$\frac{I_{in}}{I_o} = \frac{V_o}{V_{in}} = \frac{1}{1 - D} \qquad (4.43)$$

As with the buck converter, the input-output current and voltage ratios are equivalent to a dc transformer with a transformer mode ratio equal to $1/(1 - D)$, as shown in Fig. 4.18.

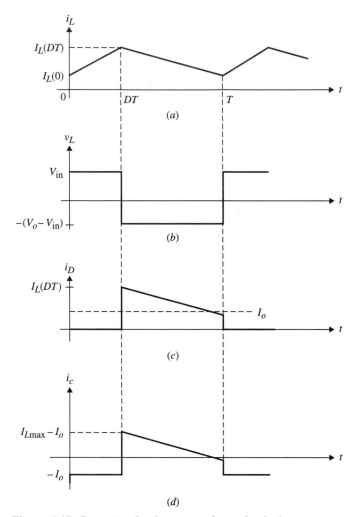

Figure 4.17 Current and voltage waveforms for the boost converter.

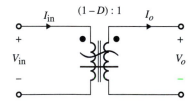

Figure 4.18 Equivalent transformer circuit representation for the boost converter.

Using Eqs. (4.38) and (4.42), we can solve for the maximum and minimum inductor current values:

$$I_L(0) = I_{L\min} = V_{in}\left(\frac{1}{R(1-D)^2} - \frac{DT}{2L}\right) \tag{4.44a}$$

$$I_L(DT) = I_{L\max} = V_{in}\left(\frac{1}{R(1-D)^2} + \frac{DT}{2L}\right) \tag{4.44b}$$

For positive values of $I_{L\max}$ and $I_{L\min}$, the converter will operate in the continuous conduction mode. To solve for the minimum critical inductor value that will keep the converter in the ccm, we set $I_{L\min}$ to zero:

$$I_{L\min} = 0$$

Under this boundary condition, the critical inductor value is given by

$$L_{\text{crit}} = \frac{RT}{2}(1-D)^2 D \qquad (4.45)$$

Output Ripple Voltage

It is clear from Fig. 4.17 that when the diode is reverse biased, the capacitor current is the same as the load current. Since we assume the load current is purely dc, the capacitor current is given by

$$i_c = -I_o \qquad\qquad 0 \le t < DT$$
$$i_c = i_L - I_o \qquad DT \le t < T$$

The capacitor current waveform is shown in Fig. 4.17(d) and redrawn in Fig. 4.19 along with the capacitor voltage waveform. Mathematical expressions for i_c can be obtained directly from this figure.

The current $i_c(t)$ is expressed mathematically as

$$i_c(t) = -\frac{\Delta I}{(1-D)T}(t-DT) + I_c(DT) \qquad DT \le t \le T \qquad (4.46)$$

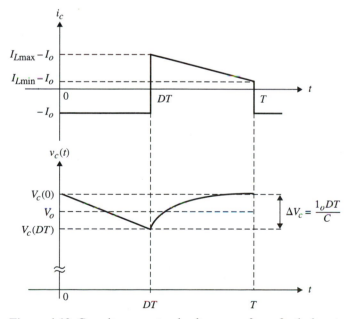

Figure 4.19 Capacitor current and voltage waveforms for the boost converter, assuming $I_{L\min} > I_o$.

where $I_c(DT)$ is the initial $i_c(t)$ at $t = DT$. The capacitor voltage for $0 \le t < DT$ is given by

$$v_c(t) = \frac{1}{C}\int_0^t -I_o \, dt + V_c(0)$$

$$= -\frac{I_o}{C}t + V_c(0)$$

(4.47)

where $V_c(0)$ is the initial capacitor voltage at $t = 0$.
    At $t = DT$ we have

$$V_c(DT) = -\frac{I_o}{C}DT + V_c(0)$$

(4.48)

Since the average capacitor voltage is V_o, we can solve for $V_c(0)$ and $V_c(DT)$ as follows:

$$V_c(0) = V_o + \frac{I_o DT}{2C}$$

$$V_c(DT) = V_o - \frac{I_o DT}{2C}$$

(4.49)

and the capacitor voltage variation is given by

$$\Delta V_c = V_c(0) - V_c(DT) = \frac{I_o DT}{C}$$

(4.50)

For $DT \le t < T$ the capacitor voltage is given by

$$v_c(t) = \frac{1}{C}\int_{DT}^T \left[\frac{-\Delta I}{(1-D)T}(t-DT) + I_c(DT)\right]dt + V_c(DT)$$

$$= -\frac{\Delta I}{2C(1-D)T}(t-DT)^2 + \frac{I_c(DT)(1-D)T}{C} + V_c(DT)$$

(4.51)

The output ripple voltage is given by

$$|\Delta V_o| = |\Delta V_c| = I_o\frac{DT}{C} = \frac{V_o DT}{RC}$$

Then the voltage ripple is given by

$$\frac{\Delta V_o}{V_o} = \frac{DT}{RC}$$

$$= \frac{D}{RCf}$$

(4.52)

EXAMPLE 4.4

Sketch the current waveforms for i_L, i_{in}, i_D, i_o, and i_c for the boost converter with the following parameters: $L = 1.8 \text{ mH}$, $V_{in} = 50 \text{ V}$, $V_o = 120 \text{ V}$, $R = 20 \ \Omega$, $C = 147 \ \mu\text{F}$, and $f = 15 \text{ kHz}$. Also sketch the voltage waveforms for v_L, v_{sw}, v_c, and v_D.

SOLUTION   In order to sketch the waveforms, we need to find D, the maximum and minimum inductor currents, and the average output current.

The duty cycle is given by

$$\frac{V_o}{V_{in}} = \frac{1}{1-D} = \frac{120}{50}$$

which yields $D = 0.58$

Using $R = 20 \ \Omega$ and $T = 66.67 \ \mu s$, the maximum and minimum inductor currents are given by

$$I_{L\max} = V_{in}\left(\frac{1}{(1-D)^2 R} + \frac{DT}{2L}\right) = 14.94 \text{ A}$$

and

$$I_{L\min} = V_{in}\left(\frac{1}{(1-D)^2 R} - \frac{DT}{2L}\right) = 13.86 \text{ A}$$

The average input and output currents are given by

$$I_{in} = \frac{I_{L\max} + I_{L\min}}{2} = 14.4 \text{ A}$$

$$I_o = I_{in}(1-D) = 6 \text{ A}$$

The capacitor peak currents are given by

$$I_{c\max} = I_{L\max} - I_o = 8.94 \text{ A}$$

$$I_{c\min} = I_{L\min} - I_o = 7.86 \text{ A}$$

Hence,

$$\Delta I_c = (I_{c\max} - I_{c\min}) = 1.074 \text{ A}$$

Notice this value must be equal to ΔI_L.

The capacitor voltage is given by

$$v_c(t)\big|_{t=0} = 120 \text{ V}$$

$$v_c(t)\big|_{t=DT} = \frac{-I_o}{C}DT + 120 \text{ V}$$

$$= \frac{-5.95 \text{ A}}{147 \ \mu F}(0.58)(66.67 \ \mu s) + 120 \text{ V} = 118.43 \text{ V}$$

Hence, the ripple is 1.57 V.

EXAMPLE 4.5

Design a boost converter with the following specifications: $P_o = 27 \text{ W}$, $V_o = 40 \text{ V}$, $V_{in} = 28 \text{ V}$, $\Delta V_o / V_o = 2\%$, $f_s = 35 \text{ kHz}$.

SOLUTION First let's determine the duty cycle, D:

$$D = 1 - \frac{V_{in}}{V_o} = 1 - \frac{28}{40} = 0.3$$

For continuous conduction mode, the inductance minimum value is given by

$$L_{\text{crit}} = \frac{RT}{2}(1-D)^2 D$$

where

$$T = 28.57 \ \mu s$$

$$R = \frac{V_o^2}{P_o} = \frac{(40)^2}{27} = 59.26 \ \Omega$$

$$L_{crit} = 124.44 \ \mu H$$

We choose $L = 200 \ \mu H$ since L should be greater than L_{crit} for ccm operation. The output ripple voltage is

$$\frac{\Delta V_o}{V_0} = \frac{D}{RCf}$$

$$0.02 = \frac{0.3}{(59.26)C(35 \ kHz)}$$

$$C = 7.23 \ \mu F$$

Determine the average and rms current values for the diode and transistor in Example 4.4.

ANSWER   8.4 A, 11 A, 6 A, 9.28 A

4.2.3   The Buck-Boost Converter

The third possible converter type is obtained by interchanging the diode and the inductor of the buck converter to realize the design of Fig. 4.20. This converter is known as a *buck-boost converter* since its voltage gain can be less than, equal to, or

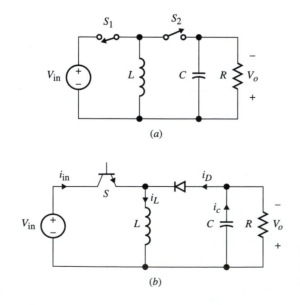

Figure 4.20 Buck-boost converter. (*a*) Switch implementation. (*b*) Transistor-diode implementation.

greater than 1. Unlike the buck and boost converters, this converter gives a negative output voltage when used without isolation.

In this analysis, we follow assumptions made in performing the steady-state analysis for the buck and boost topologies. When the switch is on, the diode is reverse biased with the equivalent circuit of mode 1 as shown in Fig. 4.20(a); Fig. 4.21(b) gives the equivalent circuit for mode 2 when the transistor is off.

When the transistor is turned on, the inductor current starts charging from the source voltage, V_{in}, while the diode D is reverse biased. The voltage across the inductor is V_{in}, and i_L is given by

$$i_L(t) = \frac{1}{L}V_{in}(t) + I_L(0) \tag{4.53}$$

where $I_L(0)$ is the initial inductor value that corresponds to the minimum inductor current value. Evaluating Eq. (4.53) at $t = DT$, when the switch is turned off, we obtain the maximum inductor current. This yields the following relation:

$$I_{L\max} - I_{L\min} = \frac{1}{L}V_{in}DT \tag{4.54}$$

Similarly, in mode 2 the inductor current is given by

$$i_L(t) = -\frac{V_o}{L}(t - DT) + I_L(DT) \tag{4.55}$$

Evaluating Eq. (4.55) at $t = T$, and since we assume steady-state operation, we use $I_L(0) = I_L(T)$ to obtain

$$I_L(0) = \frac{-V_o}{L}(T - DT) + I_L(DT) \tag{4.56a}$$

$$I_{L\max} - I_{L\min} = \frac{V_o}{L}(1 - D)T \tag{4.56b}$$

Equating Eqs. (4.54) and (4.56b), we obtain the following voltage conversion ratio:

$$\frac{V_o}{V_{in}} = \frac{D}{1 - D} \tag{4.57}$$

It is clear from this equation that the output voltage can be either smaller or greater than the input voltage:

$$D > 0.5 \qquad \text{Boost}$$
$$D < 0.5 \qquad \text{Buck}$$
$$D = 0.5 \qquad \text{Unity gain}$$

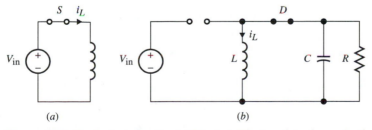

(a)   (b)

Figure 4.21 Equivalent circuits. (a) Mode 1: The transistor is conducting. (b) Mode 2: The transistor is not conducting.

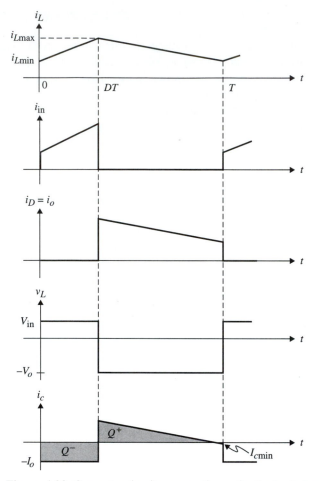

Figure 4.22 Current and voltage waveforms for the buck-boost converter.

Figure 4.22 gives typical current and voltage waveforms for the buck-boost converter. From Fig. 4.22, it is clear that the average input current is given by

$$I_{in} = \frac{I_{Lmax} + I_{Lmin}}{2}D \tag{4.58a}$$

and the average output current is given by

$$I_o = \frac{I_{Lmax} + I_{Lmin}}{2}(1 - D) \tag{4.58b}$$

Since the conservation of power must hold, we use $I_{in}V_{in} = I_o V_o$ to obtain

$$\frac{I_{in}}{I_o} = \frac{V_o}{V_{in}} = \frac{D}{1 - D} \tag{4.59}$$

To solve for I_{Lmax} and I_{Lmin} in terms of the converter components, we substitute for $I_o = V_o/R$. Hence, from Eqs. (4.56) and (4.58b) we obtain

$$I_{Lmax} = V_{in}\left[\frac{D}{R(1 - D)^2} + \frac{DT}{2L}\right] \tag{4.60a}$$

$$I_{L\min} = V_{in}\left[\frac{D}{R(1-D)^2} - \frac{DT}{2L}\right] \qquad (4.60b)$$

The critical inductor value that keeps the converter in the ~~dem~~ CCM mode can be obtained by setting $I_{L\min} = 0$ in the above equation to yield

$$L_{\text{crit}} = \frac{RT(1-D)^2}{2} \qquad (4.61)$$

We notice that for the same frequency of operation and load resistance, the buck converter has the highest critical inductor limit when compared to the boost and buck-boost, whereas the boost converter has the smallest L_{crit}, resulting in a wider range of inductor design.

Output Voltage Ripple

It can be shown that the capacitor current of the buck-boost converter is the same as that of the boost. Hence, the capacitor voltage is given by

$$\Delta V_c(t) = \frac{V_o DT}{RC}$$

$$\frac{\Delta V_o}{V_o} = \frac{DT}{RC} = \frac{D}{RCf} \qquad (4.62)$$

EXAMPLE 4.6

(a) Sketch the current and voltage waveforms for a buck-boost converter with the following parameters: $V_{in} = 40$ V, $V_o = 60$ V, $D = 0.6$, $R = 20$ Ω, $L = 750$ μH and $T_s = 200$ μs. (b) Find the rms value for i_c.

SOLUTION

(a) Let $T = 200$ μs.

$$D = \frac{1}{1 + \dfrac{V_{in}}{V_o}} = \frac{1}{1 + \dfrac{40}{60}} = 0.6$$

$$\begin{aligned}
I_{L\max} &= \frac{V_{in}D}{R(1-D)^2} + \frac{V_{in}DT}{2L} \\[2mm]
&= \frac{40(0.6)}{20(1-0.6)^2} + \frac{40(0.6)(200\,\mu s)}{2(750\,\mu H)} = 10.7 \text{ A}
\end{aligned}$$

$$\begin{aligned}
I_{L\min} &= \frac{V_{in}D}{R(1-D)^2} - \frac{V_{in}DT}{2L} \\[2mm]
&= \frac{40(0.6)}{20(1-0.6)^2} - \frac{40(0.6)(200\,\mu s)}{2(750\,\mu H)} = 4.3 \text{ A}
\end{aligned}$$

$$\begin{aligned}
I_{in} &= \frac{I_{L\max} + I_{L\min}}{2}D \\[2mm]
&= \frac{10.7 + 4.3}{2}(0.6) = 4.5 \text{ A}
\end{aligned}$$

$$P_o = V_{in}I_{in}$$
$$= (40)(4.5) = 180 \text{ W}$$

Having found P_o, we can now solve for the average output current:

$$I_D = I_o = \frac{180 \text{ W}}{60 \text{ V}} = 3 \text{ A}$$

Hence, we can calculate the values for $I_{c\max}$ and $I_{c\min}$:

$$I_{c\max} = I_{L\max} - I_o = 7.7 \text{ A}$$
$$I_{c\min} = I_{L\min} - I_o = 1.3 \text{ A}$$

$$I_{crms} = \sqrt{\frac{1}{T}\left[\int_0^{DT} 9\,dt + \int_{DT}^{T}\left(\frac{-V_o}{L}(t-DT)+I_{c\max}\right)^2 dt\right]}$$

$$= 3.856 \text{ A}$$

EXERCISE 4.7

Consider a buck-boost converter that supplies 75 W at $I_o = 5$ A from a 37 V dc source. Let $T = 130$ μs and $L = 250$ μH. Determine: (a) the duty ratio, D, (b) $I_{L\max}$ and $I_{L\min}$, (c) average input current, (d) average diode and transistor currents, and (e) the rms value of the capacitor current.

ANSWER   0.29, 9.8 A, 4.25 A, 2.03 A, 5 A, 3.464 A.

EXERCISE 4.8

Derive the output voltage ripple formula of Eq. (4.62) for a buck-boost converter.

EXERCISE 4.9

Determine D, $I_{L\max}$, $I_{L\min}$, I_{in}, I_D, I_{sw}, and I_{crms} for a buck-boost converter whose capacitor current waveform is shown in Fig. E4.9. Assume $L = 120$ μH.

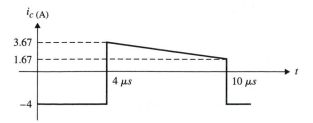

Figure E4.9   Waveform for Exercise 4.9.

ANSWER   0.4, 7.67 A, 5.67 A, 2.67 A, 4 A, 2.67 A, 3.3 A

Under certain input and output voltage conditions, none of the preceding three basic converter topologies is suitable. For example, suppose the input voltage varies between 8 and 18 V while the output voltage is desired to be maintained fixed at +12 V. It is clear that the voltage gain varies between 2/3 and 3/2 while the output is constant at +12 V. Even though the buck-boost converter allows the voltage gain to be less or greater than 1, it provides only a negative output voltage polarity. One topology that is capable of providing a positive output voltage is known as the *Single-Ended Primary Inductance Converter* (SEPIC), shown in Fig. 4.23.

4.2.4   Fourth-Order Converters

It is possible to cascade or cascode more than one of the basic topologies—buck, boost, and buck-boost—to form new topologies that have attractive features that a single topology does not. Figure 4.24(*a*) and (*b*) shows a block diagram of two converters

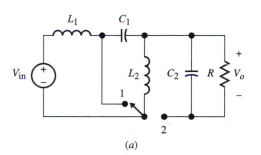

(*a*)

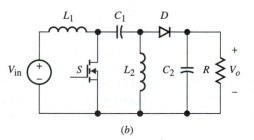

(*b*)

Figure 4.23 Single-ended primary inductance converter (SEPIC). (*a*) Single-switch, double-throw implementation. (*b*) MOSFET-diode implementation.

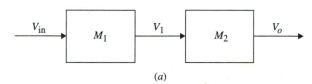

(*a*)

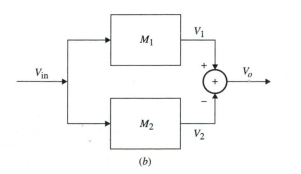

(*b*)

Figure 4.24 Combinations of basic converter topologies. (*a*) Cascade. (*b*) Cascode.

connected in series (cascade) and in parallel (cascode), respectively. The parallel connection is also known as a differential configuration.

The overall voltage gains for the Fig. 4.24(*a*) and (*b*) converter arrangements are given in Eqs. (4.63) and (4.64), respectively.

$$M = \frac{V_o}{V_{in}} = M_1 M_2 \tag{4.63}$$

$$M = \frac{V_o}{V_{in}} = M_1 - M_2 \tag{4.64}$$

where, in Fig. 4.24(*a*), M_1 and M_2 are given by

$$M_1 = \frac{V_1}{V_{in}} \quad M_2 = \frac{V_o}{V_1}$$

and in Fig. 4.24(*b*) we have

$$M_1 = \frac{V_1}{V_{in}} \quad M_2 = \frac{V_2}{V_{in}}$$

Figure 4.25(*a*) and (*b*) shows the block diagram and circuit implementation for the two possible series cascadings of the boost converter with a buck converter. Regardless of the sequence, the voltage gain for both converters should be $M_1 M_2 = D/(1-D)$.

The equivalent circuit representation for the cascade of the boost and buck converters is shown in Fig. 4.26(*a*). After careful circuit manipulation, it can be shown that the equivalent one-switch implementation of Fig. 4.26(*a*) is as shown in Fig. 4.26(*b*).

To illustrate how Fig. 4.26(*b*) is obtained from Fig. 4.26(*a*), we assume i_{L1} and i_{L2} are constant current sources as shown in Fig. 4.27(*a*). Figure 4.27(*b*) shows the equivalent circuit when S is in positions 1 and 2. Next we redraw the portion of the output circuit as shown in Fig. 4.27(*c*), which is redrawn in Fig. 4.27(*d*). The two modes of Fig. 4.27(*d*) are represented in the equivalent circuit shown in Fig. 4.27(*e*).

Similarly, the buck and boost cascade is shown in Fig. 4.28(*a*). The circuit can be simplified through several straightforward steps as shown in Fig. 4.28(*b*)–(*d*). Notice the capacitor in Fig. 4.28(*a*) is removed since regardless the state of S_1 and S_2, the capacitor average current is always zero. In Fig. 4.28(*c*) we combine L_1 and L_2 in series. Finally, Fig. 4.28(*d*) is obtained using similar steps for the buck-boost cascade.

EXAMPLE 4.7

Figure 4.29(*a*) shows a cascade of a buck and a Cuk converter. Show that the voltage gain is given by

$$\frac{V_o}{V_{in}} = \frac{D^2}{1-D} \tag{4.65}$$

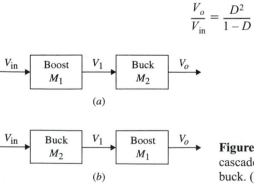

(*a*)

(*b*)

Figure 4.25 Block diagram representation for the cascade of a boost and a buck converter. (*a*) Boost-buck. (*b*) Buck-boost.

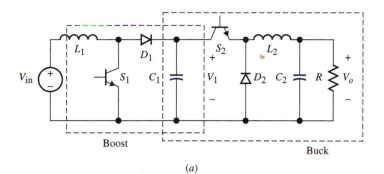

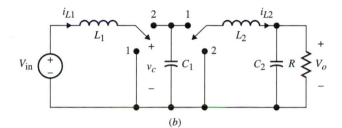

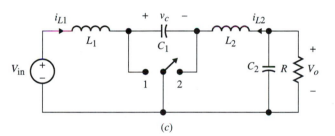

Figure 4.26 (*a*) Boost-buck cascade. (*b*) Two-switch implementation. (*c*) One-switch equivalent circuit implementation.

SOLUTION During mode 1, when S_1 and S_2 are in position 1, and during mode 2, when S_1 and S_2 are in position 2, the inductor voltages are given by

Mode 1 (*DT*):
$$v_{L1} = V_{in}$$
$$v_{L2} = -V_o + V_c$$

Mode 2 ((1–*D*)*T*):
$$v_{L1} = -V_c$$
$$v_{L2} = -V_o$$

Apply the volt-second balance principle to v_{L1} and v_{L2}, to obtain

$$v_{L1}: \quad DV_{in} - (1-D)V_c = 0$$

$$M_1 = \frac{V_c}{V_{in}} = \frac{D}{1-D}$$

$$V_{L2}: \quad D(-V_o + V_c) - (1-D)V_o = 0$$

$$M_2 = \frac{V_o}{V_c} = D$$

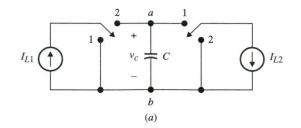

(a)

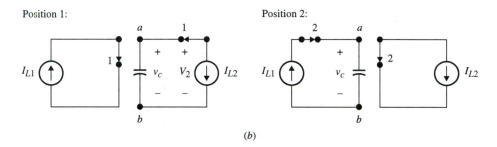

(b)

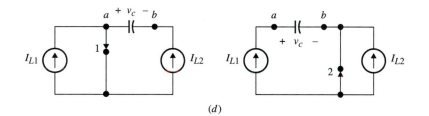

(c)

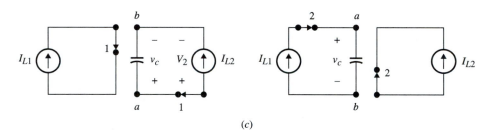

(d)

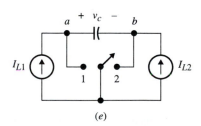
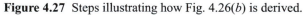

(e)

Figure 4.27 Steps illustrating how Fig. 4.26(b) is derived.

Therefore, the total voltage gain is

$$\frac{V_o}{V_c} = M_1 M_2 = \frac{D^2}{1 - D}$$

Another straightforward way to find the gain equation is to realize that the average voltage between a and a' is DV_{in} and use this voltage as an input to the Cuk converter. The switch implementation for Fig. 4.29(a) is shown in Fig. 4.29(b).

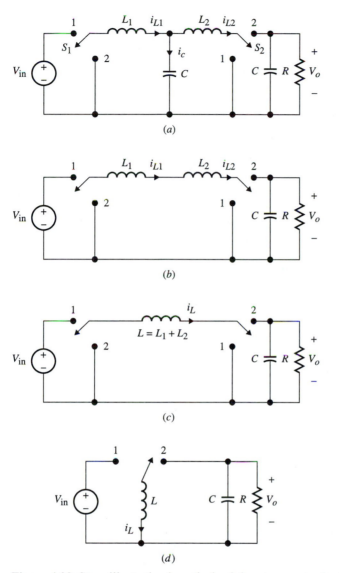

Figure 4.28 Steps illustrating how the buck-boost converter is obtained from the cascade configuration.

EXERCISE 4.10

Show that the cascaded buck and boost converters of Fig. 4.28(c) produce the same voltage gain conversion as that of Fig. 4.28(a), i.e., $V_o/V_{in} = D/(1-D)$, when S_1 and S_2 are single-pole, double-through switches; assume S_1 and S_2 are synchronized in position 1 for the DT interval and in position 2 for the $(1-D)T$ interval.

Figure 4.26(b) is known as a Cuk converter, whose voltage gain is $D/(1-D)$, as will be discussed in detail shortly. This concept can be extended to other cascaded topologies. The generalized fourth-order switched-mode voltage-to-voltage converter is given in Fig. 4.30. Components 1, 2, 3, 4, and 5 consist of a switch, a diode, two inductors, and one capacitor. By disallowing capacitor loops and inductor cut-sets, the

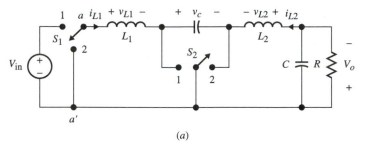

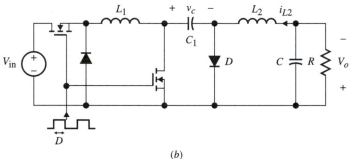

Figure 4.29 (a) Cascade configuration of buck and Cuk converters. (b) Diode-switch implementation.

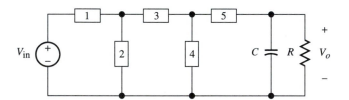

Figure 4.30 Generalized representation of fourth-order voltage-to-voltage converter.

total number of physically realizable topologies can be reduced. For example, all possible topologies with component 1 representing a switch are shown in Fig. 4.31.

The topology of Fig. 4.31(e) is a buck converter with an additional output LC filter. Figure 4.31(c) is a buck-boost converter with an additional output LC filter. The topologies in Fig. 4.31(b) and (f) are physically unrealizable since the average output current in each of the capacitors is zero; hence, no power is delivered to the load. On the other hand, Fig. 4.31(d) does not have an average input current.

The Cuk Converter

It has been shown that other converter topologies can be obtained by combining some of the three topologies discussed earlier. One cascade combination of a buck and a boost converter is known as a Cuk converter, given in Fig. 4.26(b) and redrawn in Fig. 4.32, named after its inventor, Slobodan Cuk from the California Institute of Technology. Figure 4.32(c) shows the Cuk converter with a magnetically coupled inductor representation.

The front end of the converter is a boost, and the back end of it is a buck. Hence, we may refer to the Cuk converter as a boost-buck converter. Unlike the previous converters, this converter requires two switches and uses two inductors, and a capacitor to store and transfer energy from the input to the output, resulting in a higher level of

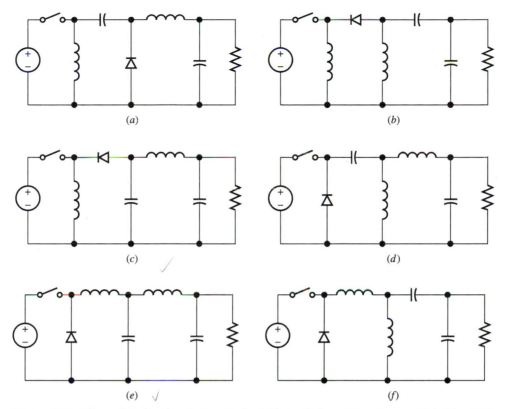

Figure 4.31 All possible fourth-order topologies with a switch used as component 1 in Fig. 4.30.

complexity. Like the buck-boost, the gain of the Cuk converter can be less than, equal to, or greater than 1, with a positive output polarity. The major advantage of this converter is that the input inductor, L_1, and the output inductor, L_2, can be coupled on one magnetic core structure such that with the proper core gap design, the input and output switching currents can be made zero. Other possible combinations of converters are shown in Fig. 4.33(a) and (b), which represent a boost cascade with an LC output filter and a buck cascade with an LC input filter, respectively.

The analysis of the Cuk converter can be carried out the same way as for the other converter topologies. There are two modes of operations: mode 1 when the switch is on, and mode 2 when the switch is off.

Mode 1   Mode 1 starts when the transistor is turned on at $t = 0$ (Fig. 4.34(a)).

The inductor L_1 voltage is given by

$$v_{L1} = V_{\text{in}}$$

$$L_1 \frac{di_{L1}}{dt} = V_{\text{in}} \tag{4.66}$$

resulting in the following current relation:

$$i_{L1}(t) = \frac{V_{\text{in}}}{L_1} t + I_{L1}(0) \tag{4.67}$$

where $I_{L1}(0)$ is the initial current value.

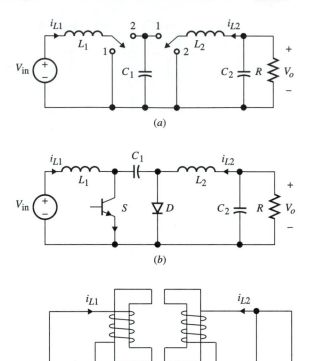

Figure 4.32 Cuk converter with magnetically coupled inductors. (*a*) Two-switch implementation. (*b*) Transistor-diode implementation. (*c*) Core implementation.

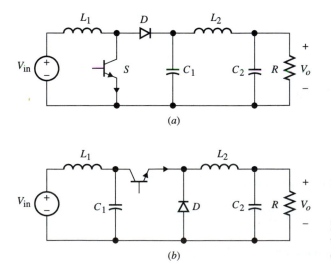

Figure 4.33 (*a*) Boost cascade with *LC* output filter. (*b*) Buck cascade with *LC* input filter.

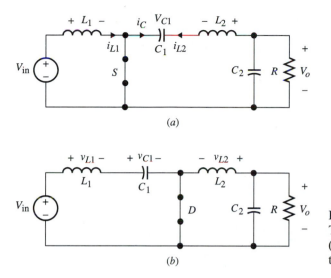

Figure 4.34 (a) Mode 1: Transistor is turned on. (b) Mode 2: Transistor is turned off.

The voltage across L_2 is given by

$$v_{L2} = v_{c1} + V_o$$

$$= L_2 \frac{di_{L2}}{dt}$$

Since $i_{L2} = -i_{c1} = -C(dv_{c1}/dt)$, the above equation yields

$$L_2 C_1 \frac{d^2 v_{c1}}{dt^2} + v_{c1} = -V_o$$

If we assume the average voltage across C_1 has no ripple, then its average value, V_{c1}, is given by

$$V_{c1} = V_{in} - V_o \qquad (4.68)$$

This equation is obtained by using KVL from the output through C_1, L_1, L_2, and V_{in} and setting the average inductor voltages to zero.

$$v_{L2} = V_{in}$$

$$i_{L2} = \frac{V_{in}}{L_2} t + I_{L2}(0)$$

At $t = DT$,

$$I_{L1}(DT) = \frac{V_{in}}{L_1} DT + I_{L1}(0) \qquad (4.69a)$$

Similarly,

$$I_{L2}(DT) = \frac{V_{in}}{L_2} DT + I_{L2}(0) \qquad (4.69b)$$

Mode 2   When the transistor is turned off at $t = DT$, D turns on and the equivalent circuit is given by Fig. 4.34(b). Similar analysis shows

$$v_{L1} = V_o$$

and the inductor currents are given by

$$i_{L1}(t) = \frac{V_o}{L_1}(t - DT) + I_{L1}(DT)$$

$$i_{L2}(t) = \frac{V_o}{L_2}(t - DT) + I_{L2}(DT)$$

At $t = T$,

$$i_{L1}(T) = \frac{V_o}{L_1}(1 - D)T + I_{L1}(DT) \tag{4.70a}$$

$$i_{L2}(T) = \frac{V_o}{L_2}(1 - D)T + I_{L2}(DT) \tag{4.70b}$$

Since $I_{L1}(T) = I_{L1}(0)$ and $I_{L2}(T) = I_{L2}(0)$, from Eqs. (4.69) and (4.70), we obtain

$$\frac{-V_{in}}{L_1}DT = \frac{V_o}{L_1}(1 - D)T$$

$$\frac{V_o}{V_{in}} = \frac{-D}{1 - D} \tag{4.71}$$

The current and voltage waveforms for the Cuk converter are shown in Fig. 4.35.

The average input current is the same as the average inductor current $i_{L1}(t)$, given by

$$I_{in} = \frac{I_{L1max} + I_{L1min}}{2} \tag{4.72}$$

and the average output current is the same as the average inductor current $i_{L2}(t)$:

$$I_o = \frac{I_{L2max} + I_{L2min}}{2} \tag{4.73}$$

Since the average input power and output power are equal, we can obtain from the above equations the following:

$$I_{L2}(DT) = \left[\frac{D}{(1 - D)R} + \frac{DT}{2L_2}\right]V_{in} \tag{4.74a}$$

$$I_{L2}(0) = \left[\frac{D}{(1 - D)R} - \frac{DT}{2L_2}\right]V_{in} \tag{4.74b}$$

Similarly, we obtain

$$I_{L1}(DT) = \left[\frac{D^2}{(1 - D)^2R} + \frac{DT}{2L_1}\right]V_{in} \tag{4.75a}$$

$$I_{L1}(0) = \left[\frac{D^2}{(1 - D)^2R} - \frac{DT}{2L_1}\right]V_{in} \tag{4.75b}$$

For continuous input current $i_{L1}(t)$, we set $I_{L1}(0) = 0$ to obtain the critical value of L_1 as follows:

$$L_{1crit} = \frac{(1 - D)^2RT}{2D} \tag{4.76}$$

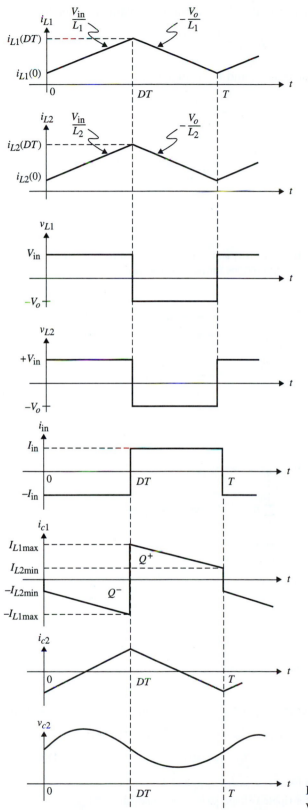

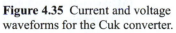

Figure 4.35 Current and voltage waveforms for the Cuk converter.

Similarly, for continuous current in L_2, the minimum value is given by

$$L_{2\text{crit}} = \frac{(1-D)RT}{2} \tag{4.77}$$

The ripple voltage across C_1 and C_2 is given by

$$\frac{\Delta V_{c1}}{V_o} = \frac{D}{RC_1 f}$$

$$\frac{\Delta V_{c2}}{V_o} = \frac{1-D}{8L_2 C_2 f^2} \tag{4.78}$$

EXERCISE 4.11

The converters shown in Fig. E4.11(a) and (b) are for a SEPIC and a converter known as a Zeta. Derive the voltage gain expressions for both converters.

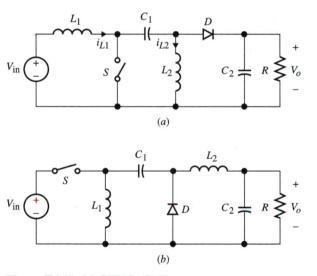

(a)

(b)

Figure E4.11 (a) SEPIC. (b) Zeta.

ANSWER $\dfrac{D}{1-D}$

EXERCISE 4.12

Consider the SEPIC converter in Fig. E4.11(a) operating in ccm with the following parameters:

$$D = 0.52 \qquad V_{\text{in}} = 112 \text{ V} \qquad f_s = 110 \text{ kHz}$$

$$R = 12 \ \Omega \qquad L_1 = L_2 = 50 \ \mu\text{H} \qquad C_1 = C_2 = C = 147 \mu\text{F}$$

(a) Sketch the waveforms for i_{L1}, i_{L2}, i_D, i_{C1}, and i_{C2}.
(b) Determine the capacitor ripple voltage $|\Delta V_{C1}/V_o|$ and $|\Delta V_{C2}/V_o|$.

ANSWER $\dfrac{\Delta V_{C1}}{V_o} = \dfrac{\Delta V_{C2}}{V_D} = \dfrac{D}{RCf}$

EXERCISE 4.13

Consider the PWM converter shown in Fig. E4.13 with S_1 and S_2 turning on and off simultaneously. Derive the expression for V_o/V_{in}.

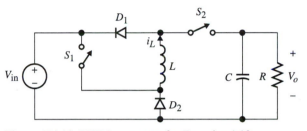

Figure E4.13 PWM converter for Exercise 4.13.

ANSWER $(2D-1)/D$

4.2.5 Bipolar Output Voltage Converters

All single-switch converter topologies discussed so far produce a unipolar average output voltage and a unipolar average output current as shown in Fig. 4.36, resulting in unidirectional average power flow from the source to the load.

In some converter topologies, like bridge converters, the output voltage can be of two polarities, depending on the duty cycle. Figure 4.37(a) and (b) shows two ways to implement a converter that produces bipolar output voltage.

Furthermore, depending on the switching sequence, the average output voltage and current can be unidirectional. Figure 4.37(c) shows the switch-diode implementation of Fig. 4.37(b). It can be shown that the voltage gain is given by

$$\frac{V_o}{V_{in}} = 2D - 1 \tag{4.79}$$

Figure 4.38 shows a plot of M vs. D for several converters.

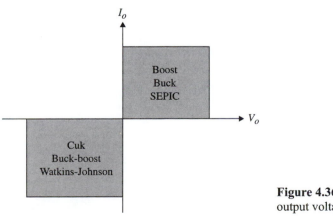

Figure 4.36 Average output current vs. output voltage.

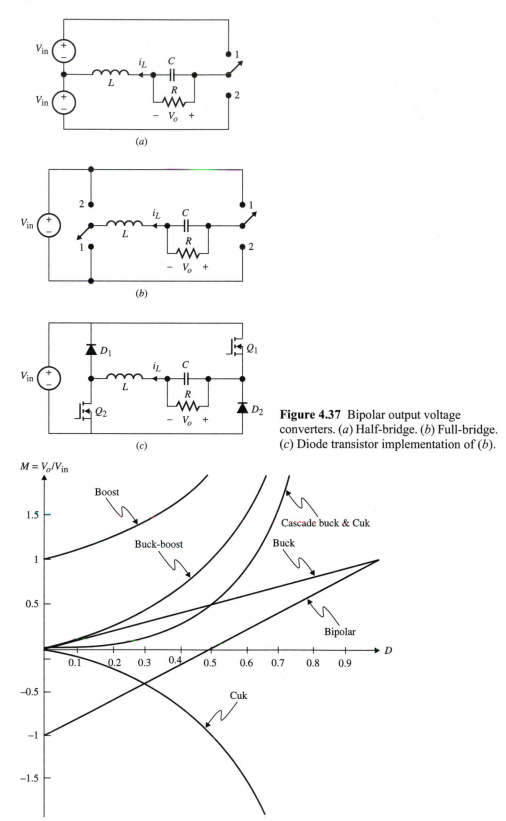

Figure 4.37 Bipolar output voltage converters. (*a*) Half-bridge. (*b*) Full-bridge. (*c*) Diode transistor implementation of (*b*).

Figure 4.38 M vs. D for various converters.

EXERCISE 4.14

The converter shown in Fig. E4.14 is known as a Watkins-Johnson converter and is capable of producing a bipolar output voltage. Assume S_1 and S_2 are thrown in positions 1 and 2 simultaneously for the DT and $(1-D)T$ time intervals, respectively. Derive the expression for the voltage gain V_o/V_{in}.

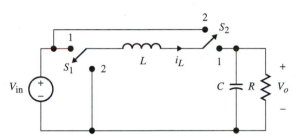

Figure E4.14 Watkins-Johnson converter.

ANSWER $\dfrac{V_o}{V_{in}} = \dfrac{2D-1}{D}$

4.3   DISCONTINUOUS CONDUCTION MODE

Unlike the continuous conduction mode (ccm), in the discontinuous conduction mode (dcm), the minimum inductor current in each of the three basic topologies is zero. Hence, there exists a short interval in which the inductor current is zero, i.e., discontinuous. dcm operation is frequently encountered in dc-to-dc converters since these converters normally operate under open load conditions. Both the steady-state conversion ratio and the closed-loop dynamic charge significantly change. It will be shown shortly that under dcm the converter voltage gain is a function of not only D, but also the load, switching frequency, and circuit components.

As in the ccm case, the steady-state condition will be assumed when it comes to deriving the expression for the voltage gain. In this section we will present the steady-state analysis for the buck, boost, and buck-boost topologies operated in dcm. Unlike the steady-state analysis in ccm, the analysis for dcm requires solving for the time interval during which the inductor current becomes zero. The basic analysis procedure for each converter is the same.

4.3.1   The Buck Converter

The inductor current waveform under dcm operation is shown in Fig. 4.39. Recall the inductor current equations for the buck converter:

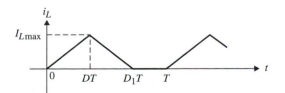

Figure 4.39 Inductor current waveform under dcm operation.

$$i_L(t) = \frac{1}{L}(V_{in} - V_o)t + i_L(0) \qquad\qquad 0 \le t < DT \qquad\qquad (4.80a)$$

$$i_L(t) = \frac{-1}{L}V_o(t - DT) + i_L(DT) \qquad\qquad DT \le t < T \qquad\qquad (4.80b)$$

The edge of discontinuity occurs when $i_L(T) = i_L(0) = 0$. This value occurs when $L = L_{crit}$. The waveform for $i_L(t)$ at the edge of discontinuity is shown in Fig. 4.40. These waveforms correspond to the equivalent circuits given in Fig. 4.41(a), (b), and (c) when the switch(es) are on, off, and both off, respectively.

If the converter inductor becomes less than L_{crit}, there exists a dead time greater than zero, as shown in Fig. 4.39. It is shown that D_1 is the duty ratio at which the inductor current becomes zero.

At $t = DT$, Eq. (4.80a) gives

$$i_L(DT) = \frac{1}{L}(V_{in} - V_o)(DT) \qquad\qquad (4.81)$$

Notice the initial inductor current at $t = 0$ is zero, $i_L(0) = 0$.

At $t = D_1 T$, Eq. (4.80b) gives

$$i_L(D_1 T) = \frac{-1}{L}V_o(D_1 T - DT) + i_L(DT) = 0 \qquad\qquad (4.82)$$

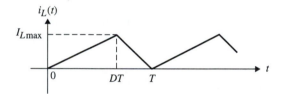

Figure 4.40 Inductor current waveform at the edge of the discontinuity.

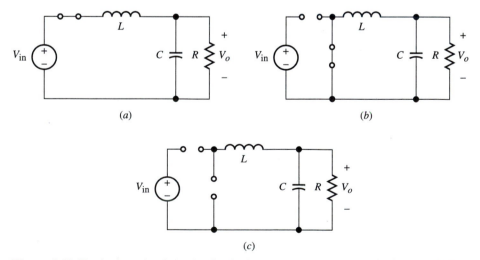

Figure 4.41 Equivalent circuit modes for the buck converter operating in dcm. (a) Switch is on. (b) Switch is off. (c) $i_L = 0$.

From the above two equations, we obtain

$$\frac{V_o}{V_{in}} = \frac{D}{D_1} \qquad (4.83)$$

Using the above equation, we solve for $i_L(DT)$,

$$i_L(DT) = \frac{2V_o^2}{DV_{in}R} \qquad (4.84)$$

Solve for D_1 from Eqs. (4.83) and (4.84) to obtain

$$D_1 = \frac{2V_oL}{DRTV_{in}} + D \qquad (4.85)$$

Thus, the voltage gain in terms of D and the circuit parameters is given by

$$\frac{V_o}{V_{in}} = \frac{D}{\dfrac{2V_oL}{DRTV_{in}} + D} \qquad (4.86)$$

Let the voltage gain M be equal to V_o/V_{in}. We must solve for the voltage gain in terms of the circuit components. From Eq. (4.86), we solve for M to yield

$$M = \frac{D^2RT}{4L}\left[\sqrt{\frac{8L}{D^2RT}+1}-1\right] \qquad (4.87)$$

We can also express D_1 as follows:

$$D_1 = \frac{1}{\dfrac{DRT}{4L}\left[\sqrt{\dfrac{8L}{D^2RT}+1}-1\right]} \qquad (4.88)$$

The voltage conversion characteristics of the buck converter can be expressed in terms of the normalized time constant, τ_n, which is given by

$$\tau_n = \frac{\tau}{T} \qquad (4.89)$$

where $\tau = L/R$ and T is the switching period.

The voltage gain can be expressed as

$$M = \frac{D^2}{4\tau_n}\left[\sqrt{\frac{8\tau_n}{D^2}+1}-1\right] \qquad (4.90)$$

The characteristic curves for M vs. D under different values of the normalized time constant are shown in Fig. 4.42.

The normalized relation for D_1 is given by,

$$D_1 = \frac{1}{\dfrac{D}{4\tau_n}\left[\sqrt{\dfrac{8\tau_n}{D^2}+1}-1\right]} \qquad (4.91)$$

The characteristic curves for M vs. D_1 under different values of the normalized time constants are shown in Fig. 4.43.

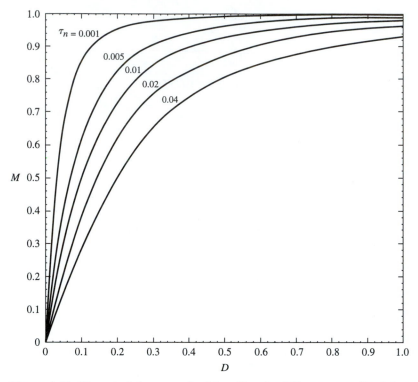

Figure 4.42 Characteristic curves for M vs. D under different normalized time constants.

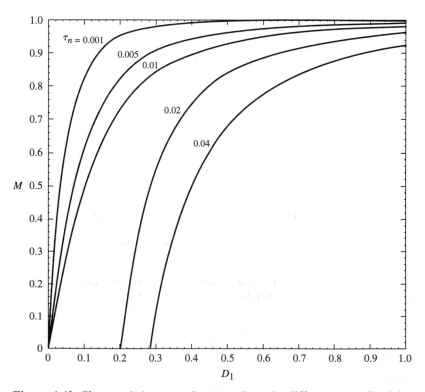

Figure 4.43 Characteristic curves for M vs. D_1 under different normalized time constants.

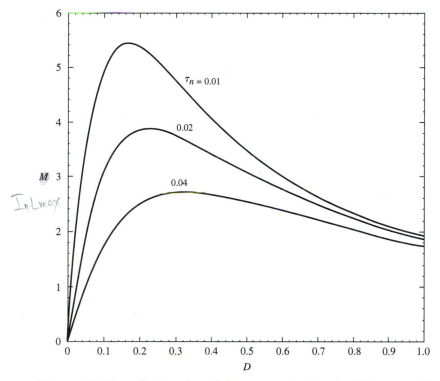

Figure 4.44 Normalized maximum inductor current vs. D under different normalized time constants.

The maximum inductor current of Eq. (4.81) can be expressed in terms of τ_n:

$$I_{L\max} = -\frac{V_{in}TD}{L}\left[\frac{D^2}{4\tau_n}\left[\sqrt{\frac{8\tau_n}{D^2}+1}-1\right]-1\right] \qquad (4.92)$$

Normalizing the current by V_{in}/R,

$$I_{nL\max} = \frac{I_{L\max}R}{V_{in}}$$

$$I_{nL\max} = -\frac{D^3}{4\tau_n^2}\left[\sqrt{\frac{8\tau_n}{D^2}+1}-1\right]+\frac{D}{\tau_n} \qquad (4.93)$$

The plot for $I_{nL\max}$ vs. D under different normalized time constants is given in Fig. 4.44.

EXAMPLE 4.8

Consider a buck converter with a dc voltage source of 80 V and a load resistance equal to 18 Ω. It is required that this converter deliver at least 100 W to the load. Assume the switching frequency is 150 kHz. Determine: (a) the inductor critical value, L_{crit}, (b) the voltage gain for $L = 0.1L_{crit}$ and $10L_{crit}$, (c) D_1 for $L = 0.1L_{crit}$, and (d) the maximum inductor current at $t = DT$.

SOLUTION

(a) The output voltage, V_o, is given by

$$V_o = \sqrt{P_oR} = \sqrt{(100)(18)} = 42.43 \text{ V}$$

For the given $V_{in} = 80$ V, the duty ratio under ccm is 0.53. Under this condition, with $T = 6.67$ μs the inductor critical value is

$$L_{crit} = \frac{RT}{2}(1-D)$$

(b) If the inductor is chosen to be less than $L = 0.1L_{crit} = 2.812$ μH, we have

$$M = \frac{V_o}{V_{in}} = \frac{D^2RT}{4L}\left[\sqrt{\frac{8L}{D^2RT}+1}-1\right] = 0.873$$

For $L = 10L_{crit}$,

$$M = \frac{V_o}{V_{in}} = \frac{42.43}{80} = 0.53$$

The new output voltage is $V_o = 0.873(80) = 69.8$ V.

(c) D_1 is given by

$$D_1 = \frac{D}{M} = \frac{0.53}{0.873} = 0.61$$

(d) The maximum inductor current at $t = DT$ is

$$I_{Lmax} = -\frac{V_{in}TD}{L}(M-1) = 12.73 \text{ A}$$

The time intervals are given as $DT = 3.53$ μs and 4.2 μs with

$$I_o = I_L = \frac{V_o}{R} = \frac{69.8 \text{ V}}{18} = 3.88 \text{ A} \qquad P_o = (3.88)(69.8) = 270.8 \text{ W}$$

$$I_{in} = \frac{I_L(DT)}{2}D = \frac{(12.73)(0.53)}{2} = 3.37 \text{ A} \quad P_{in} = (3.37)(80) = 269.8 \text{ W}$$

The plot for i_L is shown in Fig. 4.45.

EXAMPLE 4.9

Consider a buck converter with the following parameters: $V_{in} = 80$ V, $R_o = 18$ Ω, $P_o = 100$ W, $L = 0.4$ mH, $f_s = 150$ kHz.

(a) Determine the mode of operation.

(b) Determine the range of R_o for the converter to remain in the ccm.

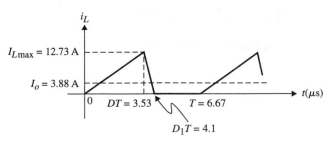

Figure 4.45 Inductor current waveform for Example 4.8 under dcm.

SOLUTION

(a) $V_o = \sqrt{P_o R_o} = 42.43$ V
Assuming continuous mode of operation,

$$D = \frac{V_o}{V_{\text{in}}} = \frac{42.43 \text{ V}}{80 \text{ V}} = 0.53$$

The critical inductor value is given by

$$L_{\text{crit}} = \frac{R_o T}{2}(1 - D) = 28.21 \ \mu\text{H}$$

Since $L > L_{\text{crit}}$, the converter is operating in the ccm.

(b) Notice that the inductor current tends to decrease faster as the load resistance increases.
Recall,

$$I_L(DT) = I_{L\max} = \left[\frac{1}{R} + \frac{(1-D)T}{2L}\right]DV_{\text{in}}$$

$$I_L(0) = I_{L\min} = \left[\frac{1}{R} - \frac{(1-D)T}{2L}\right]DV_{\text{in}}$$

For $I_{L\min} = 0$, we have

$$0 = \left[\frac{1}{R} - \frac{(1-0.53)6.67 \ \mu\text{s}}{2(0.4 \text{ mH})}\right](0.53)(80)$$

$$R = \frac{2(0.4 \ \mu\text{H})}{(1-0.53)(6.67 \ \mu\text{s})} = 255.2 \ \Omega$$

For $R < 255.2 \ \Omega$ ccm (heavy load)
For $R = 255.2 \ \Omega$ dcm (boundary)
For $R > 255.2 \ \Omega$ dcm (light load)

4.3.2 The Boost Converter

The boost converter can also be analyzed to obtain the voltage gain, M, the time at which the inductor current reads zero, D_1, and the maximum inductor current, $I_{L\max}$. For the on state in the time interval $0 \le t \le DT$, we have

$$i_L = \frac{V_{\text{in}}}{L}t \qquad (4.94)$$

In the dcm $I_L(0)$ equals zero, and the maximum inductor current occurs at $t = DT$, and is given by

$$I_{L\max} = I_L(DT) = \frac{V_{\text{in}}}{L}DT \qquad (4.95)$$

For the *off* state in the time interval $DT \le t$, we have

$$i_L(t) = \frac{V_{\text{in}} - V_o}{L}[t - DT] + i_L(DT)$$

In the dcm, $i_L(D_1 T) = 0$; hence,

$$0 = \frac{V_{\text{in}} - V_o}{L}(D_1 - D)T + i_L(DT) \qquad (4.96)$$

From Eqs. (4.95) and (4.96), we obtain the voltage gain:

$$\frac{V_o}{V_{in}} = M = \frac{D_1}{D_1 - D} \tag{4.97}$$

To solve for D_1 in terms of the circuit parameters, we use conservation of power to obtain another relation. The average input power, P_{in}, equals $I_{in}V_{in}$, where $I_{in} = \frac{1}{2}I_{Lmax}D_1$; thus,

$$P_{in} = \frac{1}{2}I_{Lmax}V_{in}D_1 \tag{4.98}$$

The average output is given by

$$P_o = \frac{V_o^2}{R} \tag{4.99}$$

Equating P_{in} and P_o, from Eqs. (4.98) and (4.99) we obtain

$$I_{Lmax} = \frac{2V_o^2}{RV_{in}D_1} \tag{4.100}$$

From Eqs. (4.97) and (4.100), and using $I_{in} = \frac{1}{2}I_{Lmax}D_1$, we obtain

$$M = \frac{1}{2}\left[1 + \sqrt{1 + \frac{2RTD^2}{L}}\right] \tag{4.101}$$

In terms of τ_n,

$$M = \frac{1}{2}\left[1 + \sqrt{1 + \frac{2D^2}{\tau_n}}\right] \tag{4.102}$$

The duty ratio D_1 is obtained from Eq. (4.97) in terms of τ_n:

$$D_1 = \frac{\tau_n}{D} + D + \sqrt{\frac{\tau_n^2}{D^2} + 2\tau_n} \tag{4.103}$$

Setting $D_1 = 1$ and solving for L, we obtain

$$L = L_{crit} = \frac{RTD}{2}(1 - D)^2$$

The normalized maximum inductor current is given by

$$I_{nLmax} = \frac{D}{\tau_n} \tag{4.104}$$

The characteristic curves for M vs. D, M vs. D_1, and D vs. I_{nLmax} under different normalized time constants are shown in Figs. 4.46, 4.47, and 4.48, respectively.

EXAMPLE 4.10

Consider a boost converter that supplies 4 A to an 80 V output with $V_{in} = 60$ V and $L = 67$ μH. Assume the inductor current is discontinuous and $f_s = 100$ kHz. Determine the operation mode and the maximum inductor that can be used to achieve the dcm mode of operation.

SOLUTION   The voltage gain is 1.33 and $\tau = L/R = 3.35 \times 10^{-6}$ s , and $\tau_n = 0.335$. The duty ratio is

$$D = \frac{M-1}{M} = \frac{1.33 - 1}{1.33} = 0.25$$

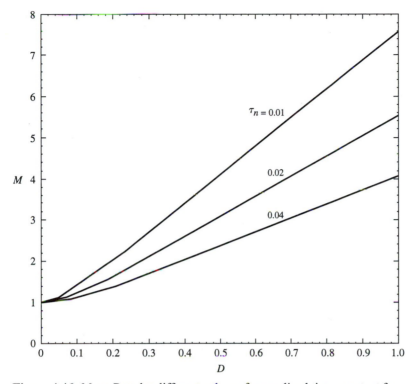

Figure 4.46 M vs. D under different values of normalized time constant for the boost converter.

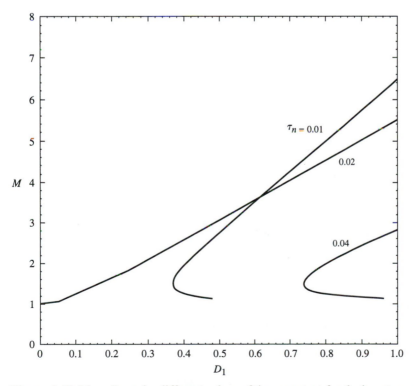

Figure 4.47 M vs. D_1 under different values of time constant for the boost converter.

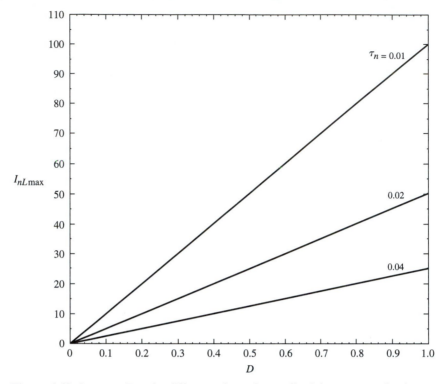

Figure 4.48 $I_{nL\text{max}}$ vs. D under different values of normalized time constant for the boost converter.

Thus, we obtain the value for D_1 as

$$D_1 = \frac{\tau_n}{D} + D + \sqrt{\frac{\tau_n^2}{D^2} + 2\tau_n}$$

$$D_1 = \frac{0.335}{0.248} + 0.248 + \sqrt{\frac{(0.335)^2}{(0.248)^2} + 2(0.335)} = 3.18$$

Since $D_1 > 1$, the converter must be operating in the ccm under the specified values.

The maximum inductor value to maintain is

$$L < L_{\text{crit}} = \frac{RTD}{2}(1-D)^2$$

$$L \leq 0.125 \text{ mH for dcm}$$

4.3.3 The Buck-Boost Converter

As with the other two converters, the derivation of the gain equation for the buck-boost operating in dcm is straightforward. Similar analysis shows the following relations for M and D_1.

$$D_1 = D\left(1 + \frac{1}{M}\right) \tag{4.105}$$

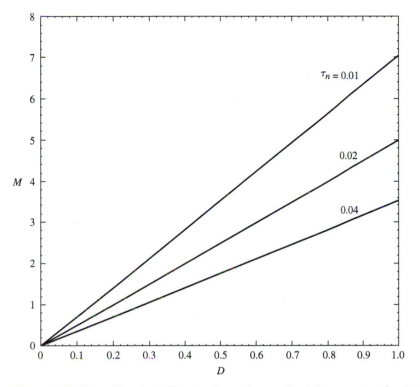

Figure 4.49 M vs. D under different values of normalized time constant for the buck-boost converter.

$$M = \frac{D}{\sqrt{2\tau_n}}$$ (4.106)

The characteristic curves for M vs. D under different values of normalized time constant are shown in Fig. 4.49. For the M vs. D_1 curve, refer to Fig. 4.50.

The normalized maximum inductor current is given by

$$I_{nL\max} = \frac{MD}{\tau_n}$$ (4.107)

Figure 4.51 shows $I_{nL\max}$ vs. D under different normalized time constants.

EXAMPLE 4.11

Consider a buck-boost converter with the following values: $V_o = 12$ V, $P_o = 25$ W, $V_{in} = 20$ V, and $f = 100$ kHz .

(a) Design the converter so that it will operate in ccm.

(b) Repeat part (a) for dcm.

(c) Find the maximum inductor current under both ccm and dcm.

(d) If the load resistance increases by 50% (i.e., the load current changes from 2.08 A to 1.39 A), determine the mode of operation for the two converters and then the maximum inductor current. Sketch the new inductor currents.

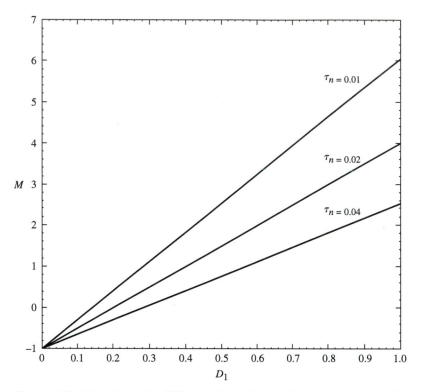

Figure 4.50 M vs. D_1 under different values of normalized time constant for the buck-boost converter.

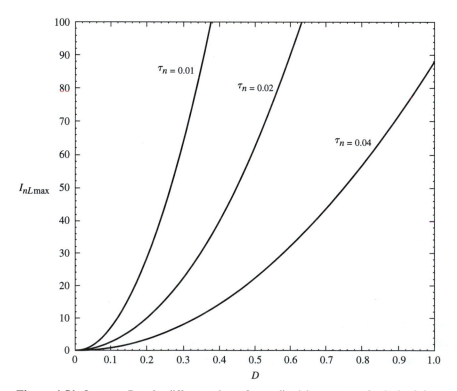

Figure 4.51 $I_{nL\max}$ vs. D under different values of normalized time constant for the buck-boost converter.

SOLUTION

(a) The load resistance at 25 W and 12 V output is given by

$$R = \frac{V_o^2}{P_o} = \frac{12^2}{25} = 5.76 \; \Omega$$

The duty ratio, D, under ccm is 0.375.

For ccm, the inductor must be larger than the critical value given by

$$L_{crit} = \left(\frac{RT}{2}\right)(1-D)^2 = 0.01125 \text{ mH} .$$

For ccm, we choose $L = 0.1$ mH.

(b) For dcm, we choose L smaller than L_{crit}. Let $L = 0.005$ mH.

$$\tau_n = \frac{\tau}{T} = \frac{0.868 \; \mu s}{10 \; \mu s} = 0.087$$

Hence, for $M = 0.6$ and $\tau_n = 0.087$ we obtain $D = 0.25$. Solve for D_1 to yield $D_1 = 0.67$.

(c) The maximum inductor current for ccm is given by

$$I_{L.max} = \frac{DV_{in}}{R(1-D)^2} + \frac{V_{in}DT}{2L} = 3.71 \text{ A}$$

The minimum inductor current is given by

$$I_{L\min} = 3.33 - 0.371 \approx 3 \text{ A}$$

For dcm, we have

$$I_{L\max} = \frac{1}{L}V_{in}DT = 10 \text{ A}$$

(d) If the load resistance changes by +50% for the designs, the new modes of operations can be obtained by finding the new value of τ_n or determining the new L.

(i) For the ccm:

$$R_{new} = 5.76 + \tfrac{1}{2}(5.76) = 8.64 \; \Omega$$

The new L_{crit} is given by

$$L_{crit} = 0.017 \text{ mH}$$

Hence, the converter will remain in the ccm.

$$I_{L\max} = 2.6 \text{ mA}$$

$$I_{L\min} = 1.85 \text{ mA}$$

(ii) For the dcm the new τ is $0.005/8.64 = 0.58 \; \mu s$; then the normalized τ_n is given by

$$\tau_n = \frac{\tau}{T} = \frac{0.58}{10} = 0.058$$

For $M = 0.6$ and $\tau_n = 0.058$, we have $D = 0.204$, and D_1 is obtained from

$$D_1 = D + \sqrt{2\tau_n} = 0.544$$

The maximum inductor current is given by

$$I_{L\max} = \frac{1}{L}V_{in}DT = 8.165 \text{ A}$$

Table 4.1 shows the summary of equations for the three basic converters.

Table 4.1   Summary of ccm Equations

	ccm		dcm	
Converter type	$\dfrac{V_o}{V_{in}}$	$\dfrac{V_o}{V_{in}}$	D_1	$\dfrac{V_o}{V_{in}}$ (in terms of D and τ_n)
Buck	D	$\dfrac{D}{D_1}$	$\dfrac{1}{\dfrac{D}{4\tau_n}\left[\sqrt{\dfrac{8\tau_n}{D}+1}-1\right]}$	$\dfrac{D^2}{4\tau_n}\left[\sqrt{\dfrac{8\tau_n}{D^2}+1}-1\right]$
Boost	$\dfrac{1}{1-D}$	$\dfrac{D_1}{D_1-D}$	$D_1=\dfrac{\tau_n}{D}+D+\sqrt{\dfrac{\tau_n^2}{D^2}+2\tau_n}$	$\dfrac{1}{2}\left[1+\sqrt{1+\dfrac{2D^2}{\tau_n}}\right]$
Buck-boost	$\dfrac{D}{1-D}$	$\dfrac{D_1}{D_1-D}$	$D+\sqrt{2\tau_n}$	$D\sqrt{\dfrac{1}{2\tau_n}}$

EXERCISE 4.15

Consider a boost converter that delivers power to the load with the inductor voltage waveform shown in Fig. E4.15. Assume $R = 176\ \Omega$. Determine L.

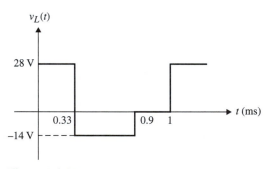

Figure E4.15

ANSWER   10.59 mH

4.4   THE EFFECTS OF CONVERTER NONIDEALITIES

The analysis thus far has been based on the assumption that all components, switching devices, and diodes are ideal. Depending on the application and power levels, the inclusion of some parasitic effects of both components and devices is very important to the design for acceptable performance. In this section, we are going to study the second-order nonideality effects on output voltage and efficiency. The following nonideal characteristics will be investigated:

1. Inductor resistance (r_L)
2. Transistor and diode voltage drops (V_Q, V_D)
3. Switching and conduction losses (r_{sw})

Other nonidealities include the equal series resistance of the output capacitor, r_{ESR}, which is important to include since its presence affects the design of the closed-loop compensator to stabilize the converter.

4.4.1   Inductor Resistance

The Buck Converter

To study the effect of inductor resistance on buck converter performance, we assume the inductor has a finite resistance, r_L, in series as shown in Fig. 4.52. The source of this resistive loss for a practical inductor are the core and copper losses.

When the switch is on, the inductor voltage is given by

$$V_{in} = r_L i_L + L\frac{di_L}{dt} + V_o$$

The first-order differential equation is obtained as

$$\frac{di_L}{dt} + \frac{i_L}{\tau} = \frac{1}{L}(V_{in} - V_o)$$

where $\tau = L/r_L$.

The general solution for i_L is given by

$$i_L(t) = (I_i - I_f)e^{-t/\tau} + I_f \tag{4.108}$$

The initial, I_i, and final, I_f, values must be determined in order to solve for $i_L(t)$.

At $t = DT$ we have

$$i_L(DT) = (I_i - I_f)e^{-DT/\tau} + I_f$$

and the final value, I_f, is given by

$$I_f = \frac{V_{in} - V_o}{r_L}$$

A typical sketch of $i_L(t)$ including inductor resistance is shown in Fig. 4.53.

Let I_i be the minimum inductor current at $t = 0$, $I_{L\min}$, and $I_{L\max}$ be the maximum at $t = DT$. Hence, using the above value for I_f, we obtain the following relation:

$$I_{L\max} = \left[I_{L\min} - \left(\frac{V_{in} - V_o}{r_L}\right)\right]e^{-DT/\tau} + \frac{V_{in} - V_o}{r_L} \tag{4.109}$$

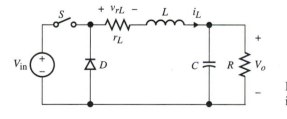

Figure 4.52 Buck converter with inductor resistance.

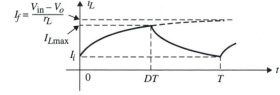

Figure 4.53 Inductor current including r_L.

When the diode is conducting, the inductor equation is given by

$$\frac{di_L}{dt} + \frac{r_L}{L} i_L = -V_o$$

The solution for i_L is given by

$$i_L(t) = (I_i - I_f) e^{(t-DT)/\tau} + I_f$$

The final value, I_f, is given by

$$I_f = \frac{-V_o}{r_L}$$

and the initial value, I_i, is $I_{L\max}$. Hence, $i_L(t)$ is given by

$$I_L(t) = \left(I_{L\max} + \frac{V_o}{r_L} \right) e^{-(t-DT)/\tau} - \frac{V_o}{r_L} \tag{4.110}$$

Evaluating $i_L(t)$ at $t = T$, we obtain

$$I_{L\min} = \left(I_{L\max} + \frac{V_o}{r_L} \right) e^{-T(1-D)/\tau} - \frac{V_o}{r_L} \tag{4.111}$$

where $i_L(t) = I_{L\min}$.

To solve for $i_L(t)$, expressions for $I_{L\min}$ and $I_{L\max}$ must be obtained from Eqs. (4.109) and (4.111) in terms of the circuit components.

By replacing the exponential function by its first two linear terms,

$$e^{-DT/\tau} \approx 1 - \frac{DT}{\tau} \tag{4.112a}$$

$$e^{-T(1-D)/\tau} \approx 1 - \frac{T}{\tau} + \frac{DT}{\tau} \tag{4.112b}$$

Eqs. (4.109) and (4.111) may be written as follows:

$$I_{L\min} = \left(I_{L\max} + \frac{V_o}{r_L} \right) \left(1 - \frac{T}{\tau} + \frac{DT}{\tau} \right) - \frac{V_o}{r_L} \tag{4.113a}$$

$$I_{L\max} = \left(I_{L\min} + \frac{(V_{in} - V_o)}{r_L} \right) \left(1 - \frac{DT}{\tau} \right) + \frac{V_{in} - V_o}{r_L} \tag{4.113b}$$

Further, it can be shown that the following equation may be obtained:

$$I_{L\max} \approx I_{L\min} \approx \frac{(V_{in}D - V_o)}{r_L} \tag{4.114}$$

Equation (4.114) suggests that the inductor ripple is approximated at zero.

Using the relation

$$I_o = \frac{I_{L\min} + I_{L\max}}{2} = \frac{V_o}{R}$$

we obtain

$$\frac{V_o}{V_{in}} = \frac{D}{1 + \frac{r_L}{R}} \tag{4.115}$$

Notice as $r_L \longrightarrow 0$, the gain becomes D.

Another way to solve for the voltage gain is to assume the time constant τ is very large compared to the switching period, and the voltage across r_L is small compared to the input voltage V_{in}; hence, an approximation V_{rL} is calculated using the average value of inductor current. The plot for M vs. D for different r_L/R values is shown in Fig. 4.54.

The Boost Converter

To study the effect of inductor resistance on the boost converter performance, we again assume the inductor has a finite resistance, r_L, in series as shown in Fig. 4.55. The inductor current waveform is shown in Fig. 4.56.

Similar analysis gives the following inductor current equations:

$$i_L(t) = \left(I_{Lmin} + \frac{V_{in}}{r_L} \right) e^{-t/\tau} - \frac{V_{in}}{r_L} \qquad\qquad 0 \le t < DT$$

$$i_L(t) = \left(I_{Lmax} - \frac{V_{in} - V_o}{r_L} \right) e^{-(t-DT)/\tau} + \frac{V_{in} - V_o}{r_L} \qquad DT \le t < T$$

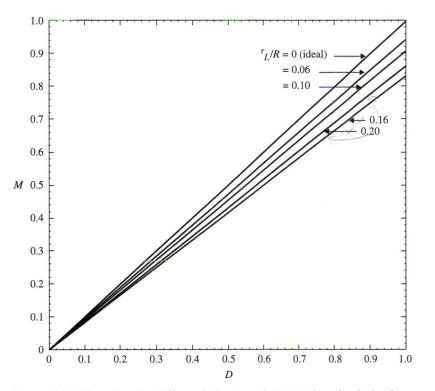

Figure 4.54 M vs. D under different inductor resistance values for the buck converter.

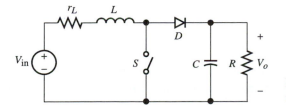

Figure 4.55 Boost converter with inductor resistance.

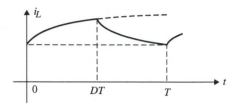

Figure 4.56 Boost inductor current.

where $i_L(0) = I_{L\min}$ and $i_L(DT) = I_{L\max}$. Using the same approximation we have applied to the buck converter, the maximum and minimum inductor currents can be approximated as follows:

$$I_{L\max} \approx I_{L\min} \approx -\frac{1}{r_L}[V_{\text{in}} - (1-D)V_o]$$

Using the average output current, which is given by,

$$I_o = \left(\frac{I_{L\max} + I_{L\min}}{2}\right)(1-D)$$

we obtain the equation for the voltage gain:

$$\frac{V_o}{V_{\text{in}}} = \frac{1}{(1-D) + \dfrac{r_L}{R}\dfrac{1}{1-D}} \tag{4.116}$$

The plot of Eq. (4.116) is shown in Fig. 4.57.

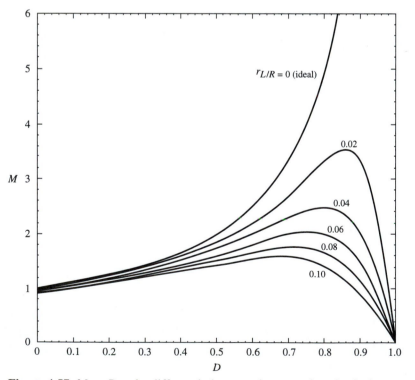

Figure 4.57 M vs. D under different inductor resistance values for the boost converter.

EXERCISE 4.16

Consider the buck-boost converter with inductor resistance shown in Fig. E4.16. Show that the voltage gain equation can be approximated by the following relation:

$$\frac{V_o}{V_{in}} = \frac{D}{(1-D) + \dfrac{r_L}{R(1-D)}} \qquad (4.117)$$

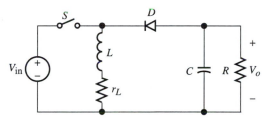

Figure E4.16 Buck-boost converter with inductor resistance.

whose plot is shown in Fig. 4.58.

4.4.2   Transistor and Diode Voltage Drop

As an illustration, let us consider the buck converter. Let us assume that when the transistor is on, a nonzero voltage drop across it, V_Q, is present; and when the diode is

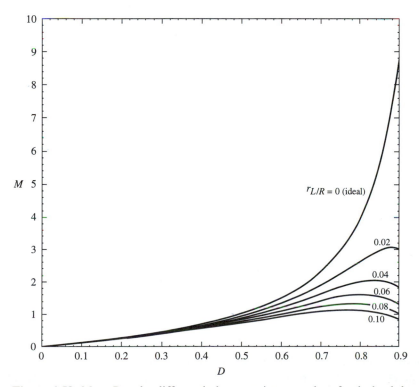

Figure 4.58 M vs. D under different inductor resistance values for the buck-boost converter.

turned on, a voltage V_D appears across it. Hence, the voltage across the inductor while the transistor and diode are conducting is given by

$$\text{Switch on:} \quad v_L = V_{in} - V_o - V_Q \qquad 0 \leq t < DT$$

$$\text{Switch off:} \quad v_L = - V_o - V_D \qquad DT \leq t < T$$

Hence, the inductor currents are given by

$$i_L(t) = \frac{1}{L}(V_{in} - V_o - V_Q)t + I_{L\min} \qquad 0 \leq t \leq DT$$

$$i_L(t) = \frac{1}{L}(- V_o - V_D)(t - DT) + I_{L\max} \qquad DT \leq t \leq T$$

Evaluating the above equations at $t = DT$ and T, we obtain

$$i_{L\max} = \frac{1}{L}(V_{in} - V_o - V_Q)DT + I_{L\min}$$

$$i_{L\min} = \frac{1}{L}(- V_o - V_D)(1 - D)T + I_{L\max}$$

Using this input and output average power, the voltage conversion is given by

$$\frac{V_o}{V_{in}} = D - D\frac{V_Q}{V_{in}} - \frac{V_D}{V_{in}}(1 - D)$$

If we normalize voltages by V_{in}, we obtain,

$$M = D\left(1 - V_{nQ} - V_{nD}\left(\frac{1}{D} - 1\right)\right) \qquad (4.118)$$

where V_{nQ} and V_{nD} are the normalized transistor and diode voltage drops.

Derive the normalized voltage gain expression for the boost converter by including V_{nQ} and V_{nD}.

ANSWER $\quad M = 1 - V_{nD} - \dfrac{D}{1 - D}V_{nQ} + \dfrac{D}{1 - D}$

Show that gain for the buck-boost converter is given by the following equation when including the diode and switch voltage drops.

$$M = \frac{(1 - V_{nQ})D - (1 - D)V_{nD}}{1 - D} \qquad (4.119)$$

4.4.3   Switch Resistance

To study the effect of switching resistance on the converter's performance, we assume the inductor has a finite resistance, r_{sw}, as shown in Fig. 4.59. The resultant inductor current is shown in Fig. 4.60.

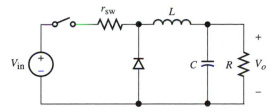

Figure 4.59 Buck converter with switching resistance.

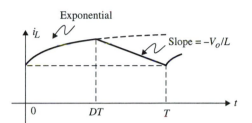

Figure 4.60 Inductor current for Fig. 4.59.

The inductor current equations are given for the *on* and *off* times of the switch as follows:

$$i_L(t) = \left(I_{L\min} - \frac{V_{in} - V_o}{r_{sw}}\right)e^{-t/\tau} + \frac{V_{in} - V_o}{r_{sw}} \qquad 0 \le t < DT$$

$$i_L(t) = \frac{V_o}{L}(t - DT) + I_{L\max} \qquad\qquad DT \le t < T$$

where τ is the time constant and equals L/r_{sw}.

Using the same approximation applied to the inductor resistance, we obtain the following expression for $I_{L\max}$ and $I_{L\min}$:

$$I_{L\max} \approx \frac{1 - DV_{in} - V_o}{Dr_{sw}}$$

$$I_{L\min} = -\frac{V_o}{L}(1 - D)T + \frac{DV_{in} - V_o}{Dr_{sw}}$$

$$I_o = \frac{V_o}{R} = \frac{I_{L\max} + I_{L\min}}{2}$$

Hence, the voltage gain is given by

$$M = \frac{1}{\dfrac{1}{D} + \dfrac{1 - D}{2\tau/T} + \dfrac{r_{sw}}{R}} \tag{4.120}$$

Again assuming $\tau/T \gg 1 - D$, we obtain

$$M = \frac{D}{1 + D\dfrac{r_{sw}}{R}} \tag{4.121}$$

Typically, for a MOSFET, $r_{sw} = r_{DS(ON)} = 0.15\ \Omega$.

A plot of M vs. D for the buck converter under different values of r_{sw}/R is shown in Fig. 4.61. Figures 4.62 and 4.63 show the plots for M vs. D under different values of r_{sw}/R for the boost and buck-boost, respectively.

EXERCISE 4.19

Show that the gain expressions for the boost and buck-boost converters including the switch resistance, r_{sw}, are

$$\frac{V_o}{V_{in}} = \frac{1}{1 - D + \left(\dfrac{D}{1-D}\right)\dfrac{r_{sw}}{R}} \tag{4.122}$$

$$\frac{V_o}{V_{in}} = \frac{D}{1 - D + \left(\dfrac{D}{1-D}\right)\dfrac{r_{sw}}{R}} \tag{4.123}$$

EXAMPLE 4.12

A buck converter is modeled by including a switch resistance, r_{sw}, an inductor resistance, r_L, and a diode voltage drop, V_D. Assume $V_{in} = 50$ V, $V_D = 0.9$ V, $V_o = 20$ V, $R = 4 \ \Omega$, $r_{sw} = 0.08 \ \Omega$, $r_L = 0.06 \ \Omega$.

(a) Derive the relation for V_o/V_{in} that includes the above effects.

(b) Find the duty cycle, D.

(c) Find the efficiency, $\eta = P_o/P_{in}$.

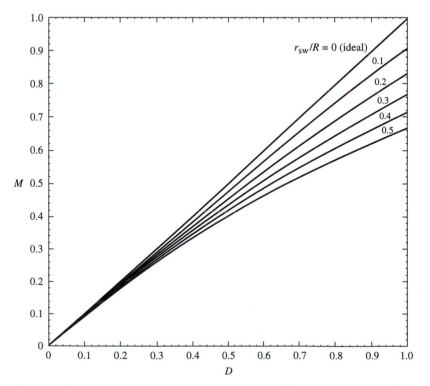

Figure 4.61 M vs. D for the buck converter under different values of r_{sw}/R.

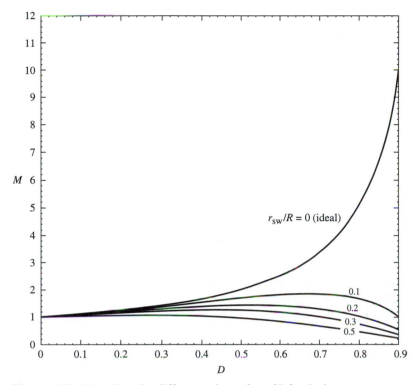

Figure 4.62 M vs. D under different values of r_{sw}/R for the boost converter.

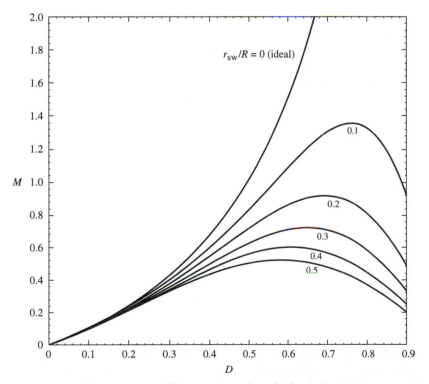

Figure 4.63 M vs. D under different values of r_{sw}/R for the buck-boost converter.

SOLUTION

(a) The inductor current relations are given by

$$I_{L\max} - I_{L\min} = \left(\frac{V_{in} - I_L(r_L + r_{sw}) - V_o}{L}\right)DT$$

$$I_{L\min} - I_{L\max} = \left(\frac{V_o - I_L r_L + V_D}{L}\right)(1 - D)T$$

$$I_L = \frac{V_o}{R}$$

These result in

$$\frac{V_o}{V_{in}} = \frac{D\left(1 + \dfrac{V_D}{V_{in}}\right) - \dfrac{V_D}{V_{in}}}{1 + \dfrac{r_L}{R} + D\dfrac{r_{sw}}{R}} \tag{4.124}$$

(b) Substituting for $V_{in} = 50$ V, $V_D = 0.9$ V, $V_o = 20$ V, $R = 4\ \Omega$, $r_{sw} = 0.08\ \Omega$, $r_L = 0.06\ \Omega$, we obtain $D = 0.42$.

(c) The power loss in the inductor and switch resistors is given by

$$P_{loss} = (I_{in})^2 r_{sw} + (I_L)^2 r_L$$

$$I_{in} = DI_o = D\frac{V_o}{R} = (0.42)\frac{20}{4} = 2.1$$

$$I_o = 5 \text{ A}$$

$$P_{loss} = (2.1)^2(0.08) + (5)^2(0.06) = 0.353 + 1.5 = 1.853 \text{ W}$$

$$P_{in} = V_{in}I_{in} = (50)(2.1) = 105 \text{ W}$$

$$P_o = V_o I_L = (20)(5) = 100 \text{ W}$$

The total loss is 5 W, and the efficiency, η, is given by

$$\eta = \frac{100}{105}100\% = 95.2\%$$

EXERCISE 4.20

Derive the voltage gain equation for the buck-boost converter by including the switch resistance, r_{sw}, and the inductor resistance, r_L.

ANSWER

$$\frac{V_o}{V_{in}} = \frac{D}{(1 - D) + \dfrac{r_L}{R(1 - D)} + \left(\dfrac{D}{1 - D}\right)\dfrac{r_{sw}}{R}}$$

---

dc Transformer Model

We can develop the dc transformer equivalent circuit model by including the converter nonidealities discussed above. Such linear circuit models are very useful when it comes to solving for voltage conversion, average current gain, and efficiency. This is because linear circuit analysis techniques that are well understood by design engineers can be employed in solving these models.

By inspection, it can be seen that the dc transformer equivalent circuit for the buck converter of Fig. 4.64 is given in Fig. 4.65. The derivation of the model is quite simple. All we need to do is to write the dc equation relating V_{in}, V_o, I_{in}, D, and I_o during modes 1 and 2. For example, for the buck converter of Fig. 4.64, we have the following relations for $v_L(t)$:

$$v_L(t) = V_{in} - V_Q - (r_L + r_{on}D)I_o - V_o \qquad 0 \le t < DT \qquad (4.125a)$$

$$v_L(t) = -V_D - V_o - (r_D(1-D) + r_L)I_o \qquad DT \le t < T \qquad (4.125b)$$

Under dc conditions, the average inductor voltage is zero,

$$\frac{1}{T}\int_0^T v_L(t)dt = 0$$

This integral yields the following dc equation:

$$D[V_{in} - V_o - V_Q - (r_L + r_{on}D)I_o] + (1-D)[-V_D - V_o - (r_D(1-D) + r_L)I_o] = 0$$

Rearrange the terms in terms of V_{in}, V_o/D, and DI_o to obtain

$$V_{in} = \frac{V_o}{D} + V_Q + \frac{1-D}{D}V_D + \left[r_{on} + \left(\frac{1-D}{D}\right)^2 r_D + \frac{r_L}{D^2}\right]DI_o \qquad (4.125)$$

It can be easily seen that the above equation can be represented by the dc equivalent model shown in Fig. 4.66(a). The simplified equivalent circuit is shown in Fig. 4.66(b), where

$$r_{eq} = r_{on} + \left(\frac{1-D}{D}\right)^2 r_D + \frac{r_L}{D^2}$$

$$V_{eq} = V_{in} + V_Q + \frac{1-D}{D}V_D$$

$$n = D$$

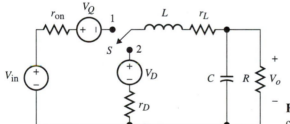

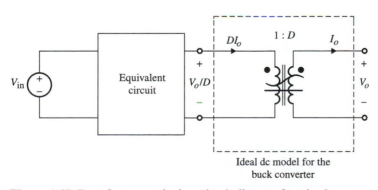

Figure 4.64 Buck converter with component nonidealities.

Ideal dc model for the
buck converter

Figure 4.65 Transformer equivalent circuit diagram for a buck converter.

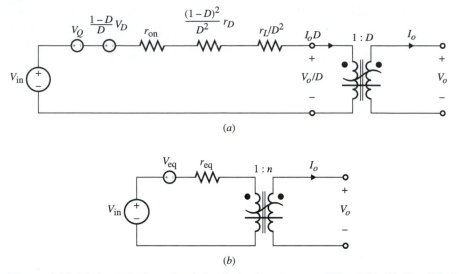

(a)

(b)

Figure 4.66 (a) dc equivalent circuit for the buck converter of Fig. 4.64. (b) Simplified equivalent circuit.

Table 4.2   dc Equivalent Circuit Model

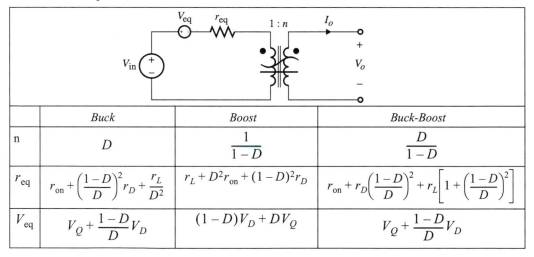

		Buck	Boost	Buck-Boost
n		D	$\dfrac{1}{1-D}$	$\dfrac{D}{1-D}$
r_{eq}		$r_{\text{on}} + \left(\dfrac{1-D}{D}\right)^2 r_D + \dfrac{r_L}{D^2}$	$r_L + D^2 r_{\text{on}} + (1-D)^2 r_D$	$r_{\text{on}} + r_D\left(\dfrac{1-D}{D}\right)^2 + r_L\left[1 + \left(\dfrac{1-D}{D}\right)^2\right]$
V_{eq}		$V_Q + \dfrac{1-D}{D}V_D$	$(1-D)V_D + DV_Q$	$V_Q + \dfrac{1-D}{D}V_D$

Similarly, we can obtain the dc equivalent circuits for the boost and buck-boost. Table 4.2 shows the values for V_{eq}, r_{eq}, and the transformer ratio for the buck, boost, and buck-boost.

EXERCISE 4.21

Show that the transformer equivalent circuit model for a boost converter that includes r_L, r_{on}, V_D, and V_Q is as shown in Fig. 4.66(a) with $n = 1/(1-D)$, $r_{\text{eq}} = r_L + D^2 r_{\text{on}}$, and $V_{\text{eq}} = (1-D)V_D + DV_Q$.

4.5   SWITCH UTILIZATION FACTOR

It is possible to evaluate the use of the power switch in the buck, boost, buck-boost, and Weinberg converters by investigating the average switch power-handling capability with respect to the average power delivered to the load. Figure 4.67 shows a block diagram representation for any switch-mode power converter, where

$$I_o = \text{average load current}$$

$$V_o = \text{average load voltage}$$

$$I_{sw,max} = \text{maximum current through the switch when turned on}$$

$$V_{sw,max} = \text{maximum voltage across the switch when turned off}$$

Let us define the switch maximum power as

$$P_{sw,max} = V_{sw,max}I_{sw,max}$$

and the average output power is given by

$$P_o = V_oI_o$$

Then we define the switch utilization factor (K_{sw}) as the ratio of $P_{sw,max}$ and P_o:

$$K_{sw} = \frac{P_o}{P_{sw,max}} \tag{4.127}$$

EXAMPLE 4.13

Determine K_{sw} for the buck converter.

SOLUTION   The maximum switch current and voltage expressions are given by Eqs. (4.128) and (4.129), respectively,

$$I_{sw,max} = DV_{in}\left[\frac{1}{R} + \frac{(1-D)T}{2L}\right] \tag{4.128}$$

$$V_{sw,max} = V_{in} \tag{4.129}$$

and the maximum switch power is given by

$$P_{sw,max} = I_{sw,max}V_{sw,max}$$

$$= DV_{in}^2\left[\frac{1}{R} + \frac{(1-D)T}{2L}\right]$$

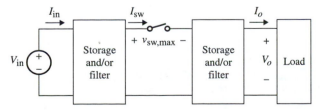

Figure 4.67 Block diagram for switch-mode power converter.

The average output power is given by

$$P_o = \frac{V_o^2}{R}$$

Therefore, the switch utilization factor, K_{sw}, is

$$K_{sw} = \frac{V_o^2/R}{DV_{in}^2\left[\dfrac{1}{R} + \dfrac{(1-D)T}{2L}\right]}$$

$$= \frac{D}{1 + \dfrac{(1-D)RT}{2L}}$$

Using the normalized time constant, $\tau_n = L/RT$, K_{sw} can be expressed as

$$K_{sw} = \frac{D}{1 + \dfrac{(1-D)}{2\tau_n}}$$

Figure 4.68 shows a plot of K_{sw} vs. D under different values of τ_n for the buck converter.

Similarly, it can be shown that the switch utilization factor for the boost and buck-boost converters are given in Eqs. (4.131) and (4.132), respectively.

$$K_{sw} = \frac{1}{\dfrac{1}{1-D} + \dfrac{(1-D)D}{2\tau_n}} \tag{4.131}$$

$$K_{sw} = \frac{D}{\dfrac{1}{1-D} + \dfrac{(1-D)}{2\tau_n}} \tag{4.132}$$

Figures 4.69 and 4.70 show the plots of K_{sw} vs. D under different values of τ_n for the boost and buck-boost, respectively.

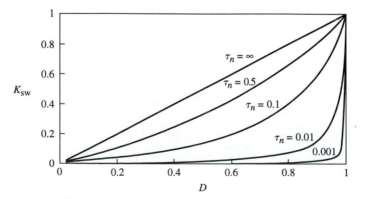

Figure 4.68 K_{sw} vs. D under different values of τ_n for the buck converter.

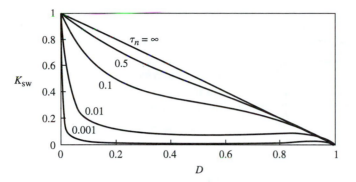

Figure 4.69 K_{sw} vs. D under different values of τ_n for the boost.

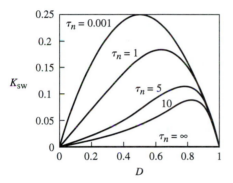

Figure 4.70 K_{sw} vs. D under different values of τ_n for the buck-boost.

EXERCISE 4.22

Show that the switch utilization factor for the boost and the buck-boost are given in Eqs. (4.131) and (4.132).

PROBLEMS

Continuous Conduction Mode

Buck Converter

4.1 Consider the following specifications for a buck converter:

$$V_{in} = 80 \text{ V}$$

$$R_o = 9 \ \Omega$$

$$P_o = 100 \text{ W}$$

$$f_s = 150 \text{ kHz}$$

Determine:

(a) The inductor value at the boundary condition (critical value L_{crit})

(b) The maximum inductor current value for $L = 10L_{crit}$

(c) The diode rms and average current values

4.2 Derive the expressions for $v_c(0)$ and $v_c(DT)$ for a buck converter operating in ccm whose current waveform is drawn in Fig. P4.2.

4.3 Consider the buck converter with $V_{in} = 25$ V, $V_o = 12$ V at $I_o = 2$ A, $f_s = 50$ kHz. Determine:

(a) The duty cycle, D

(b) L_{crit}

(c) I_{Lmin}, I_{Lmax}, $I_{o,ave}$, $I_{in,ave}$ for $L = 100L_{crit}$

(d) Output voltage ripple for $C = 0.47 \ \mu F$

(e) C for $|V_c| = 2\%$ of V_o

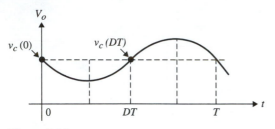

Figure P4.2

4.4 Derive the expressions for the inductor and capacitor rms currents for the buck. Find these values for Problem 4.3.

D4.5 (a) Design a buck converter for the following specifications:
$$\Delta V_o/V_o = 0.1\%, \qquad V_{in} = 40 \text{ V},$$
$$P_{o,ave} = 20 \text{ W}, f_s = 75 \text{ kHz}, D = 0.5$$
(b) What is the inductor current's peak-to-peak value?

(c) What are the diode and switch rms current values?

D4.6 Consider a buck converter with $V_{in} = 48$ V, $V_o = 27$ V at $I_o = 8$ A, and $f_s = 50$ kHz. Design for L and C such that the peak value of the inductor current does not exceed 15% of its average value, and the capacitor peak voltage does not exceed 2% of its average value.

D4.7 Consider the buck converter of Problem 4.6 by assuming the input voltage varies by ±20% and the output current varies from 10 A to 6 A. Redesign for L and C using the same specifications with V_{in}(nominal) = 48 V, $V_o = 27$ V, and $f_s = 50$ kHz.

4.8 Determine the diode and switch rms and peak current and voltage ratings for the design of Problem 4.6.

4.9 Determine the circuit parameters including V_{in}, D, f_s, C, L, and I_{Lmax} and I_{Lmin} for a buck converter whose capacitor voltage is given in Fig. P4.9.

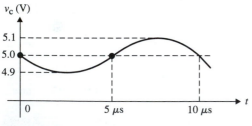

Figure P4.9

D4.10 Design the buck converter to operate in ccm with the following specifications: $V_{in,min} = 22$ V,

$V_{in,max} = 48$ V, $V_{in,nom} = 32$ V, $V_o = 12$ V,
$I_{o,max} = 4$ A, $I_{o,min} = 0.5$ A, $f_s = 50$ kHz, and
$\Delta V_o = 120$ mV.

Boost Converter

D4.11 Design a boost converter with the following specifications:
$$V_{in} = 28 \text{ V}, V_o = 48 \text{ V}, P_o = 100 \text{ W},$$
$$f_s = 110 \text{ kHz}, 2\% \text{ output voltage ripple}$$

4.12 Sketch the inductor current for the boost converter shown in Fig. P4.12. Assume ccm operation with $D = 0.4$.

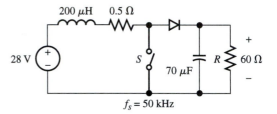

$f_s = 50$ kHz

Figure P4.12

D4.13 Design a boost converter to deliver 80 V at 4 A from a 60 V source with an output ripple voltage not to exceed 1%. Assume $f_s = 20$ kHz.

4.14 The source for a boost converter has an internal resistance R_s as shown in Fig. P4.14. Derive expressions for:

(a) V_o/V_{in}

(b) The efficiency, $\eta = \dfrac{P_{o,ave}}{P_{in,ave}}$

(c) The duty cycle at which the output voltage is maximized

Your expressions should be given as a function of R/R_s and D.

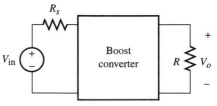

Figure P4.14

D4.15 Design a boost converter for the following requirements: $P_o = 20$ W, $V_{in} = 20$ V, $D = 0.35$, and 1% output voltage ripple.

4.16 Determine ΔV_o for a boost converter with $V_{in} = 12$ V, $V_o = 15$ V at $I_o = 250$ mA. Use $L = 150$ μH, $C = 470$ μF, and $f_s = 20$ kHz. Sketch the capacitor voltage and inductor current.

4.17 Derive the capacitor and inductor current rms values for Problem 4.16.

4.18 Derive the expressions for V_o/V_{in} for the boost by including r_L and r_{sw}.

4.19 Derive the expressions for V_o/V_{in} for the boost by including V_D and V_Q.

D4.20 Design a boost converter to operate in ccm with the following specifications:

$$V_{in, min} = 90 \text{ V}, \ V_{in, max} = 150 \text{ V},$$

$$V_{in, nom} = 120 \text{ V}, \ V_o = 152 \text{ V}, \ I_{o, max} = 2 \text{ A},$$

$$I_{o, min} = 0.2 \text{ A}, f_s = 50 \text{ kHz}, \Delta V_o/V_o = 1\%$$

Buck-Boost Converter

D4.21 Consider a buck-boost converter with the following specifications:

$$V_o = 12 \text{ V}, P_o = 25 \text{ W}, V_{in} = 24 \text{ V},$$

$$f_s = 100 \text{ kHz}, \Delta V_o/V_o = 1\%$$

(a) Design the converter so it operates in the ccm.

(b) Find the maximum inductor current under both operating modes.

(c) If the load changes by 50% (lighter load, R increases), determine the new mode of operation.

(d) Sketch $i_c(t)$ and $v_c(t)$ for parts (a) and (b).

4.22 Consider the following values for a buck-boost converter:

$$V_{in} = 30 \text{ V}, D = 0.25, R = 2 \ \Omega, L = 330 \ \mu\text{H},$$

$$f_s = 10 \text{ kHz}$$

(a) Determine the mode of operation.

(b) Sketch i_L.

(c) Determine V_o.

(d) Determine C for a 1% output ripple.

(e) At what value of D does the transition between ccm and dcm occur?

4.23 Consider a buck-boost converter that supplies 100 W to $V_o = 30$ V from a 50–30 V dc source. Let $T = 100$ μs and $L = 800$ μH.

(a) Determine the range of D for the given range of source voltage.

(b) Determine the average input current, diode current, I_{Lmax}, and I_{Lmin}, under $V_{in} = 35$ V.

(c) Determine the rms value of the diode current and the capacitor current under $V_{in} = 35$ V.

(d) Design for C so that the output ripple voltage is limited to 2% of V_o under $V_{in} = 30$ V.

(e) If the load resistor is changed so that the converter load current is decreased by 10%, what is the new duty cycle when $V_{in} = 35$ V?

4.24 For stability considerations, sometimes it is required that the equivalent series resistance (ESR) of the output capacitor be included in modeling the power stage of a PWM converter. Fig. P4.24 shows

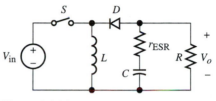

Figure P4.24

a buck-boost converter including r_{ESR}. Derive the expression for the output ripple.

4.25 Repeat Problem 4.4 for a buck-boost converter.

D4.26 Redesign Problem 4.5 for a buck-boost converter.

D4.27 Redesign Problem 4.13 for a buck-boost converter.

4.28 Derive the expression for V_o/V_{in} for the buck-boost converter by including r_L.

4.29 Derive the expression V_o/V_{in} for the buck-boost by including V_D and V_Q.

4.30 Derive the expression for V_o/V_{in} for the buck-boost converter by including r_{sw}.

Fourth-Order Converters

4.31 Derive the expression for V_o/V_{in} for the Zeta converter by including r_L.

4.32 Derive the expression for V_o/V_{in} for the Cuk converter by including V_D and V_Q.

4.33 Derive the expression for V_o/V_{in} for the Cuk converter by including r_{sw}.

4.34 Determine the output voltage ripple equation for the Cuk and Zeta converters.

4.35 Consider the Cuk converter of Fig. P4.35.

(a) Sketch the waveforms for i_{L1}, i_{L2}, i_D, and i_s assuming constant voltages across C_1 and C_2.

(b) Determine the rms current values in the diode and switch.

(c) What is the ripple voltage across C_1 and C_2?

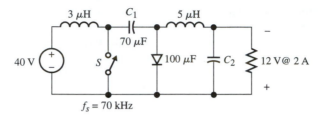

Figure P4.35

4.36 Derive the voltage gain equation for the single-ended primary inductance converter (SEPIC) shown in Fig. E4.11(a).

4.37 Derive the rms expressions for the inductor and diode currents for the Cuk converter.

D4.38 (a) Derive the voltage gain expression for the SEPIC converter operating in the dcm.

(b) Design the converter for the following specifications:

$$V_{in} = 45 \text{ V}, \ V_o = 15 \text{ V}, \ P_o = 25 \text{ W},$$
$$f_s = 50 \text{ kHz}$$

(i) The maximum inductor ripple current should not exceed 15% of its average value.

(ii) The maximum voltage ripple across C_1 should not exceed 5% of its average value.

(iii) The maximum output voltage ripple is less than 0.1%.

4.39 Develop an equivalent circuit for the cascaded boost and buck-boost converters.

4.40 Derive the voltage gain expression for the cascaded arrangement in Problem 4.39.

Discontinuous Conduction Mode

4.41 Consider the buck converter of Problem 4.1.

(a) Determine the voltage gain for $L = 0.1L_{crit}$.

(b) Determine D_1.

4.42 Derive the ripple voltage for the buck converter operating in the dcm.

4.43 Derive the ripple voltage for the buck-boost converter operating in the dcm.

4.44 Consider a boost converter with $L = 67 \ \mu H$, $V_{in} = 60 \text{ V}$, $f_s = 20 \text{ kHz}$, $C = 10 \ \mu F$, $V_o = 80$ V at $I_o = 4$ A.

(a) Determine the mode of operation.

(b) Repeat part (a) for $I_o = 2$ A.

(c) Sketch i_e, V_D, and V_c for parts (a) and (b).

(d) Determine D and D_1 for $V_o = 80$ V at (i) $I_o = 4$ A and (ii) $I_o = 2$ A.

(e) Repeat part (d) for $V_{in} = 65$ V.

4.45 Determine the expressions for the diode and transistor currents of the buck-boost operating in the dcm.

D4.46 Design the buck-boost converter to operate in the dcm with the following specifications:

$$V_{in} = 18 \text{ V}, \ V_o = -38 \text{ V}, \ I_{o,max} = 2 \text{ A},$$
$$I_{o,min} = 0.2 \text{ A}, f_s = 95 \text{ kHz}, \Delta V_o = 180 \text{ mV}.$$

4.47 Derive the following expressions for the buck-boost converter operating in the dcm:

$$M = \frac{D}{\sqrt{2\tau_n}}$$

$$D_1 = D + \sqrt{2\tau_n}$$

where

$$\tau_n = \frac{\tau}{T}$$

$$\tau = L/R$$

4.48 Derive the expression for the voltage ripple for the boost converter operating in dcm. Also give an expression for the capacitor voltage at $t = 0$ and $t = DT$ in terms of the circuit parameters.

4.49 Repeat Problem 4.21 under dcm operation.

4.50 Derive the voltage gain for the Cuk converter operating in the dcm.

4.51 Derive the expressions for M and D_1 for the Cuk converter when it operates in the dcm. (dcm operation for the Cuk converter is assumed only when both inductor currents are discontinuous.)

4.52 (a) Derive the expression for ΔV_o (peak-to-peak) for a buck converter operating in the dcm.

(b) Determine ΔV_o for a boost converter for $V_{in} = 12$ V, $V_o = 15$ V, and $I_o = 250$ mA. Use $L = 150 \ \mu H$, $C = 470 \ \mu F$, and $f_s = 20$ kHz.

Other Converter Topologies

4.53 Derive the expression for the voltage gain for the two-switch PWM converter shown in Fig. P4.53.

S_1 and S_2 are switched simultaneously. Assume $RC \gg T$. Compare this converter to the single-switch buck-boost converter.

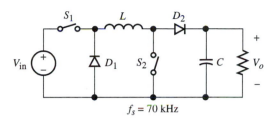

Figure P4.53

4.54 It is possible to develop a basic switched-mode dc converter by using constant input and output current sources. Such converters are called *current inverters*. Figure P4.54 shows the three basic current converters: buck, boost, and buck-boost. Assume L/R is very large compared to the switching frequency so that I_o is assumed constant. Show that the current gains for these topologies are given by, respectively,

(a) $\dfrac{I_o}{I_{in}} = (1-D)$

(b) $\dfrac{I_o}{I_{in}} = \dfrac{1}{D}$

(c) $\dfrac{I_o}{I_{in}} = \dfrac{1-D}{D}$

4.55 Show that the critical capacitor value that will produce a continuous capacitor voltage for the buck converter in Fig. P4.54(a) is given by

$$C_{crit} \geq \dfrac{D^2 T}{2R}$$

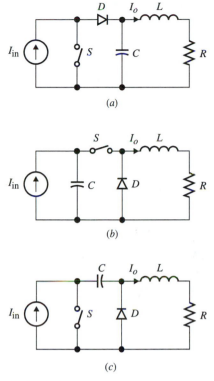

Figure P4.54

4.56 Consider the circuits shown in Fig. P4.56 with V_o and I_o constants, i.e., $RC \gg T$ and $L/R \gg T$. The input current and voltage squarewaves are as shown. Derive the expressions for the inductor and capacitor ripples in terms of the circuit parameters.

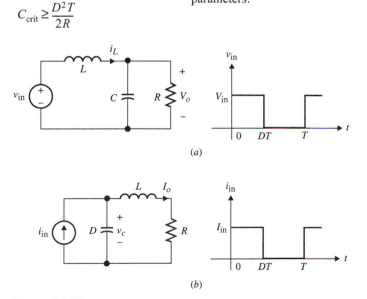

Figure P4.56

4.57 Repeat Problem 4.56 for Fig. P4.57, where $1 \le \beta < 2$.

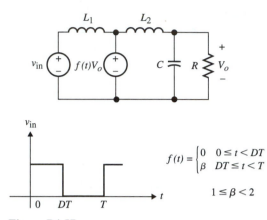

$$f(t) = \begin{cases} 0 & 0 \le t < DT \\ \beta & DT \le t < T \end{cases}$$

$$1 \le \beta < 2$$

Figure P4.57

4.58 Derive the dc and fundamental components of the Fourier series for the input current in the buck converter.

Additional General Problems

4.59 Derive the rms expressions for the inductor and diode currents for the buck converter.

4.60 Derive the rms expressions for the inductor and diode currents for the boost converter.

4.61 Derive the rms expressions for the inductor and diode currents for the buck-boost converter.

4.62 Consider the bidirectional double-switch buck converter shown in Fig. P4.62. Assume S_1 and S_2

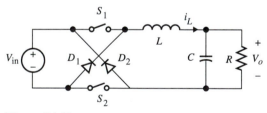

Figure P4.62

are turned on and off simultaneously at duty ratio D. Show that the voltage gain is given by

$$\frac{V_o}{V_{in}} = 2D - 1$$

4.63 Derive the ripple voltage expression for the bidirectional converter of Problem 4.62.

D4.64 Design the three buck-boost converters needed to be used in the dc distributed power system in Fig. P4.64.

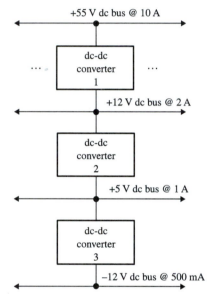

Figure P4.64

4.65 Show that the nth harmonic peak value for the diode voltage of the buck converter is given by

$$v_{D,n} = \frac{V_{in}\sqrt{2}}{n\pi} \sin nD\pi$$

4.66 The converter shown in Fig. P4.66 is known as a noninverting buck-boost converter. Derive the expression for V_o/V_{in}. Assume S_1 and S_2 are turned on and off simultaneously.

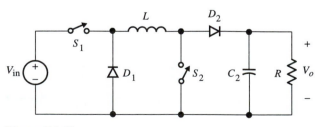

Figure P4.66

D4.67 Design a switched-mode power supply that is capable of supplying +12 V output voltage with an output power of 25 W to 375 W from a 56 V dc source. Assume the dc input comes from a variable dc source supply that changes by ±25%. The converter should not produce more than ±2% output ripple, given the peak and rms rating values for the power devices. Let $f_s = 75$ kHz.

D4.68 Design a switched-mode converter that delivers 25 W output power to a regulated output voltage of +12 V. Assume the dc source is obtained from an ac-dc SCR rectifier whose dc unregulated output vary between 8 V and 18 V. It is desired to limit the output voltage ripple to 1% and the switching frequency to 100 kHz.

4.69 Figure P4.69 shows another way to represent a buck-boost converter using a single-pole, double-through switch. Assume C_1 and C_2 are large enough that their voltages are constant. Derive the expression for V_o/V_{in} and I_o/I_{in}. Assume in steady state the switch is in position 1 for the interval DT and in position 2 for the interval $(1 - D)T$.

4.70 Figure P4.70 shows the circuit representation for cascading a buck-boost and a buck converter. Show that the total voltage gain is $D^2/(1-D)$.

4.71 Determine the voltage gain for the circuit shown in Fig. P4.71. What converters are put in cascade to produce this topology?

4.72 Consider a boost converter with its dc source coming from an unregulated 124 V dc input that varies by ±18% and delivers power to the load between 75 W and 225 W. It is desired to regulate the output voltage to 72 V. Assume the power switch has an on-resistance of 0.5 Ω, and the power diode has a 0.7 V voltage drop with a 0.25 Ω forward resistance. Also assume the inductor loss can be modeled by a discrete resistance of 0.1 Ω. Determine the range of the duty cycle D needed to be employed to maintain a constant output voltage. Assume ccm operation.

4.73 Figure P4.73 shows a two-quadrant boost converter topology that produces bipolar output voltage. Derive the expression for V_o/V_{in} assuming Q_1 and Q_2 are synchronized switches during turn-on for DT and turn-off for $(1-D)T$.

4.74 The circuit shown in Fig. P4.74 is known as an inverse SEPIC converter. Derive the voltage gain equation V_o/V_{in}. The switch is in position 1 during the DT interval and in position 2 during the $(1-D)T$ interval.

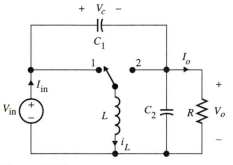

Figure P4.69

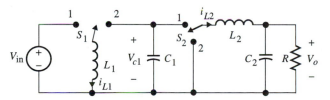

Figure P4.70

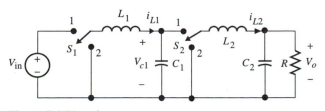

Figure P4.71

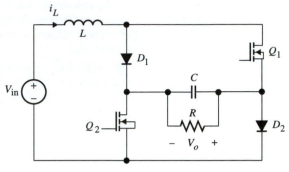

Figure P4.73

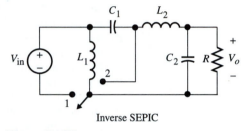

Inverse SEPIC

Figure P4.74

Chapter 5

Isolated Switch-Mode dc-to-dc Converters

INTRODUCTION

The discussion in Chapter 4 showed that the four basic converter topologies have their output conversions determined by the duty ratio, D, and that they consist of a single input and a single output with a common reference point. For applications in which the output voltage does not differ from the input voltage by a large factor, those topologies can be used. However, for many applications the input voltage is normally derived from an off-line half- or full-wave rectifier, and the output voltage is a very small fraction of the rectified dc input voltage. As a result, transformers must be added between the power stage and the output for the purpose of voltage scaling. Unlike line-frequency transformers, these transformers are high frequency and much smaller in size and weight. In addition, transformers are used in switched-mode converters for electrical isolation between the input and output, reduction of stresses in switching devices, and provision of multi-output connections. With isolation transformers, the output voltage polarity reversal does not become a design restriction. Also, in some applications, system isolation may be required by the regulatory body. However, the benefits obtained by adding isolation transformers come with a price. The major drawbacks include high converter volume and weight, reduced efficiency, and added circuit complexity to limit the effect of leakage inductance and avoid core saturation.

In this chapter, we will discuss some widely used high-frequency switched-mode dc-dc converters: the flyback, forward, push-pull, half-bridge, and full-bridge converters. It will be shown that the flyback converter is based on the boost converter, and the forward converter is based on the buck converter. In analyzing these converters, we will use the transformer model by including the magnetizing inductance. In all the dc-dc converters discussed in Chapter 4, the role of the inductor is to store energy as it comes from the source in one portion of the switching cycle and release it to the load in the other portion of the cycle. As the converter reaches the steady state, the mechanism of energy storage and release reaches equilibrium and the net energy stored in the inductor over one switching cycle is zero. In other words, energy in the inductor does not build up from one cycle to the next. Otherwise, if energy build-up in the inductor were allowed, soon the inductor would reach saturation. The issue of saturation is important not only for the inductor, but also for the transformer. Unlike converter topologies presented in Chapter 4, the topologies in this chapter require more than one switch, since the transformer must be magnetized and demagnetized repeatedly within a switching cycle, to avoid saturation. The conversion can be done unidirectionally, in which the magnetization current is positive, or bidirectionally, in which the magnetization current is positive and negative. Finally, we consider only the continuous conduction mode of operation for the isolated converters, since their dcm analysis is very similar to the dcm analysis presented in the previous chapter.

5.1 TRANSFORMER CIRCUIT CONFIGURATIONS

Transformer Model

An ideal transformer has no leakage inductances, an infinite magnetizing inductance, no copper and core losses, the ability to pass all signal frequencies without any power loss, and the ability to provide any level of current and voltage ratio transformation. Figure 5.1(a) shows an ideal transformer with its current and voltage relations. Figure 5.1(b) shows the equivalent transformer circuit including only the magnetic inductance, L_m, reflected in the primary side.

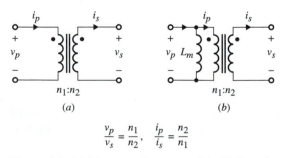

$$\frac{v_p}{v_s} = \frac{n_1}{n_2}, \quad \frac{i_p}{i_s} = \frac{n_2}{n_1}$$

Figure 5.1 (a) Ideal transformer model. (b) Transformer equivalent circuit including the magnetizing inductance.

Circuit Configurations

Throughout this chapter, as far as the primary side is concerned, two configurations of transformers are used: non-center-tap and center-tap. The non-center-tap configuration has three possible connections, known as single-ended, half-bridge, and full-bridge, as shown in Fig. 5.2(a), (b), and (c), respectively. The center-tap configuration is known as *push-pull* and is shown in Fig. 5.3.

As far as the transformer's secondary side is concerned, two configurations of transformer-rectifier connections are normally used: center-tap full-wave and bridge full-wave, as shown in Fig. 5.4(a) and (b), respectively. The center-tap arrangement results in one diode forward voltage drop, compared to two diode voltage drops for the bridge configuration.

These configurations are the most popular ones used in today's switch-mode power supply design. Since we have a dc source voltage across S_1 and S_2 of Figs. 5.2(b) and 5.3, the two switches are not allowed to close simultaneously. Hence, in a normal operation, S_1 and S_2 in these two configurations switch alternately, each having the same duty cycle. This will prevent transformer core saturation. In the full-wave bridge

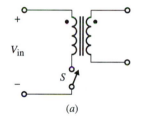

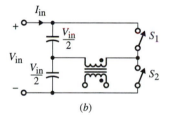

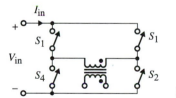

Figure 5.2 Transformer configurations: (a) single-ended, (b) half-bridge, and (c) full-bridge.

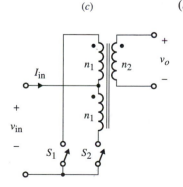

Figure 5.3 Push-pull transformer configuration.

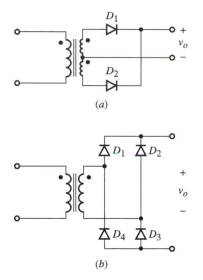

(a)

(b)

Figure 5.4 Transformer configurations in the secondary (load) side: (a) center-tap full-wave, and (b) bridge full-wave.

configuration, S_1 and S_3 are closed simultaneously in the first half cycle, and in the second half cycle switches S_2 and S_4 are closed simultaneously.

The difference between the two-switch transformer connections of Fig. 5.2(b) and Fig. 5.3 is that the switches in the former connection must be able to sustain a maximum voltage of V_{in} compared to $2V_{in}$ for the latter connection. However, in the push-pull configuration, each switch carries the average input current, I_{in}, whereas in the half-bridge configuration, each switch must carry twice the average input current. This is why the push-pull configuration is chosen when the source voltage is low. In the full-bridge case, each switch must be able to block V_{in} and allow a peak current equal to I_{in}. This makes it attractive for high-power applications. The only disadvantage of the full-bridge configuration is that it requires two floating driving circuits.

EXAMPLE 5.1

Consider the push-pull converter with a current source load shown in Fig. 5.5. If S_1 and S_2 are turned on and off according to the waveforms shown, determine the current and voltage stresses for S_1 and S_2 in terms of the average input current, I_{in}, and voltage V_{in}.

SOLUTION   Consider the first half cycle, when S_1 is closed. The input voltage is applied to the upper primary winding of the transformer. An equal amount of voltage is also induced on the lower

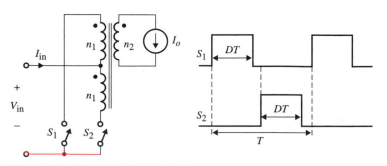

Figure 5.5 Circuit for Example 5.1.

Table 5.1   Comparison between Transformer Configurations

Transformer configuration	Maximum voltage stress across each switch	Maximum current stress through each switch	Core excitation and reset mechanism	Power level applications
Single-ended Forward	$\geq 2V_{in}* =$ $(1 + n_3/n_1)V_{in}$	$\geq 2I_{in}*$ $\geq I_{in}/D$	Unidirectional complex reset circuit	Low power <500 W
Flyback	$\geq 2V_{in}$ $= V_{in}(1-D)$	$2I_{in}$ $\geq I_{in}/D$	Unidirectional and core reset not necessary	<200 W
Half-bridge	V_{in}	$2I_{in}$ $\geq I_{in}/D**$	Bidirectional and core reset not necessary	High input voltage, medium power with low source current <800 W
Full-bridge	V_{in}	I_{in} $\geq I_{in}/D**$	Bidirectional and core reset simple	High power >800 W
Push-pull	$2V_{in}$	I_{in} $\geq I_{in}/D**$	Bidirectional and core reset simple	Medium power with low source voltage

$*I_{in}$ and V_{in} are average values.
$**D < 0.5.$

primary winding. The polarity of this voltage is in the direction of the input voltage. As a result, S_2 is subjected to twice the input voltage, i.e., $V_{stress} = 2V_{in}$. S_1 is also subjected to the same voltage stress when S_2 is turned on.

Neglecting the transformer's magnetizing current, the maximum switch current can be obtained as $I_{max} = I_{stress} > I_{in,ave}/2D = n_2 I_o/2n_1 D$ $(0 < D < 0.5)$. Depending on the value of D, the current stress on each switch can be much larger than the average input current.

The primary-switch connections for the isolated switched-mode converters are compared in Table 5.1.

5.2   BUCK-DERIVED ISOLATED CONVERTERS

Depending on the location at which the isolation transformer is inserted in the power stage of the basic buck converter, and on the type of isolation transformer configuration, several buck-derived converter topologies are possible. These topologies vary in complexity and features. In this section we will consider some of the more popular buck-derived converters: single-ended forward, half- and full-bridge, push-pull, and Weinberg converters. Other topologies are given as exercises at the end of this chapter.

ac Transformer Insertion

Let us consider isolating the input and output voltages of the buck converter by inserting an ac transformer. Figure 5.6(a) shows four places where physical transformer

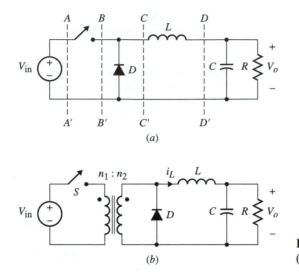

Figure 5.6 (*a*) Buck converter.
(*b*) Isolated buck converter.

insertion is possible. Locations AA' and DD' are not possible since the transformer's primary and secondary voltages would be dc V_{in} and V_o, respectively. The obvious location is either BB' or CC' since the voltage at either location is an ac squarewave voltage.

It is clear that the isolated buck-derived converter of Fig. 5.6(*b*) will not function properly in steady state. This is because the average voltage across the transformer is positive (nonzero), which results in a continuous increase in the magnetization current that will eventually lead to transformer saturation. Let us further investigate the problem of transformer magnetizing inductance saturation.

When the switch is turned on at $t = 0$, the diode becomes reverse biased as shown in Fig. 5.7(*a*). The magnetizing and output inductor currents are given by

$$i_{Lm} = \frac{V_{in}}{L_m}t + I_{Lm}(0)$$

$$i_L(t) = \frac{n_1}{n_2}i_p = \frac{\frac{n_2}{n_1}V_{in} - V_o}{L}t + I_L(0)$$

where $I_L(0)$ and $I_{Lm}(0)$ are the initial output and magnetizing inductor currents, respectively.

At $t = DT$, S is switched off (Fig. 5.7(*b*)), and D turns on. Hence, the magnetizing current stays constant at $I_{Lm}(DT)$, which is given by

$$I_{Lm}(DT) = \frac{DT}{L_m}V_{in} + I_{Lm}(0)$$

The diode current is given by

$$i_D(t) = i_L(t) + I_{Lm}(DT)\frac{n_1}{n_2}$$

The inductor current $i_L(t)$ for $t \geq DT$ is given by

$$i_L(t) = \frac{-V_o}{L}(t - DT) + I_L(DT)$$

where

$$I_L(DT) = \frac{\frac{n_1}{n_2}V_{in} - V_o}{L}DT + I_L(0)$$

At $t = T$, when the switch is turned on again, it causes i_{Lm} to increase starting at a higher initial value, $I_{Lm}(DT)$, as shown for three cycles in Fig. 5.7(c).

One way to avoid the problem of transformer saturation is to add another diode as shown in Fig. 5.8 to produce a negative voltage across the primary side when the switch is switched off. However, this converter as shown provides a path for the magnetizing current to reset to zero when the switch is open. As a result, a third winding, known as "catch" winding, is added to allow i_{Lm} to discharge to zero, preventing magnetic flux build-up from one cycle to the next. This converter is known as a single-ended forward converter, to be discussed in the next section.

The isolation of the buck-boost converter can be carried out in a similar manner. Figure 5.9(a) shows the buck-boost converter with a transformer insertion as shown in Fig. 5.9(b). Depending on the transformer windings, negative or positive output polarities can be obtained as shown in Fig. 5.9(b) and (c), respectively. Figure 5.9(c) is known as a single-ended flyback converter, which will be discussed later in the chapter.

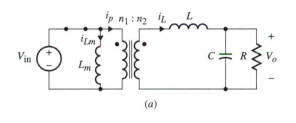

(a)

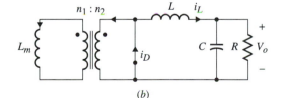

(b)

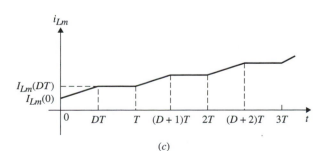

(c)

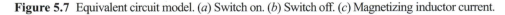

Figure 5.7 Equivalent circuit model. (a) Switch on. (b) Switch off. (c) Magnetizing inductor current.

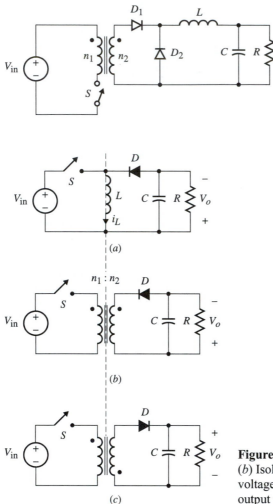

Figure 5.8 Isolated buck-derived converter known as a single-ended forward converter.

(a)

(b)

(c)

Figure 5.9 (a) Buck-boost converter. (b) Isolated converter with negative output voltage. (c) Isolated converter with positive output voltage.

5.2.1   Single-Ended Forward Converter

In this section we will analyze Fig. 5.8, which shows the simplest isolated dc-to-dc converter utilizing one switch and two diodes. As stated earlier, this circuit is commonly referred to as a *forward converter.* It can be shown that the design of Fig. 5.8 will not work properly since its magnetizing current will not be allowed to reset to zero, causing the magnetizing current to continuously increase linearly until it finally saturates the core. A more practical forward converter must include a transformer core-resetting circuit as shown in Fig. 5.10(a). The additional winding n_r is known as *tertiary winding.* Figure 5.10(b) and (c) shows two alternative ways to draw Fig. 5.10(a).

To allow for a zero average voltage across the transformer primary winding, the maximum duty cycle is 50%. However, if the winding ratio $n_r/n_1 < 1$, then it is possible to have a duty cycle that exceeds 50%. This will result in a voltage stress across the switch that exceeds $2V_{in}$.

For illustration purposes, we will analyze the converters of Figs. 5.8 and 5.10(a). If an ideal transformer is assumed in Fig. 5.8, then the steady state is quite simple. Assume the switch is turned on for the period DT and off for the period $(1-D)T$, resulting in the two modes of operation shown in Fig. 5.11(a) and (b), respectively.

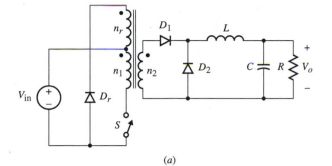

(a)

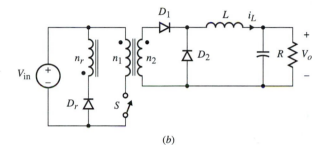

(b)

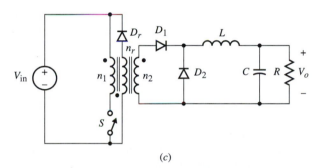

(c)

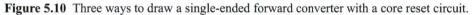

Figure 5.10 Three ways to draw a single-ended forward converter with a core reset circuit.

When the switch is turned on initially at $t = 0$, the initial inductor current is $I_L(0)$, and v_L is given by

$$v_L = \frac{n_2}{n_1} V_{in} - V_o$$

$$= L \frac{di_L}{dt}$$

The inductor current is given by

$$i_L(t) = \frac{\frac{n_2}{n_1} V_{in} - V_o}{L} t + I_L(0) \qquad 0 \le t < DT$$

At $t = DT$, the switch turns off, forcing D_1 to reverse-bias and D_2 to become forward biased, resulting in the inductor current given by

$$i_L(t) = \frac{-V_o}{L}(t - DT) + I_L(DT)$$

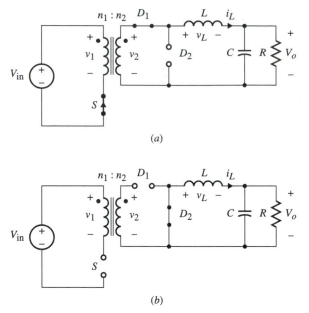

Figure 5.11 Modes of operation. (*a*) Mode 1: *S* is on. (*b*) Mode 2: *S* is off.

where $I_L(DT)$ is the inductor current at $t = DT$, when the switch is turned off. From the above equations, we obtain the following voltage gain relation:

$$\frac{V_o}{V_{in}} = \frac{n_2}{n_1}D$$

Sketches of the waveforms for v_1, v_2, v_L, i_L, i_{D1}, and i_{D2} are shown in Fig. 5.12.

For proper operation that will allow the inductor current to reach steady state, the relation $(n_2/n_1)V_{in} > V_o$ must hold, which provides a step-down operation. The capacitor voltage ripple is similar to that for the nonisolated buck converter.

Next we carry out the analysis by assuming the transformer has a finite magnetizing inductance, L_m, as shown in the equivalent circuit given in Fig. 5.13. With careful investigation of the circuit, it is clear that the magnetizing current $i_m(t)$ has no place to discharge its value when the switch is off. This will cause the converter to fail. As a result, a core-resetting mechanism as mentioned above must be used. Figure 5.14 shows the equivalent circuit for the forward converter by including the core reset circuit given in Fig. 5.10(*b*).

The operation of this converter can be easily explained by assuming that before the switch is turned on again in a new cycle, the magnetizing current, i_m, has reached zero, i.e., the transformer core is being reset. The energy is delivered to the load during the period the switch is on, and the core resetting takes place during the *off* time. It will be shown that $D \le 50\%$ for $n_r \ge n_1$ to allow time for the magnetic core flux to reset during the *off* switch time. In steady state, this converter has three modes of operation, as follows.

The first mode starts when *S* is turned on at $t = 0$, causing voltage across the primary equal to V_{in}. This will force D_1 to turn on and D_2 to reverse-bias. Since the reset winding has the opposite polarity of the primary, the diode D_r becomes reverse biased as shown in Fig. 5.15(*a*).

From Fig. 5.15(*a*), the following voltage equations are obtained:

$$v_L = v_2 - V_o, \; v_1 = V_{in}, \; v_2 = \frac{n_2}{n_1}v_1, \; v_r = \frac{n_r}{n_1}V_{in}, \; v_{Dr} = \left(1 + \frac{n_r}{n_1}\right)V_{in} \qquad (5.1)$$

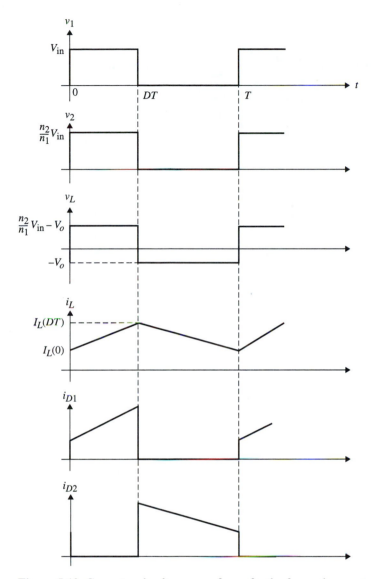

Figure 5.12 Current and voltage waveforms for the forward converter.

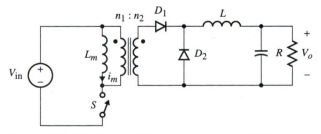

Figure 5.13 Single-ended forward converter with the magnetizing inductance.

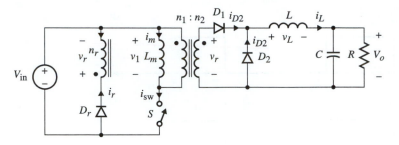

Figure 5.14 Forward converter with core reset circuit n_r–D_r, including the magnetizing inductance.

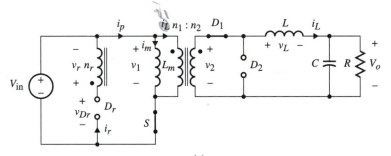

(a)

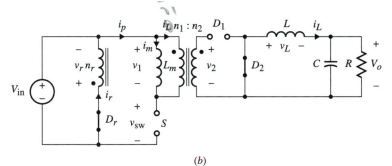

(b)

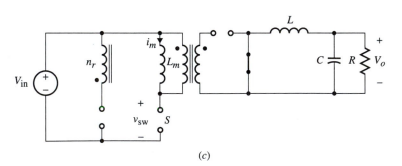

(c)

Figure 5.15 Modes of operation. (a) Mode 1: $0 \leq t < DT$. (b) Mode 2: $DT < t < D_1 T$. (c) Mode 3: $D_1 T \leq t < T$.

The current equations are given by

$$L_m \frac{di_m}{dt} = V_{in} \tag{5.2a}$$

$$L \frac{di_L}{dt} = \frac{n_2}{n_1} V_{in} - V_o \tag{5.2b}$$

$$i_r = 0 \tag{5.2c}$$

$$i_1 = \frac{n_2}{n_1} i_L \tag{5.2d}$$

$$i_p = i_m + i_L \tag{5.2e}$$

From Eqs. (5.2a) and (5.2b), the following relations are obtained:

$$i_m(t) = \frac{V_{in}}{L_m} t \tag{5.3}$$

$$i_L(t) = \frac{\frac{n_2}{n_1} V_{in} - V_o}{L} t + I_L(0) \tag{5.4}$$

where $I_L(0)$ is the initial output inductor current and $I_m(0) = 0$ since the core has been reset prior to turning on the switch. Since $L_m \gg L$, the slope of the magnetizing current is much smaller than the slope of i_L, as shown in Fig. 5.16. At $t = DT$, the switch is turned off and the circuit enters mode 2, shown in Fig. 5.15(b). At the instant S opens, the transformer primary current, i_p, becomes zero, turning off D_1 and forcing i_L to go through D_2. At this point, i_m now is forced to flow in the n_1 winding, which forces D_r to turn on to carry the reflected current through n_r.

The voltage equations in this mode are given by

$$v_L = -V_o, \; v_r = -V_{in}, \; v_1 = -\frac{n_1}{n_r} V_{in}, \; v_2 = \frac{n_2}{n_1} v_1, \; V_{Dr} = 0, \; v_{sw} = \left(1 + \frac{n_r}{n_1}\right) V_{in} \tag{5.5}$$

and the current equations are as follows:

$$L_m \frac{di_m}{dt} = -\frac{n_1}{n_r} V_{in} \tag{5.6a}$$

$$L \frac{di_L}{dt} = -V_o \tag{5.6b}$$

$$i_r = \frac{n_1}{n_r} i_m \tag{5.6c}$$

$$i_m = -i_L \tag{5.6d}$$

The waveforms are shown in Fig. 5.16. At $t = D_1 T$, the magnetizing inductor current becomes zero, since it represents the smaller portion of i_p. The magnetizing and inductor currents in this time interval are given by

$$i_m(t) = -\frac{n_1}{n_r} \frac{V_{in}(t - DT)}{L_m} + I_m(DT) \tag{5.7}$$

$$i_L(t) = -\frac{V_o}{L}(t - DT) + I_L(DT) \tag{5.8}$$

where $I_m(DT) = (V_{in}/L_m)DT$. Setting Eq. (5.7) to zero at $t = D_1$, we obtain

$$D_1 = \left(1 + \frac{n_r}{n_1}\right) D \tag{5.9}$$

At $t = D_1 T$, $i_m = 0$, causing D_r to become reverse biased, as shown in Fig. 5.15(c).

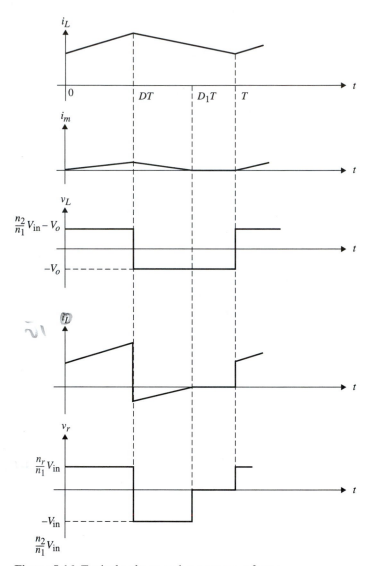

Figure 5.16 Typical voltage and current waveforms.

The voltage and current equations are given by

$$v_L = -V_o, \; v_r = v_1 = v_2 = 0, \; v_{Dr} = V_{in}, \; v_{sw} = V_{in}$$

$i_m = i_r = i_p = i_L = 0$, and $i_L(t)$ is the same as given in Eq. (5.8).

For the core to be fully magnetized, i_m must reach zero; therefore, D_1 must not be greater than 1.

$$D_1 \le 1$$

Hence, from Eq. (5.9) we restrict D by the following relation:

$$D \le \frac{1}{1 + \dfrac{n_r}{n_1}} = \frac{n_1}{n_1 + n_r}$$

The maximum duty cycle of 50% occurs when $n_r = n_1$.

EXAMPLE 5.2

Consider the forward converter of Fig. 5.14 with an input voltage of 50 V, an output voltage of 35 V, and $n_1/n_2 = 1$ and $n_1/n_r = 0.25$. With a frequency of 35 kHz, an inductance of 180 μH, and a minimum inductor current of 1.1 A, calculate the duty cycle and the maximum inductor current in this converter.

SOLUTION   The duty cycle is given in the following equation:

$$D = \frac{V_o}{V_{in}} = \frac{35 \text{ V}}{50 \text{ V}} = 0.7$$

$$T = \frac{1}{f} = \frac{1}{35 \text{ kHz}} = 2.86 \times 10^{-5}\text{s}$$

Calculating for the maximum inductor current,

$$i_{L,max} = \frac{V_{in} - V_o}{L}DT + i_{L,min}$$

$$= \frac{50 \text{ V} - 35 \text{ V}}{180 \ \mu\text{H}}(0.7)(2.86 \times 10^{-5}) + 1.1$$

$$= 2.77 \text{ A}$$

EXERCISE 5.1

Figure E5.1 shows a two-switch forward converter.

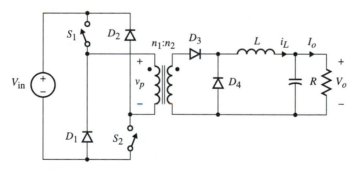

Figure E5.1 Two-switch forward converter.

(a) Show that the voltage gain expression is given by

$$\frac{V_o}{V_{in}} = \frac{n_2}{n_1}D$$

S_1 and S_2 are switched simultaneously.

(b) Find the critical inductor value (L_{crit}) to maintain a continuous conduction mode of operation. Assume $n_1 = n_2 = 1$, $V_{in} = 80$ V, $V_o = 45$ V @ $I_o = 5$ A, $f_s = 50$ kHz.

(c) Compare the one-switch, single-ended topology with this two-switch topology.

ANSWER   (b) 39.4 μH

EXERCISE 5.2

Find the rms current values for diodes D_3 and D_4 in Fig. E5.1 using the values of part (b) of Exercise 5.1 with $L = 10L_{crit}$.

ANSWER   3.76 A, 3.31 A

EXERCISE 5.3

Determine the efficiency of the converter described in Exercise 5.1 by assuming that the output rectifier diodes have a 1 V forward voltage drop and a 1.5 Ω conduction resistance.

ANSWER   84.1%

EXERCISE 5.4

Determine the rms value of the inductor current i_L for Example 5.2.

ANSWER   1.99 A

5.2.2   Half-Bridge Converter

Another way to avoid the transformer saturation problem is to use half- and full-bridge converter topologies to generate symmetrical ac waveforms at the primary side of the transformer. In this way the core flux is excited bidirectionally, resulting in a better utilization of the core, which in turn results in an increased power rating.

Figure 5.17(a) shows the circuit topologies for the half-bridge converter with the center-tap output rectifier configuration. The filtering capacitors are relatively large and used as voltage dividers, resulting in a $V_{in}/2$ applied voltage across each primary winding. As stated before, in the half-bridge converter, S_1 and S_2 are switched on and off in complementary fashion but with equal conduction periods. Since the input voltage is not allowed to be shorted, S_1 and S_2 are normally designed such that there exists a dead time during which both switches are off. This results in a duty ratio less than 50%. Key current and voltage waveforms are shown in Fig. 5.17(b).

The voltage gain for the half-bridge converter is the same as the gain for the push-pull converter, to be discussed in a later section. The filtering capacitors C_1 and C_2 are used to divide the input voltage, so each has $V_{in}/2$. Unlike the push-pull converter, the maximum blocking voltage for each switch of the half-bridge converter is V_{in}, rather than $2V_{in}$.

Finally, the switches S_1 and S_2 are implemented using bidirectional semiconductor devices (i.e., MOSFETs with anti-parallel diodes) to provide conduction paths for the inductor leakage currents that exist due to the nonideal transformers.

EXERCISE 5.5

(a) Show that the voltage gain for the half-bridge converter of Fig. 5.17(a) is given by

$$\frac{V_o}{V_{in}} = \frac{n_2}{2n_1}D$$

(b) Find the expression for the maximum diode current.

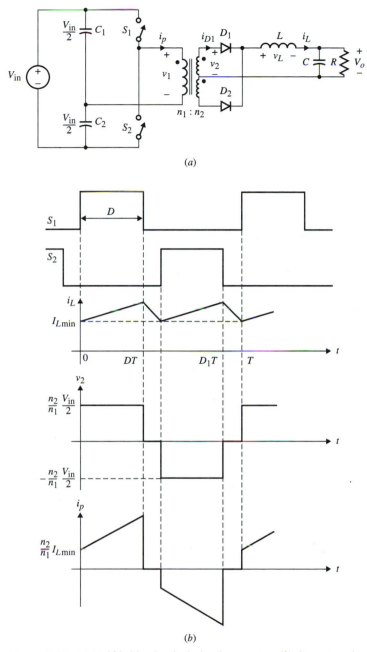

(a)

(b)

Figure 5.17 (a) Half-bridge buck-derived converter. (b) Current and voltage waveforms.

ANSWER (b)

$$I_o + \frac{V_o(1-D)T_s}{L}$$

EXERCISE 5.6

Determine the average load current and the rms inductor current for the half-bridge converter of Fig. 5.17(a) with the following specifications:

$$V_{in} = 135 \text{ V}, \; V_o = 12 \text{ V}, \; f_s = 100 \text{ kHz}, \; R = 2 \; \Omega,$$

$$n_2 = 13, \; n_1 = 39, \; L = 20 \; \mu\text{H}$$

ANSWER   6 A, 6.01 A

5.2.3   Full-Bridge Converter

Figure 5.18 shows the buck-derived full-bridge converter with a full-wave center-tap output transformer.

We must note that the push-pull, half-bridge, and full-bridge converters can use the full-bridge rectifier at the output side. Unlike the half-bridge converter, the full-bridge converter is used in high-input-voltage applications, since the power switching devices are required to block only V_{in}.

EXAMPLE 5.3

Design the full-bridge dc-dc converter with the center-tap output transformer as shown in Fig. 5.18 with the following specifications: $V_{in} = 480 \text{ V}$, $V_o = 600 \text{ V}$ @ $I_o = 10 \text{ A}$, $f_s = 50 \text{ kHz}$.

SOLUTION   The load resistor is obtained from $R = V_o/I_o = 600/10 = 60 \; \Omega$. If we choose $n_2 = 2n_1$, then

$$D = \frac{V_o n_1}{V_{in} n_2} = \frac{600}{480} \frac{1}{2} = 0.625$$

Since L_{crit} for the buck converter is expressed as

$$L_{crit} = \left(\frac{1-D}{4}\right) RT$$

we have $L_{crit} = 112.5 \; \mu\text{H}$. Choose $L > 10L_{crit} = 1.125 \text{ mH}$. If output ripple is less than 1%, according to

$$\frac{\Delta V_o}{V_o} = \frac{1-D}{8LC(2f)^2}$$

C can be determined as $C = 0.42 \; \mu\text{F}$. We can choose $C = 10 \; \mu\text{F}$.

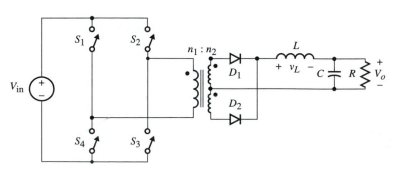

Figure 5.18 Full-bridge converter.

5.2.4   Push-Pull Converter

The circuit configuration for the push-pull converter is shown in Fig. 5.19(a). The circuit uses two active switches. It uses the transformer for voltage scaling and electrical isolation, and the output inductor is used for energy storage. Hence, unlike the design of the transformer for the single-ended converter, where care must be taken in selecting the core material and geometry to design for proper magnetizing inductance, in the push-pull the transformer is used as an ideal element. Since S_1 and S_2 share the current, the push-pull converter is used for higher-power applications compared to the single-ended converters. Figure 5.19(b) illustrates the switching waveforms for S_1 and S_2 with a dead time during which both switches are open.

During this dead time, the load current is carried by the two output diodes, D_1 and D_2. The maximum duty cycle for both switches is 0.5. As can be noticed, the switching frequency of the converter is twice the switching frequency of each switch acting alone. The converter's basic operation is straightforward and similar to the analysis of the half-bridge converter. When S_1 is on, the possible primary voltage causes D_1 to conduct and D_2 to turn off, resulting in the equivalent circuit shown in Fig. 5.20(a).

The converter voltages are given by

$$v_{s1} = v_{s2} = \frac{n_2}{n_1} V_{in}$$

$$v_{p1} = v_{p2} = + V_{in} \tag{5.10}$$

$$v_L = \frac{n_2}{n_1} V_{in} - V_o$$

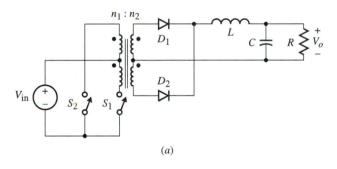

(a)

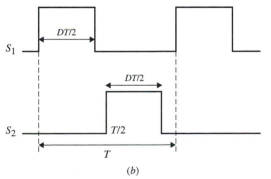

(b)

Figure 5.19 (a) Push-pull converter. (b) Switching waveforms for S_1 and S_2.

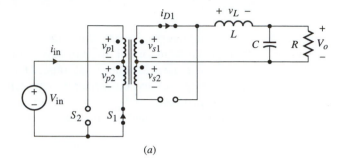

(a)

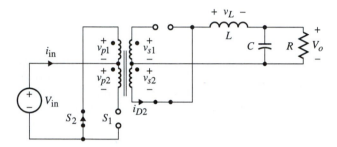

(b)

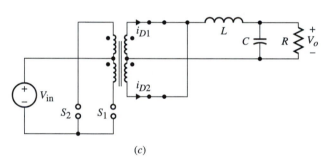

(c)

Figure 5.20 (a) Equivalent circuit when S_1 is on and S_2 is off. (b) Equivalent circuit when S_1 is off and S_2 is on. (c) Equivalent circuit when S_1 is off and S_2 is off.

This circuit is equivalent to a buck converter with a dc input of $(n_2/n_1)V_{in}$. All voltage and current waveforms are similar to those for the buck converter.

Similarly, when S_1 is off and S_2 is on, the equivalent circuit is shown in Fig. 5.20(b). The voltages are given by

$$v_{s1} = v_{s2} = -\frac{n_2}{n_1}V_{in} \tag{5.11}$$

$$v_{p1} = v_{p2} = -V_{in}$$

The voltage across the inductor is given by

$$v_L = -v_{s2} - V_o = \frac{n_2}{n_1}V_{in} - V_o \tag{5.12}$$

Notice that for both modes, whether S_1 or S_2 is on, the circuit is similar to the buck converter when the main switch is on.

Finally, consider the case when both S_1 and S_2 are off. The equivalent circuit is shown in Fig. 5.20(c). The voltages v_{s1} and v_{s2} are both zero, and the inductor voltage is $-V_o$. Hence, the inductor current starts discharging with a slope of $-V_o/L$. This mode is similar to the buck converter when the main switch is off. Therefore, the voltage gain of the push-pull converter is given by

$$\frac{V_o}{V_{in}} = \frac{n_2}{n_1}D \tag{5.13}$$

where D is the duty cycle for either switch, which ranges between 0 and 0.5.

We observe that because of the presence of the transformer, each of S_1 and S_2 should be able to withstand a reverse voltage of at least $2V_{in}$. Also, a peak reverse diode voltage for each of D_1 and D_2 is $2(n_2/n_1)V_{in}$. One disadvantage of the push-pull converter is the imbalance of the voltages applied across the transformer primaries, resulting in unequal switch current. This in turn results in a nonzero magnetizing inductance current at the end of each switching cycle. This eventually will lead to a transformer saturation problem. This problem is caused by a mismatch in the transistor characteristics, such as switching times and voltage drops. To avoid this problem, push-pull converters are designed with not only a voltage control loop (duty cycle control), but also a current loop (current-programmed control) that prevents the transformer from reaching saturation.

Key waveforms for the push-pull converter are shown in Fig. 5.21 with no magnetizing inductance included.

EXAMPLE 5.4

Another boost-derived converter that uses an isolation transformer is the push-pull arrangement shown in Fig. 5.22(a).

The input inductor, L, is very large so that the input current, I_{in}, is assumed constant. These types of converters are useful in high-output-voltage applications. Since I_{in} is continuous, there should be an overlap in the conduction times of S_1 and S_2 as shown in the switching waveforms in Fig. 5.19(b), where δ is the overlapping conduction time. Derive the voltage gain expression for the converter and compare it with the push-pull converter shown earlier. Figure 5.22(c) shows the waveforms for the primary voltage and input current.

SOLUTION Let us assume the first mode begins at $t = 0$, when both S_1 and S_2 are on, resulting in the transformer primary voltage equaling zero since D_1 and D_2 are off. The voltage across L equals the input voltage,

$$v_L = V_{in}$$

$$= L\frac{di_{in}}{dt}$$

while i_{in} is given by

$$i_{in}(t) = \frac{V_{in}}{L}t + I_{in}(0)$$

$I_{in}(0)$ is the initial inductor current value in L.

At $t = t_1$, S_2 is turned off, forcing D_2 to conduct. The voltage across the inductor is given by

$$v_L = V_{in} - \frac{n_1}{n_2}V_o$$

$$= L\frac{di_{in}}{dt}$$

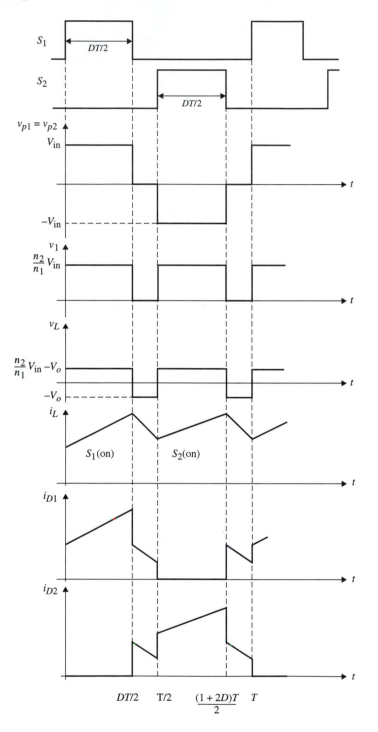

Figure 5.21 Voltage and current waveforms for the push-pull converter of Fig. 5.19(*a*).

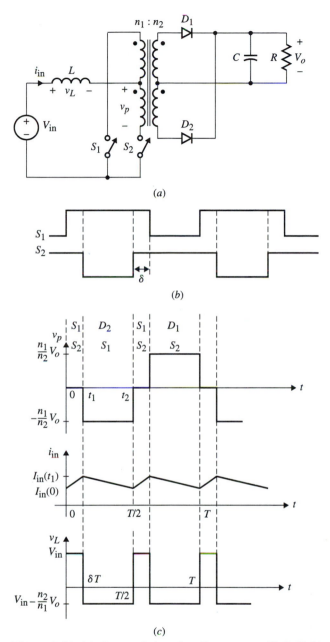

Figure 5.22 (*a*) Current-fed push-pull converter. (*b*) Switching waveforms. (*c*) Converter waveforms.

The inductor current is given by

$$i_{in}(t) = \frac{V_{in} - \frac{n_1}{n_2}V_o}{L}(t - t_1) + I_{in}(t_1)$$

The inductor current discharges at a rate of $(V_{in} - (n_1/n_2)V_o)/L$. At $t = t_2$, S_2 turns on again, resulting in $v_p = 0$. The next two modes are similar to the first two modes, except that when S_2 is on, D_1 conducts.

The gain equation is obtained by applying the volt-second balance to L as follows:

$$V_{in}t_1 = -\left(V_{in} - \frac{n_1}{n_2}V_o\right)(t_2 - t_1)$$

Since $t_1 = \delta T$ and $t_2 = T/2$,

$$V_{in}\delta = \left(\frac{n_1}{n_2}V_o - V_{in}\right)\left(\frac{1}{2} - \delta\right)$$

Solving for V_o/V_{in}, we obtain

$$\frac{V_o}{V_{in}} = \frac{n_2}{n_1}\frac{1}{1 - 2\delta}$$

If we let DT represent the *on* time, $(1 - D)T$ is the *off* time of S_1; then we have

$$\delta = D - \frac{1}{2}$$

Hence, the gain is given by

$$\frac{V_o}{V_{in}} = \frac{n_2}{n_1}\frac{1}{2(1 - D)}$$

EXERCISE 5.7

Calculate the diode rms and average current values for the push-pull converter of Fig. 5.19(a) whose current i_L is given in Fig. E5.7. Determine the average output power if the load resistance is 10 Ω.

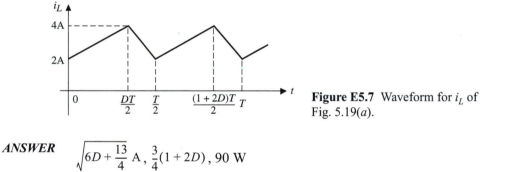

Figure E5.7 Waveform for i_L of Fig. 5.19(a).

ANSWER $\sqrt{6D + \dfrac{13}{4}}$ A , $\dfrac{3}{4}(1 + 2D)$, 90 W

5.3   BOOST-DERIVED ISOLATED CONVERTERS

In this section we will discuss several well-known boost-derived isolated converters. All of these converters have the same function of the basic boost converter discussed in Chapter 4. We begin with the simplest: the single-ended one-switch boost-derived converter known as the *flyback converter.*

5.3.1   Single-Ended Flyback Converter

The circuit topology for the flyback converter is shown in Fig. 5.23(a). This converter is one of the most common isolated switch-mode converters.

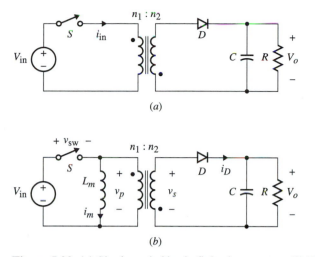

Figure 5.23 (*a*) Single-ended basic flyback converter. (*b*) Flyback converter with magnetizing inductor.

The transformer shown in this topology serves as a step-up/step-down to the input voltage, reverses output voltage polarity, provides electrical isolation, and provides energy storage during the operation. Since all the energy obtained from the source is first stored in the transformer and then passed on to the load, this converter is also known as an energy storage converter. Commercial flyback converters are normally designed with several multicoil output transformers. During the turn-on period, energy is stored in the magnetic inductor and transferred to the output side during the turn-off period. In order for the diode to conduct only during the *off* period in which energy is transferred to the output, the polarities of the transformer windings are reversed, as shown in the figure. One popular application for the flyback converter is in television screens, in which high output voltage is required. This can be obtained by using a high transformer turn ratio, n_2/n_1. Unlike the forward converter, the flyback converter doesn't need an output inductor; it uses only one diode and does not suffer from a core saturation problem. When operated in dcm, it uses a relatively small, magnetizing inductance. To understand the role of the magnetizing inductance, we replace the transformer by a simple model that includes the magnetizing inductor, as shown in Fig. 5.23(*b*).

Typical waveforms for the flyback converter of Fig. 5.23(*b*) operating in the continuous conduction mode are shown in Fig. 5.24. When the transistor is turned on, the primary voltage v_p becomes equal to the source voltage V_{in}, and the diode D is turned off by the negative polarity of $(n_2/n_1)V_{in} + V_o$. The magnetizing inductance, L_m, starts charging linearly with slope V_{in}/L_m. Note that this mode of operation is like the boost converter, in which the inductor L stores energy during the *on* time. When S is turned off, the diode is forced to carry the magnetizing inductor current through the secondary winding. The drawback of this topology is the transformer's primary-side leakage inductance, whose energy must be dissipated when the transistor is turned off. For this reason, in order to reduce the stress on the switching transistor, a snubber circuit is normally added.

The analysis of this circuit consists of two modes of operation as follows.

Mode 1 (*S* is turned on at $t = 0$ and *D* is off). The *on*-state equivalent circuit model is shown in Fig. 5.25(*a*). The voltage across L_m is V_{in}, yielding the following equation for $i_m(t)$:

$$i_m(t) = \frac{V_{in}}{L_m}t + I_m(0) \tag{5.14}$$

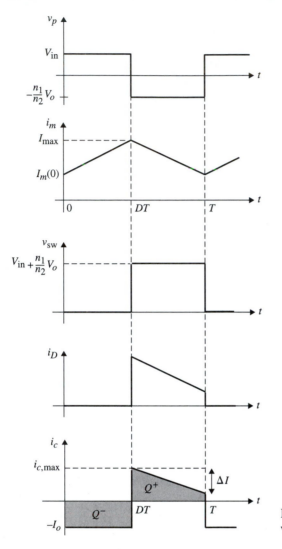

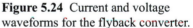

Figure 5.24 Current and voltage waveforms for the flyback converter.

where $I_m(0)$ is the initial value of the magnetizing current when the transistor is turned on. From this equation, we notice that the inductor current, i_m, will linearly charge to $t = DT$ when the transistor is turned off to enter mode 2.

Mode 2 (S is turned off and D is on). This mode of operation starts at $t = DT$, when the transistor is turned off. To maintain the continuity of i_m, the diode D turns on. The equivalent circuit model is shown in Fig. 5.25(b).

The voltage across the magnetizing inductance is nV_o, where n is n_1/n_2. In terms of i_m we have

$$v_p = -nV_o = L_m \frac{di_m}{dt}$$

Integrating this equation from DT to t, we obtain

$$i_m(t) = \frac{-V_o n}{L_m}(t - DT) + I_m(DT) \tag{5.15}$$

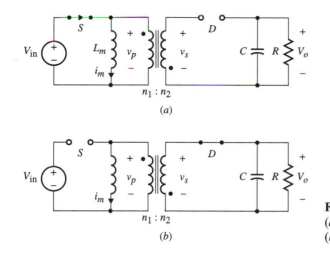

Figure 5.25 Modes of operation.
(a) Mode 1: S is on and D is off.
(b) Mode 2: S is off and D is on.

where $I_m(DT)$ is the magnetizing current value at $t = DT$, when the diode D starts conducting. Evaluating Eq. (5.15) at $t = DT$ and at $t = T$, respectively, and using $i_m(T) = i_m(0)$, we obtain the following relations:

$$I_m(DT) - I_m(0) = \frac{nV_o}{L_m}(1 - D)T \tag{5.16a}$$

$$I_m(DT) - I_m(0) = \frac{V_{in}}{L_m}DT \tag{5.16b}$$

Equating Eqs. (5.16a) and (5.16b), we obtain the following conversion ratio:

$$\frac{V_o}{V_{in}} = \frac{D}{n(1 - D)} \tag{5.17}$$

This gain equation is similar to the buck-boost converter gain when $n = 1$. All relevant current and voltage waveforms are shown in Fig. 5.24. The average output current is the same as the average diode current, which is given

$$I_o = \frac{n(I_m(DT) + I_m(0))}{2}(1 - D)$$

and the average input current is given by

$$I_{in} = \frac{I_m(DT) + I_m(0)}{2}D$$

Hence, the current conversion ratio is given by

$$\frac{I_o}{I_{in}} = \frac{n(1 - D)}{D} \tag{5.18}$$

This equation can also be obtained by equating the average input and output powers by replacing $I_o = V_o/R$; thus we obtain the following relation:

$$I_m(DT) + I_m(0) = \frac{2V_o}{nR(1 - D)}$$

From the above relation, we obtain the minimum and maximum current values of i_m as follows:

$$I_m(0) = \frac{V_{in}D}{n^2R(1-D)^2} - \frac{V_{in}DT}{2L_m} \tag{5.19a}$$

$$I_m(DT) = \frac{V_{in}D}{n^2R(1-D)^2} + \frac{V_{in}DT}{2L_m} \tag{5.19b}$$

When setting $I_m(0) = 0$, we obtain the critical value of the magnetizing inductance for the continuous conduction mode of operation.

$$L_{crit} = \frac{n^2(1-D)^2RT}{2} \tag{5.20}$$

If the inductor current is allowed to reach zero, i.e., $L_m < L_{crit}$, then the converter will operate in dcm and the core becomes fully demagnetized in each cycle.

EXAMPLE 5.5

Sketch the waveform for i_c and determine the expression for the output voltage ripple for the flyback converter of Fig. 5.23(b).

SOLUTION   The capacitor current equals the ac ripple in the diode current, i_D. So i_c is the same as the diode current minus the average output current:

$$i_c = i_D - I_o$$

The current i_c is obtained by simply shifting i_D down by I_o as shown in Fig. 5.24. It is clear that the waveform is similar to the buck-boost capacitor current. The total charge, ΔQ, during the DT interval is given by

$$\Delta Q^- = (DT)(I_o)$$

$$C\Delta V_c = (DT)(I_o)$$

$$\therefore \Delta V_c = \frac{DTI_o}{C}$$

$$\frac{\Delta V_c}{V_o} = \frac{D}{RCf}$$

The peak capacitor current is given by

$$I_{c,max} = I_{Lm}(DT) - I_o$$

Substitute for $I_{Lm}(DT)$ from Eq. (5.19b) and simplify to obtain

$$I_{c,max} = I_o\left[\frac{1-n(1-D)}{n(1-D)} + \frac{nTR(1-D)}{2L_m}\right]$$

EXERCISE 5.8

Determine the voltage gain V_o/V_{in} for the flyback converter operating in the discontinuous conduction mode. Compare this expression with the voltage gain for the continuous conduction mode.

ANSWER
$$\frac{V_o}{V_{in}} = D\sqrt{\frac{RT}{2nL_m}}$$

EXERCISE 5.9

Consider the flyback converter of Fig. 5.23(*b*) with $V_o = 48$ V at $I_o = 1$ A, $V_{in} = 18$ V, $f = 150$ kHz , $n = 0.3$. Find the maximum L_m so that the converter operates in dcm.

ANSWER 4.44 μH

5.3.2 Half-Bridge Converter

Because of the presence of an input inductance in series with the dc voltage source, the circuit implementation of the half-bridge converter requires a center-tap transformer. Figure 5.26(*a*) and (*b*) shows a discrete inductor and a center-tap transformer implementation of the half-bridge boost-derived converter.

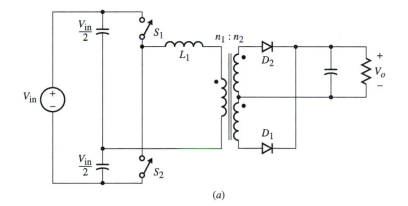

(*a*)

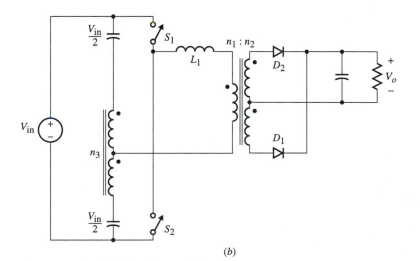

(*b*)

Figure 5.26 Half-bridge boost-derived converter. (*a*) Single-inductor implementation. (*b*) Center-tap transformer implementation.

Because switches S_1 and S_2 are switched alternately, the topology of Fig. 5.26(a) is not suitable due to the abrupt change in the primary current. This problem does not exist for Fig. 5.26(b), in which a center-tap transformer is added. It can be shown that the voltage gain is given by

$$\frac{V_o}{V_{in}} = \frac{1}{2}\left(\frac{n_2}{n_1}\right)\frac{1}{1-D} \tag{5.21}$$

5.3.3 Full-Bridge Converter

Figure 5.27 shows the full-bridge boost-derived converter with a full-wave rectifier in the output side. This converter is also known as a double-ended converter.

The main advantage of the full-bridge configuration is that its core material is better utilized compared to the single-ended case, in which the magnetic field density changes from 0 to $+B_{sat}$. For the full-bridge configuration the core excitation varies between $+B_{sat}$ and $-B_{sat}$. There are several switching sequences for Fig. 5.27 that will result in a duty cycle above 50%. Figure 5.28 shows two possible switching sequences.

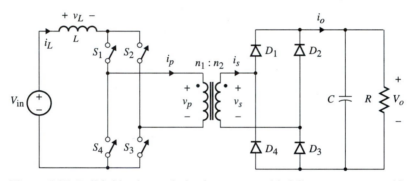

Figure 5.27 Full-bridge boost-derived converter with full-wave output rectifier.

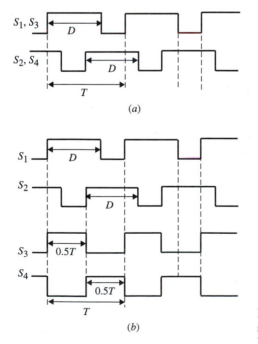

Figure 5.28 Two possible switching sequences for the full-bridge converter of Fig. 5.27.

In Fig. 5.28(a), each of the two diagonal pairs of switches are switched simultaneously with $D > 0.5$ for each switch. With this switching sequence, there exist two intervals during which all four switches are on at the same time. During these intervals, the inductor current is divided almost equally between the switch legs S_1-S_4 and S_2-S_3, reducing their rms current values. In the switching sequence of Fig. 5.28(b), two of the switches operate with a fixed duty cycle and the other two switches operate with a varying duty cycle. Of course, the driving circuit for Fig. 5.28(a) is more complex since unlike Fig. 5.28(b), where only two switches are controlled, the switching sequence of Fig. 5.28(a) is required to control four switches. Next we consider the analysis of the circuit shown in Fig. 5.27 when operated under the switching sequence of Fig. 5.28(a). It can be shown that there are four modes of operation during one switching period, as shown in Fig. 5.29.

The inductor current during mode 1 is given by

$$i_L(t) = \frac{V_{in}}{L}t + I_L(0) \tag{5.22}$$

where $I_L(0)$ is the initial inductor current when S_1 and S_3 are turned on, while S_2 and S_4 were on initially. This current is divided equally through S_1-S_4 and S_2-S_3, with $v_s = v_p = 0$ and $i_o = 0$.

At $t = t_1$, S_2 and S_4 are turned off simultaneously, forcing $i_L(t)$ to flow through the primary, through diodes D_1 and D_3, and then to the load. The equivalent circuit is shown in Fig. 5.29(b).

The inductor current is obtained from v_L:

$$v_L(t) = V_{in} - \frac{n_1}{n_2}V_o \qquad 0 \leq t < t_1 \quad t_1 \leq t < t_2$$

Therefore, the inductor current is given by

$$i_L(t) = \frac{V_{in} - \frac{n_1}{n_2}V_o}{L}(t - t_1) + I_L(t_1) \tag{5.23}$$

In this mode v_p and v_s are given by

$$v_s = V_o \qquad v_p = \frac{n_1}{n_2}V_o$$

Mode 3 starts at $t = t_2$, when S_2 and S_4 are turned on again, resulting in a similar equivalent circuit to that in mode 1. The current equation is given by

$$i_L(t) = \frac{V_{in}}{L}(t - t_2) + I_L(t_2) \tag{5.24}$$

At $t = t_3$, we have

$$I_L(t_3) = \frac{V_{in}}{L}(t_3 - t_2) + I_L(t_2)$$

Finally, mode 4 starts at $t = t_3$, when S_1 and S_3 are turned off simultaneously. The voltages are given as follows:

$$v_L = V_{in} + v_p \qquad v_p = \frac{n_1}{n_2}v_s \qquad v_s = -V_o$$

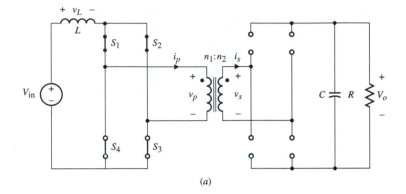

(a)

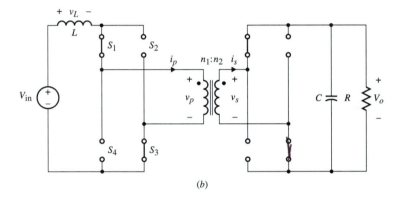

(b)

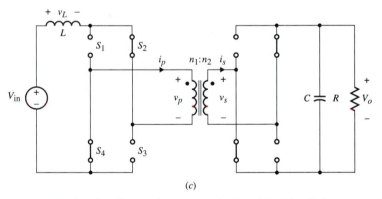

(c)

Figure 5.29 Modes of operation. (*a*) Modes 1 and 3: All switches are on. (*b*) Mode 2: S_1 and S_3 are on. (*c*) Mode 4: S_2 and S_4 are on.

Hence, the currents are given as

$$i_L(t) = \frac{V_{in} - \frac{n_1}{n_2}V_o}{L}(t - t_3) + I_L(t_3)$$

$$i_p(t) = -i_L$$

$$i_s(t) = \frac{n_1}{n_2}i_p$$

The current and voltage waveforms are shown in Fig. 5.30.

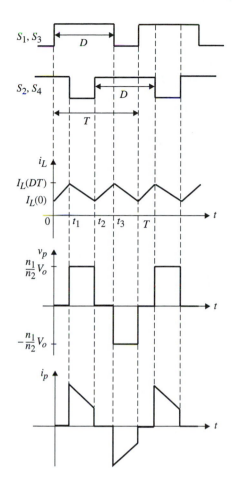

Figure 5.30 Current and voltage waveforms for Fig. 5.27.

From the volt-second principle, we have

$$t_1 V_{in} = (t_1 - t_2)\left(V_{in} - \frac{n_1}{n_2} V_o\right)$$

From Fig. 5.29, $t_1 = (D - \frac{1}{2})T$ and $(t_2 - t_1) = (1 - D)T$; hence, we have

$$(D - \tfrac{1}{2})T V_{in} = -(1 - D)T\left(V_{in} - \frac{n_1}{n_2} V_o\right) \tag{5.25}$$

Solving Eq. (5.25) for the voltage gain, we obtain

$$\frac{V_o}{V_{in}} = \frac{n_2}{n_1}\frac{1}{2(1 - D)} \tag{5.26}$$

EXAMPLE 5.6

(a) Derive the voltage gain expression for Fig. 5.27 by using the switching sequence given in Fig. 5.28(*b*).

(b) What is the rms current in each switch?

SOLUTION

(a) Although there is a change in the switching sequence between Fig. 5.28(a) and (b) in switches S_3 and S_4, careful investigation of the circuit operation modes indicates that conditions governing the charging and discharging of the inductor are exactly the same for both switching sequences. Therefore, the expression according to the volt-second balance as obtained above also applies here, and the voltage gain expression for the switching sequence given in Fig. 5.28(b) is the same as that given in Eq. (5.26).

(b)

$$I_{s,\text{rms}} = \sqrt{\frac{I_L^2}{2} + \frac{\Delta I_L^2}{24}} \qquad (5.27)$$

where

$$I_L = \frac{n_2}{n_1}\frac{I_o}{2(1-D)}$$

is the average inductor current and

$$\Delta I_L = \frac{V_{\text{in}}}{L}\left(D - \frac{1}{2}\right)T \qquad (5.28)$$

is the inductor current ripple.

EXERCISE 5.10

Derive the expressions for $i_L(0)$ and $i_L(DT)$ in Fig. 5.30 and determine the inductor's critical value, L_{crit}.

ANSWER $I_L(0) = \dfrac{n_2}{n_1}\dfrac{I_o}{2(1-D)} - \dfrac{V_{\text{in}}}{2L}\left(D - \dfrac{1}{2}\right)T, \qquad I_L(DT) = \dfrac{n_2}{n_1}\dfrac{I_o}{2(1-D)} + \dfrac{V_{\text{in}}}{2L}\left(D - \dfrac{1}{2}\right)T,$

$$L_{\text{crit}} = 2\left(\frac{n_1}{n_2}\right)^2 R(1-D)^2\left(D - \frac{1}{2}\right)T$$

EXERCISE 5.11

Show that the primary rms current of Fig. 5.27 is given by the following equation:

$$I_{p,\text{rms}} = \sqrt{2(1-D)\left(I_L^2 + \tfrac{1}{12}DI^2\right)}$$

5.4 OTHER ISOLATED CONVERTERS[1]

In addition to the circuits presented thus far, there are many other isolated converters that are covered in the open literature. Here we will consider two examples: the Cuk and the Weinberg converters.

5.4.1 Isolated Cuk Converter

Figure 5.31 shows the isolated Cuk converter. The converter storage capacitor is split into two series capacitors, C_1 and C_2, as shown. The output voltage remains negative since the transformer's primary and secondary windings have the same polarity. Let

---

[1]This section may be omitted without loss of continuity.

us derive the expression for the voltage gain. Assume the converter operates in the continuous conduction mode. Like the push-pull and bridge converters, the entire core *B-H* loop is utilized in both directions.

Figure 5.32(*a*) and (*b*) shows the two equivalent circuits for mode 1 when *S* is on and mode 2 when *S* is off.

Mode 1 (during *DT* interval):

$$v_{L1} = V_{in}$$

$$v_{L2} = -V_o - V_{c2} - \frac{n_2}{n_1}V_{c1}$$

Mode 2 (during $(1-D)T$ interval):

$$v_{L1} = V_{in} + V_{c1} + \frac{n_1}{n_2}V_{c2}$$

$$v_{L2} = -V_o$$

Applying the volt-second principle to v_{L1} and v_{L2}, we obtain

v_{L1}:

$$DV_{in} = -(1-D)\left[V_{in} + V_{c1} + \frac{n_1}{n_2}V_{c2} \right]$$

$$V_{in}\frac{-D}{1-D} = V_{in} + V_{c1} + \frac{n_1}{n_2}V_{c2}$$

$$V_{in}\frac{-1}{1-D} = V_{c1} + \frac{n_1}{n_2}V_{c2} \tag{5.29}$$

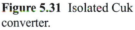

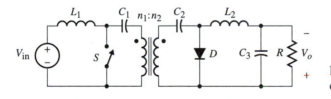

Figure 5.31 Isolated Cuk converter.

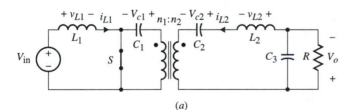

(*a*)

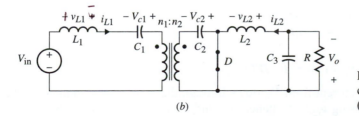

(*b*)

Figure 5.32 Isolated Cuk converter. (*a*) Mode 1. (*b*) Mode 2.

v_{L2}:

$$D\left[-V_o - V_{c2} - \frac{n_2}{n_1}V_{c1}\right] = +(1-D)V_o$$

$$-D\left(V_{c2} + \frac{n_2}{n_1}V_{c1}\right) = [+(1-D)+D]V_o$$

$$V_{c2} + \frac{n_2}{n_1}V_{c1} = \frac{-V_o}{D}$$

$$V_{c1} + \frac{n_1}{n_2}V_{c2} = -\frac{n_1}{n_2}\frac{V_o}{D} \tag{5.30}$$

From Eqs. (5.29) and (5.30), we obtain

$$\frac{V_o}{V_{in}} = \frac{D}{1-D}\frac{n_2}{n_1}$$

EXERCISE 5.12

Consider the Cuk converter shown in Fig. E5.12, which provides a positive output voltage. Assume ideal components to derive the voltage gain equation V_o/V_{in}, and sketch the waveforms for i_{L1}, i_{L2}, i_D, v_{sw}, v_D, and v_{c1}.

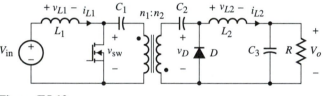

Figure E5.12

5.4.2 Weinberg Converter

In this section, the basic topology and operation of another buck-derived push-pull converter, known as the Weinberg converter, will be discussed. There are several versions of this topology that perform the same function. The steady-state analysis will be presented in this section for the converter operating in ccm. The original proposed Weinberg dc-to-dc topology is shown in Fig. 5.33(a). A more simplified equivalent circuit can be obtained. In this analysis we will assume the n-turn-ratio transformer has a magnetizing inductance, L_m, and the m-turn-ratio transformer is ideal.

The switching waveforms are similar to those for the push-pull converter and are shown in Fig. 5.33(b). When switches S_1 and S_2 are turned on and off alternately, the following cases are investigated:

1. Mode 1: S_1 on, S_2 off
2. Mode 2: S_1 off, S_2 on
3. Mode 3: S_1 off, S_2 off

Mode 1: S_1 on, S_2 off ($0 < t \leq DT$)

In this case, the current from V_{in} flows through the magnetizing inductor of the input transformer, causing $v_{s2} > 0$; hence, D_1 and D_4 are turned off. With S_1 on, $v_{s1'} > 0$

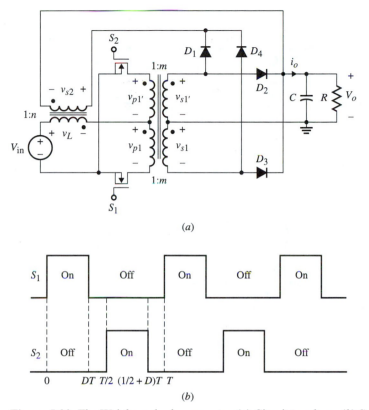

Figure 5.33 The Weinberg dc-dc converter. (a) Circuit topology. (b) Switching waveforms for the power switches S_1 and S_2.

and $v_{s1} < 0$, causing D_3 to reverse-bias and D_2 to be forward biased. The equivalent circuit model is shown in Fig. 5.34(a).

From Fig. 5.34 we have $v_L = V_{in} - V_o/m$, yielding the following equation for i_L:

$$i_L(t) = \frac{V_{in} - V_o/m}{L}t + I_L(0) \tag{5.31}$$

where $I_L(0) = I_{L\min}$ is the initial inductor current at $t = 0$. Since the center-tap transformer has the same ratio, m, diode D_3 blocks twice the output voltage, i.e.,

$$v_{D3} = -2V_o$$

Mode 2: S_1 off, S_2 on $(T/2 < t \le (0.5 + D)T)$

At $t = T/2$, S_2 is turned on while S_1 is kept off (S_1 is turned off in the previous interval at $t = DT$). In this case since the current from V_{in} flows through the magnetizing inductor, diodes D_1 and D_4 are turned off. The output diode, D_2, is off because S_2 is on, while D_3 is conducting to carry the inductor current.

From Fig. 5.34(b), the inductor voltage is $v_L = V_{in} - V_o/m$ and i_L is given by

$$i_L(t) = \frac{V_{in} - V_o/m}{L}(t - T/2) + I_L(T/2) \tag{5.32}$$

The blocking voltage of D_2 is $-2V_o$, and this case is identical to the case in mode 1. Since we have a symmetric circuit operation, these two modes are similar when either switch is on and the other one is off.

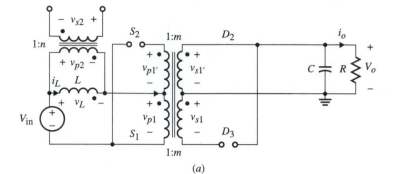

(a)

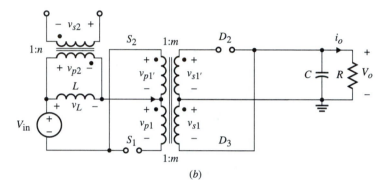

(b)

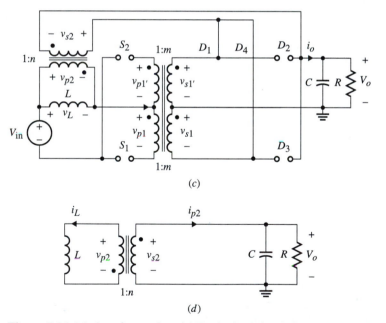

(c)

(d)

Figure 5.34 Modes of operation. (*a*) Equivalent circuit for mode 1, when S_1 is on and S_2 is off.
(*b*) Equivalent circuit for mode 2, when S_2 is on and S_1 is off. (*c*) Equivalent circuit for mode 3,
when both S_1 and S_2 are off. (*d*) Simplified equivalent circuit of mode 3.

Mode 3: S_1 off, S_2 off ($DT < t \leq T/2$ and $(0.5 + D)T < t \leq T$)

In this case, since the two switches are turned off, D_1 and D_4 are turned on, while D_2
and D_3 are turned off. The equivalent circuit is shown in Fig. 5.34(*c*), and the simplified equivalent circuit is shown in Fig. 5.34(*d*).

From Fig. 5.34(*d*), the inductor voltage is $-V_o/n$, and the inductor current for the first interval is given by

$$i_L(t) = -\frac{V_o}{nL}(t-(DT)) + I_L(DT) \tag{5.33}$$

where $I_L(DT) = I_{L\max}$ is the initial inductor current at $t = DT$.

EXERCISE 5.13

Show that the average output and input currents for the Weinberg converter of Fig. 5.33(*a*) are given by Eqs. (5.34) and (5.35), respectively.

$$I_{o,\text{ave}} = \frac{(I_{L\max} + I_{L\min})}{2}\left(\frac{D}{m} + \frac{1-D}{n}\right) \tag{5.34}$$

$$I_{\text{in,ave}} = \frac{(I_{L\max} + I_{L\min})D}{2} \tag{5.35}$$

EXAMPLE 5.7

It can be shown that the two-switch Weinberg converter of Fig. 5.33(*a*) can be simplified by using a one-switch equivalent circuit model as shown in Fig. 5.35. Derive the inductor current and voltage gain equation, and draw the key current and voltage waveforms for the simplified converter of Fig. 5.35.

SOLUTION   Figure 5.36(*a*) and (*b*) shows the two equivalent circuit modes. Figure 5.36(*c*) shows the switching waveforms for the simplified circuit of Fig. 5.35.
   The inductor currents for the two modes are given by

$$i_L(t) = \frac{V_{in} - V_o/m}{L}t + I_L(0) \qquad 0 < t \le DT \tag{5.36a}$$

$$i_L(t) = -\frac{V_o}{nL}(t-(DT)) + I_L(DT) \qquad DT < t \le T \tag{5.36b}$$

It can be shown that the voltage gain equation is

$$\frac{V_o}{V_{in}} = \frac{D}{\dfrac{D}{m} + \dfrac{1-D}{n}} \tag{5.37}$$

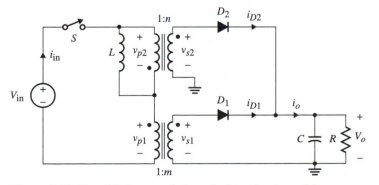

Figure 5.35 Simplified one-switch equivalent circuit model.

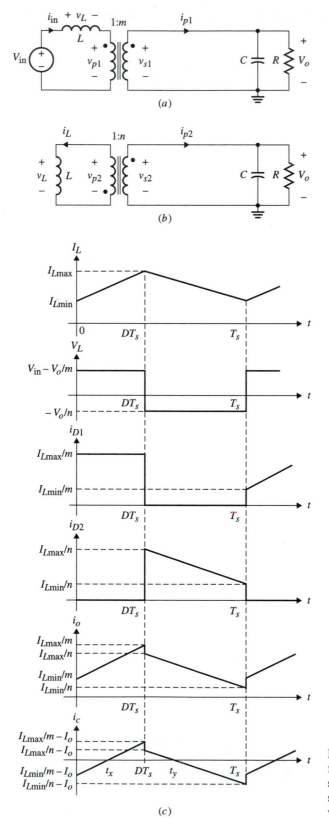

(a)

(b)

(c)

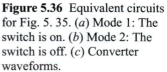

Figure 5.36 Equivalent circuits for Fig. 5. 35. (*a*) Mode 1: The switch is on. (*b*) Mode 2: The switch is off. (*c*) Converter waveforms.

If $m = n$, then the voltage gain becomes the gain of the buck converter. Therefore,

$$\frac{V_o}{V_{in}} = D \tag{5.38}$$

It can be also shown that I_{Lmax} and I_{Lmin} are given by

$$I_{Lmax} = \frac{V_o^2}{RDV_{in}} + \frac{V_o}{2nL}(1-D)T \tag{5.39a}$$

$$I_{Lmin} = \frac{V_o^2}{RDV_{in}} - \frac{V_o}{2nL}(1-D)T \tag{5.39b}$$

For ccm operation, $I_{Lmin} > 0$, and to obtain the critical inductance set $I_{Lmin} = 0$:

$$L_{crit} > \frac{RDV_{in}}{2nV_o}(1-D)T \tag{5.40}$$

The normalized output capacitor ripple voltage is given by

$$\Delta V_{nc} = \frac{n^2 \tau_n}{2\tau_{nRC}}\left[\left(\frac{I_{n,max}}{n}-1\right)^2 + \left(\frac{mD}{n(1-D)}\right)\left(1-\frac{I_{n,min}}{m}\right)^2\right] \tag{5.41}$$

where $V_{nc} = V_c/V_{in}$, $\tau_n = \tau/T = L/RT$, $\tau_{nRC} = RC/T$, $I_{n,max} = I_{max}/I_o$, and $I_{n,min} = I_{min}/I_o$.

EXAMPLE 5.8

Consider a dc-dc converter unit system that uses the Weinberg converter with an input dc equal to 125 V and an output of 48 V. The design has the following specifications:

Output power $P_o = 2$ kW
Switching frequency $f = 40$ kHz
Output ripple voltage $\Delta V_c/V_o = 0.2\%$

Determine the following:

(a) The value of n where $D = 0.7$ and $m = 0.5$
(b) The critical inductance value
(c) The maximum and minimum inductor currents
(d) The average input and output currents
(e) The capacitor current value

SOLUTION

(a) From the gain relation, we have

$$\frac{V_o}{V_{in}} = \frac{D}{\dfrac{D}{m} + \dfrac{1-D}{n}}$$

$$\frac{48}{125} = \frac{0.7}{\dfrac{0.7}{0.5} + \dfrac{1-0.7}{n}}$$

$$n \approx 0.7$$

(b) The critical inductance value is given by

$$L_{crit} > \frac{RDV_{in}}{nV_o}(1-D)T_s$$

$R = V_o^2/P = 1.152 \; \Omega$ and $T_s = 1/f = 25 \; \mu s$; therefore,

$$L_{crit} = \frac{(1.152)(0.7)(125)}{2(48)(0.7)}(1-0.7)(25 \; \mu s)$$

$$L_{crit} = 11.25 \; \mu H$$

Select $L = 30 \; \mu H$.

(c) The maximum inductor current is

$$I_{Lmax} = \frac{V_o^2}{RDV_{in}} + \frac{V_o}{2nL}(1-D)T_s$$

$$I_{Lmax} = \frac{(48)^2}{(1.152)(0.7)(125)} + \frac{48}{2(0.7)(30 \; \mu H)}(1-0.7)(25 \; \mu s)$$

$$I_{Lmax} = 31.4 \; A$$

and the minimum value is given by

$$I_{Lmin} = \frac{V_o^2}{RDV_{in}} - \frac{V_o}{2nL}(1-D)T_s$$

$$I_{Lmin} = 14.3 \; A$$

(d) From Equation (5.34) we have

$$I_{o,ave} = \frac{(I_{Lmax}+I_{Lmin})}{2}\left(\frac{D}{m}+\frac{1-D}{n}\right)$$

$$I_{o,ave} = \frac{(31.4+14.3)}{2}\left(\frac{0.7}{0.5}+\frac{1-0.7}{0.7}\right)$$

$$I_{o,ave} = 41.78 \; A$$

and from Equation (5.35) we have

$$I_{in,ave} = \frac{(I_{Lmax}+I_{Lmin})D}{2}$$

$$I_{in,ave} = \frac{(31.4+14.3)}{2}(0.7)$$

$$I_{in,ave} = 16 \; A$$

(e) The times at which the capacitor current becomes zero are given by

$$t_x = \frac{(I_o - I_{Lmin}/m)}{\frac{1}{L}\left(\frac{V_{in}}{m}-\frac{V_o}{m^2}\right)}$$

$$t_x = \frac{(41.78 - 14.3/0.5)}{\frac{1}{30 \; \mu H}\left(\frac{125}{0.5}-\frac{48}{0.5^2}\right)}$$

$$t_x = 6.83 \; \mu s$$

and

$$t_y = \frac{nL}{V_o}(I_{L\max} - nI_o) + DT_s$$

$$t_y = \frac{(0.7)(30~\mu H)}{48}(31.4 - 0.7 \times 41.78) + (0.7)(25~\mu s)$$

$$t_y = 18.44~\mu s$$

Recall

$$\tau_n = \frac{\tau}{T_s} = \frac{L/R}{T_s} = 1.04 \quad \text{and} \quad \tau_{nRC} = \frac{RC}{T_s}$$

Therefore,

$$\Delta V_{nc} = \frac{n^2 \tau_n}{2\tau_{nRC}}\left\{\left(\frac{I_{n,\max}}{n} - 1\right)^2 + \left(\frac{mD}{n(1-D)}\right)\left(1 - \frac{I_{n,\min}}{m}\right)^2\right\}$$

$$\frac{0.2}{100} = \frac{(0.7)^2(1.04)(25~\mu s)}{2(1.152)C}[4.9\times10^{-3} + (0.67)(0.25)]$$

The value for the capacitor is obtained as

$$C = 476.6~\mu F$$

5.5 MULTI-OUTPUT CONVERTERS

As stated earlier, one of the advantages of using an isolated transformer is to allow for multiple outputs at the secondary side of the output transformer. This is very common in today's power supplies used in components to provide several voltage levels at different currents and polarities. Figure 5.37(a) and (b) shows three-output examples of a half-bridge buck-derived converter and a single-ended flyback converter, respectively.

EXERCISE 5.14

Design the half-bridge converter of Fig. 5.37(a) to provide three output voltages with $V_{o1} = 5$ V at 6 A, $V_{o2} = 37$ V at 0.5 A, and $V_{o3} = 12$ V at 2 A. Design the transformer turn ratio and the converter components such that the voltage ripple for any output does not exceed 0.1%. Assume $V_{in} = 160$ VDC and a switching frequency of 100 kHz.

EXERCISE 5.15

Determine the rms currents for the power switches and transformer windings of Example 5.6.

ANSWER 0.877 A, 1.24 A, 3.24 A, 0.534 A, 1.34 A

EXERCISE 5.16

Design the multi-output converter of Fig. 5.37(b) for the following output specifications: $V_{o1} = +5$ V at 4 A, $V_{o2} = +12$ V at 0.5 A, $V_{o3} = -12$ V at 0.3 A. The ripple voltage does not exceed 100 mV peak-to-peak for each output. Use $V_{in} = 185$ V, $f_s = 50$ kHz, and $D = 0.5$.

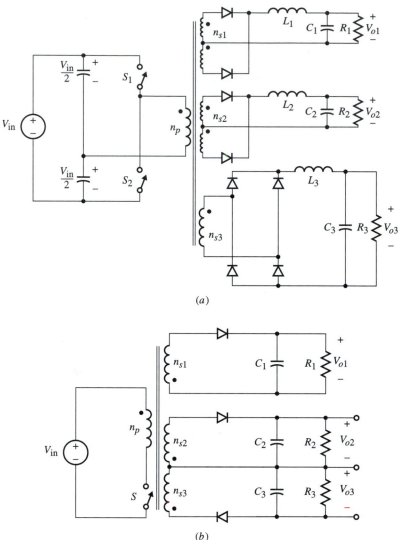

(a)

(b)

Figure 5.37 (a) Three-output half-bridge buck-derived converter. (b) Three-output flyback converter.

ANSWER $n_{s1}/n_p = 1/37$, $n_{s2}/n_p = n_{s3}/n_p = 12/185$, $L_m > 2.925$ mH, $C_1 > 400\ \mu$F, $C_2 > 50\ \mu$F, $C_3 > 30\ \mu$F

     In practice, normally the output voltage with the highest power is controlled, with the other outputs using linear regulators to stabilize their voltages. This is why the open-loop outputs are designed for a higher voltage than specified.

PROBLEMS

Forward Converter

5.1 Consider the single-ended forward converter of Fig. P5.1 including a leakage inductance of the power transformer. Discuss the effect of L_k on the circuit operation.

D5.2 Design the forward converter of Fig. 5.8 that needs to supply 400 W at 15 A load current. The output ripple should not exceed 1%.

5.3 Derive the expression for the output ripple voltage for the forward converter of Fig. 5.8.

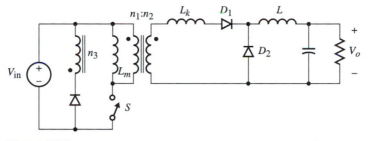

Figure P5.1

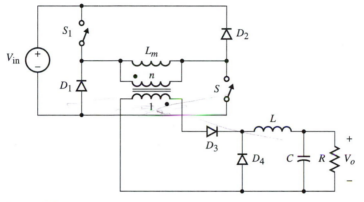

Figure P5.4

5.4 The circuit given in Fig. P5.4 is another type of forward converter. Sketch the primary voltage and diode current waveforms for the circuit.

5.5 Replace D_1 and D_2 by S_3 and S_4, respectively, in Problem 5.4. Derive the voltage gain if the four switches have the switching waveforms in Fig. P5.5.

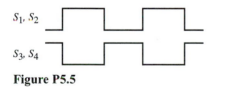

Figure P5.5

5.6 Consider the single-ended forward converter given in Fig. P5.6, which uses a regular and a zener diode to provide a negative voltage damp across the primary in order to reset the core to avoid transformer saturation.

(a) Discuss the operation of the circuit.

(b) Sketch the current waveforms i_{D1}, i_{D2}, i_L, i_{D3}, i_{sw}, and i_p.

Full-Bridge

D5.7 Design a full-bridge dc-dc converter with a center-tap output transformer with the following

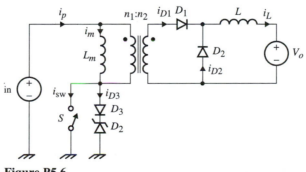

Figure P5.6

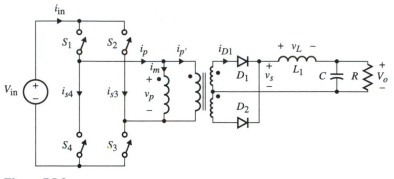

Figure P5.8

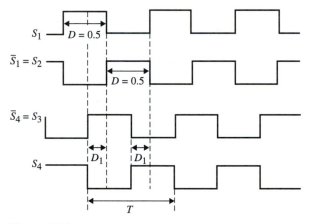

Figure P5.9

specifications: $V_{in} = 480$ V, $V_o = 240$ V at $I_o = 20$ A, $f_s = 75$ kHz.

5.8 Sketch the labeled current and voltage waveforms for the double-ended buck-derived converter shown in Fig. P5.8. Assume the switching sequence for S_1-S_3 and S_2-S_4 is the same as that in Fig. 5.19(b). Assume $RC \gg T$.

5.9 Repeat Example 5.6 by using the switching sequence in Fig. P5.9. Derive the voltage gain in terms of the phase shift period D_1.

Push-Pull Converter

5.10 Derive the output voltage ripple for the push-pull converter.

Isolated SEPIC and Cuk Converter

5.11 Derive the voltage gain equation for the isolated SEPIC converter shown in Fig. P5.11.

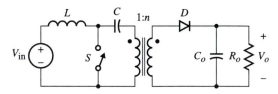

Figure P5.11

5.12 Derive the voltage gain expression for the isolated Cuk converter shown in Fig. P5.12.

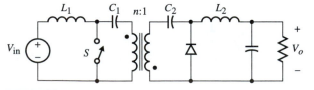

Figure P5.12

Weinberg Converter

5.13 Repeat Problem 5.3 for the Weinberg converter.

5.14 Show that the voltage gain equation for the Weinberg converter of Fig. 5.35 is given by

$$\frac{V_o}{V_{in}} = \frac{D}{\dfrac{D}{m} + \dfrac{1-D}{n}}$$

5.15 Show that t_x and t_y of Fig. 5.36 are given by

$$t_x = \frac{(I_o - I_{L\min}/m)}{\dfrac{1}{L}\left(\dfrac{V_{in}}{m} - \dfrac{V_o}{m^2}\right)}$$

$$t_y = \frac{nL}{V_o}(I_{L\max} - nI_o) + DT$$

For the special case when $I_{n,\max}/n = 1$, show that $t_y = DT_s$, t_x is the same as above, and ΔV_{nc} is given by

$$\Delta V_{nc} = \frac{n^2 \tau_n}{2\tau_{nRC}}\left(\frac{mD}{n(1-D)}\right)\left(1 - \frac{I_{n,\min}}{m}\right)^2$$

Design Problems

Multi-Output Converters

D5.16 Design the four-output buck-boost converter shown in Fig. P5.16 to operate in ccm with the following specifications: $V_{in,\min} = 28$ V, $V_{in,\max} = 48$ V, $V_{in,nom} = 36$ V, $V_{o1} = 5$ V at $I_o = 10$ A, $V_{o2} = 5$ V at $I_o = 1$ A, $V_{o3} = 12$ V at $I_o = 2$ A, $V_{o4} = -12$ V at $I_o = 3$ A, $f_s = 70$ kHz, and $\Delta V_o/V_o = 1\%$ for each output.

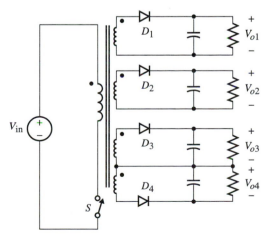

Figure P5.16

D5.17 Design the two-output forward converter shown in Fig. P5.17 to operate in ccm with the following specifications: $V_{in,\min} = 78$ V, $V_{in,\max} = 110$ V, $V_{in,nom} = 92$ V, $V_{o1} = 5$ V at $I_o = 15$ A, $V_{o2} = 12$ V at $I_o = 5$ A, $f_s = 70$ kHz, and $\Delta V_o/V_o = 1.5\%$ for each output.

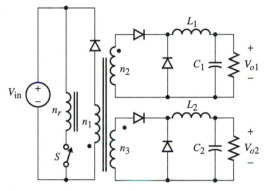

Figure P5.17

D5.18 Design the two-output, two-switch forward converter shown in Fig. P5.18 to operate in ccm with the following specifications: $V_{in,\min} = 210$ V, $V_{in,\max} = 480$ V, $V_{in,nom} = 370$ V, $V_{o1} = 5$ V at $I_o = 1.5$ A, $V_{o2} = 28$ V at $I_o = 5$ A, $f_s = 85$ kHz, $\Delta V_{o1}/V_{o1} = 0.1\%$, $\Delta V_{o2}/V_{o2} = 0.5\%$, $\Delta I_{o1}/I_{o1} = 15\%$, $\Delta I_{o2}/I_{o2} = 10\%$.

Single-Output Converters

D5.19 Design the simplified Weinberg converter shown in Fig. P5.19 to operate in ccm with the following specifications: $V_{in,nom} = 48$ V, $V_o = 22$ V at $I_o = 3$ A to 0.3 A, $f_s = 120$ kHz, and $\Delta V_o/V_o = 1\%$.

D5.20 Design the single-ended forward converter of Fig. 5.8 to operate in ccm with the following specifications: $V_{in,nom} = 28$ V, $V_o = 12$ V at $I_o = 4$ A to 1 A, $f_s = 125$ kHz, and $\Delta V_o/V_o = 1\%$. Assume the diode forward voltage drop is 1 V.

D5.21 Redesign Problem 5.20 using the push-pull converter of Fig. 5.19(a).

D5.22 Design the full-bridge converter of Fig. 5.18 to operate in ccm with the following specifications: $V_{in,nom} = 150$ V, $V_o = 12$ V at $I_o = 2$ A, $f_s = 50$ kHz, $\Delta V_o/V_o = 0.5\%$, $\Delta I_o/I_o = 5\%$.

D5.23 Design the half-bridge converter of Fig. 5.17(a) to operate in ccm with the following specifications: $V_{in,nom} = 185$ V, $V_o = 28$ V at $I_o = 2$ A, $f_s = 100$ kHz, $\Delta V_o/V_o = 0.1\%$, $\Delta I_o/I_o = 5\%$.

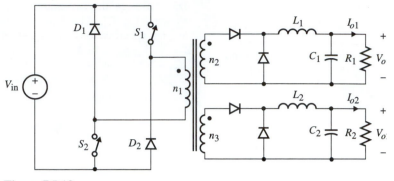

Figure P5.18

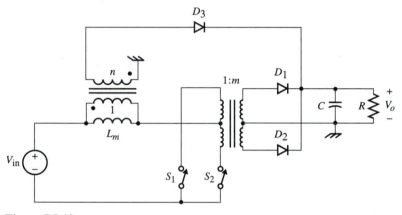

Figure P5.19

D5.24 Design the forward converter that uses a tertiary winding shown in Fig. 5.14 with the following specifications: $V_{in,nom} = 285$ V $\pm 20\%$, $V_o = 12$ V at $I_o = 2$ A to 200 mA, $f_s = 120$ kHz, $\Delta V_o = 100$ mV. Determine the diode and transistor voltage and current ratings.

D5.25 Repeat Problem 5.24 with the same specifications except for an additional output voltage with the following specifications: $V_{o2} = 5$ V at 4 A to 500 mA and the output voltage ripple does not exceed 50 mV. The circuit configuration is shown in Fig. P5.25.

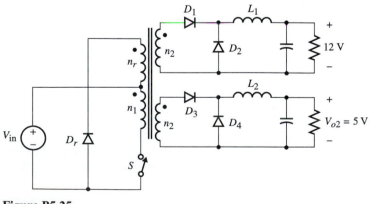

Figure P5.25

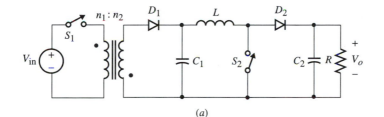

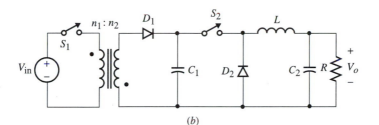

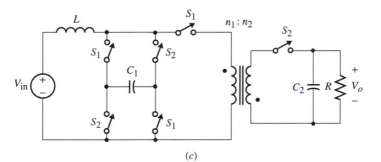

Figure P5.26

Additional Topologies

5.26 For the three circuits given in Fig. P5.26(a), (b), and (c), show that the voltage gains are given by the following equations, respectively. Assume S_1 and S_2 are switched simultaneously for (a) and (b), and S_1 and S_2 are switched complementarily for (c). Assume C_1 and C_2 are very large.

(a) $\dfrac{V_o}{V_{in}} = \dfrac{D}{(1-D)^2}\dfrac{n_2}{n_1}$

(b) $\dfrac{V_o}{V_{in}} = \dfrac{D^2}{(1-D)}\dfrac{n_2}{n_1}$

(c) $\dfrac{V_o}{V_{in}} = \dfrac{-D}{(1-D)(1-2D)}\dfrac{n_2}{n_1}$

5.27 The converter shown in Fig. P5.27(a) is another version of the forward converter. Assume ideal diodes. L_m is the transformer magnetizing inductance. The switching waveforms for S_1, S_2, S_3, and S_4 are shown in Fig. P5.27(b). Assume $RC \gg T$.

(a) Sketch the current and voltage waveforms labeled on the figure.

(b) Derive the voltage gain expression, V_o/V_{in}.

5.28 Draw the waveforms for the branch currents and voltages for the push-pull converter including a finite magnetic inductance as shown in Fig. P5.28.

5.29 Show that if the single-ended flyback converter of Fig. 5.23(b) is operating in dcm (magnetizing current is discontinuous), then the voltage conversion ratio is given by

$$\frac{V_o}{V_{in}} = \frac{n_2}{n_1}D\sqrt{\frac{1}{2\tau}}$$

where $\tau_n = L/RT$.

5.30 Consider the circuit given in Fig. P5.30(a) with the switching sequence given in Fig. P5.30(b).

(a) Sketch the waveforms for i_L, v_s, i_{D1}, and i_{D4}.

(b) Derive the expression for V_o/V_{in}.

(c) What is the expression for the critical value of L to maintain ccm operation?

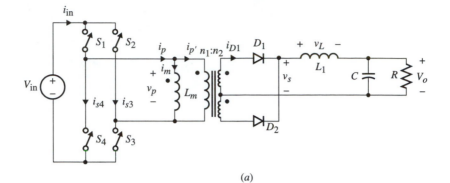

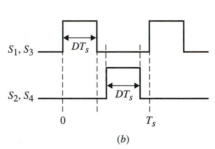

(a)

(b)

Figure P5.27

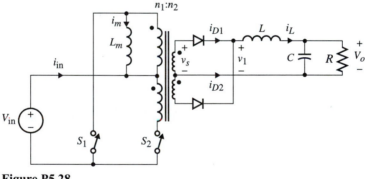

Figure P5.28

5.31 There are several core reset schemes available in today's dc-dc converters. Figure P5.31(a) and (b) shows two possible core-resetting circuits used in high-power applications. The winding n_r is the reset winding.

(a) Discuss the operation of each of the circuits and compare them to each other.

(b) Draw the waveforms for i_{in}, v_{sw}, i_o, v_o, and i_{D1}.

5.32 Derive the voltage gain equation for the isolated double-switch, single-ended dc-dc converter in Fig. P5.32. S_1 and S_2 are switched simultaneously.

5.33 Consider the current-fed push-pull converter given in Fig. P5.33. Derive the current gain expression I_o/I_{in}.

5.34 The converter shown in Fig. P5.34 is an isolated version of a converter given in Chapter 4. Assume the magnetizing inductance, L_m, is finite. Derive the expression for V_o/V_{in}.

5.35 Figure P5.35 shows the isolated version of the SEPIC converter, where a transformer is inserted in place of L_2. Assume the transformer has a magnetizing inductance, L_m, and a turn ratio of $n_1:n_2$.

(a) Sketch the waveforms for i_{L1}, i_{Lm}(magnetizing inductance), i_{c1}, i_D, v_p, v_s, and v_{sw}.

(b) Show that the voltage gain is given by

$$\frac{V_o}{V_{in}} = \frac{n_2}{n_1}\frac{D}{1-D}$$

where D is the duty cycle of the switch.

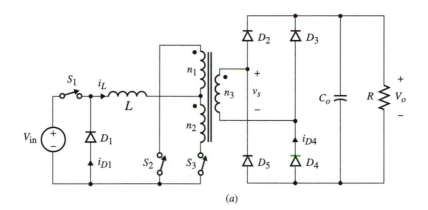

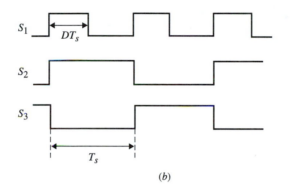

Figure P5.30

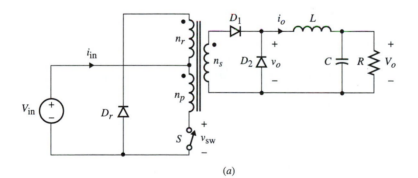

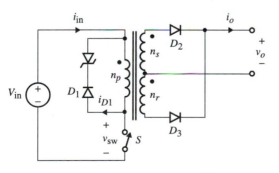

Figure P5.31

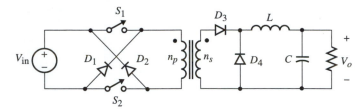

Figure P5.32

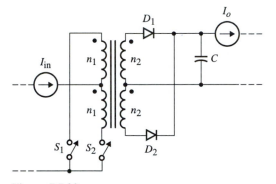

Figure P5.33

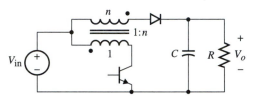

Figure P5.34

5.36 The converter in Fig. P5.36(a) and (b) is known as the isolated Watkins-Johnson converter. Assume S_1 and S_2 are switched on and off as shown in Fig. P5.36(c). Assume finite magnetizing inductance. Derive the expressions for V_o/V_{in} for each converter.

5.37 Draw the waveforms for the voltage and current variables in the push-pull converter shown in Fig. P5.37. Unlike Fig. 5.16(a), this converter is assumed to have finite magnetizing inductance.

D5.38 Consider a flyback converter that draws its input current from a dc voltage source varying between 185 V and 275 V and that produces an output voltage of +12 V at an average load current of 10 A.

(a) Determine the value of the transformer ratio and magnetizing inductance L_m.

(b) Determine the range of the duty cycle needed to maintain a constant output voltage. Assume $f = 85$ kHz and a maximum duty cycle of 0.6.

D5.39 Repeat the design in Problem 5.38 by assuming the power switch and diode have 3 V and 0.7 V voltage drops when conducting, respectively.

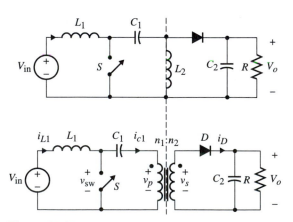

Figure P5.35

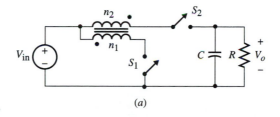

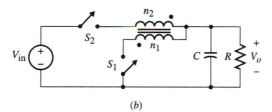

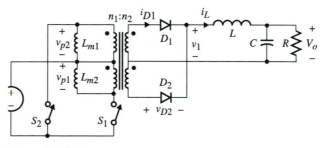

Figure P5.36

Figure P5.37

Chapter **6**

Soft-Switching dc-dc Converters

A new class of dc-to-dc converters, known in the literature as *soft-switching resonant converters,* has been thoroughly investigated in recent years. Soft switching means that one or more power switches in a dc-dc converter have either the turn-on or turn-off switching losses eliminated. This is in contrast to hard switching, where both turn-on and turn-off of the power switches are done at high current and high voltage levels. One approach is to create a full-resonance phenomenon within the converter through series or parallel combinations of resonant components. Such converters are generally known as resonant converters. Another approach is to use a conventional PWM converter—buck, boost, buck-boost, Cuk, SEPIC—and replace the switch with a resonant switch that accomplishes the loss elimination. Because of the nature of the PWM circuit, resonance occurs for a shorter time interval compared to the full-resonance case. This class of converters, combining resonance and PWM, is appropriately known as quasi-resonance converters. In this chapter, our focus will be on the latter method, mainly using the resonance PWM switch to achieve soft switching. For simplicity, here we use the term soft switching to refer to dc-dc converters, quasi-resonance converters, and other topologies that employ resonance to reduce switching losses. Two major techniques are employed to achieve soft switching: zero-current switching (ZCS) and zero-voltage switching (ZVS). This chapter will focus on ZCS and ZVS types of PWM resonant switches and their steady-state analyses.

6.1 TYPES OF DC-DC CONVERTERS

As shown in previous chapters, linear-mode and switch-mode converters have been used widely in the design of commercial dc-to-dc power supplies. Linear power supplies offer the designer four major advantages: simplicity in design, no electrical noise in the output, fast dynamic response time, and low cost. Their applications, however, are limited due to several disadvantages: (1) The input voltage is at least 2 or 3 V higher than the output voltage because the circuit can only be used as a step-down regulator, (2) each regulator is limited to only one output, and (3) efficiency is low compared to other switching regulators (30% to 60% for an output voltage less than 20 V).

High-frequency pulse-width-modulation (PWM) switching regulators overcome all the linear regulators' shortcomings: (1) They have higher efficiency (>90%); (2) power transistors operate at their most efficient points—cutoff and saturation—allowing for power densities of around 40 to 50 W/in^3; (3) multi-output applications are possible; and (4) the size and the cost are much lower, especially at high power levels.

However, PWM switching converters still have several limitations, among them (1) greater circuit complexity compared to the linear power supplies, (2) high electromagnetic interference (EMI), and (3) switching speeds below 100 kHz because of high stress levels on power semiconductor devices.

A third generation of power converters was introduced in the late 1980s. This group of topologies is known as soft-switching resonant converters. Compared with linear and switched-mode converters, the potential advantage of the soft-switching resonant converters is reduced power losses, thus achieving high switching frequency and high power density while maintaining high efficiency. Moreover, due to the higher switching frequency, such converters exhibit faster transient responses. Today's soft-switching techniques are used in the design of both high-frequency dc-to-dc conversion and high-frequency dc-to-ac inversion. Only the first application is discussed in this text.

The Resonance Concept

Like switched-mode dc-to-dc converters, resonant converters are used to convert dc to dc through an additional stage: the resonant stage, in which the dc signal is converted to a high-frequency ac signal. The advantages of the resonant converter include the natural commutation of power switches, resulting in low switching power dissipation and reduced component stresses, which in turn results in increased power efficiency and increased switching frequency; and higher operating frequencies, resulting in reduced size and weight of equipment and in faster response, and hence a possible reduction in EMI problems.

Since the size and weight of the magnetic components (inductors and transformers) and the capacitors in a converter are inversely proportional to the converter's switching frequency, many power converters have been designed for

Figure 6.1 Typical block diagram of soft-switching dc-to-dc converter.

progressively higher frequencies in order to reduce the size and weight and to obtain fast converter transients. In recent years, the market demand for wide applications that require variable-speed drives, highly regulated power supplies, and uninterruptible power supplies, as well as the desire for smaller size and lighter weight, has increased.

There are many soft-switching techniques available in the literature to improve the switching behavior of dc-to-dc resonant converters. At the time of this writing, intensive research in soft switching is under way to further improve the efficiency through increased switching frequency of power electronic circuits.

From a circuit standpoint, a dc-to-dc resonant converter can be described by three major circuit blocks as shown in Fig. 6.1: the dc-to-ac input inversion circuit, the resonant energy buffer tank circuit, and the ac-to-dc output rectifying circuit. Typically, the dc-to-ac inversion is achieved by using various types of switching network topologies. The resonant tank, which serves as an energy buffer between the input and the output, is normally synthesized by using a lossless frequency-selective network. The purpose of that network is to regulate the energy flow from the source to the load. Finally, the ac-to-dc conversion is achieved by incorporating rectifier circuits at the output section of the converter.

Resonant versus Conventional PWM

For many years, high-efficiency power-processing circuits have been achieved by operating power semiconductor devices in the switching mode, whereby switching devices are operated in either the on or off state, as in the PWM method. In PWM converters, the switching of semiconductor devices normally occurs at high current levels. Therefore, when switching at high frequencies these converters are associated with high power dissipation in their switching devices. Furthermore, PWM converters suffer from EMI caused by high-frequency harmonic components associated with their quasi-square switching current and/or voltage waveforms. Unfortunately, even though the technological advancements of PWM switch-mode converters has resulted in faster switching, their operating frequency is limited by the factors mentioned above.

In the resonant technique, switching losses in the semiconductor devices are avoided due to the fact that the current through or voltage across the switching device at the switching point is equal to or near zero. This reduction in switching losses allows the designer to attain a higher operating frequency without sacrificing the converter's efficiency. Compared to the PWM converters, the resonant converters show the promise of achieving the design of small-size and low-weight converters. Currently, resonant power converters operating in the range of a few megahertz are available. Another advantage of resonant converters over PWM converters is the decrease in the harmonic content in the converter voltage and current waveforms. Therefore, when the resonant and PWM converters are operated at the same power level and frequency, the resonant converter can be expected to have lower harmonic emission.

6.2   CLASSIFICATION OF SOFT-SWITCHING RESONANT CONVERTERS

The literature is very rich with resonant power electronic circuits used in applications such as dc-to-dc and dc-to-ac resonant converters. To date, there exists no general classification of resonant converter topologies. In fact, more and more resonant topologies continue to be introduced in the open literature. Several types of dc-to-dc converters employing additional resonant stages that have been explored thus far may be summarized as follows:

- Quasi-resonant converters (single-ended)
  - Zero-current switching (ZCS)
  - Zero-voltage switching (ZVS)
- Full-resonance converters (conventional)
  - Series resonant converter (SRC)
  - Parallel resonant converter (PRC)
- Quasi-squarewave (QSW) converters
  - Zero-current switching (ZCS)
  - Zero-voltage switching (ZVS)
- Zero-clamped topologies
  - Zero-clamped-voltage (CV)
  - Zero-clamped-current (CC)
- Class E resonant converters
- dc link resonant inverters
- Multi-resonant converters
  - Zero-current switching (ZCS)
  - Zero-voltage switching (ZVS)
- Zero Transition Topologies
  - Zero-voltage transition (ZVT).
  - Zero-current transition (ZCT).

Many other variations of soft-switching topologies that exist today are beyond the scope of this text. Since this text targets senior undergraduate electrical engineering students, we focus mainly on the quasi-resonant type of PWM converters.

6.3   ADVANTAGES AND DISADVANTAGES OF ZCS AND ZVS

The major advantage of ZCS and ZVS quasi-resonant converters is that the power switch is turned on and off at zero voltage and zero current, respectively. In ZCS topologies the rectifying diode has ZVS, whereas in ZVS topologies the rectifying diode has ZCS. A second advantage is that both ZVS and ZCS converters utilize transformer leakage inductors and diode junction capacitors and the output parasitic capacitor of the power switch.

The major disadvantage of the ZVS and ZCS techniques is that they require variable-frequency control to regulate the output. This is undesirable since it complicates the control circuit and generates unwanted EMI harmonics, especially under wide load variations. In ZCS, the power switch turns off at zero current, but at turn-on the converter still suffers from the capacitor turn-on loss caused by the output capacitor of the power switch.

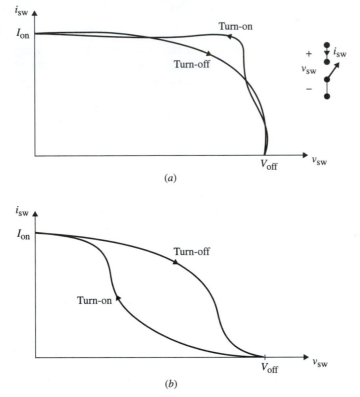

(a)

(b)

Figure 6.2 Switching loci. (a) Without snubber circuit. (b) With snubber circuit.

Switching Loci

Most regulator converter switches need to turn on or turn off the full load current at a high voltage, resulting in what is known as hard switching. Figure 6.2(a) and (b) shows typical switching loci for a hard-switching converter without and with a snubber circuit, respectively.

In a soft-switching converter topology, an LC resonant network is added to shape the switching device's voltage or current waveform into a quasi-sinewave in such a way that a zero voltage or a zero current condition is created. This technique eliminates the turn-on or turn-off loss associated with the charging or discharging of the energy stored in the MOSFET's parasitic junction capacitors. Figure 6.3(a) and (b) shows typical switching loci for the ZVS at turn-on and the ZCS at turn-off cases.

Switching Losses

As the frequency of operation increases, the switching losses also increase. As shown in Chapter 3, there are two types of switching losses:

1. At turn-off, the power transformer leakage inductance produces high di/dt, which results in a high voltage spike across it.
2. At turn-on, the switching loss is mainly caused by the dissipation of energy stored in the output parasitic capacitor of the power switch.

Figure 6.4(a) and (b) shows typical switching waveforms at turn-off and turn-on, respectively.

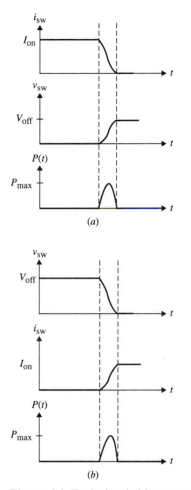

(a)

(b)

Figure 6.4 Typical switching current, voltage, and power loss waveforms at (a) turn-off and (b) turn-on.

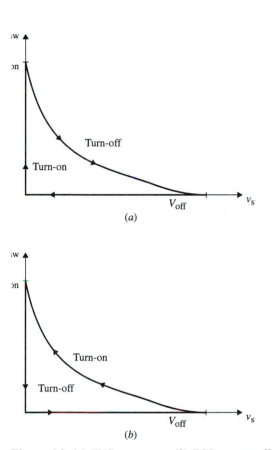

Figure 6.3 (a) ZVS at turn-on. (b) ZCS at turn-off.

6.4   ZERO-CURRENT SWITCHING TOPOLOGIES

6.4.1   The Resonant Switch

In this section, we will present one class of PWM converters that was introduced in the late 1980s based on the concept of using conventional PWM switching along with an LC tank circuit. Depending on the inductor-capacitor arrangements, there are two possible resonant switch configurations. The switch is either L-type or an M-type[1] and can be implemented as half-wave or full-wave, which corresponds to whether the switch current is unidirectional or bidirectional, respectively. Figure 6.5 shows the L-type switch in both the half- and full-wave implementations. The M-type switch is shown in Fig. 6.6.

The three conventional converter topologies–buck, boost, and buck-boost—will be analyzed here. In all of these topologies, the LC tank forms the resonant tank that causes ZCS to occur. The buck, boost, and buck-boost converters are shown in Fig. 6.7(a), (b), and (c), respectively.

---

[1]The notations *L-type* and *M-type* were used by the original authors.

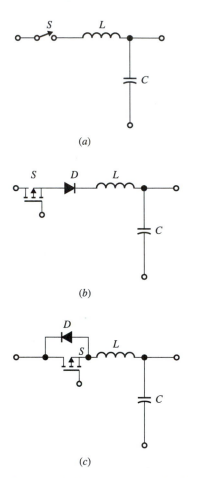

Figure 6.5 Resonant switch. (*a*) L-type switch. (*b*) Half-wave implementation. (*c*) Full-wave implementation.

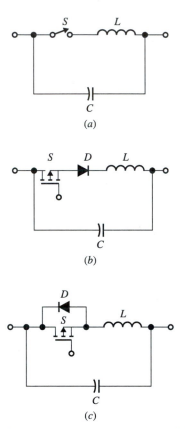

Figure 6.6 Resonant switch. (*a*) M-type switch. (*b*) Half-wave implementation. (*c*) Full-wave implementation.

The detailed steady-state analyses of these conventional converters was presented in Chapter 4. There exist two ccm modes of operation (as far as energy transfer is concerned): one mode during which energy is transferred from the source to the storage inductor, and a second mode during which energy is transferred from the storage inductor to the load. A low-pass *LC* filter is used in all three topologies to filter the fundamental frequency from the harmonics. Using the two types of switch arrangements, it is possible to convert the three topologies into quasi-resonant converters.

6.4.2 Steady-State Analysis of Quasi-Resonant Converters

To simplify the steady-state analysis of the converters, some assumptions need to be made:

1. The filtering components L_o, L_{in}, L_F, and C_o are very large compared to the resonant components L and C.
2. The output filter L_o-C_o-R is treated as a constant current source, I_o.
3. The output filter C_o-R is treated as a constant voltage source, V_o.
4. Switching devices and diodes are ideal.
5. Reactive circuit components are ideal.

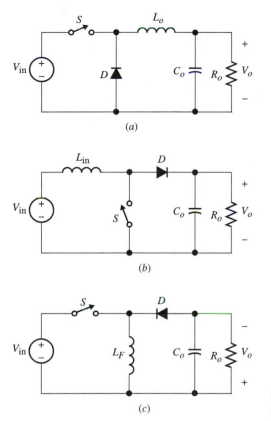

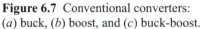

Figure 6.7 Conventional converters: (a) buck, (b) boost, and (c) buck-boost.

The Buck Resonant Converter

Replacing the switch in Fig. 6.7(a) by the resonant-type switch of Fig. 6.5(a), we obtain a quasi-resonant PWM buck converter, as shown in Fig. 6.8(a). Its simplified equivalent circuit is shown in Fig. 6.8(b). It can be shown that there are four modes of operation under the steady-state condition.

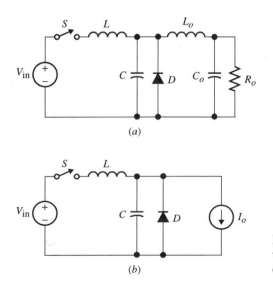

Figure 6.8 (a) Conventional buck converter with L-type resonant switch. (b) Simplified equivalent circuit.

Mode I $(0 \leq t < t_1)$ Mode I starts at $t = 0$, when S is turned on. Since the switch was off prior to $t = 0$, it is clear that the diode must have been on for $t < 0$ to carry the output inductor current. Hence, we assume for $t > 0$, both S and D are on. The output current is equal to the constant current source, I_o, as shown in Fig. 6.9(a). In this mode, the capacitor voltage, v_c, is zero and the input voltage is equal to the inductor voltage as given by

$$V_{in} = L\frac{di_L}{dt} \tag{6.1}$$

Integrating Eq. (6.1) from 0 to t, the inductor current, i_L, is given by

$$i_L(t) = \frac{V_{in}}{L}t \tag{6.2}$$

Equation (6.2) assumes a zero initial condition for i_L.

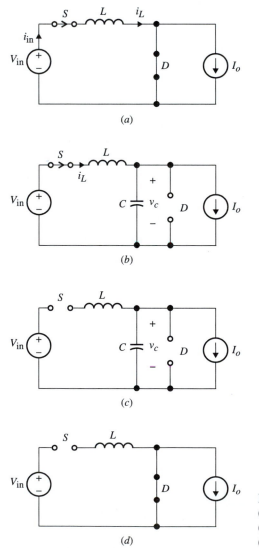

(a)

(b)

(c)

(d)

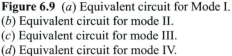

Figure 6.9 (*a*) Equivalent circuit for Mode I. (*b*) Equivalent circuit for mode II. (*c*) Equivalent circuit for mode III. (*d*) Equivalent circuit for mode IV.

As long as the inductor current is less than I_o, the diode will continue conducting and the capacitor voltage remains at zero. At time t_1, the inductor current becomes equal to I_o, the diode stops conducting, and the circuit enters mode II. Evaluating Eq. (6.2) at $t = t_1$, we have

$$I_o = \frac{V_{in}}{L}t_1 \qquad (6.3)$$

Hence, the time interval $\Delta t_1 = t_1$ is given by

$$\Delta t_1 = t_1 = \frac{LI_o}{V_{in}} \qquad (6.4)$$

This is the inductor current charging state.

Mode II $(t_1 \leq t < t_2)$ Mode II starts at t_1, when the diode is open-circuited as shown in Fig. 6.9(b), resulting in a resonant stage between L and C. During the time between t_1 and t_2, the switch remains on, but the diode is off. The initial capacitor voltage remains zero, but the initial inductor current changes to I_o.

The first-order differential equations that represent this mode are

$$C\frac{dv_c}{dt} = i_L - I_o \qquad (6.5a)$$

$$L\frac{di_L}{dt} = V_{in} - v_c \qquad (6.5b)$$

Next we express the inductor current in a second-order differential equation for $t \geq t_1$ as given by

$$\frac{d^2i_L}{dt^2} + \frac{1}{LC}i_L = \frac{I_o}{LC} \qquad (6.6)$$

From Eq. (6.6), the general solution for $i_L(t)$ is given by

$$i_L(t) = A_1 \sin \omega_o(t - t_1) + A_2 \cos \omega_o(t - t_1) + A_3 \qquad (6.7)$$

where the resonant angular frequency is defined by

$$\omega_o = \sqrt{\frac{1}{LC}}$$

Now we evaluate the constants A_1, A_2, and A_3. Recall at $t = t_1$, $i_L(t_1) = I_o$. Using this value, Eq. (6.7) becomes

$$A_2 + A_3 = I_o$$

The other relation in terms of the constants is obtained from the following equation:

$$L\frac{di_L(t)}{dt} = V_{in} - v_c(t)$$

Substituting for i_L from Eq. (6.7), and since the capacitor voltage is zero at t_1, we have

$$A_1 = \frac{V_{in}}{L\omega_o}$$

By taking the derivative of i_L again where $L(di_L^2(t_1)/dt^2) = 0$, we find A_2 to equal zero. Therefore, $A_3 = I_o$.

By solving the above equations, i_L and v_c are given by

$$i_L(t) = I_o + \frac{V_{in}}{Z_o}\sin\omega_o(t - t_1) \tag{6.8}$$

$$v_c(t) = V_{in}[1 - \cos\omega_o(t - t_1)] \tag{6.9}$$

where $Z_o = \sqrt{L/C}$ is known as the characteristic impedance.

This mode will last until $t = t_2$, when the transistor turns off because the inductor current reaches zero. The peak inductor current occurs at $t = t'_{max}$, where $\omega_o(t'_{max} - t_1) = \pi/2$. When $v_c(t) = V_{in}$, the peak inductor current is $I_o + V_{in}/Z_o$. Moreover, the peak capacitor voltage occurs at $t = t''_{max}$, where $\omega_o(t''_{max} - t_1) = \pi$. When $i_L(t) = I_o$, the peak capacitor voltage is $2\,V_{in}$.

The time interval in this mode can be derived at $t = t_2$ by setting $i_L(t_2) = 0$,

$$i_L(t_2) = I_o + \frac{V_{in}}{Z_o}\sin\omega_o(t_2 - t_1) \tag{6.10}$$

$$= 0$$

Therefore,

$$\Delta t_2 = t_2 - t_1 = \frac{1}{\omega_o}\sin^{-1}\frac{-Z_o I_o}{V_{in}} \tag{6.11}$$

It should be noted that the angle $\omega_o(t_2 - t_1)$ is in the third quadrant; hence, Eq. (6.11) may be written as

$$t_2 - t_1 = \frac{T_s}{2} + \frac{1}{\omega_o}\sin^{-1}\frac{Z_o I_o}{V_{in}} = \frac{1}{\omega_o}\left(\pi + \sin^{-1}\frac{Z_o I_o}{V_o}\right)$$

Mode III starts at $t = t_2$, when the switch is turned off.

Mode III ($t_2 \le t < t_3$)   At t_2 the inductor current becomes zero, and the capacitor linearly discharges from $v_c(t_2)$ to zero during t_2 to t_3. The diode remains off since its voltage is negative, as shown in Fig. 6.9(c).

Now the initial value of $i_L(t_2)$ is zero and the initial value of the capacitor at $t = t_2$ is $V_c(t_2)$. The inductor has no current going through it when the switch is off, so the capacitor current equals I_o, as given by

$$i_c = C\frac{dv_c}{dt} = -I_o \tag{6.12}$$

The capacitor voltage $v_c(t)$ is obtained by integrating Eq. (6.12) from t_2 to t with $V_c(t_2)$ as the initial value, resulting in

$$v_c(t) = \frac{-I_o}{C}(t - t_2) + V_c(t_2) \tag{6.13}$$

The initial value $V_c(t_2)$ is obtained from Eq. (6.9) for the previous mode, to yield

$$V_c(t_2) = V_{in}[1 - \cos\omega_o(t_2 - t_1)] \tag{6.14}$$

Substituting Eq. (6.14) into Eq. (6.13), we obtain

$$v_c(t) = \frac{-I_o}{C}(t - t_2) + V_{in}[1 - \cos\omega(t_2 - t_1)] \tag{6.15}$$

At $t = t_3$, the capacitor voltage becomes zero, and the equation for this time interval is given by

$$\Delta t_3 = t_3 - t_2 = \frac{C}{I_o}V_{in}[1 - \cos \omega_o(t_2 - t_1)] \tag{6.16}$$

At this point, the diode turns on and the circuit enters mode IV. In this mode, the capacitor voltage and inductor current remain zero until the switch is turned on again to repeat mode I.

Mode IV $(t_3 \le t < t_4)$ In this mode the switch remains off, but the diode starts conducting at $t = t_3$. Mode IV will continue as long as the switch is off, and the output current starts the free-wheeling stage through the diode. The inductor current and the capacitor voltage are zero when the switch closes.

$$i_L(t) = 0$$
$$v_c(t) = 0$$

Therefore, there will be no power transfer during this mode. By turning on the switch at $t = T_s$, the cycle will repeat these four modes. The dead time Δt_4 is given by

$$\Delta t_4 = T_s - \Delta t_1 - \Delta t_2 - \Delta t_3 \tag{6.17}$$

Figure 6.10 shows the steady-state waveforms for v_c and i_L for the buck converter with L-type switch.

Voltage Gain   In this section, a detailed derivation for the expression for the voltage gain, $M = V_o/V_{in}$, in terms of the circuit parameters will be given.

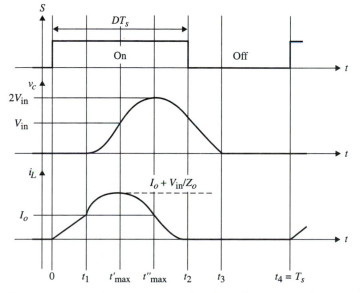

Figure 6.10 Steady-state current and voltage waveforms of buck L-type.

The average output voltage, V_o, can be obtained by evaluating the following integral.

$$V_o = \frac{1}{T_s} \int_0^{T_s} v_c(t)\, dt \tag{6.18}$$

Substitute for $v_c(t)$ from Eqs. (6.9) and (6.13) for the intervals $(t_2 - t_1)$ and $(t_3 - t_2)$, respectively, into Eq. (6.18) to yield

$$V_o = \frac{1}{T_s}\left[\int_{t_1}^{t_2} V_{in}(1 - \cos\omega_o(t - t_1))\, dt + \int_{t_2}^{t_3}\left(\frac{-I_o}{C}(t - t_2) + V_c(t_2) \right) dt \right]$$

The voltage gain ratio is given by

$$\frac{V_o}{V_{in}} = \frac{1}{T_s}\left[(t_2 - t_1) - \frac{\sin\omega_o(t_2 - t_1)}{\omega_o} - \frac{I_o}{V_{in}C}\frac{(t_3 - t_2)^2}{2} + V_c(t_2)(t_3 - t_2) \right] \tag{6.19}$$

Substitute for $(t_2 - t_1)$, $(t_3 - t_2)$, and $V_c(t_2)$ from Eqs. (6.11), (6.16), and (6.14), respectively, into Eq. (6.19) to yield a closed-form expression for M in terms of the circuit parameters. However, for illustration purposes, we will show that the same expression can be obtained from the law of power conservation. The conservation of energy per switching cycle states that since the converter is assumed to be ideal, then the average input and output powers should be equal. Next we evaluate the average input and output powers and then equate them since the converter is assumed to be ideal.

The total input energy over one switching cycle is given by

$$E_{in} = \int_0^{T_s} i_{in} V_{in}\, dt \tag{6.20}$$

Since i_{in} is equal to $i_L(t)$, Eq. (6.20) is rewritten as

$$E_{in} = \int_0^{t_1} i_L(t) V_{in}\, dt + \int_{t_1}^{t_2} i_L(t) V_{in}\, dt \tag{6.21}$$

Substituting for $i_L(t)$ from Eqs. (6.2) and (6.8) into the integrals, respectively, Eq. (6.21) becomes

$$E_{in} = V_{in}\left\{ \frac{V_{in}}{2L}t_1^2 + I_o(t_2 - t_1) + \frac{V_{in}}{Z_o\omega_o}[\cos\omega_o(t_2 - t_1) - 1] \right\} \tag{6.22}$$

Substituting for $\cos\omega_o(t_2 - t_1) = 1 - [I_o(t_3 - t_2)]/CV_{in}$ from Eq. (6.16) for mode III, Eq. (6.22) becomes

$$E_{in} = V_{in}\left\{ \frac{t_1}{2}I_o + I_o(t_2 - t_1) + \frac{V_{in}}{Z_o\omega_o}\left[\frac{I_o(t_3 - t_2)}{CV_{in}} \right] \right\} \tag{6.23}$$

With $Z_o\omega_o = 1/C$, Eq. (6.23) gives

$$E_{in} = V_{in}I_o\left[\frac{t_1}{2} + (t_2 - t_1) + (t_3 - t_2) \right] \tag{6.24}$$

The output energy over one cycle is obtained by evaluating Eq. (6.25):

$$E_o = \int_0^{T_s} I_o V_o\, dt = I_o V_o T_s \tag{6.25}$$

From the conservation of energy theory, equating the input and output energy expressions from Eqs. (6.24) and (6.25), we have

$$I_o V_o T_s = V_{in} I_o \left[\frac{t_1}{2} + (t_2 - t_1) + (t_3 - t_2) \right]$$

(6.26)

From Eq. (6.26) the voltage gain is expressed by

$$\frac{V_o}{V_{in}} = \frac{1}{T_s} \left[\frac{t_1}{2} + (t_2 - t_1) + (t_3 - t_2) \right]$$

(6.27)

Substituting for t_1, $(t_2 - t_1)$, and $(t_3 - t_2)$ from Eqs. (6.4), (6.11), and (6.16), respectively, into Eq. (6.27), the voltage gain becomes

$$\frac{V_o}{V_{in}} = \frac{1}{T_s} \left\{ \frac{L I_o}{2 V_{in}} + \frac{1}{\omega_o} \sin^{-1} \frac{-Z_o I_o}{V_{in}} + \frac{C V_{in}}{I_o} [1 - \cos \omega_o (t_2 - t_1)] \right\}$$

(6.28)

To simplify and generalize the gain equation, the following normalized parameters are defined:

$$M = \frac{V_o}{V_{in}} \quad \text{normalized output voltage (voltage gain)}$$

(6.29a)

$$Q = \frac{R_o}{Z_o} \quad \text{normalized load}$$

(6.29b)

$$I_o = \frac{V_o}{R_o} \quad \text{average output current}$$

(6.29c)

$$f_{ns} = \frac{f_s}{f_o} \quad \text{normalized switching frequency}$$

(6.29d)

By substituting Eqs. (6.29) into Eq. (6.28), the final voltage gain is simplified to

$$M = \frac{f_{ns}}{2\pi} \left[\frac{M}{2Q} + \alpha + \frac{Q}{M} (1 - \cos \alpha) \right]$$

(6.30)

where

$$\alpha = \sin^{-1} \left(\frac{-M}{Q} \right)$$

(6.31)

A plot of the control characteristic curve of M vs. f_{ns} under various normalized loads is given in Fig. 6.11.

EXAMPLE 6.1

Consider the following specifications for the ZCS buck converter of Fig. 6.8(a). Assume the parameters are $V_{in} = 25$ V, $V_o = 12$ V, $I_o = 1$ A, $f_s = 250$ kHz. Design for the resonant tank parameters L and C and calculate the peak inductor current and peak capacitor voltage. Determine the time interval for each mode.

SOLUTION   The voltage gain is $M = V_o / V_{in} = 12/25 = 0.48$. Let us select $f_{ns} = 0.4$. Next we determine Q from either the control characteristic curve of Fig. 6.11 or from the gain equation of

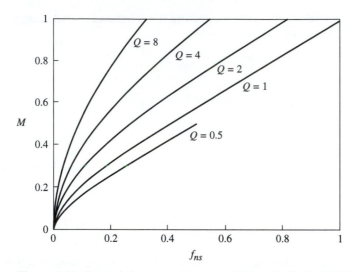

Figure 6.11 Control characteristic curve of M vs. f_{ns} for the ZCS buck converter.

Eq. (6.30). This results in Q approximately equal to 1. Since $R_o = V_o/I_o$, the characteristic impedance is given by

$$Z_o = \frac{R_o}{Q} = 12\ \Omega$$

$$= \sqrt{\frac{L}{C}} = 12\ \Omega \tag{6.32}$$

The second equation in terms of L and C is obtained from f_o. From the normalized switching frequency, f_o may be given by

$$f_o = \frac{f_s}{f_{ns}}$$

$$= \frac{f_s}{0.4} = 625\ \text{kHz}$$

In terms of the angular frequency, ω_o, we have

$$\omega_o = 2\pi f_o = \sqrt{\frac{1}{LC}} \tag{6.33}$$

Solving Eqs. (6.32) and (6.33) for L and C, we obtain

$$L = \frac{Z_o}{\omega_o} = \frac{12\ \Omega}{2\pi \times 625 \times 10^3\ \text{rad/s}}$$

$$= 3.06 \times 10^{-6} \approx 3\ \mu\text{H}$$

$$C = \frac{1}{Z_o\omega_o} = \frac{1}{12 \times 2 \times \pi \times 625 \times 10^3}$$

$$\approx 0.02\ \mu\text{F}$$

The peak inductor current is given by

$$I_{L,\text{peak}} = I_o + \frac{V_{in}}{Z_o}$$

$$\approx 3\ \text{A}$$

and the peak capacitor voltage is

$$v_{c,\text{peak}} = 2V_{\text{in}}$$

$$= 50 \text{ V}$$

The time intervals are calculated from the following expressions:

$$t_1 = \frac{I_o L}{V_{\text{in}}} = \frac{1 \text{ A} \times (3 \times 10^{-6} \text{ H})}{25 \text{ V}} \approx 0.122 \ \mu s$$

$$t_2 = t_1 + \frac{1}{\omega_o} \sin^{-1}\left(\frac{-Z_o I_o}{V_{\text{in}}}\right)$$

$$= 0.122 + \frac{1}{2\pi \times 625 \times 10^3} \sin^{-1}\left(\frac{-12 \times 1}{25}\right)$$

$$\approx 0.795 \ \mu s$$

$$t_3 = t_2 + \frac{CV_{\text{in}}(1 - \cos\omega_o(t_2 - t_1))}{I_o}$$

$$= 0.795 + \frac{(0.02 \times 25 \times 10^{-6})(1 - \cos(2\pi \times 0.625 \times 0.67))}{1 \text{ A}}$$

$$\approx 1.79 \ \mu s$$

For t'_{max} we have

$$\omega_o(t'_{\text{max}} - t_1) = \frac{\pi}{2}$$

$$t'_{\text{max}} = \frac{\pi/2}{\omega_o} + t_1$$

$$= 0.4 \ \mu s + 0.122 \ \mu s$$

$$= 0.522 \ \mu s$$

and finally, $t_4 = 4 \ \mu s = T_s$.

EXERCISE 6.1

Consider the ZCS buck converter with the following parameters: $V_{\text{in}} = 40$ V, $V_o = 28.7$ V @ $I_o = 0.6$ A, $f_s = 100$ kHz, $L = 15$ μH, and $C = 60$ nF. Determine the peak inductor current and the time at which the capacitor reaches its maximum voltage of 40 V.

ANSWER   3.1 A, 3.2 μs

EXERCISE 6.2

Figure 6.12(a) shows another type of quasi-resonant buck converter that uses an M-type resonant switch arrangement. Its simplified circuit is shown in Fig. 6.12(b). Derive the voltage gain equation, $M = V_o/V_{\text{in}}$, for the steady-state waveforms shown in Fig. 6.12(c).

The ZCS Boost Converter

Let us consider the boost quasi-resonant converter with an M-type switch as shown in Fig. 6.13(a), with its equivalent circuit shown in Fig. 6.13(b). Here we assume the input

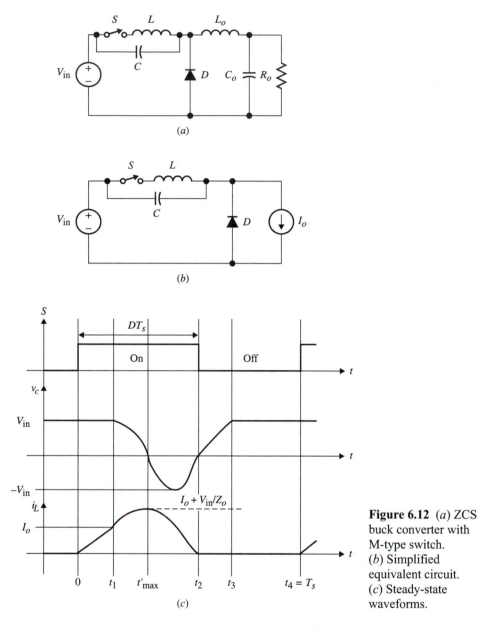

Figure 6.12 (*a*) ZCS buck converter with M-type switch. (*b*) Simplified equivalent circuit. (*c*) Steady-state waveforms.

current is constant and the load voltage is constant. The following analysis assumes half-wave operation.

Mode I $(0 \leq t < t_1)$ We first assume that the switch and the diode are both on in mode I, as shown in Fig. 6.14(*a*). The output voltage is given by

$$V_o = L\frac{di_L}{dt} \tag{6.34}$$

The initial inductor current and capacitor voltage are given by

$$i_L(0) = 0$$
$$v_c(0) = V_o$$

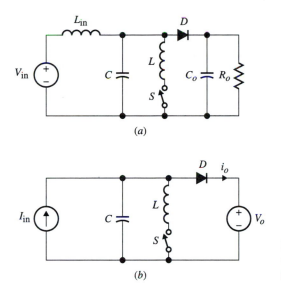

Figure 6.13 (*a*) ZCS boost converter with M-type switch. (*b*) Simplified equivalent circuit.

Integrating Eq. (6.34), the inductor current becomes

$$i_L(t) = \frac{V_o}{L}t + i_L(0) = \frac{V_o}{L}t \tag{6.35}$$

When the resonant inductor current reaches the input current, I_{in}, the diode turns off; hence, we have

$$\frac{V_o}{L}t_1 = I_{in}$$

with t_1 given by

$$t_1 = \frac{I_{in}L}{V_o} \tag{6.36}$$

At $t = t_1$, the diode turns off since $i_L = I_{in}$, and the converter enters mode II.

Mode II ($t_1 \leq t < t_2$)   The switch remains closed, but the diode is off at t_1 in mode II as shown in Fig. 6.14(*b*). This is a resonant mode during which the capacitor voltage starts decreasing resonantly from its initial value of V_o. When $i_L = I_{in}$, the capacitor reaches its negative peak. At $t = t_2$, i_L equals zero, and the switch turns off (i.e., switching at zero current).

The initial conditions are given by

$$v_c(t_1) = V_o \quad \text{and} \quad i_L(t_1) = I_o$$

From Fig. 6.14(*b*), the first derivatives for i_L and v_c are given by

$$L\frac{di_L}{dt} = v_c(t)$$

$$C\frac{dv_c}{dt} = I_{in} - i_L(t)$$

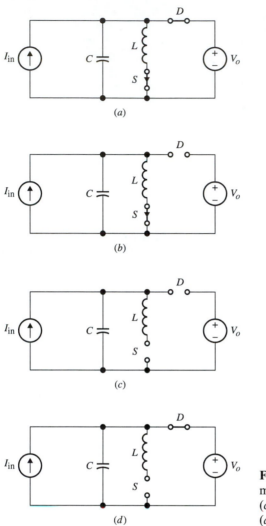

Figure 6.14 (*a*) Equivalent circuit for mode I. (*b*) Equivalent circuit for mode II. (*c*) Equivalent circuit for mode III. (*d*) Equivalent circuit for mode IV.

Next the two first-order differential equations need to be solved for this mode. Using the same solution technique used with the buck converter, the expressions for $i_L(t)$ and $v_c(t)$ are given by

$$i_L(t) = I_{in} + \frac{V_o}{Z_o} \sin \omega_o (t - t_1) \tag{6.37}$$

$$v_c(t) = V_o \cos \omega_o (t - t_1) \tag{6.38}$$

where $\omega_o = 1/\sqrt{LC}$.

At $t = t_2$, $i_L(t_2) = 0$, and the time interval can be obtained from evaluating Eq. (6.37) at $t = t_2$, to yield

$$(t_2 - t_1) = \frac{1}{\omega_o} \sin^{-1}\left(-\frac{I_{in}Z_o}{V_o}\right)$$

$$= \frac{1}{\omega_o}\left[\pi + \sin^{-1}\left(\frac{I_{in}Z_o}{V_o}\right)\right] \tag{6.39}$$

Mode III $(t_2 \leq t < t_3)$ Mode III starts at t_2, and the switch and the diode are both open, as shown in Fig. 6.14(c). Since v_c is constant, the capacitor starts charging up by the input current source. The capacitor voltage is given by

$$v_c(t) = \frac{1}{C} \int_{t_2}^{t} I_{in} \, dt$$

$$= \frac{I_{in}}{C}(t - t_2) + v_c(t_2)$$

(6.40)

The diode begins conducting at $t = t_3$ when the capacitor voltage is equal to the output voltage, i.e., $v_c(t_3) = V_o$. Equation (6.40) becomes

$$v_c(t_3) = V_o = \frac{I_{in}}{C}(t_3 - t_2) + v_c(t_2)$$

so the time interval in this period can be expressed as

$$t_3 - t_2 = \frac{I_{in}}{C}[V_o - v_c(t_2)]$$

(6.41)

where $v_c(t_2)$ may be obtained from Eq. (6.38).

Mode IV $(t_3 \leq t < t_4)$ At t_3 the capacitor voltage is clamped to the output voltage and the diode starts conducting again, but the switch remains open. This condition remains as long as the switch is open. The cycle of the mode will repeat at the time of T_s, when S is turned on again.

Typical steady-state waveforms are shown in Fig. 6.15.

Voltage Gain   As we did for the buck converter, we apply conservation of energy per switching cycle to express the voltage gain, $M = V_o/V_{in}$, in terms of the circuit parameter.

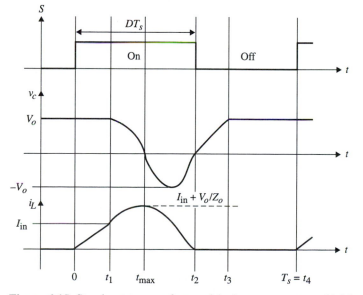

Figure 6.15 Steady-state waveforms of the boost converter with M-type switch.

The input energy is given by

$$E_{in} = V_{in} I_{in} T_s \tag{6.42}$$

and the output energy by

$$E_o = \int_0^{T_s} i_o V_o \, dt \tag{6.43}$$

The output current equals $i_o = I_{in} - i_L$ and $i_o = I_{in}$ for intervals $0 \le t < t_1$ and $t_3 \le t < T_s$, respectively. Therefore, E_o becomes

$$E_o = \int_0^{t_1} (I_{in} - i_L) V_o \, dt + \int_{t_3}^{T_s} I_{in} V_o \, dt \tag{6.44}$$

The input current is obtained from the conservation of output power as

$$I_{in} = \frac{V_o^2}{V_{in} R_o}$$

Substituting for the input current and evaluating Eq. (6.44), the output energy becomes

$$
\begin{aligned}
E_o &= V_o \int_0^{t_1} \left(I_{in} - \frac{V_o}{L} t \right) dt + I_{in} V_o (T_s - t_3) \\
&= V_o \left(I_{in} t_1 - \frac{1}{2} \frac{V_o}{L} t_1^2 \right) + I_{in} V_o (T_s - t_3)
\end{aligned}
\tag{6.45}
$$

If we substitute for $t_1 = I_{in} L / V_o$ and $(T_s - t_3) = T_s - [t_1 + (t_2 - t_1) + (t_3 - t_2)]$, and use the equations for $(t_2 - t_1)$ and $(t_3 - t_2)$ from Eqs. (6.39) and (6.41), Eq. (6.45) becomes

$$E_o = -\frac{1}{2} I_{in}^2 L + V_o I_{in} \left[T_s - \frac{\alpha}{\omega_o} - \frac{C}{I_{in}} V_o (1 - \cos \alpha) \right] \tag{6.46}$$

Following similar steps as for the quasi-resonant buck converter, it can be shown that the voltage gain expression is given by

$$\frac{M-1}{M} = \frac{f_{ns}}{2\pi} \left[\frac{M}{2Q} + \alpha + \frac{Q}{M} (1 - \cos \alpha) \right] \tag{6.47}$$

where, α, M, I_o, and f_{ns} are given in Eqs. (6.29).

Using Eq. (6.47), Fig. 6.16 shows the characteristic curve for M vs. f_{ns} as a function of the normalized load.

EXAMPLE 6.2

Design a boost ZCS converter for the following parameters: $V_{in} = 20$ V, $V_o = 40$ V, $P_o = 20$ W, $f_s = 250$ kHz.

SOLUTION   The voltage gain is $M = V_o / V_{in} = 40/20 = 2$. Let us select $f_{ns} = 0.38$. From the characteristic curve of Fig. 6.16, Q can be approximated as 6.0.

The characteristic impedance is given by

$$Z_o = \frac{R_o}{Q} = \frac{V_o^2 / P_o}{Q} = \frac{80\ \Omega}{6} = 13.33\ \Omega \tag{6.48}$$

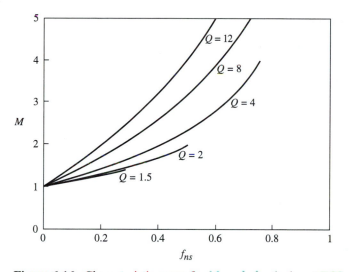

Figure 6.16   Characteristic curve for M vs. f_{ns} for the boost ZCS converter.

and the resonant frequency is

$$f_o = \frac{f_s}{f_{ns}}$$

$$= \frac{250 \text{ kHz}}{0.38} = 657.89 \text{ kHz} \tag{6.49}$$

Solve Eqs. (6.48) and (6.49) for L and C:

$$L = \frac{Z_o}{2\pi f_o} = \frac{13.33 \; \Omega}{2\pi \times 657.89 \times 10^3} = 3.22 \times 10^{-6} \text{ H}$$

$$C = \frac{1}{Z_o \omega_o} = \frac{1}{(13.33)(2\pi \times 657.89 \times 10^3)} = 18.14 \text{ nF}$$

To limit the input ripple current and the output voltage, we set

$$L_o = 100L = 322 \times 10^{-6} \text{ H}$$

$$C_o = 100C = 1.8 \times 10^{-6} \text{ F}$$

EXAMPLE 6.3

Design a boost converter with ZCS, with the following design parameters: $V_{in} = 25$ V, $P_o = 30$ W at $I_o = 0.5$ A, and $f_s = 100$ kHz. Assume the output voltage ripple $\Delta V_o / V_o$ is 0.2%.

SOLUTION   The load resistance, $R_o = P_o / I_o^2 = 30/(0.5)^2 = 120 \; \Omega$.

$$M = \frac{V_o}{V_{in}} = \frac{60}{25} = 2.4$$

From the characteristic curve of Fig. 6.15, we approximate Q at 6 when we assume $f_n = 0.58$. Hence, $f_o = 100$ kHz/0.58 = 172.4 kHz.

The characteristic impedance is obtained from

$$Q = \frac{R_o}{Z_o} = \frac{120}{Z_o} = 6 \quad \text{and} \quad Z_o = 20 \ \Omega$$

Hence, $\sqrt{L/C} = 20 \ \Omega$.

$$\omega_o = 2\pi(172 \times 10^3) = 1080.7 \times 10^3 \text{ rad/s}$$

$$\sqrt{\frac{1}{LC}} = 1080.7 \times 10^3 \text{ rad/s}$$

Solving for C and L,

$$C = 46.27 \text{ nF}$$

$$L = 18.51 \ \mu\text{H}$$

From $\Delta V_o/V_o = 0.2\%$, C_o can be obtained from the ripple voltage equation for the conventional boost converter, which is given by

$$\frac{\Delta V_o}{V_o} = \frac{D}{f_s R_o C_o}$$

Solving the above equation for C_o we obtain

$$C_o = \frac{D}{f_s R_o (\Delta V_o/V_o)}$$

$$= \frac{((10 - 6.147 - 0.193)/10) \times 10^{-6}}{(100 \times 10^3)(120)(0.2/100)} = 15.25 \ \mu\text{F}$$

The time intervals are given by

$$t_1 = \frac{L I_{\text{in}}}{V_o} = \frac{18.51 \times 10^{-3} \times 1.2}{60} = 0.370 \ \mu\text{s}$$

$$t_2 - t_1 = \frac{1}{\omega_o} \sin^{-1}\left[\frac{-Z_o I_{\text{in}}}{V_o}\right]$$

$$= \frac{1}{1080.7 \times 10^3} \sin^{-1}\left[-\frac{20 \times 1.2}{60}\right] = 3.29 \ \mu\text{s}$$

$$t_3 - t_2 = \frac{1}{\omega_n} \frac{V_o}{Z_o I_{\text{in}}}(1 - \cos\alpha)$$

$$= \frac{1}{1080.7 \times 10^3} \frac{60}{20 \times 1.2}(1 - \cos 3.553)$$

$$= 0.193 \ \mu\text{s}$$

$$t_4 - t_3 = T - t_1 - (t_2 - t_1) - (t_3 - t_2)$$

$$= 10 - 0.370 - 3.29 - 0.193 = 6.147 \ \mu\text{s}$$

The output inductor is obtained from

$$L_{\text{crit}} = \frac{R_o}{2f_s}(1 - D^2)D$$

$$= 88.8 \ \mu\text{H}$$

Hence, we select $L_o = 890 \ \mu\text{H}$.

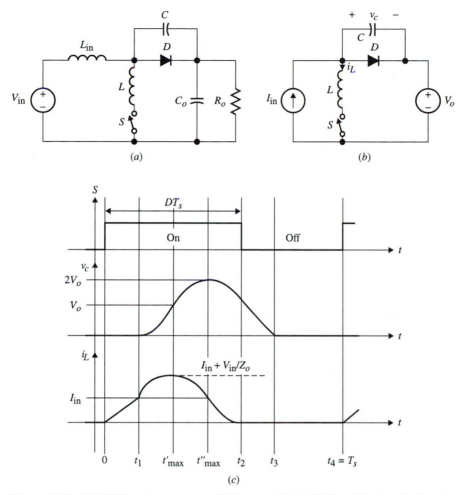

Figure 6.17 (*a*) ZCS boost converter with L-type switch. (*b*) Simplified equivalent circuit. (*c*) Steady-state waveforms.

Figure 6.17(*a*) shows the quasi-resonant boost converter using the L-type resonant switch, and the simplified circuit and steady-state waveforms are shown in Fig. 6.17(*b*) and (*c*), respectively. The reader is invited to verify these waveforms.

EXERCISE 6.3

Consider the quasi-resonant boost converter of Fig. 6.13(*a*) with the following design parameters: $V_{in} = 25$ V, $P_o = 30$ W at $I_o = 0.5$ A, and $f_s = 100$ kHz. Determine the resonant tank capacitor and inductor components, assuming $f_{ns} = 0.4$.

ANSWER $C = 42.4 \ \mu\text{F}, L = 9.55 \ \mu\text{H}$

ZCS Buck-Boost Converter

Let us consider the quasi-resonant buck-boost converter using the L-type switch as shown in Fig. 6.18(*a*); Fig. 6.18(*b*) shows the simplified equivalent circuit.

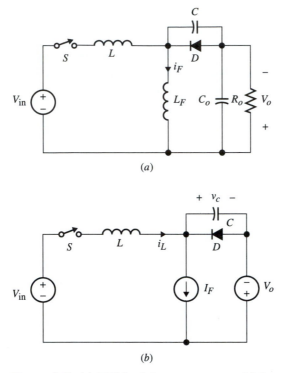

(a)

(b)

Figure 6.18 (a) ZCS buck-boost converter with L-type switch. (b) Simplified equivalent circuit.

Like the previous buck and the boost converters, the buck-boost converter has four modes of operation.

Mode I ($0 \leq t < t_1$)   Mode I starts at $t = 0$, when the switch and the diode are both conducting. According to Kirchhoff's law, the voltage equation can be written as

$$L\frac{di_L(t)}{dt} = V_{in} + V_o \tag{6.50}$$

By integrating both sides of Eq. (6.50) with the initial condition of $i_L(0) = 0$, $i_L(t)$ is given by

$$i_L(t) = \frac{V_{in} + V_o}{L}t \tag{6.51}$$

and $v_c(t) = 0$.

At $t = t_1$, the inductor current reaches I_F, forcing the output diode to stop conducting, so t_1 can be expressed as

$$t_1 = \frac{LI_F}{V_{in} + V_o} \tag{6.52}$$

Mode II $(t_1 \leq t < t_2)$ This is a resonant stage between L and C with the initial conditions given by

$$v_c(t_1) = 0$$

$$i_L(t_1) = I_F$$

Applying Kirchhoff's law in Fig. 6.19(b), the inductor current and capacitor voltage equations may be given as

$$\frac{di_L}{dt} = V_{in} + V_o - v_c(t) \tag{6.53a}$$

$$C\frac{dv_c}{dt} = i_L(t) - I_F \tag{6.53b}$$

Solving Eqs. (6.53) for $t > t_1$, we obtain

$$i_L(t) = I_F + \frac{V_{in} + V_o}{Z_o}\sin \omega_o(t - t_1) \tag{6.54}$$

$$v_c(t) = (V_{in} + V_o)[1 - \cos \omega_o(t - t_1)] \tag{6.55}$$

At $t = t_2$, the inductor current reaches zero, $i_L(t_2) = 0$, and the switch stops conducting. The time interval $(t_2 - t_1)$ is given by

$$(t_2 - t_1) = \frac{1}{\omega_o}\sin^{-1}\left(-\frac{I_F Z_o}{V_{in} + V_o}\right) \tag{6.56}$$

Mode III $(t_2 \leq t < t_3)$ Mode III starts at $t = t_2$, when the inductor current reaches zero. The switch and the diode are both off. The capacitor starts to discharge until it reaches zero, and the diode will start to conduct again at $t = t_3$. During this period, the inductor current is zero.

$$v_c(t) = \frac{-1}{C}\int_{t_2}^{t} I_F dt = \frac{-I_F}{C}(t - t_2) + V_c(t_2) \tag{6.57}$$

The diode begins to conduct at the end of this mode, $t = t_3$, because the capacitor voltage is equal to zero, yielding

$$0 = \frac{-I_F}{C}(t_3 - t_2) + V_c(t_2)$$

where $V(t_2)$ may be obtained from Eq. (6.55) by evaluating it at $t = t_2$. The expression from Eq. (6.57) for the time between t_2 and t_3 is

$$(t_3 - t_2) = \frac{C}{I_F}V_c(t_2) \tag{6.58}$$

Mode IV $(t_3 \leq t < t_4)$ Between t_3 and t_4, the switch remains off, but the diode is on. At the end of the cycle, the switch is closed again when the current becomes zero. The cycle of the modes will repeat at T_s.

The steady-state waveforms shown in Fig. 6.20 are the characteristic waveforms for the switch, v_c, and i_L.

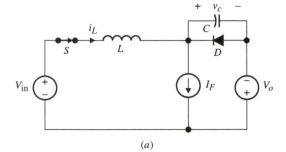

(a)

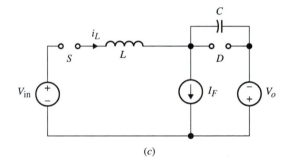

(b)

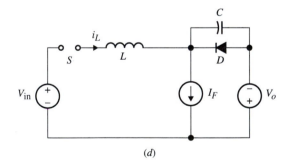

(c)

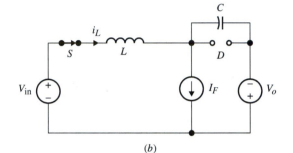

(d)

Figure 6.19 (*a*) Equivalent circuit for mode I. (*b*) Equivalent circuit for mode II. (*c*) Equivalent circuit for mode III. (*d*) Equivalent circuit for mode IV.

Voltage Gain   As before, the conservation of energy per switching cycle is used to express the voltage gain, $M = V_o/V_{in}$, in terms of the normalized circuit parameter. It can be shown that M for the buck-boost ZCS converter is given by

$$\frac{M}{1+M} = \frac{f_{ns}}{2\pi}\left[\frac{M}{2Q} + \alpha + \frac{Q}{M}(1-\cos\alpha)\right] \tag{6.59}$$

All the parameters of Eq. (6.59) are the same as before.

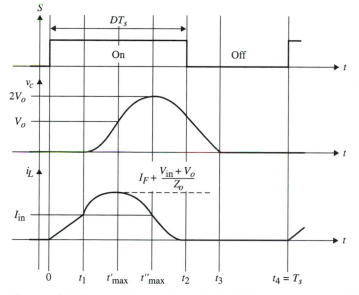

Figure 6.20 Steady-state waveforms for buck-boost converter with L-type switch.

Figure 6.21 shows the characteristic curve for M vs. f_{ns} for the ZCS buck-boost converter.

EXAMPLE 6.4

Consider a buck-boost quasi-resonant ZCS converter with the following specifications: $V_{in} = 40$ V, $P_o = 80$ W at $I_o = 4$ A, $f_s = 250$ kHz, $L_o = 0.1$ mH, and $C_o = 6$ μF. Design values for L and C, and determine the output ripple voltage, assuming $D = 0.5$.

SOLUTION The output voltage and load resistance are given by

$$V_o = \frac{80}{4} = 20 \text{ V} \quad \text{and} \quad R_o = \frac{20}{4} = 5 \text{ }\Omega$$

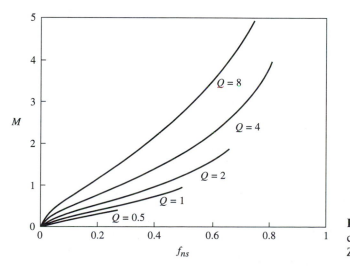

Figure 6.21 Characteristic curve for M vs. f_{ns} for the ZCS buck-boost converter.

The voltage gain is given by

$$\frac{V_o}{V_{in}} = \frac{20}{40} = 0.5$$

With $M = 0.5$ and $f_{ns} = 0.17$, we have $Q = 3$, resulting in $f_o = 250 \text{ kHz}/0.17 = 1470.6 \text{ kHz}$.

From Q, and Z_o, we have

$$Q = \frac{R_o}{Z_o} = \frac{5}{Z_o} \quad \text{and} \quad Z_o = \frac{3}{5} = 0.6 \ \Omega$$

Hence,

$$\sqrt{L/C} = 0.6 \ \Omega$$

and

$$\sqrt{1/LC} = 2\pi f_o = 2\pi \times 1470.6 \times 10^3 \text{ rad/s}$$

$$\frac{1}{C} = 0.6 \times 2\pi \times 1470.6 \times 10^3$$

From these equations C and L are given by

$$C = 180.4 \text{ nF}$$

$$L = 3^2 \times C = 1.6 \ \mu\text{H}$$

The duty cycle D is approximately 33% since the voltage gain for the buck-boost is 0.5. Hence, the voltage ripple is

$$\frac{\Delta V_o}{V_o} = \frac{D}{R_o C_o f} = \frac{0.33}{5 \times 6 \times 10^{-6} \times 250 \times 10^3} = 4.4\%$$

EXERCISE 6.4

The buck-boost converter with an M-type switch is shown in Fig. 6.22(*a*), and Fig. 6.22(*b*) is the equivalent circuit. Derive the steady-state waveforms and show that the voltage gain, $M = V_o/V_{in}$, is given by

$$\frac{M}{1+M} = \frac{f_{ns}}{2\pi}\left[\frac{M}{2Q} + \alpha + \frac{Q}{M}(1 - \cos\alpha)\right] \tag{6.60}$$

6.5   ZERO-VOLTAGE SWITCHING TOPOLOGIES

In this section, we will investigate the zero-voltage switching (ZVS) quasi-resonant converter family. Like the ZCS topologies, M-type or L-type switch arrangements can be used. In those topologies, the power switch is turned on at zero voltage (of course, turn-off also occurs at zero voltage). While the switch is off, a peak voltage will appear across it, causing the stress to be higher than in the hard-switching PWM case. In a ZVS topology, a flyback diode across the switch (body diode) is used to damp the voltage across the capacitor, which results in a zero voltage across the switch. We should point out that the capacitor across the switch can be the same as the switch's parasitic capacitor, and the flyback diode could be the same as the internal body diode of the power semiconductor switch.

Figure 6.23(*a*) shows a MOSFET switch implementation including an internal body diode and a parasitic capacitor.

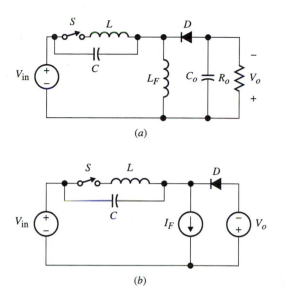

(a)

(b)

Figure 6.22 (a) ZCS buck-boost converter with an M-type switch. (b) Simplified equivalent circuit.

We will assume C_{gd} and C_{gs} are too small to be included. If the body diode, D_s, is not fast enough for the designed application or has limited power capabilities, it is possible to block it and use an external, fast flyback diode as shown in Fig. 6.23(b). D_1 is used to block D_s, and D_2 is the diode actually used to carry the reverse switch current. Both the current- and voltage-mode control methods are used in conjunction with a direct duty ratio PWM control approach to vary the on or off time of the power switch.

Over the last 15 years, many different zero-voltage resonant converter topologies have been introduced. Only the quasi-resonant soft-switching ZVS topologies will be analyzed and their control characteristic curves studied here. Regardless of their topological variations, many of these converters have several common features.

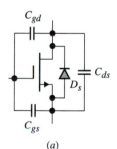

(a)

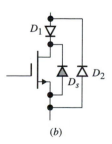

(b)

Figure 6.23 (a) MOSFET implementation. (b) MOSFET switch with fast flyback diode.

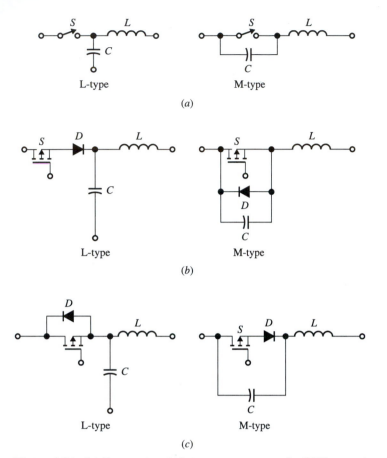

Figure 6.24   (*a*) Resonant switch arrangement types for ZVS operation. (*b*) Half-wave MOSFET implementation. (*c*) Full-wave MOSFET implementation.

6.5.1   Resonant Switch Arrangements

Next we investigate the buck, boost, and buck-boost ZVS topologies using L-type and M-type resonant switches. Figure 6.24(*a*) shows the two possible implementations using L- and M-type resonant switches. The half-wave L-type and M-type MOSFET implementations are shown in Fig 6.24(*b*), and Fig 6.24(*c*) shows the full-wave implementations for L- and M-type switches.

6.5.2   Steady-State Analyses of Quasi-Resonant Converters

As in the ZCS case, to simplify the steady-state analysis, we make the same assumptions made in Section 6.4.2.

The Buck Converter

Replacing the switch in Fig. 6.7(*a*) with the M-type switch of Fig. 6.24(*a*), we obtain the ZVS buck converter shown in Fig. 6.25(*a*). The simplified equivalent circuit is given in Fig. 6.25(*b*).

As the switch and diode are both on or off at the same time, it can be shown that under steady-state conditions there are four modes of operation. Unlike ZCS topolo-

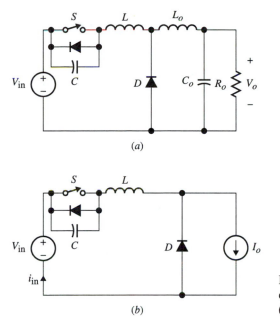

(a)

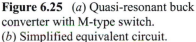

(b)

Figure 6.25   (a) Quasi-resonant buck converter with M-type switch.
(b) Simplified equivalent circuit.

gies, the switching cycle in ZVS starts with the main switch in the nonconduction state. This is in order to establish a zero-voltage condition across the switch during the resonant stage while it is open. Figure 6.26 shows the equivalent circuit modes under the steady-state condition.

Mode I ($0 \leq t < t_1$)   Assume that initially the power switch is conducting and the diode is off. Mode I starts at $t = 0$, when the switch is turned off. In this mode, since S has been closed for $t < 0$, the initial capacitor voltage, v_c, is zero, and the inductor current is I_o as shown in Fig. 6.26(a).

$$v_c(0) = 0$$

$$i_L(0) = I_o$$

Applying KCL to Fig. 6.26(a), we have

$$C \frac{dv_c}{dt} = i_L(t)$$

Since $i_L(t) = I_o$, the capacitor starts to charge according to

$$v_c(t) = \frac{1}{C} I_o t \qquad (6.61)$$

The voltage across the output diode is given by

$$v_D(t) = V_{in} - v_c(t)$$

As long as $v_c < V_{in}$, the diode remains off.

The capacitor voltage reaches the input voltage, V_{in}, at $t = t_1$, causing the diode to turn on. Hence, at $t = t_1$, we have

$$V_c(t_1) = V_{in}$$

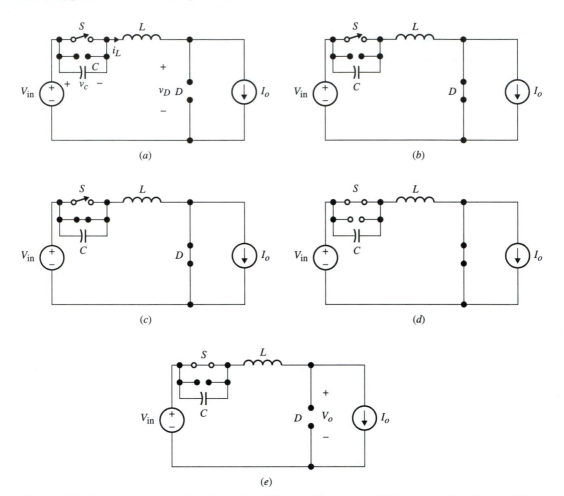

Figure 6.26 Equivalent circuits for (a) mode I, (b) mode II, (c) mode III $(t_2 \leq t < t_2')$, (d) mode III $(t_2' \leq t < t_3)$, and (e) mode IV.

and t_1 can be expressed as

$$t_1 = \frac{CV_{in}}{I_o} \tag{6.62}$$

At $t = t_1$, the circuit enters mode II. The current and voltage waveforms are shown in Fig. 6.27.

Mode II $(t_1 \leq t < t_2)$ Mode II starts at t_1, when the diode turns on, and the circuit enters the resonant stage. During the time between t_1 and t_2, the switch remains off. At $t = t_2$, the capacitor voltage tends to go negative, forcing the diode across S to turn on. The initial capacitor voltage and inductor current in this mode are given by

$$V_c(t_1) = V_{in}$$
$$I_L(t_1) = I_o$$

The expressions for the current and the voltage in the time domain are given in Eqs. (6.63) and (6.64), respectively:

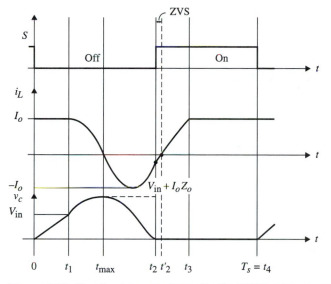

Figure 6.27 Steady-state waveforms for the ZVS buck converter.

$$i_L(t) = I_o \cos \omega_o (t - t_1) \qquad (6.63)$$

$$v_c(t) = V_{in} + I_o Z_o \sin \omega_o (t - t_1) \qquad (6.64)$$

where the resonant frequency and the characteristic impedance are defined as before.

The inductor current is zero when the capacitor voltage reaches the peak, and the capacitor starts discharging while the inductor current is a negative value. The inductor current reaches the peak when the capacitor drops to the input voltage, and at the end of the mode (i.e., at $t = t_2$), the capacitor voltage equals zero, $v_c(t_2) = 0$.

The period between t_2 and t_1 is given by

$$t_2 - t_1 = \frac{1}{\omega_o}\left[\sin^{-1}\left(\frac{+V_{in}}{I_o Z_o}\right) + \pi\right] \qquad (6.65)$$

$$= \frac{\alpha}{\omega_o}$$

and the inductor current at $t = t_2$ is

$$I_L(t_2) = I_o \cos \alpha \qquad (6.66)$$

where

$$\alpha = \omega_o(t_2 - t_1)$$

Mode III ($t_2 \le t < t_3$) At t_2 the capacitor voltage becomes zero, and the inductor current starts to charge linearly and reaches the output current at $t = t_3$. The body diode of the switch turns on at $t = t_2$ to maintain inductor current continuity, and the output diode also remains on at this point, as shown in Fig. 6.26(c). As long as the inductor current is less than I_o, the output diode will stay on. The switch may be turned on at 2 V any time after t_2 and before t'_2, when the inductor current reverses polarity.

The initial value of the capacitor voltage in mode III is zero:

$$V_c(t_2) = 0$$

The inductor voltage is equal to the input voltage,

$$L\frac{di_L}{dt} = V_{in} \tag{6.67}$$

By integrating Eq. (6.67) from t_2 to t, the inductor current can be expressed as

$$i_L(t) = \frac{V_{in}}{L}(t - t_2) + I_o \cos\alpha \tag{6.68}$$

At $t = t_3$, the inductor current reaches the output current, $i_L(t_3) = I_o$, forcing the diode to turn off. The time interval from t_3 to t_2 is

$$t_3 - t_2 = \frac{I_o L}{V_{in}}(1 - \cos\alpha) \tag{6.69}$$

Therefore, there will be no power transfer and no charge or discharge interval when the switch turns on again in the next mode.

Mode IV $(t_3 \le t < t_4)$ In mode IV, as shown in Fig. 6.26(e), the inductor current is trapped and held constant at $i_L = I_o$, with $v_c = 0$:

$$i_L = I_o$$

$$v_c = 0$$

Mode IV will continue as long as the switch is on. By turning off the switch at $t = t_4 = T_s$, the switching cycle repeats. The dead time $t_4 - t_3$ is given by

$$t_4 - t_3 = T_s - t_1 - (t_2 - t_1) - (t_3 - t_2) \tag{6.70}$$

Voltage Gain   We follow the same approach as in the ZCS case by using the energy balance concept. The input energy is given by

$$E_{in} = \int_0^{T_s} i_{in} V_{in}\, dt$$

i_{in} is the current, which is equal to $i_L(t)$. Hence, we have

$$E_{in} = \int_0^{t_1} i_L(t) V_{in}\, dt + \int_{t_1}^{t_2} i_L(t) V_{in}\, dt + \int_{t_2}^{t_3} i_L(t) V_{in}\, dt + \int_{t_3}^{T_s} i_L(t) V_{in}\, dt \tag{6.71}$$

The inductor current equals the output current in modes I and IV, and for $t_1 \le t < t_2$ and $t_2 \le t < t_3$, i_L is given in Eqs. (6.63) and (6.68), respectively. Substitute the inductor currents into Eq. (6.71) to yield Eq. (6.72):

$$E_{in} = V_{in}[I_o t_1 + I_o \sqrt{LC} \sin\omega_o(t_2 - t_1) + \frac{V_{in}}{2L}(t_3 - t_2)^2$$
$$+ I_o(t_3 - t_2)\cos\alpha + I_o(T_s - t_3)] \tag{6.72}$$

Substituting for the time intervals t_1, $(t_2 - t_1)$, $(t_3 - t_2)$, and $(T_s - t_3)$ from Eqs. (6.62), (6.65), (6.69), and (6.70), respectively, and using the normalized parameters M, Q, and ω_o, Eq. (6.72) becomes

$$E_{in} = V_{in} I_o \left[\frac{-Q}{M\omega_o} - \frac{ML}{2R_o} + T_s - \frac{\alpha}{\omega_o} + \frac{ML}{R_o}\cos\alpha - \frac{ML}{2R_o}\cos^2\alpha \right] \tag{6.73}$$

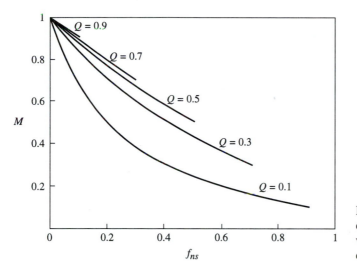

Figure 6.28 Control characteristic curve of M vs. f_{ns} for ZVS buck converter.

The output energy is expressed by

$$E_o = \int_0^{T_s} I_o V_o \, dt = I_o V_o T_s \tag{6.74}$$

Equating the input and output energy in Eqs. (6.73) and (6.74), the voltage gain expression becomes

$$M = 1 - \frac{f_{ns}}{2\pi}\left[\frac{M}{2Q} + \alpha + \frac{M}{Q}(1 - \cos\alpha)\right] \tag{6.75}$$

A plot of the control characteristic curve of M vs. f_{ns} is shown in Fig. 6.28.

The Boost Converter

In this section, we consider the quasi-resonant boost converter using the M-type switch as shown in Fig. 6.29(a), with its simplified circuit shown in Fig. 6.29(b). The four circuit modes of operation are shown in Fig. 6.30.

Mode I ($0 \leq t < t_1$)   Assume for $t < 0$, the switch is closed while D is open. At $t = 0$, the switch is turned off, allowing the capacitor to charge by the constant current I_{in} as given by

$$I_{in} = i_L(t) = i_c(t) = C\frac{dv_c}{dt} \tag{6.76}$$

Since the initial capacitor voltage equals zero, Eq. (6.76) gives the following expression for $v_c(t)$:

$$v_c(t) = \frac{I_{in}}{C}t \tag{6.77}$$

The capacitor voltage reaches the output voltage at $t = t_1$, i.e., $v_c(t_1) = V_o$, resulting in

$$t_1 = \frac{CV_o}{I_{in}} \tag{6.78}$$

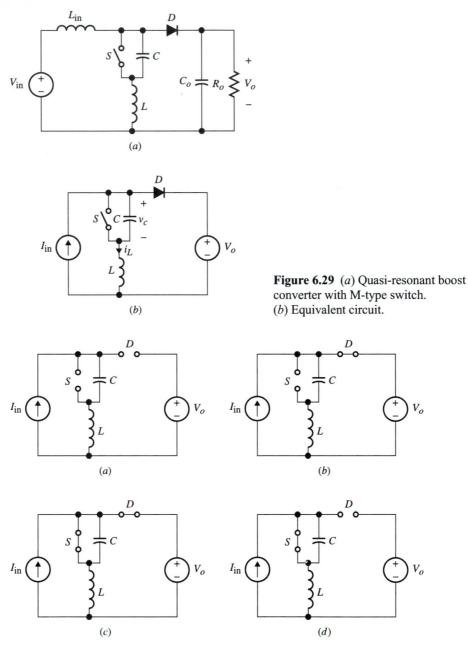

Figure 6.29 (*a*) Quasi-resonant boost converter with M-type switch. (*b*) Equivalent circuit.

Figure 6.30 Equivalent circuit modes. (*a*) Mode I. (*b*) Mode II. (*c*) Mode III. (*d*) Mode IV.

At $t = t_1$, the diode starts conducting since $v_c = V_o$, and the converter enters mode II.

Mode II ($t_1 \leq t < t_2$)   At $t = t_1$, the resonant stage begins since D is on and S is off, as shown in Fig. 6.30(*b*). When the capacitor voltage reaches the output voltage, i_L reaches the negative peak. The initial conditions are $v_c(t_1) = V_o$ and $i_L(t_1) = I_{in}$.

The expression for $v_c(t)$ is given by

$$v_c(t) = V_o + I_{in} Z_o \sin \omega_o (t - t_1) \tag{6.79}$$

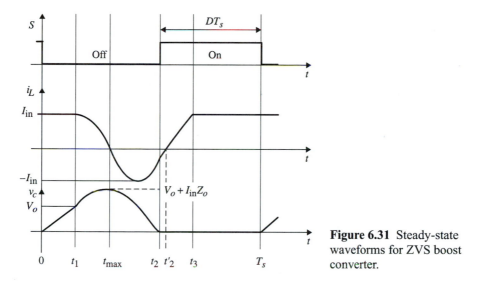

Figure 6.31 Steady-state waveforms for ZVS boost converter.

and the inductor current is

$$i_L(t) = I_{in}[1 + \cos \omega_o(t - t_1)] \tag{6.80}$$

Evaluating Eq. (6.79) at $t = t_2$ with $v_c(t_2) = 0$, the time interval between t_1 and t_2 can be found to be

$$(t_2 - t_1) = \frac{1}{\omega_o} \sin^{-1}\left(\frac{-V_o}{I_{in}Z_o}\right) \tag{6.81}$$

Mode III *($t_2 \leq t < t_3$)*   Mode III starts at t_2, when v_c reaches zero and S turns on at ZVS. The switch and the diode are both conducting, and the inductor current linearly increases to I_{in} as shown in Fig. 6.31. At $t = t_3$, the diode (anti-parallel diode) turns on, clamping the voltage across C to zero.

The initial conditions at $t = t_2$ are

$$V_c(t_2) = 0 \tag{6.82a}$$

$$I_L(t_2) = I_{in}(1 + \cos \omega_o(t_2 - t_1)) \tag{6.82b}$$

Because the capacitor voltage is zero, the inductor voltage is equal to the output voltage.

$$L\frac{di_L}{dt} = V_o \tag{6.83}$$

By integrating Eq. (6.83), the inductor current becomes

$$i_L(t) = \frac{V_o}{L}(t - t_2) + I_L(t_2) \tag{6.84}$$

To achieve ZVS, the switch can be turned on anytime after t_2 and before t'_2. At $t = t'_2$, the inductor current reverses polarity and the switch picks up the current. At $t = t_3$, i_L reaches zero, resulting in the time interval given in Eq. (6.85):

$$(t_3 - t_2) = \frac{L}{V_o} i_L(t_2) \tag{6.85}$$

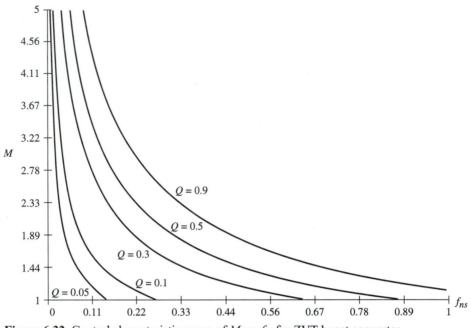

Figure 6.32 Control characteristic curve of M vs. f_{ns} for ZVT boost converter.

Substituting the initial condition into the equation, we get

$$(t_3 - t_2) = \frac{L}{V_o} I_{in}(1 + \cos \omega_o(t_2 - t_1)) \tag{6.86}$$

At $t = t_3$, the output diode turns off and the entire I_{in} current flows in the transistor and the inductor.

Mode IV $(t_3 \leq t < t_4)$ At time t_3, the inductor current reaches zero and the output diode turns off, but the switch remains closed. The cycle repeats at $t = T_s$.

Voltage Gain   As before, we use the conservation of energy per switching cycle to express the voltage gain. It can be shown that the voltage gain in terms of the normalized parameter is given in Eq. (6.87). A plot of the control characteristic curve of M vs. f_{ns} is shown in Fig. 6.32.

$$\frac{1-M}{M} = \frac{f_{ns}}{2\pi}\left[\frac{M}{2Q} + \alpha + \frac{M}{Q}(1 - \cos \alpha)\right] \tag{6.87}$$

EXAMPLE 6.5

Design a ZVS quasi-resonant boost converter for the following design parameters: $V_{in} = 30$ V, $P_o = 30$ W at $V_o = 38$ V, $f_{ns} = 0.4$, and $T_s = 4$ μs. Assume the output voltage ripple is limited to 2% at $D = 0.4$.

SOLUTION   The voltage gain is $M = V_o/V_{in} = 1.3$, and with $f_{ns} = 0.4$, we obtain $Q = 0.2$.
Using the switching frequency $f_s = 1/T_s = 1/4$ μs $= 250$ kHz, and $f_o = f_s/0.4 = 250/0.4 = 625$ kHz, the resonant frequency is obtained from

$$\omega_o = \frac{1}{\sqrt{LC}} = (2\pi)(625) \times 10^3 \text{ rad/s}$$

The second equation in terms of L and C is obtained from

$$Q = \frac{R_o}{Z_o} = \frac{R_o}{\sqrt{L/C}} = 0.2$$

The load resistance is

$$R_o = \frac{38^2}{30} = 48.13 \; \Omega$$

Substituting in the above relation for Q, we obtain

$$\sqrt{\frac{L}{C}} = \frac{48.13}{0.2} = 240.65 \; \Omega$$

Solving the above two equations for C and L, we obtain

$$C = \frac{1}{(2\pi)(625)(10^3)(240.65)} = 1.06 \; \text{nF}$$

$$L = 1.06 \times 10^{-9} \times (240.65)^2 = 61.3 \; \mu\text{H}$$

To calculate L_o and C_o, using a voltage ripple of 2%, we use the following relation:

$$\frac{D}{f_s R_o C_o} = 0.02$$

where

$$D = 0.4$$

$$C_o = \frac{0.4}{f_s R_o} \frac{100}{0.2}$$

$$= \frac{40}{0.2 \times 250 \times 10^3 \times 48.13} = 16.6 \; \mu\text{F}$$

The critical inductor value is given by

$$L_{\text{crit}} = \frac{R_o}{2f_s}(1-D)^2 D$$

$$= \frac{48.13}{2 \times 250 \times 10^3}(1-0.4)^2(0.4)$$

$$= 13.9 \; \mu\text{H}$$

To achieve a limited ripple current, it is recommended that L_o be set at about 100 times the critical inductor value. So we select $L_o = 1.4$ mH.

EXERCISE 6.5

The quasi-resonant boost converter using the L-type switch is shown in Fig. 6.33(a), and its simplified circuit is shown in Fig. 6.33(b). Derive the expression for the voltage gain.

The Buck-Boost Converter

The ZVS buck-boost converter with an M-type switch is shown in Fig. 6.34(a), with its equivalent circuit shown in Fig. 6.34(b).

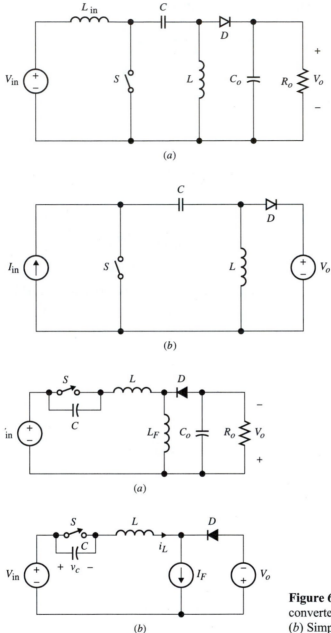

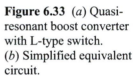

Figure 6.33 (*a*) Quasi-resonant boost converter with L-type switch. (*b*) Simplified equivalent circuit.

Figure 6.34 (*a*) ZVS buck-boost converter with M-type switch. (*b*) Simplified equivalent circuit.

Like the buck and the boost converters, the buck-boost converter steady-state operation leads to four modes of operation as shown in Fig. 6.35. Figure 6.35(*e*) shows the typical steady-state waveforms for v_c and i_L.

Following a similar analysis as before, it can be shown that the voltage gain in terms of M, Q, and f_{ns} is given in Eq. (6.88).

$$M = \frac{1}{\dfrac{f_{ns}}{2\pi}\left[\alpha + \dfrac{M}{2Q} + \dfrac{M}{Q}(1 - \cos\alpha)\right]} - 1 \qquad (6.88)$$

Figure 6.36 shows the control characteristic curve for M vs. f_{ns}.

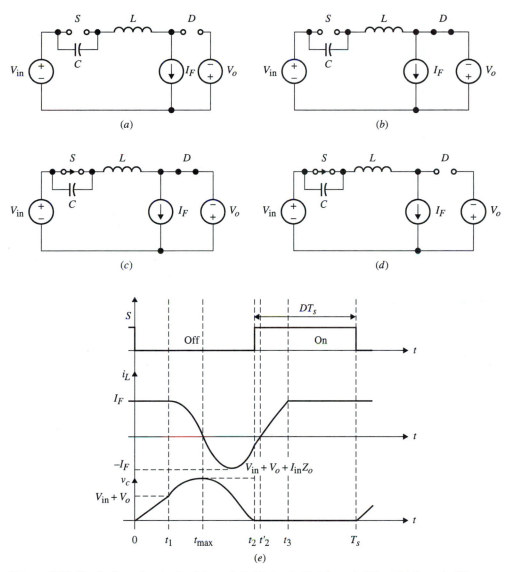

Figure 6.35 Equivalent circuits for (*a*) mode I, (*b*) mode II, (*c*) mode III, and (*d*) mode IV. (*e*) Steady-state waveforms for v_c and i_L.

6.6   GENERALIZED ANALYSIS FOR ZCS[2]

It can be shown from the preceding analysis of the quasi-resonant ZCS and ZVS PWM converters that all dc-dc converter families share the same switching network, with the orientation depending on the topology type (buck, boost, buck-boost, Cuk, Zeta, SEPIC, etc.). As a result, the switching network representation and analysis, including the switching waveforms, can be generalized for each converter family.

The generalized analysis means that in order to derive the analysis and the design curves for a family of converters, only the generalized switching cell with the generalized and normalized parameters for that family need to be analyzed. The equations for the

---

[2]This section can be skipped without loss of continuity.

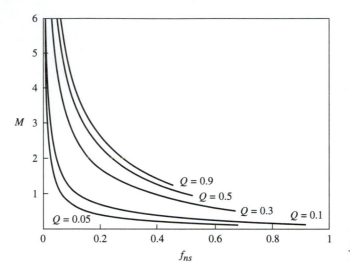

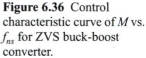

Figure 6.36 Control characteristic curve of M vs. f_{ns} for ZVS buck-boost converter.

generalized cell will be the generalized equations that describe any converter that uses this specific cell. By using generalized parameters, it is possible to generate a single *transformation table* from which the voltage ratios and other important design parameters for each converter can be obtained directly.

6.6.1   The Generalized Switching Cell

Figure 6.37(*a*) and (*b*) shows the generalized switching cells of the quasi-resonant PWM ZCS and ZVS converters, respectively. Note that these cells are all of the common-

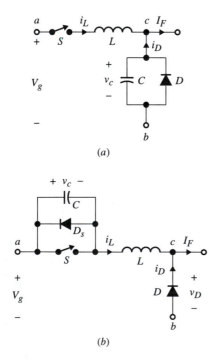

Figure 6.37 Switching cells. (*a*) ZCS quasi-resonant converter cell. (*b*) ZVS quasi-resonant converter cell.

ground, three-terminal, two-port network type. The following parameters, will be used throughout this discussion:

- The normalized cell input voltage, V_{ng}:

$$V_{ng} = \frac{V_g}{V_{in}}$$

where V_g is the switching-cell average input voltage as shown in Fig. 6.37.

- The normalized cell output current, I_{nF}:

$$I_{nF} = \frac{I_F}{I_o}$$

where I_F is the switching-cell average output current.

- The normalized filter capacitor voltage, V_{nF}:

$$V_{nF} = \frac{V_F}{V_{in}}$$

where V_F is the filter capacitor average voltage.

- The normalized filter inductor current, I_{nT}:

$$I_{nT} = \frac{I_T}{I_o}$$

where I_T is the filter inductor average current (in the ZCS QSW CC family).

- The normalized cell output average voltage, V_{nbc}:

$$V_{nbc} = \frac{V_{bc}}{V_{in}}$$

where V_{bc} is the switching-cell average output voltage.

- The normalized current entering node b in the switching cell, I_{nb}:

$$I_{nb} = \frac{I_b}{I_o}$$

where I_b is the average current entering node b.

Note that the generalized transformation table that will be presented next includes the generalized parameters V_{ng}, I_{nF}, V_{nbc}, I_{nb}, V_{nF}, and I_{nT}, which are the normalized versions of the parameters V_g, I_F, V_{bc}, I_b, V_F, and I_T. It will be noted that I_{nb}, V_{nF}, and I_{nT} have the same normalized quantity, as do V_{ng} and I_{nF}.

6.6.2 The Generalized Transformation Table

The derivation of the generalized transformation table for the converter families is beyond the scope of this book. The generalized transformation table is shown in Table 6.1.

By applying the appropriate cell to the conventional dc-dc converters, the ZCS quasi-resonant converter (QRC) family can be formed as shown in Fig. 6.38.

In the next section, the modes of operation for the ZVS QRC switching network will be discussed briefly, and the main switching waveforms will be drawn in terms of the generalized parameters.

Table 6.1 Generalized Transformation Table

	V_{ng}, I_{nF}	V_{nF}, I_{nT}, I_{nb}	V_{nbc}
Buck	1	$1 - M$	$-M$
Boost	M	1	$1 - M$
Buck-boost, Cuk, Zeta, and SEPIC	$1 + M$	1	$-M$

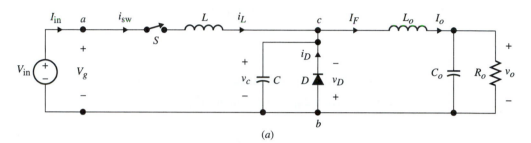

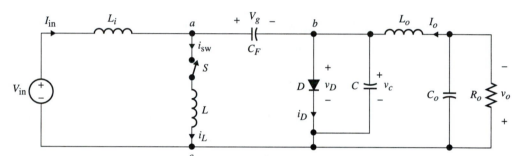

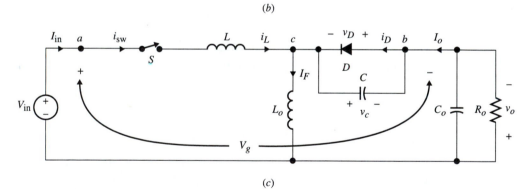

Figure 6.38 The dc-dc ZCS QRC family. (*a*) Buck. (*b*) Boost. (*c*) Buck-boost.

6.6.3 Basic Operation of the ZCS QRC Cell

Typical switching waveforms for the cell in Fig. 6.37(*a*) are shown in Fig. 6.39. Table 6.2 shows the conditions of the switches and diodes in each mode. It can be shown that there are four modes of operation.

Mode 1 $(t_0 \le t < t_1)$ It is assumed that before $t = t_0$, S was off and D was on in order to carry I_F. The resonant inductor L was carrying no current, and the resonant capacitor C voltage was zero. Mode 1 starts when S is turned on while D is on, which

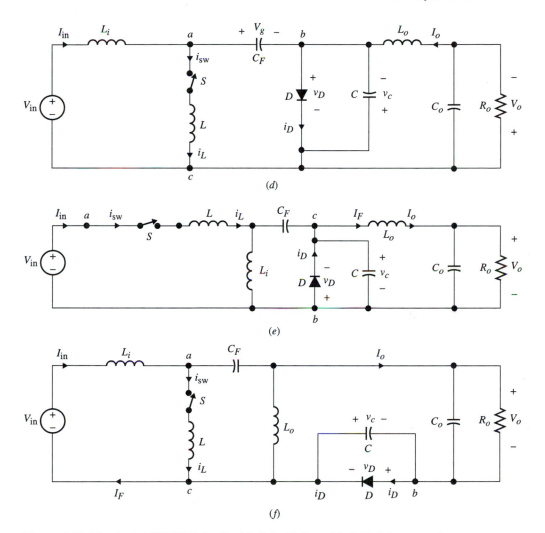

Figure 6.38 The dc-dc ZCS QRC family. (*d*) Cuk. (*e*) Zeta. (*f*) SEPIC (*continued*).

causes L to charge up linearly until the current through it becomes equal to I_F at $t = t_1$, causing D to turn off.

Mode 2 ($t_1 \leq t < t_2$) Mode 2 starts when D turns off while S is on, initiating a resonant stage between C and L until the current through L drops to zero at $t = t_2$, causing S to turn off at zero current (soft switching).

Mode 3 ($t_2 \leq t < t_3$) Mode 3 starts when S turns off at zero current. The resonant capacitor starts discharging linearly, causing the voltage across it to drop to zero again, in turn causing D to turn on at zero voltage at $t = t_3$.

Mode 4 ($t_3 \leq t < t_0 + T_s$) Mode 4 is a steady-state mode, and nothing happens in it until S is turned on again to start the next switching cycle.

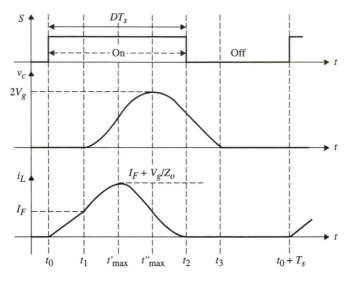

Figure 6.39 Main ZCS QRC switching-cell waveforms.

Table 6.2 Switch and Diode Conditions

	S	D
Mode 1 $(t_0 \leq t < t_1)$	On	On
Mode 2 $(t_1 \leq t < t_2)$	On	Off
Mode 3 $(t_2 \leq t < t_3)$	Off	Off
Mode 4 $(t_3 \leq t < t_0 + T_s)$	Off	On

Generalized Steady-State Analysis

From the description of the modes of operation, the equivalent circuit for each mode can be drawn as in Fig. 6.40. These equivalent circuits can be used along with the description of the modes to write the mathematical equations for each mode as follows (knowing that $v_c(t_0) = 0$ and $i_L(t_0) = 0$):

Mode 1 $(t_0 \leq t < t_1)$

$$v_c(t) = 0$$
$$i_L(t) = \frac{V_g}{L}(t - t_0) \tag{6.89}$$

with the initial conditions at $t = t_1$ given by

$$v_c(t_1) = 0$$
$$i_L(t_1) = I_F$$

Mode 2 $(t_1 \leq t < t_2)$

$$v_c(t) = V_g[1 - \cos \omega_o(t - t_1)] \tag{6.90}$$

$$i_L(t) = I_F + \frac{V_g}{Z_o} \sin \omega_o(t - t_1) \tag{6.91}$$

At $t = t_2$, we have $i_L(t_2) = 0$.

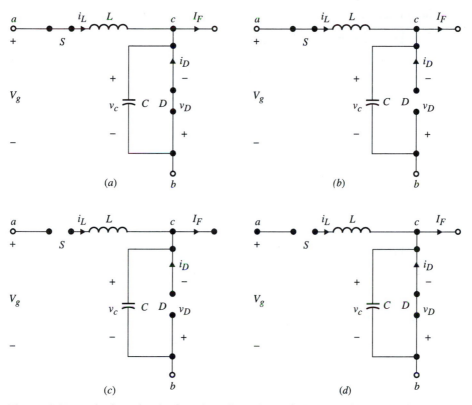

Figure 6.40 Equivalent circuits for (*a*) mode 1, (*b*) mode 2, (*c*) mode 3, and (*d*) mode 4.

Mode 3 $(t_2 \leq t < t_3)$

$$v_c(t) = -\frac{I_F}{C}(t - t_2) + V_g(1 - \cos\beta) \tag{6.92}$$

$$i_L(t) = 0 \tag{6.93}$$

At $t = t_3$, we have $v_c(t_3) = 0$.

Mode 4 $(t_3 \leq t < t_0 + T_s)$

$$v_c(t) = 0 \tag{6.94}$$

$$i_L(t) = 0 \tag{6.95}$$

Generalized Interval Equations

To simplify the analysis, the following time intervals are defined:

$$\alpha = \omega_o(t_1 - t_0)$$
$$\beta = \omega_o(t_2 - t_1)$$
$$\gamma = \omega_o(t_3 - t_2)$$
$$\delta = \omega_o((t_0 + T_s) - t_3)$$

These intervals can be derived as follows:

- From Eq. (6.89), α is given by

$$\alpha = \omega_o(t_1 - t_0) = \frac{Z_o I_F}{V_g}$$

Using the normalized parameters, we have

$$\alpha = \omega_o(t_1 - t_0) = \frac{MI_{nF}}{QV_{ng}} \tag{6.96}$$

- From Eq. (6.91), β is given by

$$\beta = \omega_o(t_2 - t_1) = \sin^{-1}\left(-\frac{MI_{nF}}{QV_{ng}}\right) \tag{6.97}$$

- From Eq. (6.92), γ is given by

$$\gamma = \omega_o(t_3 - t_2) = \frac{QV_{ng}}{MI_{nF}}(1 - \cos\beta)$$

- From Fig. 6.39 and the intervals α, β, and γ, we have δ given by

$$\delta = \omega_o((t_0 + T_s) - t_3) = \frac{2\pi}{f_{ns}} - \alpha - \beta - \gamma \tag{6.98}$$

Generalized Gain Equation

The cell output-to-input generalized gain can be found using the average output diode D voltage as follows:

$$V_{D,\text{ave}} = -V_{c,\text{ave}}$$

$$= -\frac{1}{T_s}\int_{t_0}^{t_0 + T_s} v_c(t)\, dt$$

$$= -\frac{1}{T_s}\left[V_g\left((t_2 - t_1) - \frac{\sin\beta}{\omega_o}\right) - \frac{I_F}{2C}(t_3 - t_2)^2 + V_g(1 - \cos\beta)(t_3 - t_2)\right]$$

By using the normalized parameters defined previously, we have

$$V_{nD} = \frac{f_{ns}}{2\pi}\left[\frac{MI_{nF}}{2Q}\gamma^2 - V_{ng}(\beta + \gamma - \sin\beta - \gamma\cos\beta)\right] \tag{6.99}$$

By substituting for the generalized parameters (V_{nD}, V_{ng}, and I_{nF}) from Table 6.1, we arrive at the gain equation for each converter in the family.

Generalized ZCS Condition

It can be noted from Fig. 6.39 that S can be turned off at any time after $t = t_2$. The generalized condition to achieve zero-current switching can be expressed as follows:

$$\frac{2\pi}{f_{ns}}D \geq \alpha + \beta \tag{6.100}$$

Generalized Peak Resonant Inductor Current (Peak Switch Current)

The peak resonant inductor current or peak switch current occurs at $t = t'_{max}$, when $\omega_o(t'_{max} - t_1) = \pi/2$. Using Eq. (6.92) at $t = t'_{max}$,

$$I_{nL-peak} = I_{nF} + \frac{QV_{ng}}{M} \tag{6.101}$$

The peak resonant capacitor voltage or peak diode voltage occurs at $t = t''_{max}$, when $\omega_o(t''_{max} - t_1) = \pi$. Using Eq. (6.91) at $t = t'_{max}$,

$$V_{nL-peak} = 2V_{ng}$$

Design Control Curves

By substituting the generalized parameters from Table 6.1 in the generalized equations, design equations for each topology in the family can be found and their design curves can be plotted.

Figure 6.41 shows the control characteristic curves of M vs. f_{ns} for the ZCS QRC family as an example. Figure 6.42 shows the average and the rms switch currents as a function of the voltage gain.

6.6.4   Basic Operation of the ZVS QRC Cell

Figure 6.37(b) shows the generalized ZVS QRC switching cell, which is formed by adding the resonant capacitor C (which can be considered the switch's internal capacitor) and the resonant inductor L to the conventional switching cell. As discussed before, this is also known as an M-type resonant switch.

By applying this cell to the conventional dc-dc converters in Fig. 6.7, the ZVS QRC family can be formed as shown in Fig. 6.43. In this section, the modes of operation for the ZVS QRC switching network will be discussed briefly and the main switching waveforms will be drawn.

The typical switching waveforms for the ZVS QRC cell are shown in Fig. 6.44. Table 6.3 shows the conduction status of the switches and diodes in each mode. As before, there are four modes of operation. It is assumed that before $t = t_0$, S was on and D was off. The resonant inductor L was carrying a current equal to I_F, and the resonant capacitor C voltage was zero.

Mode 1 ($t_0 \leq t < t_1$)   Mode 1 starts when S is turned on while D is off, which causes C to charge up linearly until its voltage reaches a value equal to V_g at $t = t_1$, causing D to start conducting. The resonant inductor current during this mode doesn't change and is equal to I_F.

Mode 2 ($t_1 \leq t < t_2$)   Mode 2 starts when D turns on while S is off, causing a resonant stage between C and L until the voltage across C tends to go negative, forcing the switch diode D_s to turn on at $t = t_2$. After this time, S can be turned on at zero voltage.

Mode 3 ($t_2 \leq t < t_3$)   Mode 3 starts when S is turned on at zero voltage (i.e., zero-voltage switching). The resonant inductor current starts charging up linearly until it reaches I_F, causing D to turn off at zero current at $t = t_3$.

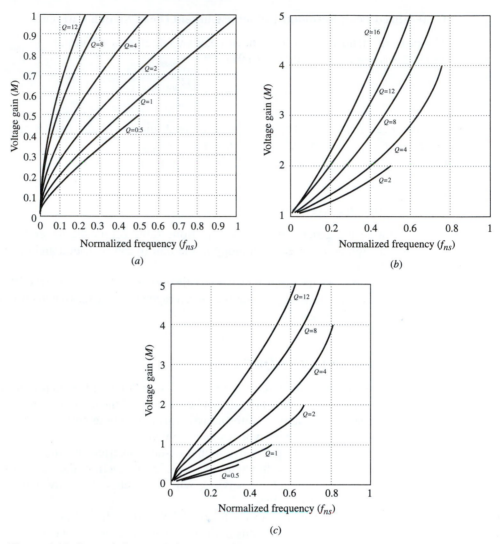

Figure 6.41 Control characteristic curves of M vs. f_{ns} for (a) ZCS QRC buck, (b) ZCS QRC boost, (c) ZCS QRC buck-boost, Cuk, Zeta, and SEPIC.

Mode 4 ($t_3 \leq t < t_0 + T_s$) Mode 4 is a steady-state mode, and nothing happens in it until S is turned off again to start the next switching cycle.

Generalized Steady-State Analysis

From the description of the modes of operation, the equivalent circuit for each mode can be drawn as shown in Figure 6.45. These equivalent circuits can be used along with the description of the modes to write the mathematical equations for each mode as follows (knowing that $v_c(t_0) = 0$ and $i_L(t_0) = I_F$):

Mode 1 ($t_0 \leq t < t_1$)

$$v_c(t) = \frac{I_F}{C}(t - t_0)$$

$$i_L(t) = I_F$$

(6.102)

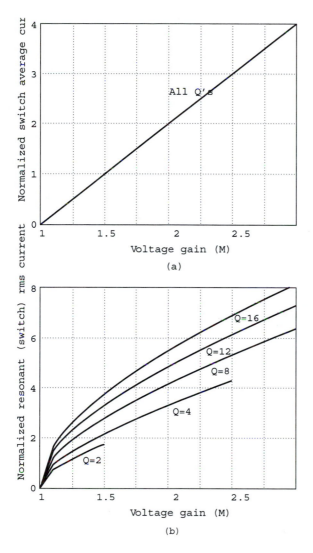

Figure 6.42 Some of the ZCS QRC boost main switch (S) normalized stresses.
(*a*) Normalized average current.
(*b*) Normalized rms current.

$$v_c(t_1) = V_g$$
$$i_L(t_1) = I_F$$

(6.103)

Mode 2 ($t_1 \leq t < t_2$)

$$v_c(t) = V_g + Z_o I_F \sin \omega_o(t - t_1)$$
$$i_L(t) = I_F \cos \omega_o(t - t_1)$$

(6.104)

$$v_c(t_2) = 0$$
$$i_L(t_2) = I_F \cos \omega_o(t_2 - t_1) = 0$$

(6.105)

Mode 3 ($t_2 \leq t < t_3$)

$$v_c(t) = 0$$

$$i_L(t) = I_F \cos \omega_o(t_2 - t_1) + \frac{V_g}{L}(t - t_2)$$

(6.106)

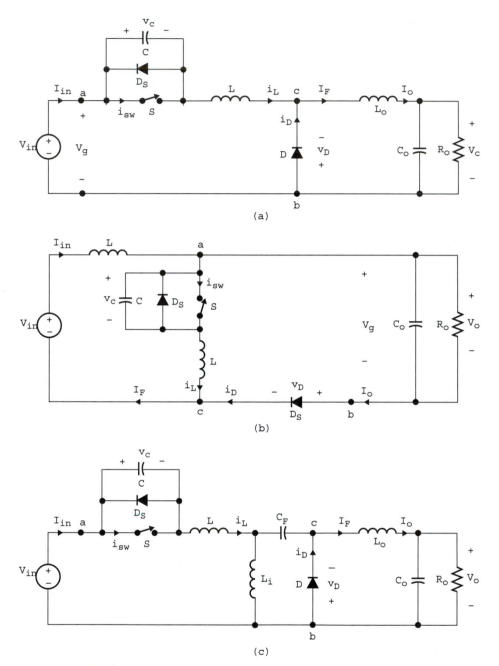

Figure 6.43 The dc-dc ZVS QRC family. (*a*) Buck. (*b*) Boost. (*c*) Buck-boost.

$$v_c(t_3) = 0$$
$$i_L(t_3) = I_F$$

(6.107)

Mode 4 ($t_3 \leq t < t_0 + T_s$)

$$v_c(t) = 0$$
$$i_L(t) = I_F$$

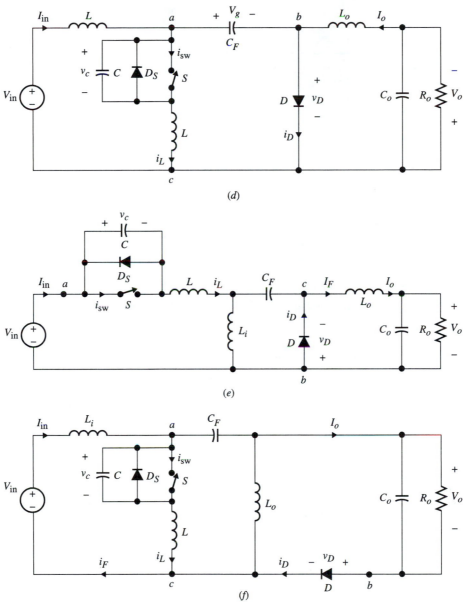

Figure 6.43 The dc-dc ZVS QRC family. (*d*) Cuk. (*e*) Zeta. (*f*) SEPIC (*continued*).

Generalized Interval Equations

To simplify the analysis, the following time intervals are defined:

$$\alpha = \omega_o(t_1 - t_0)$$
$$\beta = \omega_o(t_2 - t_1)$$
$$\gamma = \omega_o(t_3 - t_2)$$
$$\delta = \omega_o((t_0 + T_s) - t_3)$$

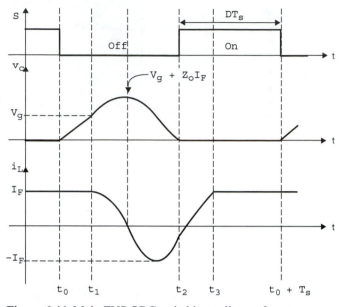

Figure 6.44 Main ZVS QRC switching-cell waveforms.

Table 6.3   Switches and Diode Conditions

	S	D	D_s
Mode 1 $(t_0 \leq t < t_1)$	Off	Off	Off
Mode 2 $(t_1 \leq t < t_2)$	Off	On	Off
Mode 3 $(t_2 \leq t < t_3)$	On	On	On
Mode 4 $(t_3 \leq t < t_0 + T_s)$	On	Off	Don't care

These intervals can be derived as follows:

- From Eqs. (6.102) and (6.103),

$$\alpha = \omega_o(t_1 - t_0) = \frac{V_g}{Z_o I_F}$$

Using the normalized parameters defined earlier,

$$\alpha = \omega_o(t_1 - t_0) = \frac{Q V_{ng}}{M I_{nF}}$$

- From Eqs. (6.104) and (6.105),

$$\beta = \omega_o(t_2 - t_1) = \sin^{-1}\left(-\frac{Q V_{ng}}{M I_{nF}}\right)$$

- From Eqs. (6.106) and (6.107),

$$\gamma = \omega_o(t_3 - t_2) = \frac{M I_{nF}}{Q V_{ng}}(1 - \cos\beta)$$

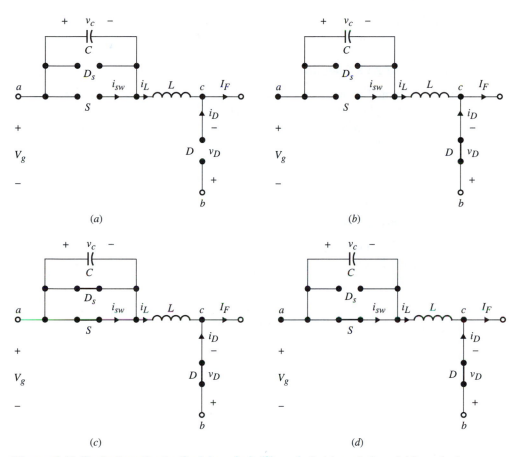

Figure 6.45 Equivalent circuits for (*a*) mode 1, (*b*) mode 2, (*c*) mode 3, and (*d*) mode 4.

- From Fig. 6.44 and the preceding intervals,

$$\delta = \omega_o((t_0 + T_s) - t_3) = \frac{2\pi}{f_{ns}} - \alpha - \beta - \gamma$$

Generalized Gain Equation

The cell output-to-input generalized gain can be found using the average output diode *D* voltage as follows:

$$V_{D,\,\text{ave}} = V_{s,\,\text{ave}} - V_g$$

$$= \left[\frac{1}{T_s} \int_{t_0}^{t_0 + T_s} v_s(t)\, dt \right] - V_g$$

$$= \frac{1}{T_s} \left[\frac{I_F}{2C}(t_1 - t_0)^2 + V_g(t_2 - t_1) - \frac{Z_o I_F}{\omega_o}(\cos \omega_o(t_2 - t_1) - 1) \right] - V_g$$

By using the normalized parameters, we have

$$V_{nD} = \frac{f_{ns}}{2\pi} \left[\frac{MI_{nF}}{2Q} \alpha^2 + V_{ng}\beta + \frac{MI_{nF}}{Q}(1 - \cos\beta) \right] - V_{ng} \qquad (6.108)$$

By substituting for the generalized parameters (V_{nD}, V_{ng}, and I_{nF}) from Table 6.1 in Eq. (6.108), we arrive at the gain equation for each converter in the family.

Generalized ZVS Condition

It can be noted from Fig. 6.43 that S must be turned on after $t = t_2$ and before $t = t_3$—in other words, before $i_L(t) = I_F$, which causes D to turn off ($i_D(t) = I_F - i_L(t)$) at $t = t_3$ causing $v_c(t)$ to start charging up linearly again. The generalized condition to achieve zero-voltage switching can be expressed as follows:

$$\alpha + \beta \le \frac{2\pi}{f_{ns}}(1 - D) \le \alpha + \beta + \gamma$$

Design Control Curves

By substituting for the generalized parameters from Table 6.1 in the generalized equations, design equations for each topology in the family can be found. Figure 6.46 shows

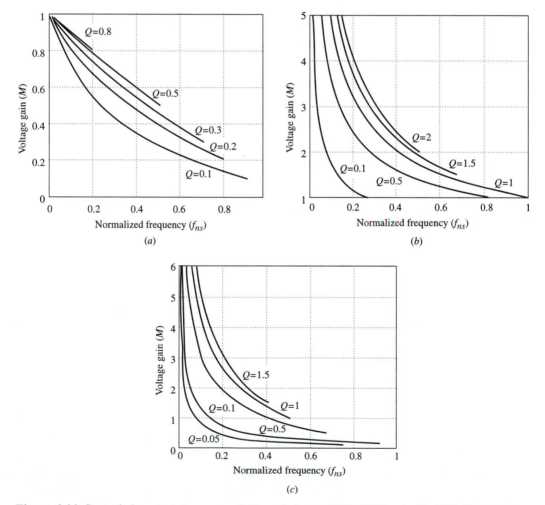

Figure 6.46 Control characteristic curves of M vs. f_{ns} for (a) ZVS QRC buck, (b) ZVS QRC boost, (c) ZVS QRC buck-boost, Cuk, Zeta, and SEPIC.

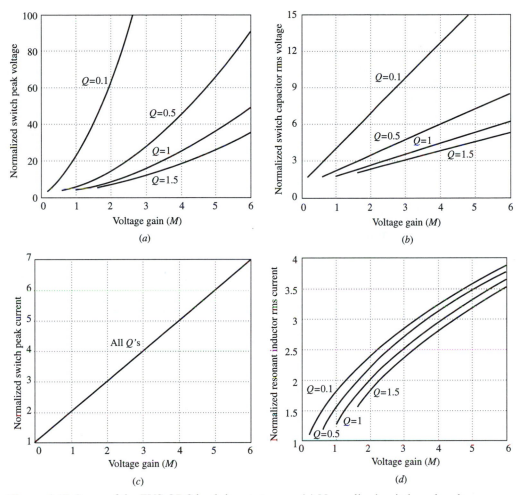

Figure 6.47 Some of the ZVS QRC buck-boost stresses. (*a*) Normalized switch peak voltage. (*b*) Normalized switch rms voltage. (*c*) Normalized switch peak current. (*d*) Normalized switch rms current.

the control characteristic curves for the ZVS QRC family. Many other curves can also be plotted; as an example, Fig. 6.47 shows the peak and the rms switch currents as a function of the voltage gain, M.

6.7   ZERO-VOLTAGE AND ZERO-CURRENT TRANSITION CONVERTERS

Traditional converters operate with a sinusoidal current through the power switches, which results in high peak and rms currents for the power transistors and high voltage stresses on the rectifier diodes. Furthermore, when the line voltage or load current varies over a wide range, quasi-resonant converters are modulated with a wide switching frequency range, making the circuit design difficult to optimize. As a compromise between the PWM and resonant techniques, various soft-switching PWM converter techniques have been proposed lately to combine the desirable features of the conventional PWM and quasi-resonant techniques without a significant increase in the circulating energy.

6.7.1   Switching Transition

To overcome the limitations of the quasi-resonant converters, zero-voltage transition (ZVT) or zero-current transition (ZCT) is the solution. Instead of using a series resonant network across the power switch, a shunt resonant network is used across the power switch. A partial resonance is created by the shunt resonant network to achieve ZCS or ZVS during the switching transition. This retains the advantages of a PWM converter because after the switching transition is over, the circuit reverts to the PWM operation mode.

The features of the ZCT PWM and ZVT PWM soft-switching converters are summarized as follows:

- Zero-current/voltage turn-off/on for the power switch
- Low voltage/current stresses of the power switch and rectifier diode
- Minimal circulating energy
- Constant-frequency operation
- Soft switching for a wide line and load range

One disadvantage is that the auxiliary switch does not operate with soft switching; it is hard-switching, but the switching loss is much lower than that of a PWM converter. Another disadvantage is that the transformer leakage inductance is not utilized, which is similar to the quasi-resonant converters. Therefore, the transformer should be designed with minimum leakage inductance.

The ZVT and ZCT converters differ from the conventional PWM converters by the introduction of a resonant branch, shown in Fig. 6.48. Figure 6.48(a) shows the

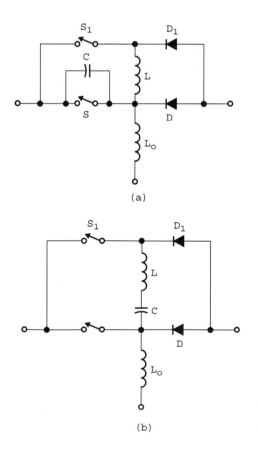

(a)

(b)

Figure 6.48 (a) ZVT PWM switching cell. (b) ZCT PWM switching cell.

ZVT PWM switching cell, and Fig. 6.48(b) shows the ZCT PWM switching cell. L is a resonant inductor, C is a resonant capacitor, S_1, is an auxiliary switch, and D_1, is an auxiliary diode.

6.7.2   The Boost ZVT PWM Converter

In this section, we consider the boost ZVT PWM, shown in Fig. 6.49, by placing the ZVT PWM switching cell shown in Fig. 6.48(a) into the conventional boost converter.

The switching cycle is divided into six modes, as shown in Fig. 6.50. These modes are defined as follows.

Mode I $(t_0 \le t < t_1)$ Mode I, shown in Fig. 6.50(a), starts at $t = t_0$, when the auxiliary switch S_1 is turned on. Since the main switch, S, and the auxiliary switch, S_1, were off prior to $t = t_0$, it is clear that the diode, D, must have been on for $t < t_0$ to carry the output current. Hence, we assume D is on and D_1 is off at $t = t_0$. So for $t > t_0$, S_1 is on. The diode current starts to decrease, and it reaches zero when the inductor current i_L increases and reaches the constant current source, I_{in}, at t_1. In this mode, the capacitor voltage, v_c, is equal to the output voltage, V_o, and also equal to the inductor voltage as given by

$$V_o = L\frac{di_L}{dt} = v_c(t)$$

From the equation, the inductor current, i_L, is given by

$$i_L(t) = \frac{V_o}{L}(t - t_0)$$

This equation assumes zero initial condition for i_L.

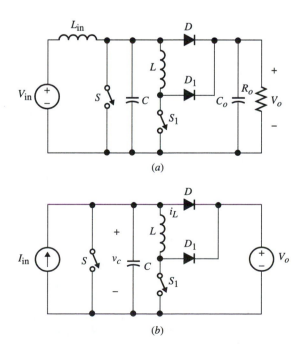

Figure 6.49 (a) Boost ZVT PWM. (b) Simplified equivalent circuit.

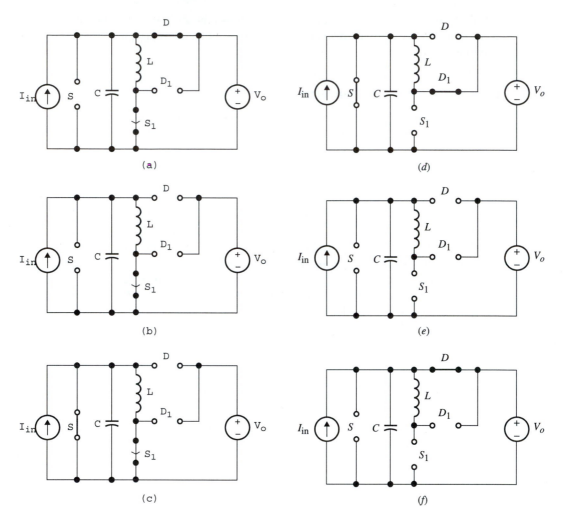

Figure 6.50 Equivalent circuits for the six modes of operation: (*a*) mode I, (*b*) mode II, (*c*) mode III, (*d*) mode IV, (*e*) mode V, and (*f*) mode VI.

As long as the inductor current is less than I_{in}, the diode will continue conducting and the capacitor voltage will remain at V_o. At time t_1, the inductor current becomes equal to I_{in}, D stops conducting, and the circuit enters mode II. From the above equation, we have

$$I_{in} = \frac{V_o}{L}(t - t_0)$$

The time interval is given by

$$(t_1 - t_0) = \frac{LI_{in}}{V_o}$$

This is the inductor current charging state.

Mode II ($t_1 \leq t < t_2$)   Mode II starts at t_1, when D is off. The circuit is shown in Fig. 6.50(*b*) and results in a resonant stage between L and C. During the time between

t_1 and t_2, the main switch, S, remains off and S_1 is still on, but both diodes are off. The initial capacitor voltage is still V_o, but the initial i_L has changed to I_{in}. The first-order differential equations that represent this mode are given by

$$C\frac{dv_c}{dt} = I_{in} - i_L(t)$$

$$L\frac{di_L}{dt} = v_c(t)$$

Equation (6.109) is obtained from the above two equations:

$$\frac{d^2 i_L}{dt^2} - \frac{1}{LC}i_L(t) = \frac{1}{LC}I_{in} \tag{6.109}$$

The solution for i_L and v_c is given by

$$i_L(t) = \frac{V_o}{Z}\sin\omega_o(t-t_1) + I_{in}$$

$$v_c(t) = V_o(2 - \cos\omega_o(t-t_1))$$

The time interval between t_1 and t_2 is given by

$$(t_2 - t_1) = \frac{1}{\omega_o}\cos^{-1}2$$

The diode voltage starts to charge up due to the decreasing capacitor voltage:

$$v_D(t) = V_o - v_c(t)$$

Substituting for v_c, the diode voltage becomes

$$v_D(t) = V_o(\cos\omega_o(t-t_1) - 1)$$

Mode III ($t_2 \le t < t_3$)   Mode III starts when the capacitor discharges to zero. The body diode of S turns on to clamp v_c to zero. In this mode the main switch, S, remains off, the auxiliary switch, S_1, is still on, and both diodes are off. Now,

$$v_c(t) = 0$$

Mode IV ($t_3 \le t < t_4$)   Mode IV starts at $t = t_3$, when the main switch, S, is turned on and the auxiliary switch, S_1, is turned off. At t_3, the initial capacitor voltage is zero, and the inductor starts linearly discharging from $i_L(t_2)$ to zero during t_3 to t_4. The diode D remains off since its voltage is negative, but D_1 turns on at $t = t_3$ to carry the inductor current.

The input voltage is equal to the inductor voltage, and the output voltage is equal to the negative inductor voltage, v_L.

$$v_L(t) = L\frac{di_L}{dt} = -V_o$$

The inductor current for $t > t_2$ is given by

$$i_L(t) = -\frac{V_o}{L}(t-t_2) + I_L(t_2)$$

Mode V ($t_4 \leq t < t_5$)   In mode V, at $t = t_4$ both switches are off, and both diodes are also off. The inductor current is zero, and the input current is going only through the capacitor:

$$I_{in} = C\frac{dv_c}{dt}$$

The capacitor voltage can be expressed as

$$v_c(t) = \frac{1}{C}I_{in}(t - t_4)$$

The capacitor is charging up from zero and will reach the output voltage at $t = t_5$. The time interval is

$$(t_5 - t_4) = \frac{V_o C}{I_{in}}$$

The circuit enters mode VI at this point.

Mode VI ($t_5 \leq t < t_6$)   When the capacitor reaches the output voltage, D starts conducting, but in this mode, both switches are still off. The diode current will equal the input current immediately. At $t = t_5$, the capacitor voltage is equal to the output voltage until the auxiliary switch is turned on again; then the cycle will repeat from mode I. The waveforms for the six modes of operation are shown in Fig. 6.51.

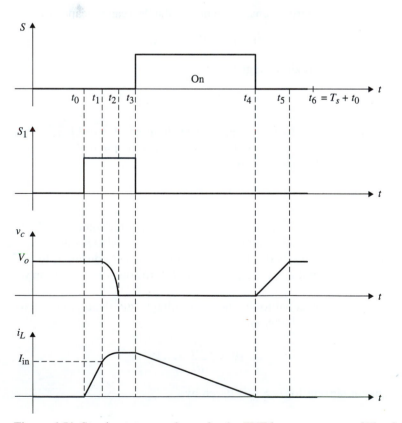

Figure 6.51 Steady-state waveforms for the ZVT boost converter of Fig. 6.49(*b*).

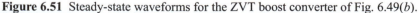

PROBLEMS

ZCS Quasi-Resonant Buck Converters

6.1 Consider the ZCS buck converter whose steady-state waveforms are shown in Fig. 6.10. Assume that the converter has the following parameters: $V_{in} = 25$ V, $I_o = 1$ A, $L = 3$ μH, $C = 0.02$ μF. Determine the time intervals at t_1, t_2, t_3.

6.2 Consider the ZCS buck converter shown in Fig. P6.2. Determine the output power.

6.3 Determine M and Q for a ZCS buck converter with an L-type switch that has the following converter components: $L = 15$ μH, $C = 60$ μF, $f_s = 100$ kHz, $V_{in} = 40$ V, and $I_o = 0.74$ A.

6.4 Consider the ZCS buck converter shown in Fig. P6.4(a) with $L = 20$ μH, $C = 5$ μF, $V_{in} = 50$ V, $V_o = 30$ V, $I_o = 20$ A. Assume that $L_o/R_o \gg 1/T$. The switching waveform of the transistor is shown in Fig. P6.4(b).

(a) Determine the minimum t_{on} in μs to achieve ZCS and determine the switching period T.

(b) Repeat part (a) by adding a diode across the transistor to allow bidirectional current flow, to produce a full-wave ZCS buck converter. Assume all values are the same as in part (a).

6.5 Determine the average output voltage for the ZCS buck converter shown in Fig. P6.5.

D6.6 Consider a ZCS buck converter with an L-type switch with $M = 0.35$ and $Q = 1$. Design for

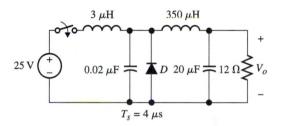

Figure P6.5

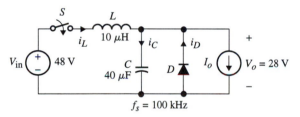

Figure P6.2

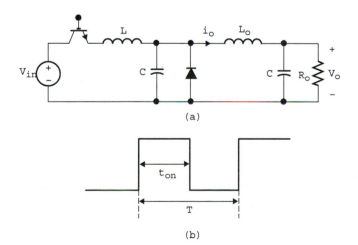

(a)

(b)

Figure P6.4

the resonant tank L and C for $V_{in} = 40$ V, $f_s = 250$ kHz, and $I_o = 0.7$ A.

D6.7 Consider the ZCS buck converter shown in Fig. P6.7 that has the following design parameters: $V_{in} = 25$ V, $V_o = 12$ V, $f_s = 250$ kHz, $I_o = 1$ A, and $f_o = 625$ kHz. Design for L and C and draw the steady-state waveforms for i_L, i, v_c, v_{sw}, v_D, i_D, i_o.

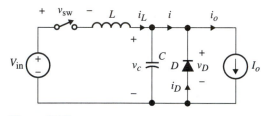

Figure P6.7

D6.8 Consider a ZCS buck converter with a unidirectional switch and the following specifications:

- Maximum load power 750 W
- Nominal output voltage 5 V
- Nominal input voltage 12 V
- Switching frequency 85 kHz
- Maximum peak resonant inductor current at twice the average load current

(a) Design for the resonant tank L and C.

(b) If the load current changes by ±50%, what are the new minimum and maximum switching frequencies required to maintain the output voltage at 5 V?

(c) Repeat part (b) for a variation of ±20% in the average input voltage.

D6.9 Given a ZCS buck converter with an L-type switch that has the following parameters: $V_{in} = 50$ V, $f_s = 100$ kHz, $I_o = 0.2$ A, $V_o = 49$ V, and $f_{ns} = 0.5$. Design for L, C, L_o, C_o, and R. Design for an output ripple current within 15% of its dc value and an output ripple voltage not to exceed 2%.

D6.10 Consider the ZCS buck converter shown in Fig. P6.10 with the following parameters:

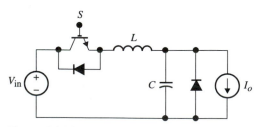

Figure P6.10

$L = 25$ μH, $C = 4.7$ μF, $V_{in} = 58$ V, and $I_o = 18$ A. Determine the output voltage and f_s to deliver an average output power equal to 580 W.

ZCS Quasi-Resonant Boost Converters

D6.11 Design a ZCS boost converter for the following parameters: $V_{in} = 20$ V, $V_o = 36$ V, $I_o = 0.7$ A, and $f_s = 250$ kHz. What is the range of f_s needed to keep V_o constant when I_o changes between 0.2 A and 1 A?

D6.12 Design a ZCS boost converter with an M-type switch with the following design parameters: $M = 2.73$ and $f_{ns} = 0.6$. It is desired to deliver a 0.5 A load current to $V_o = 68$ V.

D6.13 Design an M-type switch boost quasi-resonant converter with the following parameters: $M = 1.8$, $Q = 2.5$, $f_{ns} = 0.5$. The input voltage varies between 18 V and 26 V, and I_o varies between 0.5 A and 1 A. What is the range of f_s needed to keep the converter output voltage constant at $V_o = 38$ V?

ZCS Quasi-Resonant Buck-boost Converters

6.14 Consider the ZCS buck-boost converter with an L-type switch shown in Fig. P6.14. Find M, Q and f_s. What is the new f_s needed to keep V_o constant if the average load current changes to 1.8 A?

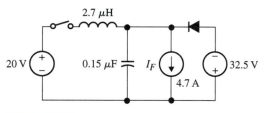

Figure P6.14

6.15 Determine the value of I_F in the ZCS buck-boost converter given in Fig. P6.15.

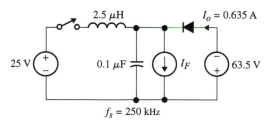

Figure P6.15

D6.16 Design a ZCS buck-boost converter with an L-type switch for the following specifications: $V_{in} = 40$ V, $V_o = 20$ V, $I_o = 4$ A, and $f_s = 250$ kHz.

D6.17 Design a ZCS buck-boost converter with L-type converter for the following specifications: $V_{in} = 30$ V, $V_o = 20$ V @ $I_o = 2.8$ A, and $f_s = 50$ kHz.

D6.18 Design a ZCS buck-boost converter with an L-type switch that operates at 150 kHz and delivers 48 W to $V_o = 47$ V from a 10 V dc source.

D6.19 Figure P6.19 shows a ZCS buck-boost with S being unidirectional.

(a) Sketch i_L and v_c.

(b) Discuss the four modes of operation.

(c) Derive the expressions for i_L and v_c in terms of the circuit parameters.

6.20 Repeat Problem 6.19 using a ZCS buck-boost L-type converter. Select any set of M, Q, f_{ns} you see fit for your design.

ZVS Quasi-Resonant Buck Converters

6.21 Derive the voltage gain expression for the M-type ZVS buck converter.

6.22 Consider the buck ZVS shown in Fig. P6.22 with an alternative way of implementing the L-type resonant switch. Assume L_o/L is very large.

(a) Discuss the modes of operation over one switching cycle.

(b) Draw the typical waveforms for i_L and v_c.

(c) Derive the expression for V_o/V_{in}.

6.23 Consider the resonant capacitor voltage and inductor current waveforms shown in Fig. P6.23 for a ZVS buck. Determine L, C, t_1, t_3, V_o, Q, f_{ns}, M, and $i_L(t_2)$.

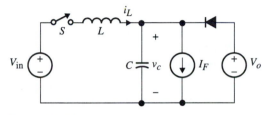

Figure P6.19

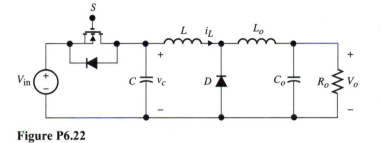

Figure P6.22

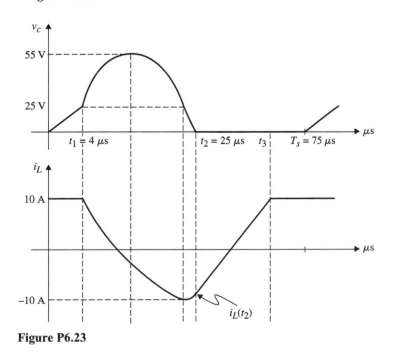

Figure P6.23

ZVS Quasi-Resonant Boost Converters

6.24 Consider the simplified ZVS boost quasi-resonant converter shown in Fig. P6.24. Assume S is a bidirectional switch.

(a) Sketch the waveforms for v_c and i_L.

(b) Discuss the four modes of operation.

(c) Derive the following expression for the voltage gain:

$$M = \frac{V_o}{V_{in}} = \frac{1}{\dfrac{f_{ns}}{2\pi}\left(\alpha - \dfrac{Q}{2M} + \dfrac{M}{Q}(1 - \cos\alpha)\right)}$$

where

$$\sin\alpha = \frac{-Q}{M}, f_{ns} = \frac{f_s}{f_o}, Q = \frac{R_o}{Z_o}$$

$$Z_o = \sqrt{L/C}, f_o = \frac{1}{2\pi\sqrt{LC}}$$

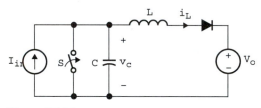

Figure P6.24

D6.25 Design a ZVS boost converter with the following specifications:

- Maximum load power 750 W
- Nominal output voltage 12 V
- Nominal input voltage 5 V
- Switching frequency 100 kHz
- Maximum peak capacitor voltage at 1.5 times the average input voltage

6.26 Figure P6.26(a) shows an isolated boost converter with switches that operates at 50% duty cycle. If it is assumed that L_{in} and C_o are large, then the input current, I_{in}, and the output voltage, V_o, may be assumed constant as shown in Fig. P6.26(b). The switching waveforms for S_1 and S_2 are shown in Fig. P6.26(c). It can be shown that there are four modes of operation in the steady state with S_2 operating at ZVS.

(a) Discuss the modes of operation.

(b) Show that $I_o/I_{in} = n(1-D)$, where I_{in} and I_o are the average input and output currents.

ZVS Quasi-Resonant Buck-Boost Converters

6.27 Determine the output voltage for the ZVS converter shown in Fig. P6.27 and sketch the waveforms for i_L, v_c, and i_D.

D6.28 Design a buck-boost ZVS converter with the following steady-state operating points: $M = 0.55, f_{ns} = 0.3, Q = 0.1$. Assume $V_{in} = 30$ V, $T_s = 10\ \mu s$, and the output current is at least 1 A.

General Soft-switching Converters

4.28 Figure P6.29 shows a zero-current transition (ZCT) buck converter.

(a) Discuss the modes of operation and show the typical waveforms for all the currents and voltages. Assume I_o is constant.

(b) Give the expression for the circulating energy in the resonant tank.

(c) What are the major features of this converter?

6.30 Consider the circuit given in Fig. P6.30(a) that operates in the ZVS mode with S_1 and S_2 operating alternatively as shown in Fig. P6.30(b). Discuss the operation of the circuit and sketch the waveforms for i_{sw}, i_{LK}, v_{c1}, v_{c2}, i_{Lm}, and v_s.

6.31 Figure P6.31(a) shows a ZVS soft-switching capacitor voltage-clamped converter. Assume $C_o \gg C_1$ and $C_o \gg C_2$. It can be shown that in the steady state there are four modes of operation. Assume S_1 and S_2 are switched as shown in Fig. 6.31(b). Sketch the waveforms for i_L, v_{c1}, v_{c2}, and i_{D2}, and discuss the modes of operation.

6.32 Figure P6.32 shows a technique that is based on the concept of clamped-current (CC) soft switching. This converter is known as the ZCS quasi-squarewave resonant converter. It is assumed that L_g is large enough so its current may be represented as a current source. S and D_s form a unidirectional switch. This topology offers several advantages. Discuss these advantages.

(a) Sketch the steady-state waveforms for i_L, i_{sw}, i_D, v_c, and v_o. Assume $I_F < I_o$ and the first mode starts at $t = t_0$ when S is on and D is on initially. It can be shown that there are four modes of operation.

(b) Show that during the resonant mode (mode 3), the resonant inductor current and resonant capacitor voltage equations are given by

$$i_L(t) = I_F(\cos\omega_o(t - t_2) + 1)$$

$$v_c(t) = V_{in} + Z_o I_F \sin\omega_o(t - t_2)$$

where $Z_o = \sqrt{L/C}$ and $t = t_2$ is the start of mode 3.

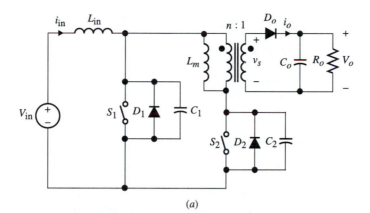

(a)

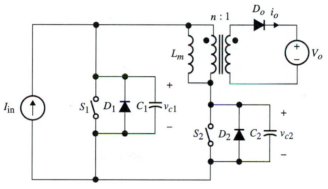

(b)

(c)

Figure P6.26

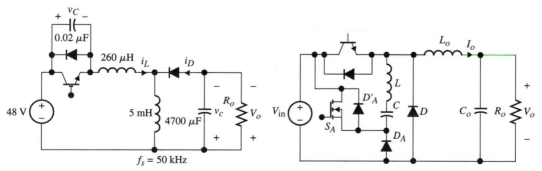

Figure P6.27

Figure P6.29

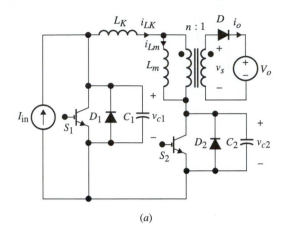

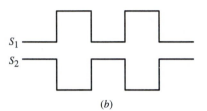

(a)

(b)

Figure P6.30

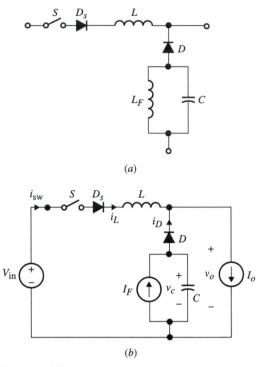

(a)

(b)

Figure P6.32

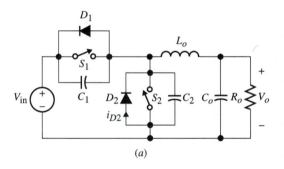

(a)

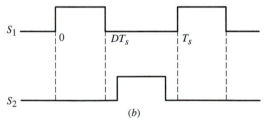

(b)

Figure P6.31

6.33 Figure P6.33 shows a soft-switching converter cell with the diode replaced by a switch S_D. By allowing the rectifier to become a controllable switch, the converter voltage gain can be controlled by the pulse width of S. This soft-switching family is known as PWM QUASI-SQUAREWAVE (PWM QSW) converters. Unlike the quasi-resonant converters, which are frequency-controlled converters, the QSW converters are PWM-controlled at constant frequency. Assume S and S_D are bidirectional switches. Discuss the four modes of operation.

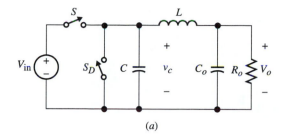

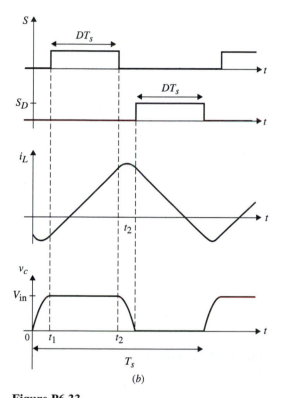

(b)

Figure P6.33

Chapter 7

Uncontrolled Diode Rectifier Circuits

INTRODUCTION

For nearly a century, rectifier circuits have been the most common power electronic circuits used to convert ac to dc. The word *rectification* is used not because these circuits produce dc, but rather because the current flows in one direction; only the *average* output signal (voltage or current) has a dc component. Moreover, since these circuits allow power to flow only from the source to the load, they are often termed *unidirectional converters*. As will be shown shortly, when rectifier circuits are used solely, their outputs consist of dc along with high-ripple ac components. To significantly reduce or eliminate the output ripple, additional filtering circuitry is added at the output. In the majority of applications, diode rectifier circuits are placed at the front end of the power electronic 60 Hz systems, interfaced with the sine-wave voltage produced by the electric utility. In dc-to-dc application, at the rectified side or the dc side, a large filter capacitor is added to reduce the rectified voltage ripple. This dc voltage maintained across the output capacitor is known as raw dc or *uncontrolled dc*.

The principal circuit operations of the various configurations of rectifier circuits are similar, whether half- or full-wave, single-phase or three-phase. Such circuits are said to be *uncontrolled* since the rectified output voltage and current are a function only of the applied excitation, with no mechanism for varying the output level. In such circuits, regardless of the configuration, diodes are almost always used to achieve rectification. This is because diodes are inexpensive, are readily available with low and high power capabilities, and have terminal characteristics that are simple and well understood. However, since diode circuits are nonlinear, i.e., their circuit topological modes vary with time, their analysis can be challenging, as illustrated in Chapter 3. To simplify the analysis it will be assumed throughout this chapter that the diodes are ideal as discussed in Chapter 2. In this chapter, we will discuss the single- and three-phase uncontrolled rectifier circuits of both the half- and full-bridge configurations as shown in the general block diagram representation in Fig. 7.1.

The analyses of rectifier circuits will include resistive, inductive, and capacitive loads, as well as loads with dc sources. We should point out that *rectification* is a term generally used for converting ac to dc with power flowing unidirectionally to the load, whereas *inversion* is a term used when dc power is converted to ac power. In diode and SCR circuits (SCR will be discussed in Chapter 8), inversion refers to the process in which power flow is from the dc side to the ac side. Another important point is that the circuits in this chapter are also known as *naturally commutating* or *line-commutating converters* since the diodes are naturally turned off and on by being connected directly to a single- or three-phase ac line supply.

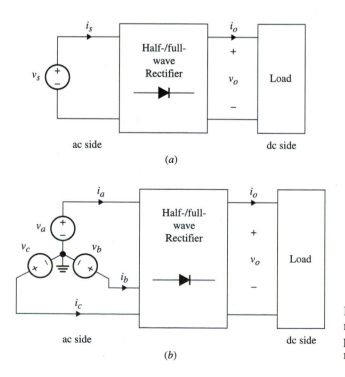

Figure 7.1 Block diagram representation for (*a*) single-phase and (*b*) three-phase rectifier circuits.

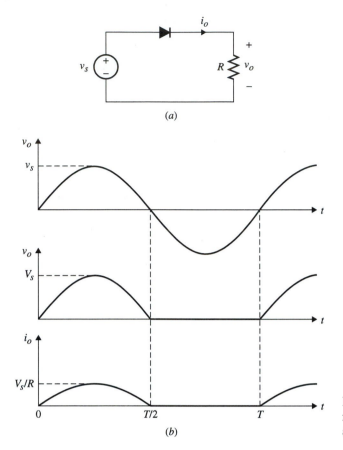

Figure 7.2 (*a*) Half-wave rectifier circuit. (*b*) Voltage and current waveforms.

7.1   SINGLE-PHASE RECTIFIER CIRCUITS

7.1.1   Resistive Load

Figure 7.2(*a*) shows the simplest single-phase, half-wave rectifier circuit under a resistive load. The diode conducts only when the source voltage v_s is positive, as shown in Fig. 7.2(*b*). Throughout this chapter and the next, the source voltage v_s is given by $v_s(t) = V_s \sin \omega t$, where V_s is the peak voltage of the source and ω is its frequency. At $t = T/2$, when v_s becomes negative, the diode ceases to conduct since it does not allow a negative current to flow through it.

The average value of the output voltage, V_o, which represents the dc component of the output signal, v_o, is calculated by evaluating the following integral over one period, T:

$$
\begin{aligned}
V_o &= \frac{1}{T} \int_0^T v_o(t)\, dt \\
&= \frac{1}{T} \int_0^{T/2} V_s \sin \omega t\ dt \\
&= \frac{V_s}{\pi}
\end{aligned}
\tag{7.1}
$$

The average load current is given by

$$I_o = \frac{V_o}{R}$$
$$= \frac{V_s}{\pi R}$$

(7.2)

We observe that the output voltage waveform contains an ac component (ripple) at the same frequency as the ac source; hence, the output dc component is only 32% of the peak applied voltage ($V_o = 0.32 V_s$). Because of their low dc output and high ripple voltages, half-wave rectifier circuits are not practical and are limited to low-voltage applications.

EXAMPLE 7.1

Consider the half-wave rectifier circuit of Fig. 7.2(a) with a resistive load of 25 Ω and a 60 Hz ac source of 110 V rms.

(a) Calculate the average values of v_o and i_o.

(b) Calculate the rms values of v_o and i_o.

(c) Calculate the average power delivered to the load.

(d) Repeat part (c) by assuming that the source has a resistance of 60 Ω.

SOLUTION

(a) The average values of v_o and i_o are given by

$$V_o = \frac{V_s}{\pi}$$
$$= \frac{\sqrt{2}(110 \text{ V})}{\pi}$$
$$= 49.52 \text{ V}$$

and

$$I_o = \frac{V_o}{R}$$
$$= 1.98 \text{ A}$$

(b) The rms value of the output voltage is expressed as

$$V_{o,\text{rms}} = \sqrt{\frac{1}{T} \int_0^T v_o^2 \, dt}$$
$$= \sqrt{\frac{1}{T} \int_0^{T/2} V_s^2 \sin^2 \omega t \, dt}$$

(7.3)

Substituting for $\sin^2 \omega t = (1 - \cos 2\omega t)/2$, Eq. (7.3) gives $V_{o,\text{rms}} = 77.78 \text{ V}$.

The rms current is given by

$$I_{o,\text{rms}} = \frac{V_{o,\text{rms}}}{R}$$
$$= 3.11 \text{ A}$$

(c) The average power delivered to the load over one cycle is obtained from the following relation:

$$P_o = \frac{1}{T}\int_0^T i_o v_o \, dt$$

$$= \frac{1}{TR}\int_0^{T/2} v_o^2 \, dt$$

$$= \frac{V_{o,\text{rms}}^2}{R}$$

$$= 242 \text{ W}$$

(d) With a 60 Ω source resistance, R_s, the average power is given by

$$P_o = \frac{R V_{o,\text{rms}}^2}{(R + R_s)^2}$$

$$= 20.93 \text{ W}$$

EXERCISE 7.1

Calculate the efficiency of the circuit of Example 7.1 for parts (c) and (d).

ANSWER   100%, 54.5%.

A more practical circuit is the full-wave rectifier shown in Fig. 7.3(a). Its typical waveforms are shown in Fig. 7.3(b).

The circuit operates as follows: When $v_s > 0$, D_1 and D_4 turn on, and D_2 and D_3 turn off. The voltage across D_2 and D_3 is $-v_s$, which keeps them off. At $t = T/2$, the negative source voltage stops D_1 and D_4 from conducting and causes D_2 and D_3 to begin conducting, until $t = T$, when the cycle repeats. The average output voltage and average output current are twice that of the half-wave rectifier under a resistive load.

EXERCISE 7.2

Calculate the rms and average diode current for $v_s = 120\sqrt{2}\sin(2\pi 60t)$ V and $R = 25\ \Omega$ in Fig. 7.2(a).

ANSWER   3.4 A rms, 2.16 A

EXERCISE 7.3

Repeat Example 7.1 for the full-wave rectifier circuit given in Fig. 7.3(a).

ANSWER   99 V, 3.96 A, 110 V rms, 4.4 A rms, 484 W, 41.87 W

7.1.2   Inductive Load

An increase in the conduction period of the load current of Fig. 7.2 can be achieved by adding an inductor in series with the load resistance as shown in Fig. 7.4(a), resulting in a half-wave rectifier circuit under an inductive load. This means the

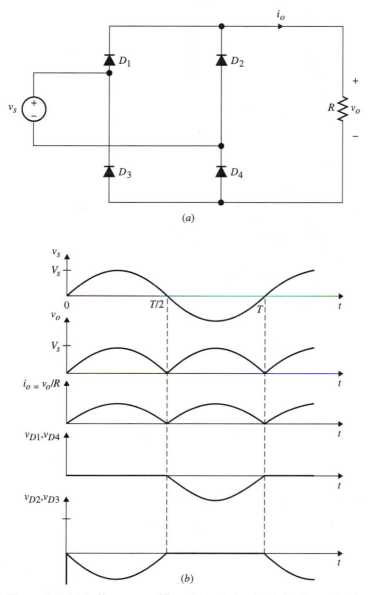

Figure 7.3 (*a*) Full-wave rectifier with resistive load. (*b*) Current and voltage waveforms.

load current flows not only during $v_s > 0$, but also for a portion of $v_s < 0$. The diode is kept in the *on* state by the inductor's voltage, which offsets the negative voltage of $v_s(t)$. The load current is present between $T/2$ and T, but never for the entire period, regardless of the inductor size. This can be easily explained by assuming that the diode conducts for the entire period. Consequently, the output voltage v_o must equal v_s, since the diode voltage is zero. This can occur only when the load current is alternating. This is clearly a contradiction, and there must be a time in which the diode stops conducting. Figure 7.4(*b*) shows the steady-state waveforms for the circuit of Fig. 7.4(*a*). The equivalent circuit mode for $t < t_2$ and $t > t_2$ are shown in Fig. 7.4(*c*) and (*d*), respectively, where t_2 is the time when D stops conducting.

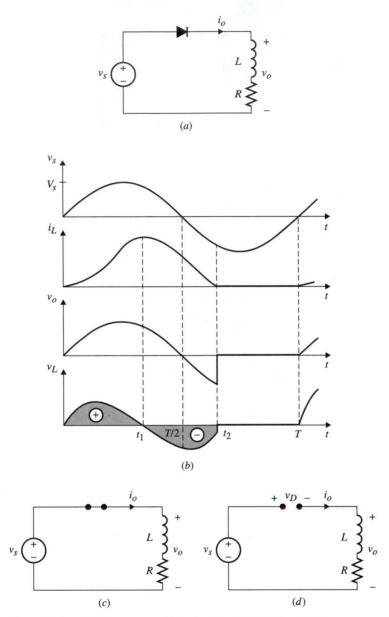

Figure 7.4 (*a*) Half-wave rectifier circuit with inductive load.
(*b*) Current and voltage waveforms. (*c*) Equivalent circuit for $0 \leq t \leq t_2$.
(*d*) Equivalent circuit for $t_2 \leq t \leq T$.

In steady state, the average value of the inductor voltage is zero. This is indicated by the equal positive and negative areas of v_L in Fig. 7.4(*b*). At $t = t_1$, the load current reaches the peak, at which $v_L = 0$. The average value of v_o is calculated from Fig. 7.4(*b*) as follows:

$$V_o = \frac{1}{T} \int_0^{t_2} V_s \sin \omega t \, dt$$

$$= \frac{V_s}{2\pi} (1 - \cos \omega t_2)$$

(7.4)

The times t_1 and t_2 can be determined from the output current equation, to be derived next. A mathematical expression for the output current, i_o, can be obtained by considering the equivalent circuit for $0 \le t \le t_2$ shown in Fig. 7.4(c). The first-order differential equation that mathematically represents this circuit is given by

$$\frac{di_o}{dt} + \frac{R}{L}i_o = \frac{1}{L}V_s \sin \omega t \tag{7.5}$$

Solving Eq. (7.5) for i_o requires obtaining both the transient, $i_{o,\text{tr}}(t)$, and steady-state, $i_{o,\text{ss}}(t)$, solutions, as given by

$$i_o(t) = i_{o,\text{tr}}(t) + i_{o,\text{ss}}(t)$$

The transient current response, also known as the *natural response*, is obtained by setting the forced excitation, v_s, to zero, resulting in the following current equation:

$$i_{o,\text{tr}}(t) = I_{o,i}e^{-t/\tau} \tag{7.6}$$

where

$I_{o,i}$ = initial value of $i_{o,\text{tr}}(t)$ at $t = 0$

τ = circuit time constant, equal to L/R

The steady-state or *forced* response is obtained by using the equivalent phasor circuit shown in Fig. 7.5. The phasor current is given by

$$I_{o,\text{ss}} = \frac{V_s \angle 0^\circ}{|Z| \angle \theta} = \frac{V_s}{|Z|} \angle -\theta \tag{7.7}$$

where

$$|Z| = \sqrt{(\omega L)^2 + R^2}$$

$$\theta = \tan^{-1}\frac{\omega L}{R}$$

In the time domain, we obtain the steady-state output current response as

$$i_{o,\text{ss}}(t) = \frac{V_s}{|Z|} \sin(\omega t - \theta) \tag{7.8}$$

The total solution for i_o is obtained by adding Eqs. (7.6) and (7.8), to yield

$$i_o(t) = I_{o,i}e^{-t/\tau} + \frac{V_s}{|Z|} \sin(\omega t - \theta) \tag{7.9}$$

The initial value, $I_{o,i}$, is obtained from evaluating Eq. (7.9) at $t = 0^+$ with $i_o(0^+) = 0$, to give

$$I_{o,i} = \frac{V_s \omega L}{|Z|^2} \tag{7.10}$$

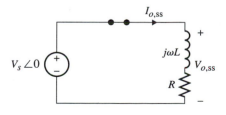

Figure 7.5 Equivalent phasor circuit for Fig. 7.4(c).

The time at which the diode stops conducting, $t = t_2$, must be determined in order to evaluate the average load voltage given in Eq. (7.4). Evaluating Eq. (7.9) at $t = t_2$, with $i_o(t_2) = 0$, we obtain the following expression in terms of t_2:

$$t_2 = -\tau \ln\left[-\sqrt{1 + \left(\frac{R}{\omega L}\right)^2} \sin(\omega t_2 - \theta) \right] \tag{7.11}$$

Since t_2 is in both sides of Eq. (7.11), a numerical analysis method is required to solve for it. By normalizing the average output voltage as $V_{no} = V_o/V_s$, we can obtain a plot of this normalized value versus a normalized load resistance (R_n), which is defined as

$$R_n = \frac{\omega L}{R} \tag{7.12}$$

In terms of R_n, the average output voltage is expressed by

$$V_{no} = \frac{1 - \cos \omega t_2}{2\pi} \tag{7.13a}$$

where

$$\omega t_2 = -R_n \ln\left[-\sqrt{1 + \frac{1}{R_n^2}} \sin(\omega t_2 - \theta) \right] \tag{7.13b}$$

$$\theta = \tan^{-1} R_n \tag{7.13c}$$

Note that, unlike the resistive half-wave rectifier circuit, this circuit experiences load regulation since the average output voltage varies with the change of the load.

One important design criterion in power converters is what is known as *load regulation*. This parameter gives a measure of how the output voltage deviates from its desired value when the load changes. If we assume the output voltage takes on a minimum value at low load, V_{ol}, and a maximum value at full load, V_{of}, then the load regulation (LR_{load}) is defined by

$$LR_{load} = \left| \frac{V_{of} - V_{ol}}{V_{of}} \right| \times 100\% \tag{7.14}$$

Another important parameter is *line regulation*, LR_{line}, which gives a measure of the output variation when the line input voltage changes. The line regulation is defined as

$$LR_{line} = \frac{V_{o,max} - V_{o,min}}{V_{o,min}} \times 100\% \tag{7.15}$$

where $V_{o,max}$ and $V_{o,min}$ are the average output voltages at maximum and minimum line voltages, respectively.

EXAMPLE 7.2

Consider the inductive-load half-wave rectifier circuit of Fig. 7.4(a) with $v_s = 120 \sin 377t$, $L = 60$ mH, and $R = 12\ \Omega$. Determine: (a) V_o, (b) I_o, (c) LR_{line} if V_s is changed by $\pm 20\%$, and (d) LR_{load} if R decreases by 50%.

SOLUTION

(a) The normalized average output voltage for $\omega L/R \approx 1.9$ is obtained from Eqs. (7.11) and (7.13a).

$$V_{no} = 0.213$$

Then we obtain the average output voltage as

$$V_o = V_s V_{no,\text{ave}}$$

$$= 25.56 \text{ V}$$

(b) The average output current is

$$I_o = \frac{V_o}{R}$$

$$= 2.13 \text{ A}$$

(c) The minimum and maximum peak input voltages are given as

$$V_{s,\min} = 96 \text{ V}$$

$$V_{s,\max} = 144 \text{ V}$$

The corresponding average outputs are

$$V_{o,\min} = 20.44 \text{ V}$$

$$V_{o,\max} = 30.67 \text{ V}$$

The line regulation is obtained from Eq. (7.15) as $LR_{\text{line}} = 50\%$.

(d) With the load resistance decreasing by 50%, the new value is $R = 6 \ \Omega$, resulting in $\omega L/R = 3.77$ and $V_{no} = 0.156$. This gives a new value of the average output voltage equal to 18.72 V. The resultant load regulation is $LR_{\text{load}} = 26.73\%$. This is considered very poor load regulation.

A more useful half-wave rectifier circuit is one in which a diode is added across the output as shown in Fig. 7.6(a). Unlike Fig. 7.4(a), the load current in Fig. 7.6(a) is permitted to be continuous and the output voltage cannot go negative.

A careful investigation of Fig. 7.6(a) shows that D_1 and D_2 cannot conduct simultaneously. For example, if we assume both diodes are on, then the average output voltage would be zero and V_s simultaneously. This is not possible. In the steady-state condition, this circuit has two topological modes of operation.

Mode 1: D_1 Is On and D_2 Is Off   Mode 1 occurs when v_s is positive, causing D_1 to become forward biased and D_2 reverse biased. The equivalent circuit model is shown in Fig. 7.6(b). The time interval for this mode is $0 \le t < T/2$. The equation that governs this mode is

$$\frac{di_o}{dt} + \frac{R}{L}i_o = \frac{1}{L}v_s(t) \tag{7.16}$$

Equation (7.16) is similar to Eq. (7.5) and has the following complete solution.

$$i_{o1}(t) = \frac{V_s}{|Z|}\sin(\omega t - \theta) + I_1 e^{-t/\tau} \tag{7.17}$$

where

$$|Z| = \sqrt{(\omega L)^2 + R^2}$$

$$\tau = \frac{L}{R}$$

$$\theta = \tan^{-1}\frac{\omega L}{R}$$

and I_1 is constant and to be determined from the initial conditions. The subscript of the load current i_{o1} denotes the current i_o of mode 1.

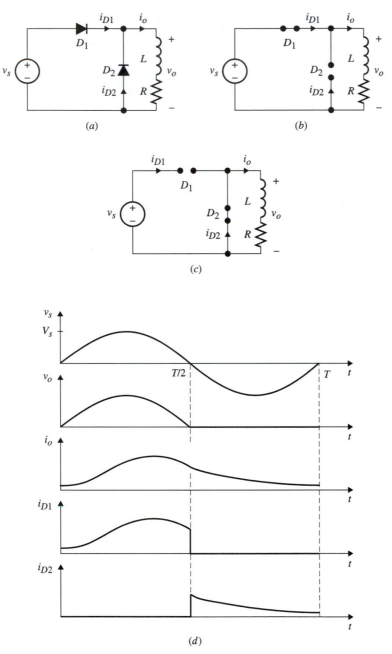

Figure 7.6 (*a*) Half-wave rectifier with a flyback diode. (*b*) Mode 1: D_1 is on and D_2 is off. (*c*) Mode 2: D_1 is off and D_2 is on. (*d*) Voltage and current waveforms.

Mode 2: D_1 Is Off and D_2 Is On Mode 2 occurs for $v_s < 0$, at which D_1 becomes reverse biased and D_2 turns on to pick up the load current. This way the load current commutates between D_1 and D_2 to maintain its continuity. The equivalent circuit is shown in Fig. 7.6(*c*), and the governing equation for this topology is given by

$$\frac{di_o}{dt} + \frac{R}{L}i_o = 0$$

The solution of this equation consists of only the transient response as shown in Eq. (7.18).

$$i_{o2}(t) = I_2 e^{-(t-T/2)/\tau} \tag{7.18}$$

where I_2 is constant and can be solved for, along with I_1, from the following relations that must hold in the steady state:

$$i_{o1}(0) = i_{o2}(T) \tag{7.19a}$$

$$i_{o1}(T/2) = i_{o2}(T/2) \tag{7.19b}$$

The left-side currents of Eqs. (7.19) are obtained from Eq. (7.17), and the right-side currents are obtained from Eq. (7.18). Consequently, we obtain the solution for I_1 and I_2 as follows:

$$I_1 = \frac{V_s}{|Z|} \sin\theta \frac{1 + e^{-T/2\tau}}{1 - e^{-T/\tau}} \tag{7.20}$$

$$I_2 = \frac{V_s}{|Z|} \sin\theta \frac{1 + e^{-T/2\tau}}{1 - e^{-T/\tau}} \tag{7.21}$$

The waveforms for i_o, i_{D1}, i_{D2}, and v_o are shown in Fig. 7.6(d).

If the load inductance is assumed infinitely large, then there will be no ripple in the load current, $i_o(t)$, Eqs. (7.17) and (7.18) become equal, and a constant load current is obtained as shown in the waveforms of Fig. 7.7.

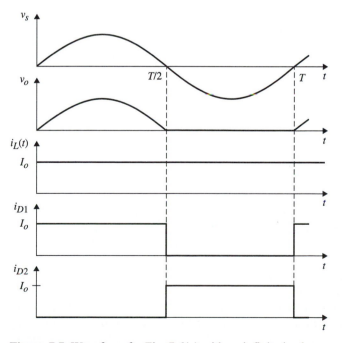

Figure 7.7 Waveform for Fig. 7.6(a) with an infinite load inductance.

The load resistance according to the following relation limits the value of the dc output current:

$$I_o = \frac{V_o}{R}$$
$$= \frac{V_s}{\pi R}$$

(7.22)

One important observation from Fig. 7.7 is that the source current is a pulsating squarewave, an unattractive feature in power electronic circuits. The above circuit is also known as a *current doubler* since the average output current is twice the average input current.

EXERCISE 7.4

Consider the circuit shown in Fig. 7.6(*a*). Determine V_o, I_o, and the inductor currents at $t = 0$ and at $t = T/2$ in the steady state by using $v_s = 220 \sin 377t$, $R = 15\ \Omega$. Use (*a*) $L = 20$ mH, (*b*) $L = 200$ mH, and (*c*) L infinitely large.

ANSWER   70 V, 4.67 A, 16 mA, 5.9 A, 3.2 A, 6 A, 4.67 A, 4.67 A

EXERCISE 7.5

Determine the rms currents of diodes D_1 and D_2 in Fig. 7.3(*a*).

ANSWER   3.4 A rms

A final observation to be made about the half-wave inductive-load rectifier circuit is that unlike in Fig. 7.4(*a*), the inductor current exists for the entire period. As a result, the rectifier circuits of Fig. 7.4(*a*) and Fig. 7.6(*a*) are said to operate in the discontinuous and continuous conduction modes, respectively.

7.1.3 Capacitive Load

In some applications in which a constant output voltage is desirable, a series inductor is replaced by a parallel capacitor as shown in the half-wave rectifier circuit of Fig. 7.8(*a*).

Like the single-diode inductive-load half-wave rectifier, the circuit of Fig. 7.8(*a*) can be analyzed by considering the two topological modes resulting from the conducting states of the diode. Let mode 1 correspond to the time interval in which the diode conducts, resulting in the equivalent circuit shown in Fig. 7.8(*b*). Figure 7.8(*c*) shows the equivalent circuit when the diode is off, representing mode 2.

The circuit will be in mode 1 as long as the capacitor voltage is lower than the input voltage, v_s. If we assume the diode starts conducting at $t = t_1$, then the next time it will turn off to enter mode 2 of operation is when $t = T/4$, at which the input voltage reaches its peak. The diode will stay off until the capacitor and the input voltages become equal again at $t = T + t_1$. Figure 7.9 shows the steady-state waveforms for the output voltage and the source current.

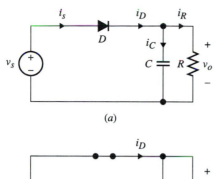

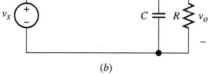

(a)

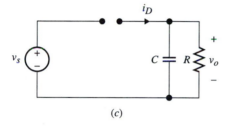

(b)

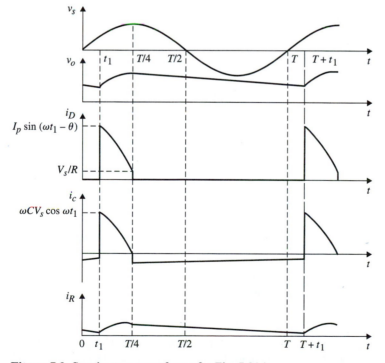

(c)

Figure 7.8 (a) Half-wave rectifier with capacitive load. (b) Equivalent circuit mode when the diode is on. (c) Equivalent circuit mode when the diode is off.

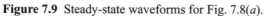

Figure 7.9 Steady-state waveforms for Fig. 7.8(a).

The analytical expressions for v_o and i_s are given as follows. For $t_1 \leq t < T/4$,

$$v_o = V_s \sin \omega t \tag{7.23}$$

$$i_s(t) = \frac{V_s}{R} \sin \omega t + \omega C V_s \cos \omega t \tag{7.24}$$

The peak input current occurs at t_1, i.e., $i_s(t_1) = I_p$, where

$$I_p = \omega C V_s \cos \omega t_1 + \frac{V_s}{R} \sin \omega t_1$$

For $T/4 \leq t < T + t_1$,

$$v_o(t) = V_s e^{-(t - T/4)/\tau} \tag{7.25}$$

$$i_s(t) = 0 \tag{7.26}$$

where the time constant is given by $\tau = RC$.

The value of $t = t_1$ can be determined by equating Eqs. (7.23) at $t = t_1$ and (7.25) at $t = t_1 + T$. This yields the following relation:

$$\sin \omega t_1 - e^{(t_1 + 3T/4)/\tau} = 0 \tag{7.27}$$

Equation (7.27) may be evaluated iteratively to obtain t_1. The average value of the output voltage is determined from the following:

$$
\begin{aligned}
V_o &= \frac{1}{T} \int_{t_1}^{T+t_1} v_o(t)\, dt \\[2mm]
&= \frac{1}{T} \left[\int_{t_1}^{T/4} V_s \sin \omega t \, dt + \int_{T/4}^{T+t_1} V_s e^{-(t - T/4)/\tau}\, dt \right] \\[2mm]
&= \frac{V_s}{2\pi} \left[\cos \omega t_1 + \tau \omega (1 - e^{-(t_1 + 3T/4)/\tau}) \right] \\[2mm]
&= \frac{V_s}{2\pi} \left[\cos \omega t_1 + \tau \omega (1 - e^{-(t_1 + 3T/4)/\tau}) \right]
\end{aligned}
$$

EXAMPLE 7.3

Consider the half-wave rectifier circuit of Fig. 7.8(a) with $v_s(t) = 20 \sin 2\pi 60 t$, $R = 10 \text{ k}\Omega$, and $C = 47 \ \mu\text{F}$.

(a) Sketch the waveforms for v_o, i_R, i_D, and i_c.

(b) Determine the output voltage ripple.

SOLUTION

(a) First we need to find the time t_1 at which the diode turns on. Using $\tau = RC = 470$ ms and $T = 16.67$ ms, through several iterations, Eq. (7.27) gives $t_1 \approx 3.48$ ms. At this time, the output voltage is 19.3 V and the resistor current is 1.93 mA.

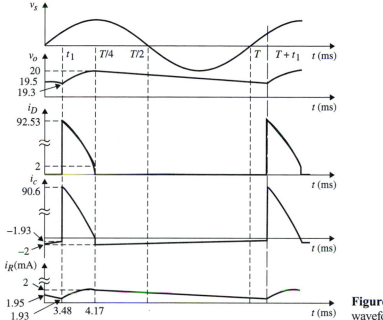

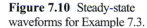

Figure 7.10 Steady-state waveforms for Example 7.3.

At $t = T/4 = 4.17$ ms, $v_o = 20$ V and $i_R = 2$ mA. When the diode turns off, $i_c = -i_R = -2$ mA. However, while the diode is conducting, i_c is given by

$$i_c(t) = \omega C V_s \cos \omega t$$

$$= 354 \cos \omega t \text{ mA}$$

At $t_1 \approx 3.48$ ms, $i_c = 354 \cos 377(3.48 \text{ ms}) = 90.6$ mA; hence, $i_D = i_R + i_c = 1.93 + 90.6 = 92.53$ mA. Finally, the output voltage at T or at $t = 0$ is obtained from Eq. (7.25) as

$$v_o = V_s e^{-(t - T/4)/\tau}$$

$$= 20 e^{-3T/4\tau}$$

$$= 20 e^{-3(16.67)/(4)(470)}$$

$$= 19.5 \text{ V}$$

The waveforms are shown in Fig. 7.10.

(b) The output voltage ripple, V_r, is given by

$$V_r = v_{o,max} - v_{o,min}$$

$$= v_o(T/4) - v_o(t_1)$$

$$= 20 - 19.3$$

$$= 0.7 \text{ V}$$

EXERCISE 7.6

Calculate the average output voltage for Example 7.3.

ANSWER   19.68 V

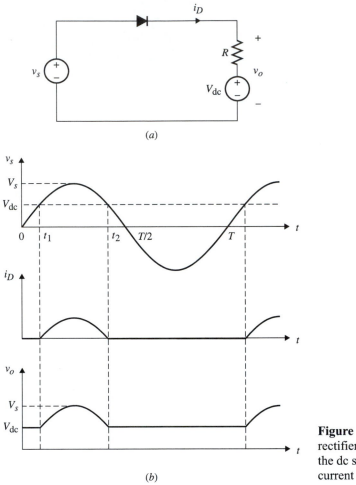

(a)

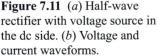

(b)

Figure 7.11 (a) Half-wave rectifier with voltage source in the dc side. (b) Voltage and current waveforms.

7.1.4   Voltage Source in the dc Side

Figure 7.11(a) shows a half-wave rectifier circuit with a voltage source in series with the load resistance. This constant voltage may be considered as an emf source located in the load side, as in the case of dc machine circuit modeling.

Let us assume that the diode starts conducting at $t = t_1$. This time occurs when $v_s = V_{dc}$. Furthermore, the diode stops conducting at $t = t_2$, when the current i_o becomes zero. Typical waveforms are shown in Fig. 7.11(b). See Problem 7.19 to derive the equation for the load current during the conduction period $t_1 \le t \le t_2$.

7.2   THE EFFECT OF AC-SIDE INDUCTANCE

7.2.1   Half-Wave Rectifier with Inductive Load

Let us consider the half-wave diode rectifier circuit by including an ac-side inductance, which is inevitable in all ac systems. Studying the effect of the line inductance on the behavior of diode circuits is important since in practical rectifier systems connected directly to the line voltage, the transformer leakage inductance and the line parasitic inductance might cause line voltage and current disturbances and could affect the average power drawn from the mains. Moreover, an external inductance might also be added in some applications in order to limit di/dt or dv/dt. For simplicity, we normally assume that the ac-side inductance is finite and may be represented by a discrete value, L_s, in series with the source voltage given in the block diagram representation in Fig. 7.1. Figure 7.12(a) shows the circuit for the half-wave rectifier including L_s and under an inductive load. In the following analysis, we will assume that $L/R \gg T/2$ so that the load current, i_o, is constant. This assumption is valid since in many applications the load inductance is very much larger than the ac-side inductance.

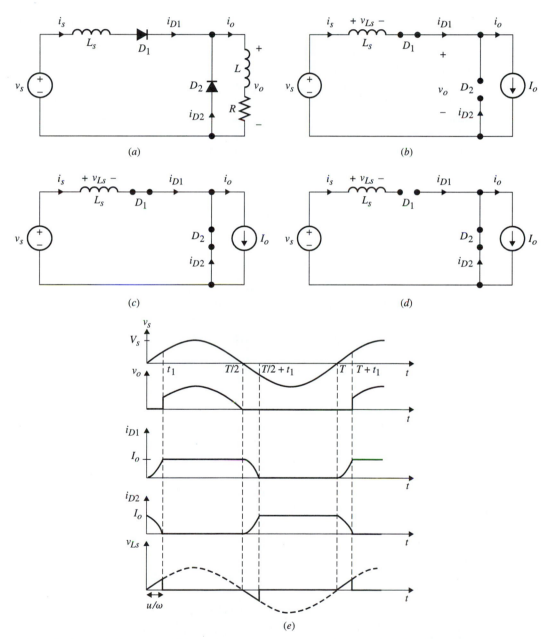

Figure 7.12 (a) Half-wave rectifier with ac-side inductance. (b) Mode 1: D_1 is on and D_2 is off during $v_s(t) > 0$. (c) Mode 2 (commutation mode): D_1 and D_2 are on while $v_s(t) < 0$. (d) Mode 3: D_1 is off and D_2 is on while $v_s(t) < 0$. (e) The waveforms for v_s, v_o, i_{D1}, i_{D2}, and v_{Ls}.

The behavior of the circuit can be easily analyzed by assuming that one of the diodes, D_1, is conducting for some time during the positive cycle of $v_s(t)$, while D_2 is off. Since the current in D_1 is constant, then the voltage across L_s is zero and the voltage across D_2 is positive. Hence, our assumption is valid. The equivalent circuit under this mode is shown in Fig. 7.12(b). During this mode, we have the following current and voltage values:

$$i_{D1} = i_s(t) = I_o \qquad i_{D2} = 0 \qquad v_{Ls} = 0 \qquad v_o = v_s$$

At $t = T/2$, $v_s(t)$ starts to become negative, causing D_1 to stop conducting. However, since the current in D_1 is the same as the inductance current, which is not allowed to change instantaneously, D_2 turns on in order to maintain the inductor current's continuity. During this overlapping time, when both diodes are conducting, $i_s(t)$ changes from $+I_o$ to zero, while $i_{D2}(t)$ changes from zero to $+I_o$. This is known as a *commutation process*, during which currents are moved from one branch of the circuit to another branch. This is why the ac-side inductance, L_s, is known as the *commutation inductance*. This circuit mode of operation is referred to as commutation mode, as shown in Fig. 7.12(c). The waveforms for v_s, v_o, i_{D1}, i_{D2}, and v_{Ls} are shown in Fig. 7.12(e).

During mode 2, the following equations hold:

$$v_{Ls} = v_s(t) = L_s \frac{di_s}{dt}$$

$$i_{D1} = i_s(t)$$

$$i_{D2} = I_o - i_{D1}$$

$$v_o = 0$$

The initial condition for $i_s(t)$ at $t = T/2$ is I_o. Using the above v_{Ls} equation with the given initial condition, we obtain the following input current integral relation:

$$i_s(t) = \frac{1}{L_s} \int_{T/2}^{t} v_{Ls}(t) \, dt + I_o \tag{7.28}$$

Substituting for $v_{Ls}(t) = v_s(t) = V_s \sin \omega t$ in the integral and solving for $i_s(t)$, we obtain

$$i_s(t) = I_o - \frac{V_s}{L_s \omega}(1 + \cos \omega t) \quad T/2 \leq t \leq t_1 + T/2 \tag{7.29}$$

At the end of the commutation period, $t = t_1 + T/2$; $i_s(t)$ becomes zero, forcing D_1 to turn off at zero current; and D_2 remains forward biased, carrying the load current as shown in the circuit of mode 3 given in Fig. 7.12(d). In this mode we have the following current and voltage equations:

$$i_{D1} = i_s(t) = 0 \quad i_{D2} = I_o \quad v_{Ls} = 0 \quad v_o = 0$$

To solve for the commutation time, t_1, or the commutation angle, $u = \omega t_1$, we evaluate $i_s(t)$ at $t = T/2 + t_1$ and set it to zero. The load current and t_1 are given by

$$I_o = \frac{V_s}{L_s \omega}(1 - \cos \omega t_1) \tag{7.30}$$

$$t_1 = \frac{1}{\omega} \cos^{-1}\left(1 - \frac{L_s \omega I_o}{V_s}\right) \tag{7.31}$$

The average output voltage is expressed by

$$V_o = \frac{1}{T} \int_{t_1}^{T/2} v_s(t)\, dt$$

$$= \frac{V_s}{\pi}\left(1 - \frac{\omega L_s I_o}{2 V_s}\right)$$

(7.32)

If we normalize the output voltage by V_s and the load current by Z_o, where $Z_o = \omega L_s$ is the characteristic impedance of the ac line, we have the following normalized output quantities:

$$V_{no} = \frac{V_o}{V_s}$$

$$= \frac{1}{\pi}\left(1 - \frac{I_{no}}{2}\right)$$

(7.33)

$$I_{no} = \frac{Z_o I_o}{V_s}$$

$$= 1 - \cos u$$

(7.34)

or u is given by

$$u = \cos^{-1}(1 - I_{no})$$

(7.35)

The maximum voltage across L_s, $V_{Ls,\max}$, occurs at $t = t_1$ and is given by

$$V_{Ls,\max} = v_s(t_1)$$

$$= V_s \sin \omega t_1$$

Normalizing $V_{Ls,\max}$ by V_s and substituting the normalized current I_{no} into the above relation, we obtain

$$V_{nLs,\max} = \sqrt{(2 - I_{no})I_{no}}$$

(7.36)

Figure 7.13 shows the curve for $V_{nLs,\max}$ vs. I_{no}, and Fig. 7.14(a) and (b) shows I_{no} vs. u and V_{no} vs. I_{no}, respectively.

EXERCISE 7.7

Consider the circuit of Fig. 7.12(a) with $L_s = 0.05$ mH, $R = 20\ \Omega$, $I_o = 1$ A, and $v_s(t) = 100 \sin 2\pi 60 t$. Find: ($a$) the average output voltage, (b) the commutation intervals in

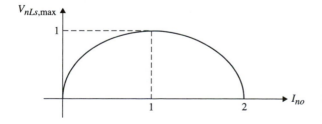

Figure 7.13 Normalized maximum inductor voltage versus normalized output current.

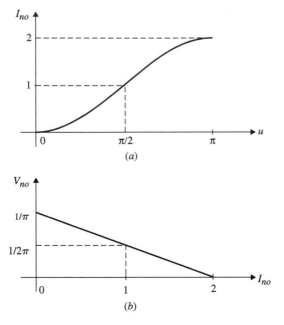

(a)

(b)

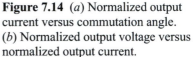

Figure 7.14 (*a*) Normalized output current versus commutation angle. (*b*) Normalized output voltage versus normalized output current.

seconds and degrees, (*c*) the rms diode currents, (*d*) the peak inverse diode voltages, and (*e*) the peak voltage across the inductor L_s.

ANSWER 31.83 V, 51.5 μs, 1.1°, 585.26 A rms, 100 V, 1.94 V

7.2.2 Half-Wave Rectifier with Capacitive Load

Consider the half-wave rectifier circuit with a capacitive load including an ac-side inductance as shown in Fig. 7.15(*a*). Again, to simplify the analysis we assume $RC \gg T/2$ so the output voltage is constant. Its equivalent circuit is given in Fig. 7.15(*b*).

Let us assume first that the diode is off while the input voltage is negative. As shown in Fig. 7.15(*a*), for the diode to conduct, its voltage must be positive. At $t = t_1$, $v_s = V_o$ and the diode turns on. The waveforms are shown in Fig. 7.15(*c*), and the voltage across L_s is given by

$$v_{Ls} = L_s \frac{di_s}{dt}$$

$$= v_s(t) - V_o$$

Since the initial inductor current value at $t = t_1$ is zero, the solution for $i_s(t)$ is given by

$$i_s(t) = \frac{V_s}{\omega L_s}(\cos \omega t_1 - \cos \omega t) - \frac{V_o}{L_s}(t - t_1) \tag{7.37}$$

The peak inductor current occurs when $v_{Ls} = 0$, i.e., when $v_s = V_o$. Let us assume this occurs at $t = t_2$.

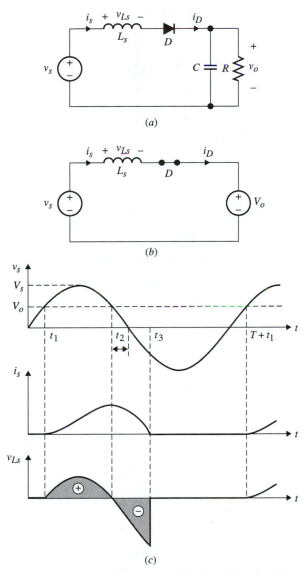

Figure 7.15 (*a*) Half-wave capacitive-load circuit with commutative inductance. (*b*) Equivalent circuit with $RC \gg T/2$. (*c*) Current and voltage waveforms.

$$I_{s,\text{peak}} = i_s(t_2)$$

$$= \frac{V_s}{\omega L_s}(\cos \omega t_1 - \cos \omega t_2) - \frac{V_o}{L_s}(t_2 - t_1) \qquad (7.38)$$

where t_2 is obtained from

$$V_o = V_s \sin \omega t_2 \qquad (7.39)$$

Also at $t = t_1$, when the diode first turns on, we have

$$V_o = V_s \sin \omega t_1 \qquad (7.40)$$

where $t_1 + t_2 = T/2$.

As long as $i_s(t)$ is positive, D remains on, and at $t = t_3$, i_s reaches zero, causing D to turn off. The current at $t = t_3$ is given by

$$i_s(t_3) = \frac{V_s}{\omega L_s}(\cos \omega t_1 - \cos \omega t_3) - \frac{V_o}{L_s}(t_3 - t_1) \tag{7.41}$$
$$= 0$$

Equations (7.39), (7.40), and (7.41) can be used to solve for t_1, t_2, and t_3 numerically. The average output current is given by

$$I_o = \frac{1}{T}\int_{t_1}^{t_3} i_s(t) \, dt$$

$$= \frac{V_s}{2\pi L_s}\left((t_3 - t_1)\cos \omega t_1 - \frac{\sin \omega t_3}{\omega} + \frac{\sin \omega t_1}{\omega}\right) - \frac{V_o}{2L_sT}(t_3 - t_1)^2$$

Normalizing I_o and V_o, and using the relation (7.39) for $t = t_2$, we obtain

$$I_{no} = \frac{\omega L_s I_o}{V_s}$$

$$= \frac{1}{2\pi}\left(\alpha\cos \omega t_1 - \sin \omega t_3 + \sin \omega t_1 - \frac{\alpha^2}{2}V_{no}\right) \tag{7.42}$$

where $\alpha = \omega(t_3 - t_1)$, which is the diode conducting time. Figure 7.16 shows I_{no} vs. α, and Fig. 7.17 shows V_{no} vs. α.

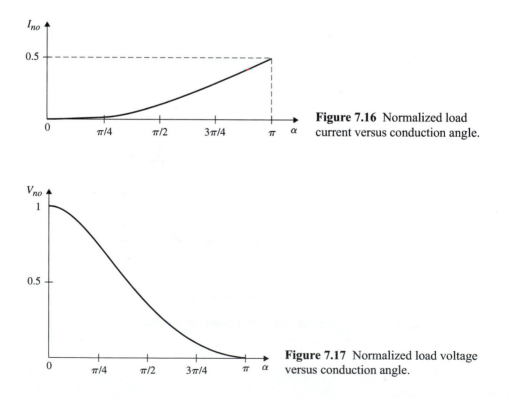

Figure 7.16 Normalized load current versus conduction angle.

Figure 7.17 Normalized load voltage versus conduction angle.

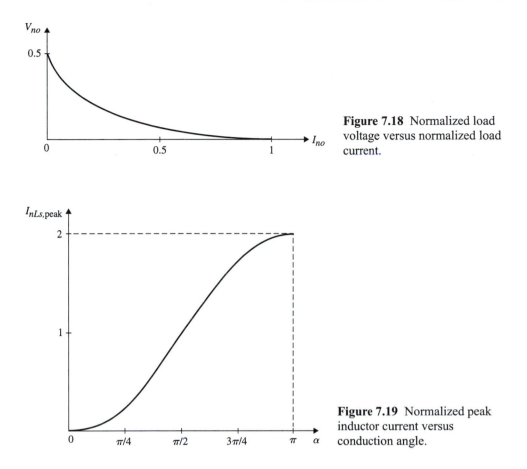

Figure 7.18 Normalized load voltage versus normalized load current.

Figure 7.19 Normalized peak inductor current versus conduction angle.

The effect of L_s on the variation of the average output voltage is shown in Fig. 7.18.

The normalized inductor peak current is given by

$$I_{nLs,\text{peak}} = \frac{\omega L_s I_{Ls,\text{peak}}}{V_s}$$

$$= (\cos \omega t_1 - \cos \omega t_2) - \alpha V_{no}$$

A plot of the normalized peak inductor current vs. α under different load conditions is shown in Fig. 7.19.

EXAMPLE 7.4

Consider the half-wave rectifier of Fig. 7.15(a) with $v_s(t) = 25 \sin 377t$, $V_o = 20$ V, and $L_s = 2$ mH. Determine V_{no}, I_{no}, α, and $I_{nLs,\text{peak}}$.

SOLUTION   First find the value of t_1:

$$t_1 = \frac{1}{\omega}\sin^{-1}\left(\frac{V_o}{V_s}\right) = 2.46 \text{ ms}$$

Through some numerical iterations we find that $t_3 = 7.66$ ms. The normalized average output current can be obtained theoretically as follows:

$$I_o = \frac{V_s}{2\pi L_s}\left((t_3 - t_1)\cos\omega t_1 - \frac{\sin\omega t_3}{\omega} + \frac{\sin\omega t_1}{\omega}\right) - \frac{V_o}{2L_s T}(t_3 - t_1)^2$$

$$= 0.99 \text{ A}$$

Hence,

$$I_{no} = \frac{\omega L_s I_o}{V_s} = 0.03$$

The other parameters are given as

$$\alpha = \omega(t_3 - t_1) = 1.96 \text{ rad/s}$$

$$V_{no} = \frac{V_o}{V_s} = 0.8 \text{ V}$$

$$I_{nLs,\text{peak}} = (\cos\omega t_1 - \cos\omega t_2) - \alpha V_{no} = 0.17$$

EXERCISE 7.8

Determine the diode rms current for Example 7.4.

ANSWER   2.078 A

7.2.3   Full-Wave Rectifier with Inductive Load

The effect of the commutation inductance on the waveforms and parameters of the full-wave rectifier shown in Fig. 7.20(a) is investigated next. As before, we will assume $L/R \gg T/2$, resulting in a constant load current, I_o. The typical waveforms are shown in Fig. 7.20(b). The behavior of the circuit can be discussed similarly to the case for the half-wave rectifier circuit of Fig. 7.6(a). We first assume that D_1 and D_4 are conducting during the positive cycle of $v_s(t)$. Since I_o is constant, the voltage across L_s is zero, and $v_o(t) = v_s(t)$ as shown in Fig. 7.21(a), which is represented as mode 1. Notice that the voltages across D_2 and D_3 are negative and equal to $-v_s(t)$. As long as $v_s(t)$ is positive, D_2 and D_3 remain off. The second topological mode starts at $t = T/2$, when $v_s(t)$ becomes negative. At this time, D_2 and D_3 start conducting while D_1 and D_4 remain on in order to maintain the continuity of the inductor current. Figure 7.21(b) shows this mode, when all diodes are conducting. The following current and voltage relations apply in this mode:

$$v_{Ls} = L_s\frac{di_s}{dt}$$

$$v_o = 0$$

$$i_{D1} = i_{D4} = \frac{I_o + i_s}{2} \tag{7.43}$$

$$i_{D2} = i_{D3} = \frac{I_o - i_s}{2}$$

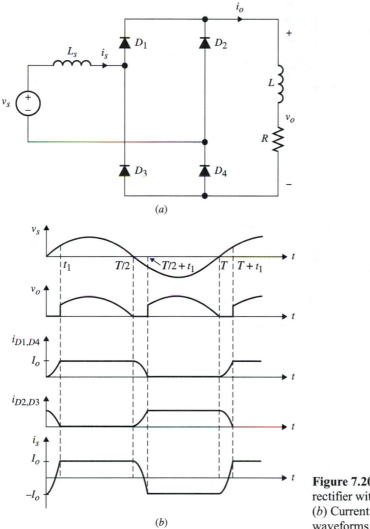

Figure 7.20 (*a*) Full-wave rectifier with ac-side inductance. (*b*) Current and voltage waveforms.

Using the initial inductor value of $i_s(T/2) = +I_o$ in solving the above equation for $i_s(t)$, we obtain

$$i_s(t) = I_o - \frac{V_s}{L_s \omega}(1 + \cos \omega t) \tag{7.44}$$

At $t = t_1 + T/2$, we have $i_s(t_1 + T/2) = -I_o$. This gives the following relation for the commutation period:

$$\cos \omega t_1 = 1 - \frac{2L_s \omega I_o}{V_s} \tag{7.45}$$

In terms of normalized values, $u = \omega t_1$ is given by

$$u = \cos^{-1}(1 - 2I_{no})$$

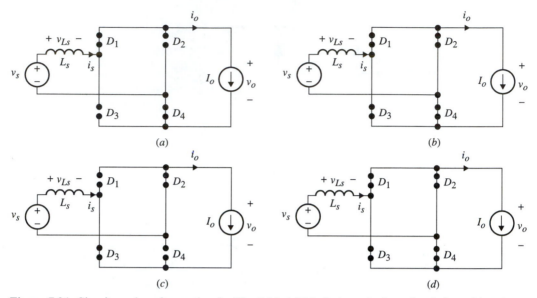

Figure 7.21 Circuit modes of operation for Fig. 7.20. (a) Mode 1: equivalent circuit for $v_s(t) > 0$. (b) Mode 2: equivalent circuit for $v_s(t) < 0$ and during diode commutation period. (c) Mode 3: equivalent circuit for $v_s(t) < 0$. (d) Mode 4: equivalent circuit for $v_s(t) > 0$ and during diode commutation period.

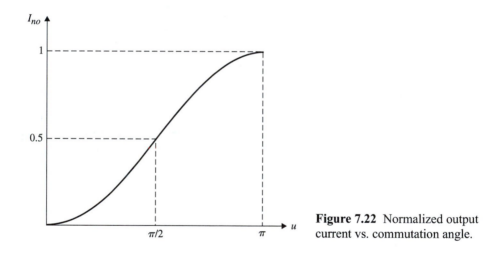

Figure 7.22 Normalized output current vs. commutation angle.

Figure 7.22 shows the curves for the normalized output current as a function of the commutation angle. The larger the commutation inductance, the less the average output current. At $t = T/2 + t_1$, the currents in D_1 and D_4 become zero and the currents in D_2 and D_3 reach I_o, as shown in the topology of mode 3 given in Fig. 7.21(c). The output voltage is equal to $v_s(t)$. This mode remains until $t = T$, when the voltage becomes positive again. Since $v_s(t) > 0$, D_1 and D_4 turn on and D_2 and D_3 remain on until the current commutates from D_2 and D_3 to D_1 and D_4 at the end of the commutation period u, and the circuit enters mode 4 as shown in Fig. 7.21(d). The current and voltage equations are the same as those for mode 2. The initial condition for $i_s(t)$ is $i_s(T) = -I_o$, and the solution for $i_s(t)$ is given by

$$i_s(t) = -I_o + \frac{V_s}{L_s \omega}(\cos \omega t - 1) \qquad (7.46)$$

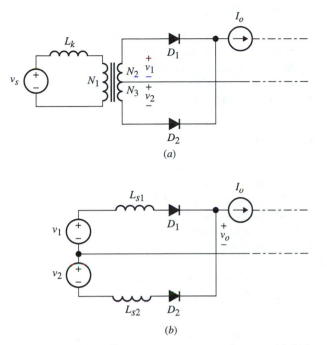

Figure 7.23 (*a*) Full-wave center-tap transformer with leakage inductance. (*b*) Simplified equivalent circuit.

We notice from the above relations that the average output voltage, inductor peak current, and rms value of $i_s(t)$ are directly related to the normalized load current.

To illustrate the source of L_s in the full-wave rectifier, we refer to Fig. 7.23. This is a center-tap transformer including a leakage inductance L_k reflected to the primary side. The equivalent circuit for Fig. 7.23(*a*) can be drawn as shown in Fig. 7.23(*b*), where the inductance is reflected to the primary side. The voltage relations are given by

$$v_1 = \frac{N_2}{N_1} v_s$$

$$v_2 = \frac{N_3}{N_1} v_s$$

where $L_{s1} = (N_1/N_2)^2 L_k$ and $L_{s2} = (N_1/N_3)^2 L_k$.

EXERCISE 7.9

Consider the center-tap full-wave rectifier of Fig. 7.23(*a*) with $N_1 = 10$, $N_2 = 5$, $N_3 = 15$, $L_k = 10~\mu H$, $I_o = 20$ A, and $v_s = 100 \sin 2\pi(60)t$. Assume the diodes have no forward voltage drops.

(a) Determine the average output voltage.

(b) Repeat part (*a*) by assuming each diode has a 1 V forward drop.

ANSWER   63.61 V, 62.63 V

7.3 THREE-PHASE RECTIFIER CIRCUITS

Many industrial applications require high power that a single-phase system is unable to provide. Three-phase diode rectifier circuits are used widely in high-power applications with low output ripple. It is important that we understand the basic concepts of a three-phase rectifier circuit. In this section, we will cover both half- and full-wave rectifier circuits under resistive and high inductive loads.

7.3.1 Three-Phase Half-Wave Rectifier

Figure 7.24 shows the general configuration for an m-phase half-wave rectifier connected to a single load. The explanation of the circuit is quite simple since all diode cathodes are connected to the same point, creating a diode-OR arrangement. At any given time, the highest anode voltage will cause its corresponding diode to conduct, with all other diodes in the reverse-bias state. In other words, the output voltage will ride on the peak voltage at all times. Figure 7.25 shows four random sine functions and the output voltage.

The half-wave three-phase resistive-load rectifier circuit is shown in Fig. 7.26(a). We assume that the three-phase voltage source is a Δ configuration with the three balance voltages given by

$$v_1 = V_s \sin \omega t \qquad v_2 = V_s \sin(\omega t - 120°) \qquad v_3 = V_s \sin(\omega t - 240°)$$

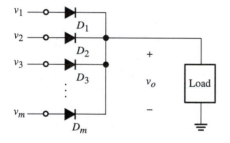

Figure 7.24 m-phase half-wave rectifier connected to a single load.

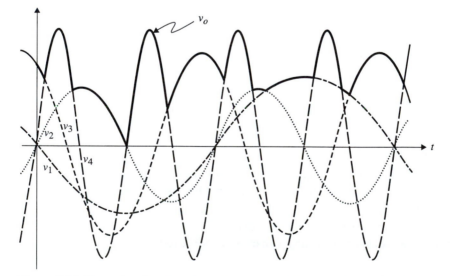

Figure 7.25 Example of random four-phase sinusoidal input voltages and the output voltage.

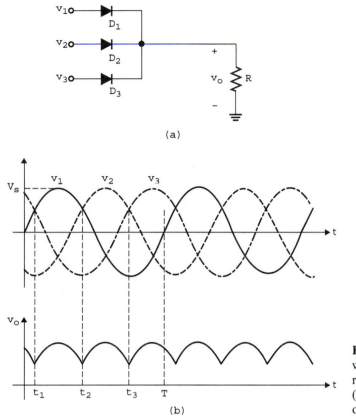

(a)

(b)

Figure 7.26 (*a*) Half-wave three-phase resistive-load rectifier. (*b*) Waveforms for the output voltage.

Figure 7.26(*b*) shows the output voltage waveform. This circuit is also known as a *three-pulse rectifier* circuit. Here, the number of pulses refers to the number of voltage peaks in a given cycle.

The circuit works as follows. At $t = t_1$, we have $v_1 = v_3$. This occurs when $t_1 = T/12$, and at $t_2 = 5T/12$ we have $v_2 = v_1$. The average output voltage is given by

$$
\begin{aligned}
V_o &= \frac{1}{T}\int_0^T v_o(t)\,dt \\[6pt]
&= \frac{3}{T}\int_{t_1}^{t_2} v_1(t)\,dt \\[6pt]
&= \frac{3}{T}\int_{T/12}^{5T/12} v_1(t)\,dt \\[6pt]
&= \frac{3\sqrt{3}}{2\pi}V_s
\end{aligned}
\tag{7.47}
$$

The source voltage arrangement of v_1, v_2, and v_3 can be a wye (Y) or a delta (Δ) connection, as shown in Fig. 7.27(*a*) and (*b*), respectively. The voltages v_a, v_b, and v_c are the balanced, positive-sequence phase voltages given by

$$
v_a = V_s \sin\omega t \qquad v_b = V_s \sin(\omega t - 120°) \qquad v_c = V_s \sin(\omega t - 240°)
$$

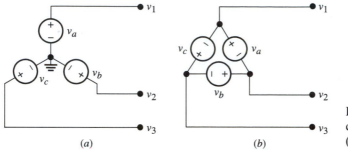

Figure 7.27 Three-phase connections: (*a*) Y and (*b*) Δ connections.

The voltage set v_1, v_2, and v_3 are line voltages. In the Δ configuration, the line-to-line voltages are the same as the phase voltages, whereas the line-to-line voltages in the Y configuration are given by

$$v_{12} = v_1 - v_2$$
$$= v_a - v_b$$
$$= v_{ab}$$
$$= \sqrt{3} V_s \sin(\omega t + 30°)$$

$$v_{23} = v_2 - v_3$$
$$= v_b - v_c$$
$$= v_{bc}$$
$$= \sqrt{3} V_s \sin(\omega t - 90°)$$

$$v_{31} = v_3 - v_1$$
$$= v_c - v_a$$
$$= v_{ca}$$
$$= \sqrt{3} V_s \sin(\omega t + 150°)$$

The phase diagram representation for the phase and the line-to-line voltages is given in Fig. 7.28.

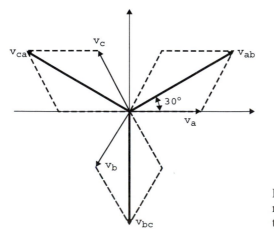

Figure 7.28 Phasor diagram representation for phase and line-to-line three-phase voltages.

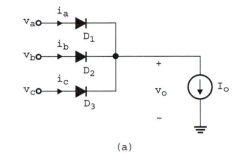

(a)

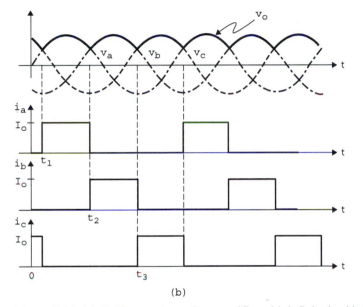

(b)

Figure 7.29 (*a*) Half-wave three-phase rectifier with infinite load inductance. (*b*) Current and voltage waveforms.

Figure 7.29(*a*) and (*b*) shows the equivalent circuit for a half-wave circuit under a highly inductive load and its waveforms, respectively.

The half-wave circuit with a capacitive load is shown in Fig. 7.30(*a*) and its output waveform is shown in Fig. 7.30(*b*). To obtain an expression for the output voltage ripple of Fig. 7.30(*b*), we will make the assumption that the load time constant, RC, is much larger than $T/3$. Let $t = t_1$ be the time at which the diode D_1 turns on when v_a is the most positive. At $t = T/4$, $v_a = V_s$ and diode D_1 stops conducting; since the capacitor voltage at this instant is equal to the peak voltage V_s, all diodes remain off and the capacitor starts discharging through R. At $t = t_1 + T/3$, v_o becomes equal to v_b and D_2 turns on. Once again, the capacitor voltage increases with v_b until it equals V_s, at which the cycle repeats. For $T/4 < t < T/3 + t_1$, the output voltage is given by

$$v_o = V_s e^{-(t - T/4)/RC}$$

At $t = t_1 + T/3$, the output voltage equals v_b, i.e.,

$$v_b(t_1 + T/3) = V_s \sin(\omega(t_1 + T/3) - 120°)$$

$$= V_s e^{-(t_1 + T/12)/RC}$$

This equation can be rewritten as follows:

$$\sin \omega t_1 = e^{-(t_1 + T/12)/RC}$$

As in the case of the single-phase rectifier circuit, we can solve for t_1 numerically. The ripple voltage, V_r, is obtained from the following relation:

$$V_r = V_s - v_b(t_1 + T/3)$$
$$= V_s(1 - \sin \omega t_1) \tag{7.48}$$

Using the approximation $t_1 \approx T/4$, the above equation becomes

$$v_o(t_1 + T/3) = V_s e^{-(T/4 + T/12)/RC}$$
$$= V_s e^{-T/3RC}$$
$$\cong V_s\left(1 - \frac{T}{3RC}\right)$$

so the ripple is given by

$$v_r = V_s - v_o(t_1 + T/3)$$
$$= \frac{T}{3RC}V_s \tag{7.49}$$
$$= \frac{1}{3fRC}V_s$$

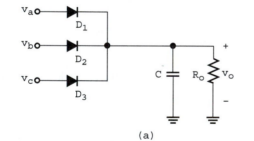

(a)

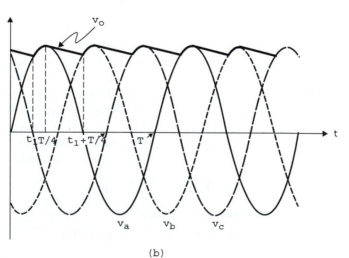

(b)

Figure 7.30 (a) Three-phase half-wave capacitive-load rectifier. (b) Output voltage waveform.

It can be shown that this ripple is one-third of the voltage ripple for the single-phase circuit.

Consider the three-phase rectifier shown in Fig. 7.30(*a*) with $f = 60$ Hz and $R_o = 10$ kΩ. Design for C so that the output voltage ripple doesn't exceed 1% of the peak voltage.

ANSWER $C \geq 55.6 \ \mu F$

It can be shown that the diode conduction time can be approximated by

$$\frac{T}{2} - t_1 - \sqrt{\frac{2V_r}{V_s}} \tag{7.50}$$

Using this expression it is possible to obtain approximate average and rms current values for the diodes.

7.3.2   Three-Phase Full-Wave Rectifier

Let us now consider the full-bridge rectifier circuit including the commutation inductance. The full-bridge rectifier is more common since it provides a high output voltage and less ripple. First let us consider the full-bridge circuit under a resistive load as shown in Fig. 7.31. Let us assume that v_a, v_b, and v_c are the three phase voltages in Fig. 7.32. The easiest way to approach the full-bridge rectifier circuit of Fig. 7.31 is to consider it as a combination of a positive commutating diode group D_1, D_2, and D_3, and a negative commutating diode group D_4, D_5, and D_6. Since no commutating inductance is included, at any given time only two diodes are conducting simultaneously—one from the positive group and the other from the negative group. To show that this is the case, let us redraw Fig. 7.31 as shown in Fig. 7.33. It is clear from Fig. 7.32 that only one diode from the upper group can be on during the positive cycle, and only one diode among the lower group can be on during the negative cycle. The output voltage, v_o, is given by

$$v_o = v_{o1} - v_{o2}$$

where v_{o1} and v_{o2} are the output voltages of the positive and negative commuting diode groups to ground, respectively.

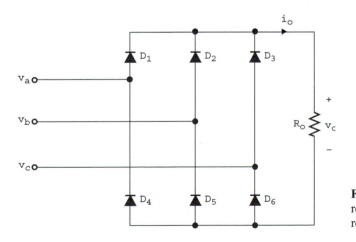

Figure 7.31 Full-bridge rectifier circuit under resistive load.

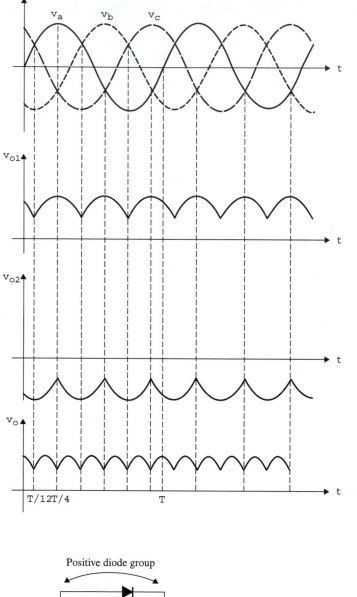

Figure 7.32 Output voltage waveforms for Fig. 7.31.

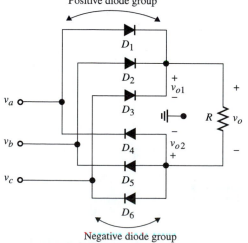

Figure 7.33 Equivalent circuit for Fig. 7.31.

The average output voltage is given by

$$V_o = \frac{1}{T/6} \int_{T/12}^{T/4} V_a \, dt$$

$$= \frac{3\sqrt{3}}{\pi} V_s$$

(7.51)

Let us consider the full-bridge rectifier under a highly inductive load using a Y-connected voltage source with $L/R \gg T/6$ as shown in Fig. 7.34(a). Table 7.1 shows all six modes of operation with the corresponding diode conduction angles and currents, where $\theta = \omega t$.

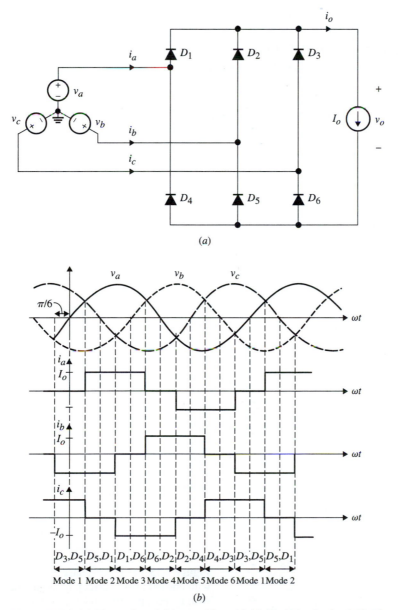

(a)

(b)

Figure 7.34 (a) Three-phase bridge rectifier with highly inductive load. (b) Phase current waveforms.

Table 7.1 Conduction Modes for Fig. 7.34

Mode	Conduction angle	Diodes conducting	i_a	i_b	i_c	Most positive absolute voltage
I	$-\pi/6 < \theta < \pi/6$	D_3, D_5	0	$-I_o$	I_o	$\lvert v_b\rvert, v_c$
II	$\pi/6 < \theta < \pi/2$	D_5, D_1	I_o	$-I_o$	0	$v_a, \lvert v_b\rvert$
III	$\pi/2 < \theta < 5\pi/6$	D_1, D_6	I_o	0	$-I_o$	$v_a, \lvert v_c\rvert$
IV	$5\pi/6 < \theta < 7\pi/6$	D_6, D_2	0	I_o	$-I_o$	$v_b, \lvert v_c\rvert$
V	$7\pi/6 < \theta < 9\pi/6$	D_2, D_4	$-I_o$	I_o	0	$\lvert v_a\rvert, v_b$
VI	$9\pi/6 < \theta < 11\pi/6$	D_4, D_3	$-I_o$	0	I_o	$\lvert v_a\rvert, v_c$

The waveforms for the output voltage and the diode and line currents are shown in Fig. 7.34(b). The output voltage is the same as that in the resistive case given in Fig. 7.31.

7.4 AC-SIDE INDUCTANCE IN THREE-PHASE RECTIFIER CIRCUITS

7.4.1 Half-Wave Rectifiers

Figure 7.35(a) shows a three-phase, half-wave rectifier circuit including an ac-side commutating inductance under a highly inductive load. The phase voltages v_a, v_b, and

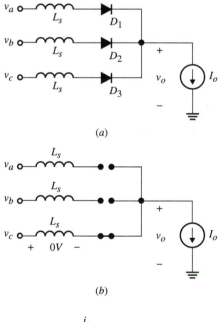

(a)

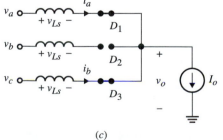

(b)

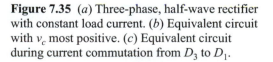

(c)

Figure 7.35 (a) Three-phase, half-wave rectifier with constant load current. (b) Equivalent circuit with v_c most positive. (c) Equivalent circuit during current commutation from D_3 to D_1.

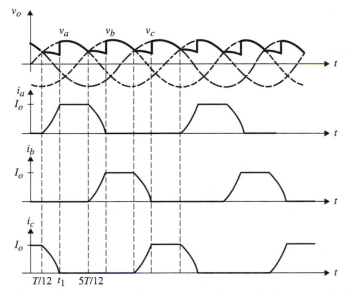

Figure 7.36 Waveforms for v_o, i_a, i_b, and i_c and the three phase voltages.

v_c are shown in Fig. 7.36. The operation of the circuit can be analyzed as follows. Prior to $\pi/12$, v_c is the most positive and D_3 is conducting, with D_1 and D_2 off, as shown in the equivalent circuit of Fig. 7.35(b). During this interval, $i_c = +I_o$ and $v_o = v_c$.

At $t = T/12$, v_a and v_c become equal and D_1 starts conducting, as in the single-phase case. Since i_c cannot change to zero instantaneously, D_3 remains on while the load current is being moved from D_3 to D_1, as shown in Fig. 7.35(c). At this instant, $i_a = 0$, $i_c = +I_o$, and v_o is the average value between v_a and v_c, to be shown next.

Assuming equal ac-side inductances, the following relations are obtained:

$$v_a = L_s \frac{di_a}{dt} + v_o \qquad (7.52a)$$

$$v_c = L_s \frac{di_c}{dt} + v_o \qquad (7.52b)$$

Since $i_a + i_c = I_o$, then $di_c/dt = -di_a/dt$, resulting in the following relation for v_o:

$$v_o = \frac{v_c + v_a}{2} \qquad (7.53)$$

Substituting for v_o in Eq. (7.53), we obtain the following differential equation for i_a:

$$\frac{di_a}{dt} = \frac{1}{2L_s}(v_a - v_c)$$

Substitute for $v_a = V_s \sin(\omega t)$ and $v_c = V_s \sin(\omega t - 240°)$ and use the initial values of $i_c(T/12) = I_o$, $i_a(T/12) = 0$. The solution for i_c is then given by

$$i_c(t) = \frac{V_s\sqrt{3}}{2L_s\omega}[1 - \cos(\omega t - 30°)]$$

The waveforms for v_o, i_a, i_b, and i_c are given in Fig. 7.36.

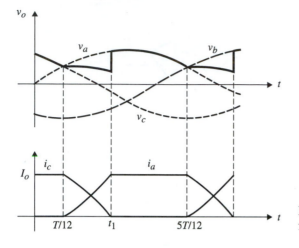

Figure 7.37 Commutation waveforms from phase c to phase a in Fig. 7.35(a).

The commutation angle is obtained by evaluating $i_a(t)$ at t_1 when the current reaches I_o, as shown in the commutating waveform from phase c to phase a in Fig. 7.37.

$$i_a(t) = \frac{V_s\sqrt{3}}{2L_s\omega}[1 - \cos(\omega t_1 - T/12)] = I_o \qquad (7.54)$$

$$i_c(t) = I_o - i_a = I_o - \frac{V_s\sqrt{3}}{2L_s\omega}[1 - \cos(\omega t - 30°)] \qquad (7.55)$$

Hence, u can be found from the following expression:

$$u = \cos^{-1}\left(1 - \frac{2L_s\omega I_o}{V_s\sqrt{3}}\right) \qquad (7.56)$$

Since the pulse width of the phase current is smaller than that of the single-phase case, the power factor for a three-phase is lower. Finally, the output voltage is given by the following integral:

$$V_o = \frac{3}{T}\left[\int_{T/12}^{t_1} v_o\,dt + \int_{t_1}^{5T/12} v_a\,dt\right] \qquad (7.57)$$

Substituting for v_o in the above equation, Eq. (7.57) can be rewritten as

$$V_o = \frac{3}{T}\left[\int_{T/12}^{t_1} \frac{v_a + v_c}{2}\,dt + \int_{t_1}^{5T/12} v_a\,dt\right]$$

$$= \frac{3}{T}\left[\int_{T/12}^{t_1} v_a\,dt - \int_{T/12}^{t_1} \frac{v_a - v_c}{2}\,dt + \int_{t_1}^{5T/12} v_a\,dt\right] \qquad (7.58)$$

$$= \frac{3}{T}\left[\int_{T/12}^{5T/12} v_a\,dt - \int_{T/12}^{t_1} \frac{v_a - v_c}{2}\,dt\right]$$

We recognize that the first integral is the average output voltage for the half-wave three-phase converter with $L_s = 0$. The second integral is the average voltage loss due to the commutation. Substituting $L_s di_a / dt$ for $(v_a - v_c)/2$ in Eq. (7.58), we have

$$V_o = V_o\big|_{L_s=0} - \frac{3}{T}\int_{T/12}^{t_1} L_s \frac{di_a}{dt}\, dt$$

$$= V_o\big|_{L_s=0} - \frac{3}{T}\int_{\pi/6}^{\pi/6+u} L_s \frac{di_a}{dt}\, d\theta \qquad (7.59)$$

$$= V_o\big|_{L_s=0} - \frac{3}{T} L_s I_o$$

Recall that the average value of the voltage when $L_s = 0$ is given by

$$V_o\big|_{L_s=0} = \frac{3\sqrt{3}\,V_s}{2\pi}$$

Then V_o of Eq. (7.59) may be written as

$$V_o = \frac{3\sqrt{3}}{2\pi} V_s \left(1 - \frac{\omega L_s I_o}{\sqrt{3}\,V_s} \right)$$

Normalizing the output voltage by V_s and I_o by $V_s/(\omega L_s)$ gives

$$V_{no} = \frac{3\sqrt{3}}{2\pi}\left(1 - \frac{I_{no}}{\sqrt{3}} \right) \qquad (7.60)$$

7.4.2   Full-Wave Bridge Rectifiers

Let us consider the full-wave, three-phase rectifier circuit including an ac-side inductance. The unidirectional source current in the half-wave circuit has a significant effect on the input power factor. This is why the bridge connection is widely used. Figure 7.38 shows the full-bridge with an ac-side commutation inductance from a three-phase Y-connected voltage source. Again, we assume a large inductive load.

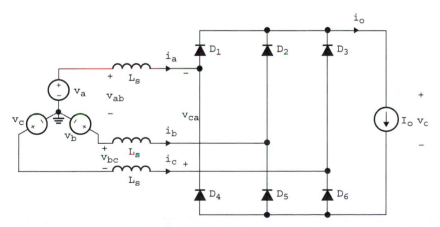

Figure 7.38 Three-phase bridge rectifier with ac-side inductance.

The three source line-to-line voltages are

$$v_{ab} = \sqrt{3}\,V_s \sin(\omega t + 30°) \quad v_{bc} = \sqrt{3}\,V_s \sin(\omega t - 90°) \quad v_{ca} = \sqrt{3}\,V_s \sin(\omega t + 150°)$$

The operation of this circuit can be discussed in a similar way to the case with $L_s = 0$, except in this case the commutation inductance will present an additional commutation angle, u, during which three diodes are on simultaneously. From mode I in Table 7.1, for $\omega t < \pi/6$, $|v_{bc}|$ is the maximum voltage with D_3 and D_5 conducting, as shown in Fig. 7.39(a). In this mode, we have

$$i_a = 0 \quad i_b = -I_o \quad i_c = I_o \quad v_o = -v_{bc} \quad v_{Ls} = 0$$

From Table 7.1, the next mode is when D_5 and D_1 turn on, since the next most positive voltage in this mode is v_{ab}. However, the current will commutate from D_3 to D_1 during a commutation period, u, during which D_3 remains on while D_1 and D_5 are conducting. Figure 7.39(b) shows the equivalent circuit during this commutation period. From mode I to mode II the currents commutate from zero to $+I_o$ in i_a; hence, D_3 remains on for a short time until its current goes to zero at $t = t_1$. In this time interval we have

$$-v_b - v_o - L_s \frac{di_c}{dt} + v_c = 0 \tag{7.61a}$$

$$v_b - v_o - L_s \frac{di_a}{dt} + v_a = 0 \tag{7.61b}$$

Since $i_a + i_c = I_o$, then $di_c/dt = -di_a/dt$. Hence, the above relation becomes

$$v_o = \frac{v_{ab} + v_{cb}}{2} = \frac{v_{ab} + v_{bc}}{2} \tag{7.62}$$

From Eqs. (7.61) and (7.62), we have the following first-order differential equations in terms of the line-to-line voltages and currents:

$$-v_{cb} - v_o - L_s \frac{di_c}{dt} = 0 \tag{7.63a}$$

$$v_{ab} - v_o - L_s \frac{di_a}{dt} = 0 \tag{7.63b}$$

Substituting for v_o in the above equations, we obtain

$$\frac{v_{cb} - v_{ab}}{2L_s} = \frac{di_c}{dt} = -\frac{di_a}{dt}$$

$$\frac{v_{ab} - v_{cb}}{2L_s} = \frac{di_a}{dt}$$

Substituting for $v_{cb} - v_{ab} = v_{ca} = \sqrt{3}\,V_s \sin(\omega t + 150°)$, and since at $t = 0$, $i_c(\pi/6) = I_o$, we have

$$i_c(t) = \frac{1}{2L_s} \int_{\pi/6}^{\omega t} \sqrt{3}\,V_s \sin(\omega t + 150°)\,d\omega t + I_o$$

At $\omega t = u + \pi/6$, we have $i_c(u + \pi/6) = 0$; hence, u is given by

$$u = \cos^{-1}\left(1 - \frac{2L_s \omega I_o}{V_s \sqrt{3}}\right) \tag{7.64}$$

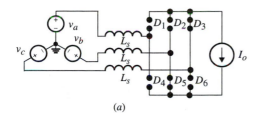

(a)

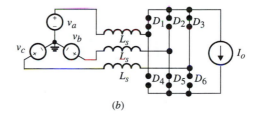

(b)

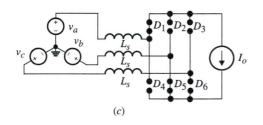

(c)

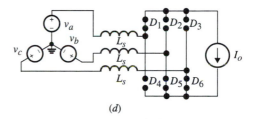

(d)

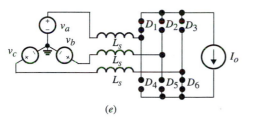

(e)

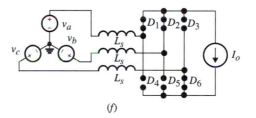

(f)

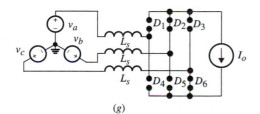

(g)

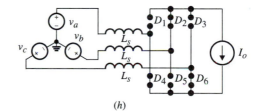

(h)

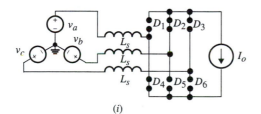

(i)

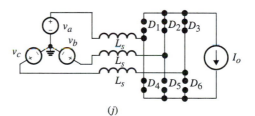

(j)

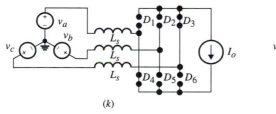

(k)

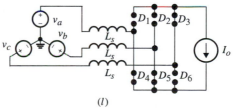

(l)

Figure 7.39 Equivalent circuit modes for mode I to mode VI. (*a*) Mode I: D_3 and D_5 are conducting. (*b*) Mode I and mode II commutation. (*c*) Mode II: D_1 and D_5 are conducting. (*d*) Mode II and mode III commutation. (*e*) Mode III: D_1 and D_6 are conducting. (*f*) Mode III and mode IV commutation. (*g*) Mode IV: D_2 and D_6 are conducting. (*h*) Mode IV and mode V commutation. (*i*) Mode V: D_2 and D_4 are conducting. (*j*) Mode V and mode VI commutation. (*k*) Mode VI: D_3 and D_4 are conducting. (*l*) Mode VI and mode I commutation.

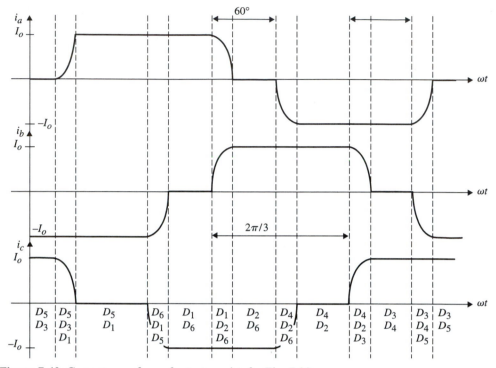

Figure 7.40 Current waveforms for i_a, i_b, and i_c for Fig. 7.38.

At the end of the commutation period u, i_c becomes zero and D_3 stops conducting. Diodes D_1 and D_5 remain on until mode III starts at $\pi/2$, when $|v_{ca}|$ becomes the most positive, turning diodes D_1 and D_6 on and in turn turning D_5 off. In this mode, the current i_b will commutate from I_o to 0 through D_5, and i_c goes from 0 to $-I_o$ through D_6. Diodes D_5 and D_6 will overlap during the commutation period u. The remaining modes and their corresponding mode-to-mode transition are shown in Fig. 7.39(a)–(l). The current waveforms of i_a, i_b, and i_c and the output voltage are shown in Fig. 7.40.

PROBLEMS

In all the problems assume ideal diodes unless stated otherwise.

Single-Phase Rectifiers

Half-Wave Rectifiers

7.1 Determine the power factor and the THD for the waveforms of Example 7.1.

7.2 Consider the half-wave rectifier in Fig. P7.2 with $v_s = 110 \sin \omega t$ V and $R = 1$ kΩ. Assume an ideal diode except the forward resistance (r_d) is 10 Ω. Sketch v_o and determine:

(a) The average output voltage

(b) The rms current in the diode

(c) The peak current in the diode

(d) The input power factor

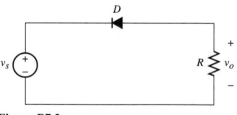

Figure P7.2

7.3 Derive Eqs. (7.9) and (7.11).

7.4 Consider the half-wave inductive-load rectifier circuit of Fig. 7.4(a) with $v_s = 110\sqrt{2}\sin\omega t$ V, $\omega = 60$ Hz, $L = 800$ mH, and $R = 100$ Ω. Find:

(a) The expression for $i_L(t)$

(b) The time t_1 when i_L becomes zero

(c) The average load current

(d) The peak inductor current

(e) The rms current in the inductor

(f) The average power delivered to the load

7.5 Repeat Problem 7.4 with the diode reversed.

7.6 Repeat Problem 7.4 by placing a free-wheeling diode in parallel with the load as shown in Fig. 7.6(a).

7.7 Repeat Example 7.2 for the circuit of Fig. 7.6(a).

D7.8 Design for L and R for the half-wave rectifier of Fig. 7.6(a) to provide a ripple current not to exceed 5% of its average value at $V_o = 20$ V. Assume $v_s = 40 \sin 120 \pi t$ V.

D7.9 The single-phase, half-wave rectifier of Fig. 7.6(a) is required to supply the load with an output ripple current not to exceed 5% of its dc value.

(a) Design for L to meet this requirement. Use $R = 20\ \Omega$, $v_s = 150 \sin 120 \pi t$ V.

(b) What is the average load voltage?

7.10 Derive Eqs. (7.20) and (7.21) for the half-wave rectifier with an L-R load.

D7.11 Design the half-wave rectifier of Fig. 7.4(a) with an L-R load to supply 25 W to the load with a minimum output ripple current of 10% and an average output of 28 V. Use $v_s = 120 \sin (120 \pi t)$.

D7.12 Design a dc power supply with an average dc output of 20 V with a 2% ripple. The power supply should be capable of providing 500 mA to the load. Use the rectifier of Fig. P7.12 for your design with $v_s = 110 \sin (2 \pi 60 t)$ V, and assume the diode's forward voltage drop is 1 V. Specify the diode's ratings, including its peak reverse voltage and rms values.

7.14 Sketch the waveforms for i_s, i_c, i_R, and v_o for Fig. 7.8(a) with $R = 10$ kΩ, $C = 100\ \mu$F, and $v_s = 10 \sin 377t$ V.

Full-Wave Rectifiers

7.15 Determine the power factor and the THD for the waveforms of Fig. 7.3(b).

7.16 Consider the full-wave rectifier with a resistive load of Fig. 7.3(a) with $v_s = 120 \sin \omega t$ and $R = 20\ \Omega$. Find:

(a) The average output voltage

(b) The average output current

(c) The rms output current

(d) The input power factor

(e) The load power

7.17 Consider a full-wave rectifier with an infinite inductive load.

(a) Show that the dc and harmonic components of the output voltage are given by

$$v_o(t) = \frac{2V_s}{\pi} - \frac{4V_s}{\pi} \sum_{n=2}^{\infty} \frac{1}{(n-1)(n+1)} \cos n \omega t$$

for $n = 2, 4, 6, \ldots$.

(b) Show that the nth harmonic of the source current is given by

$$i_s(t) = \frac{4I_o}{\pi} \sum_{n=1}^{\infty} \frac{\sin n \omega t}{n}$$

D7.18 It is required that the circuit shown in Fig. P7.18 provide an output power of 2 kW at an average load current of 20 A. Design for the transformer turns ratio and R to meet this requirement. Assume $L/R \gg (2 \pi / \omega)$. Use $v_s = 100 \sin (2 \pi 60 t)$ V.

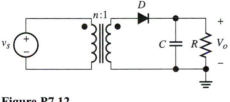

Figure P7.12

D7.13 Design the half-wave rectifier with an R-C load with a 10% output ripple voltage and that provides 10 W to the load. Assume $v_s = 20 \sin (120 \pi t)$ V.

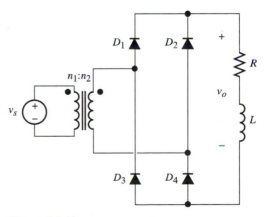

Figure P7.18

7.19 Consider the generalized-type load with a dc source representing a dc-motor-load full-wave rectifier as shown in Fig. P7.19. Assume $v_s = V_s \sin \omega t$ V and $V_{dc} < V_s$. Derive:

(a) The expression for $i_o(t)$

(b) The rms value of $i_o(t)$

(c) The average load current

(d) The ripple load current

(e) Calculate the above values for $V_s = 25$ V, $\omega = 377$ rad/s, $R = 2$ Ω, and $L = 5$ mH.

with an output power of 25 W. Design for the required transformer turns ratio n. Assume $v_s = v_s = 110\sqrt{2}\sin 377t$ V.

D7.22 Redesign Problem 7.21 by assuming $v_s(t)$ fluctuates by ±20%. Design for the worst case.

7.23 Consider the full-wave, center-tap diode rectifier circuit shown in Fig. P7.23.

(a) Sketch v_{o1} and v_{o2}.

(b) Determine the peak inverse voltage for each diode.

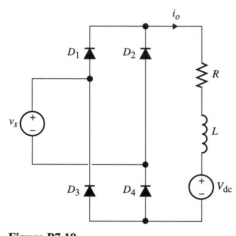

Figure P7.19

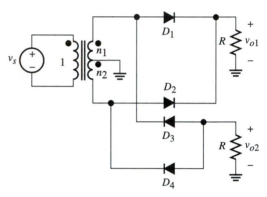

Figure P7.23

7.20 Consider the center-tap full-wave rectifier of Fig. P7.20.

(a) Sketch v_o.

(b) Determine the average output voltage.

(c) Repeat parts (a) and (b) by including each diode drop V_D.

(d) Determine the peak inverse voltage (PIV) for the two diodes.

D7.24 Design the dual center-tap full-wave rectifier of Fig. P7.24 to provide ±15 VDC from a line voltage $v_s = 155 \sin \omega t$ V, where $\omega = 377$ rad/s.

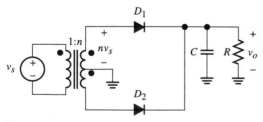

Figure P7.20

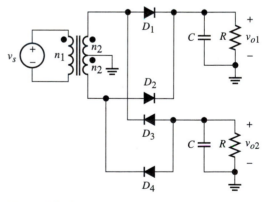

Figure P7.24

D7.21 The full-wave rectifier of Problem 7.20 is to provide an average output voltage of 48 V

7.25 Repeat Problem 7.9 by using the full-bridge rectifier given in Fig. 7.20(a) with $L_s = 0$.

7.26 (a) Calculate the output ripple voltage for the full-wave rectifier of Fig. P7.26. Use $R = 10$ kΩ and $C = 47$ μF, with $v_s = 20 \sin 377 \omega t$.

(b) Show that the expression for the average output voltage is given by

$$V_o = \frac{V_s}{2\pi}\left[\cos\omega t_1 + \tau\omega\left(1 - e^{-(t_1 + 3T/4)/\tau}\right)\right]$$

where $\tau = RC$, V_s is the peak voltage, and t_1 ($0 \leq t_1 \leq T/4$) is the time at which the diode starts conducting.

(c) If it is assumed that the ripple voltage is much smaller than the peak input voltage, i.e., $t_1 = T/4$, and $RC \gg T$, then show that $V_r = V_s/2RCf$. Apply this simplified ripple voltage to part (a).

(d) Use the approximation in part (c) to show that the average diode current is given by

$$I_{D,ave} = \frac{V_s}{2R}\left[1 + \frac{1}{\pi}\sqrt{1 - \left(1 - \frac{V_r}{V_s}\right)^2}\right]$$

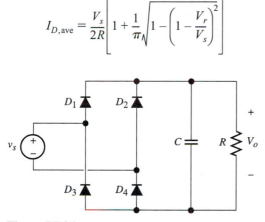

Figure P7.26

7.27 The circuit shown in Fig. P7.27 is called a voltage doubler. It is used when the line voltage of a given system changes from 110 VAC to 220 VAC or vice versa. By changing the switch position, the output voltage can be held constant under both a 110 VAC and a 220 VAC line input. Consider the voltage doubler shown. Determine the voltages v_{o1} and v_{o2} for both cases when the switch is on and off. Assume $v_s = V_s\sin\omega t$ V with $R_1C_1 \gg T$ and $R_2C_2 \gg T$ ($\omega = 2\pi/T$).

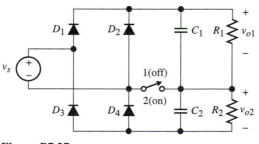

Figure P7.27

7.28 Consider the full-wave rectifier with an LC output filter shown in Fig. P7.28.

(a) Draw the equivalent circuit between a and b by assuming $L = 0$ and $C = \infty$.

(b) Repeat part (a) by assuming $L = \infty$ and $C = 0$.

(c) Calculate the power dissipated in the load in parts (a) and (b). Assume $v_s = V_s\sin\omega t$.

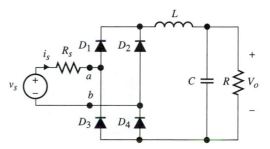

Figure P7.28

D7.29 The single-phase, full-wave rectifier with an R-C load shown in Fig. P7.26 is required to supply the load with an output ripple voltage not to exceed 5% of its dc value.

(a) Design for C to meet this requirement. Use $R = 20\ \Omega$, $v_s = 150\sin 120\pi t$ V.

(b) Calculate the average load current.

The Effect of ac-Side Inductance

Half-Wave Circuits

7.30 Consider the half-wave rectifier shown in Fig. P7.30 with a flyback diode including L_s and the diode resistances. Assume $v_s = V_s\sin\omega t$ V.

(a) Derive the expression for the power dissipated in the rectifier.

(b) Sketch i_s, i_{D2}, and v_o.

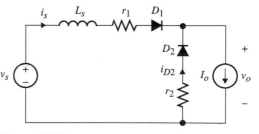

Figure P7.30

D7.31 The circuit of Fig. 7.12(a) is to be designed to deliver a 1 kW output power at $I_o = 20$ A. Assume L_s fluctuates in the range of ±20%. Use $L_s = 0.1$ mH and $v_s(t) = 120 \sin 2\pi 60t$ V. Design for the load resistance and inductance.

Full-Wave Circuits

7.32 Derive the rms expression for the input current $i_s(t)$ of a full-wave rectifier circuit with an ac-side inductance, which is redrawn in Fig. P7.32.

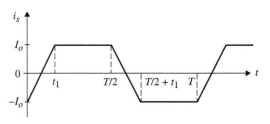

Figure P7.32

7.33 Repeat Problem 7.32 by using the approximation for $i_s(t)$ shown in Fig. P7.33.

Figure P7.33

7.34 (a) Discuss the operation of the circuit in Fig. P7.34 and sketch the waveforms for $v_o, i_{s1}, i_{s2}, v_{Ls1}, v_{Ls2}$. Assume $v_{s1} = V_{s1} \sin \omega t$ V, $v_{s2} = V_{s2} \sin \omega t$ V.

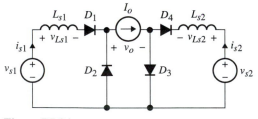

Figure P7.34

(b) Derive the expression for $v_o(t)$.
(c) What is the difference between this circuit and the full-wave bridge circuit of Fig. 7.21?

7.35 (a) Find the ratio between the rms currents for the waveforms shown in Fig. P7.35(a) and (b), representing the line currents for half-wave rectifiers with and without an ac-side inductance, respectively.
(b) Sketch the ratios as a function of $\omega t_1 = u$.

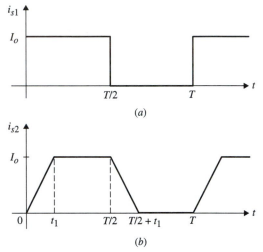

Figure P7.35

Input Power Factor

7.36 Consider the line current for the half-wave rectifier with L_s as shown in Fig. 7.12(e), which is redrawn in Fig. P7.36(a).
(a) Derive the expression for the power factor for $i_s(t)$.
(b) Repeat part (a) by using the trapezoidal approximation shown in Fig. P7.36(b).
(c) Compare the two solutions.

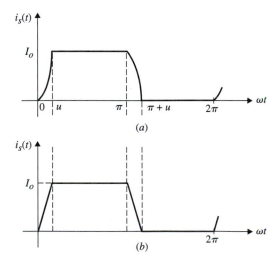

Figure P7.36

7.37 Consider the one-port converter network shown in Fig. P7.37 with $i_s = 5\sin(377t + 30°)$ A, and $v_s = 10\sin(377t - 45°)$ V. Determine:

(a) The instantaneous input power

(b) The input power factor

(c) The average input power

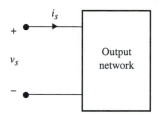

Figure P7.37

7.38 Derive the input power factor and the THD for the input current waveforms shown in Fig. P7.38. Assume the line voltage is pure sinusoidal. (See Appendix B.)

Three-Phase Rectifier Circuit

7.39 **(a)** Show that the average output voltage for the n-phase, half-wave rectifier of Fig. P7.39 is given by

$$V_o = V_s \frac{\sin(\pi/n)}{(\pi/n)}$$

where V_s is the phase peak voltage. What is the value of V_o as $n \longrightarrow \infty$?

(b) Show that the rms and average current in each diode are given by

$$I_{Di,\text{rms}} = \frac{I_o}{\sqrt{n}}$$

$$I_{D,\text{ave}} = \frac{I_o}{n}$$

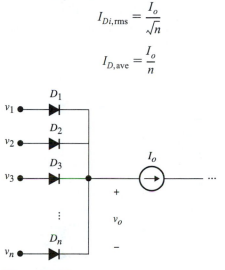

Figure P7.39

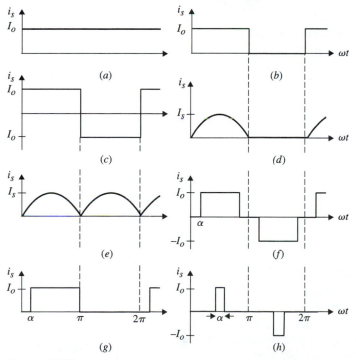

Figure P7.38

7.40 Consider the balanced three-phase rectifier circuit shown in Fig. P7.40 with the delta-connected voltage sources given below:

$$v_a = V_s \sin \omega t \quad v_b = V_s \sin(\omega t - 120°)$$
$$v_c = V_s \sin(\omega t - 240°)$$

(a) Sketch the waveforms for i_1, i_2, i_3, and v_o.

(b) Derive the expression for the average V_o.

(c) Determine the first fundamental component for the line currents.

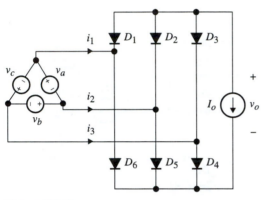

Figure P7.40

7.41 Consider the half-wave, three-phase, capacitive-load circuit of Fig. P7.41.

(a) Determine the rms value for v_o.

(b) Calculate the input power factor for each phase.

(c) Sketch v_o for $RC = T/3$.

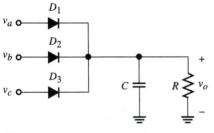

Figure P7.41

Three-Phase Rectifiers with ac-Side Inductance

7.42 Derive Eq. (7.60) for the average output voltage for the half-wave, three-phase converter.

D7.43 Design the transformer turns ratio of the circuit shown in Fig. P7.43 to deliver an average 10 A load current with a 120 V output voltage. Assume the line-to-line voltages are 580 VAC, with $\omega = 2\pi60$.

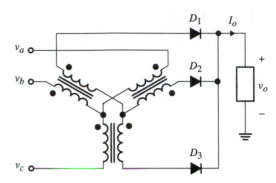

Figure P7.43

7.44 Derive Eq. (7.64) for the commutation angle u for the full-wave converter.

General Problems

7.45 Sketch the input current, i_s, and output voltage, v_o, for each of the circuits shown in Fig. P7.45. Assume $v_s(t) = V_s \sin \omega t$ V.

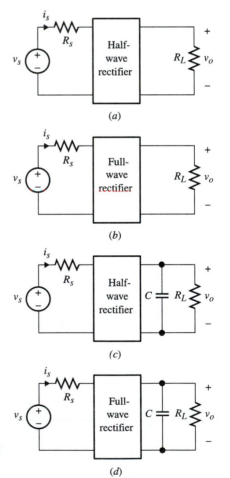

Figure P7.45

7.46 Derive the expression for the average output power and the commutation angle u for the circuit shown in Fig. P7.46. Find the average and rms values of each diode current. Assume $v_s = V_s \sin \omega t$ V.

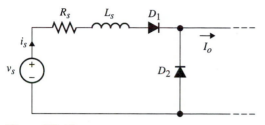

Figure P7.46

7.47 In Problem 7.46, if we assume that D_1 and D_2 have forward resistances r_{D1} and r_{D2}, respectively, show that the commutation angle is given by

$$u = \frac{\omega L_s}{r_{D1} + r_{D2}} \ln \frac{V_s + r_{D2} I_o}{(r_{D1} + r_{D2}) I_o}$$

7.48 Find the average output voltage for the circuit of Fig. P7.48. Determine the percentage change in the average output voltage if the ac-side line inductance doubles. Assume $v_s = 110\sqrt{2}\sin 377t$ V.

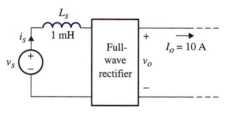

Figure P7.48

7.49 Consider the full-wave rectifier with an ac-side line inductance and a dc source in the load side as shown in Fig. P7.49.

(a) Sketch i_o, i_s, and v_{Ls}.
(b) Calculate the average output power.

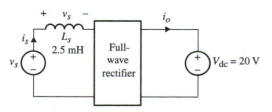

Figure P7.49

(c) Calculate the percentage change in the average output power if the ac-side line inductance doubles in value. Assume $v_s = 50 \sin 2\pi 60t$ V.

7.50 Repeat Problem 7.49 for v_s being a square-wave with ± 50 V peaks and $T = 20$ ms.

7.51 Sketch the waveforms for i_{s1}, i_{s2}, and v_s' for the circuit in Fig. P7.51. Find the THD for v_s'.

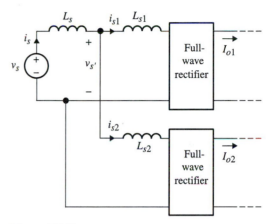

Figure P7.51

7.52 Derive the expression for the average output voltage in Fig. P7.52 in terms of the circuit parameters, assuming $v_s(t) = V_s \sin \omega t$.

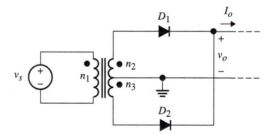

Figure P7.52

7.53 Repeat Problem 7.52 by including diode resistors r_{D1} and r_{D2} as shown in Fig. P7.53.

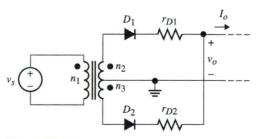

Figure P7.53

Chapter 8

Phase-Controlled Converters

INTRODUCTION

We saw in the preceding chapter how diodes can be used to rectify an ac input voltage to produce an uncontrolled dc output. These circuits—whether half-wave or full-wave configurations under resistive, inductive, or capacitive loads—have one common feature: The level of the output voltage is a function only of the circuit parameters and the peak voltage and frequency of the applied voltage source. For this reason such circuits are known as *uncontrolled rectifier* circuits. In this chapter *controlled rectifier circuits*, using the silicon-controlled rectifier (SCR) instead of the diode, will be discussed. Unlike in diode rectifier circuits, in controlled circuits the power may flow from the load side (dc side) to the source side (ac side) under some control condition. This negative direction of power flow is known as *inversion* and the circuits are known as *controlled inverter circuits*.

In controlled rectifier circuits, diodes are replaced by SCRs to allow control of the conduction period or the phase of the conducting waveform. Since the voltage or current at the output is controlled by varying or delaying the phase of the conducting waveforms, such circuits are also known as *phase-controlled converters*. As shown in Chapter 2, unlike a diode, an SCR does not turn on when only the anode-cathode voltage becomes positive; rather, an additional signal must be applied to a third terminal (gate). The source of control stems from the fact that the gate signal can be applied at *any time* in the period during which the anode-cathode voltage is positive.

Controlled SCR rectifiers have a wide range of industrial and residential applications, especially applications in which power flows in both directions. For example, in the electrochemical industry, phase-controlled rectifiers are used to control the power in electroplating to the dc side, and in rotary machines they are used to control the speed of dc and ac motors in both directions. In residential applications, medium-power phase-controlled rectifiers are used in light dimmers and variable-speed appliances.

As in Chapter 7, half- and full-wave circuit configurations under resistive, inductive, and capacitive loads will be discussed. Also, because of their wide use in high-power applications, three-phase controlled circuits will be investigated. The SCR is used in such circuits because of its high current and voltage capabilities, its simple gating circuits, and its ability to control large anode currents with a small gate signal. Throughout this chapter, it will be assumed that the gate signal used to turn on the SCR is very short in duration and does not interfere with the circuit operation, that the turn-on and turn-off times are negligible, and that the anode-cathode voltage is zero in the conducting state (*on* state) and the anode current is zero in the nonconducting state (*off* state).

8.1 BASIC PHASE CONTROL CONCEPTS

Figure 8.1 shows block diagram representations for single- and three-phase SCR-controlled circuits.

The ac side consists of the line-frequency input voltage supply, which is either single-phase or three-phase (delta or wye connection), with an inductance, L_s. As in Chapter 7, the input power factor will be defined.

The SCR circuit block consists of one or more SCRs and possibly diodes, resulting in half- and full-wave and other configurations. The control signals are externally applied gating signals to turn on the SCRs in order to control the average output voltage. There are various integrated circuits commercially available to generate gating signals for such applications. Such ICs and other gating techniques are outside the scope of this textbook. The final block is labeled *load,* which represents the dc side of the converter.

Depending on the type of load used, the load current can be either continuous or discontinuous. Both cases will be analyzed in this chapter. Under highly inductive loads, i_o will be assumed constant. This will simplify the analysis significantly.

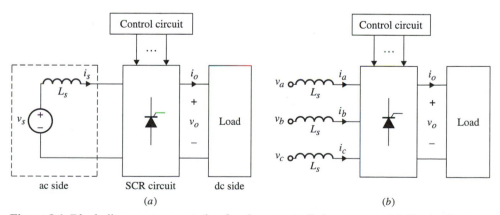

Figure 8.1 Block diagram representation for phase-controlled converters. (*a*) Single-phase. (*b*) Three-phase.

Depending on the SCR circuit configuration, the control signals, and the type of load, it is possible to produce a dc load current, I_o, and a dc load voltage, V_o, with two polarities, resulting in a four-quadrant mode of operation, as shown in Fig. 8.2.

In the two- and four-quadrant modes of operation, power can flow from the dc side to the ac side, resulting in what is known as a *phase-controlled inverter.* It will be shown later that for the inversion process to take place, the dc side must contain a type of energy source capable of delivering energy from the dc circuit to the ac supply side. The four-quadrant mode of operation is attainable if two circuits of the two-quadrant type are connected in a series or a back-to-back configuration.

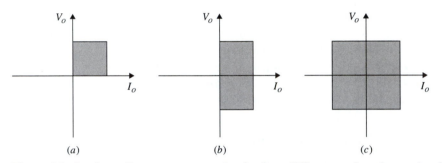

Figure 8.2 Quadrant diagram representation for three different modes of operation for Fig. 8.1. (*a*) One-quadrant mode. (*b*) Two-quadrant mode. (*c*) Four-quadrant mode.

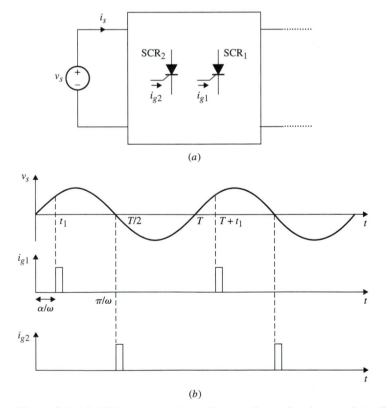

Figure 8.3 (*a*) SCR representation with gate trigger signals i_{g1} and i_{g2}. (*b*) Example with firing at t_1 and $T/2$ or angles at α and π.

Gating of the SCR is done by means of a gate trigger signal as shown in Fig. 8.3(*a*). The time t_1 is known as the firing time or the delay time. It is common to represent this instant in terms of α, which is known as the *firing angle* or the *delay angle*,

$$\alpha = \omega t_1$$

where ω is the frequency (rad/s) of the supplied voltage. For example, the delay angle for SCR$_1$ is α and for SCR$_2$ it is π, as shown in Fig. 8.3(*b*).

Since we assume an ideal SCR, the height and width of the gate trigger signal are not of importance. We should point out that time and angle notations will be used interchangeably.

8.2   HALF-WAVE CONTROLLED RECTIFIERS

In this section, we will discuss half-wave phase-controlled rectifiers under resistive and inductive loads. These circuits are mainly used in medium-power applications involving several kilowatts.

8.2.1   Resistive Load

When the diode is replaced by an SCR in the single-phase half-wave rectifier circuit, the resultant circuit is known as a half-wave phase-controlled rectifier, shown in Fig. 8.4.

When the source voltage is positive, the SCR will not conduct until the gate signal is applied at $t = t_1$, as shown in Fig. 8.5. At this time, the output voltage becomes equal to the input voltage. The output voltage, v_o, is sketched in Fig. 8.5.

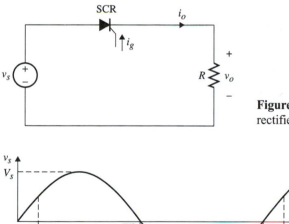

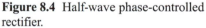

Figure 8.4 Half-wave phase-controlled rectifier.

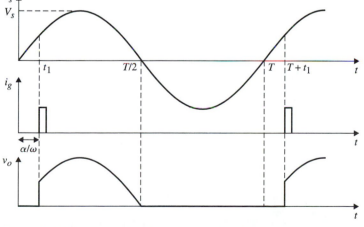

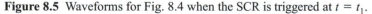

Figure 8.5 Waveforms for Fig. 8.4 when the SCR is triggered at $t = t_1$.

We must restate that once the SCR starts conducting, the gate signal can be removed. The delay angle (α) range is 0° to 180°. At $t = T/2$ the input voltage becomes negative, causing the SCR to turn off. The average output voltage is evaluated from the following equation:

$$V_o = \frac{1}{T}\int_0^T v_o\, dt = \frac{1}{T}\int_{t_1}^{T/2} v_s\, dt = \frac{V_s}{2\pi}(1 + \cos\alpha) \tag{8.1}$$

where the normalized voltage is given by

$$V_{no} = \frac{V_o}{V_s/\pi}$$
$$= \frac{1 + \cos\alpha}{2} \tag{8.2}$$

$\alpha = \omega t_1$ is the firing angle, the varying of which allows the controllability of the output voltage. Figure 8.6 shows the normalized average output voltage V_{no} as a function of the delay angle, α. This control characteristic curve will allow us to determine the rectifier firing angle for a desired normalized output.

The rms value can be evaluated from Eq. (8.3) or Eq. (8.4):

$$V_{o,\,rms} = \sqrt{\frac{1}{T}\int_{t_1}^{T/2}(V_s\sin\omega t)^2 dt} \tag{8.3}$$

$$= \frac{V_s}{2}\sqrt{\left(1 - \frac{\alpha}{\pi}\right) + \frac{\sin 2\alpha}{2\pi}} \tag{8.4}$$

The input power factor (pf) and the total harmonic distortion (THD) for Fig. 8.4 are given by Eqs. (8.5) and (8.6), respectively:

$$pf = \frac{\sqrt{2}}{2}\sqrt{\left(1 - \frac{\alpha}{\pi}\right) + \frac{\sin 2\alpha}{\pi}} \tag{8.5}$$

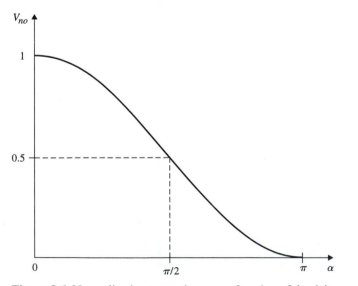

Figure 8.6 Normalized output voltage as a function of the delay angle.

$$\text{THD} = \sqrt{\frac{2}{\left(1 - \dfrac{\alpha}{\pi}\right) + \left(\dfrac{\sin 2\alpha}{2\pi}\right)} - 1} \qquad (8.6)$$

The reader is invited to verify these equations (see Problem 8.1). Since the average output voltage is positioned for $0 \le \alpha \le \pi$, and the average output current is V_o/R, this inverter operates in the first quadrant. Comparing this inverter with the uncontrolled half-wave diode rectifier circuit, we note that the average output voltage can be controlled by varying α. The phase-controlled current and voltage waveforms are related by the delay angle, α. This is why α is also known as the *angle of retard*. Both the THD and the input power factor are reduced in the phase-controlled rectifier due to the presence of the displacement angle α.

8.2.2   Inductive Load

Figure 8.7(*a*) shows a half-wave phase-controlled converter with an inductive-resistive load. The relevant waveforms are depicted in Fig. 8.7(*b*).

When the SCR is fired at $t = t_1$, the load current starts flowing until it reverses direction at $t = t_2$, at which the SCR is naturally turned off. It can be shown that the output current equation in the interval $t_1 \le t \le t_2$ is given by

$$i_o(t) = \frac{V_s}{|Z|}\left[\sin(\omega t - \theta) - \sin(\alpha - \theta)e^{-(t - t_1)/\tau}\right] \qquad (8.7)$$

where $\alpha = \omega t_1$, $\tau = L/R$, $\theta = \tan^{-1}(\omega L/R)$, and $|Z| = \sqrt{R^2 + (\omega L)^2}$.

The average value of $v_o(t)$ can be obtained by evaluating the following integral:

$$V_o = \frac{1}{T}\int_{t_1}^{t_2} V_s \sin \omega t \; dt \qquad (8.8)$$

$$= \frac{V_s}{2\pi}(\cos\alpha - \cos\beta) \qquad (8.9)$$

The numerical value of $\beta = \omega t_2$ can be obtained by evaluating Eq. (8.7) at $t = t_2$ with $i_o(t_2) = 0$.

Since the average value of the inductor voltage is zero in the steady state, the output average value is the same as the value across the load resistance. As a result, the average value of the load current, I_o, is given by

$$I_o = \frac{V_s}{2\pi R}(\cos\alpha - \cos\beta) \qquad (8.10)$$

Similarly, we normalize the average output voltage by

$$V_{no} = \frac{V_o}{V_s/\pi}$$

$$= \frac{\cos\alpha - \cos\beta}{2}$$

Figure 8.7(*c*) shows the plot of V_{no} versus α under different values of β. Notice that for $\beta = \pi$, the curve gives the resistive-load case shown in Fig. 8.6.

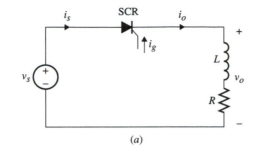

(a)

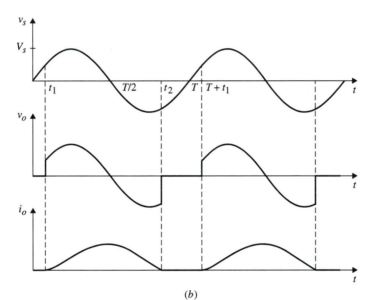

(b)

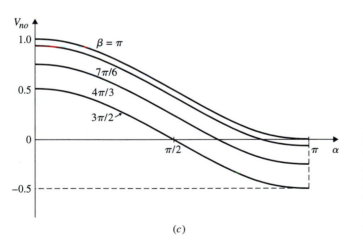

(c)

Figure 8.7 (a) Half-wave phase-controlled circuit with inductive load. (b) Relevant waveforms. (c) V_{no} versus α under different values of β.

EXERCISE 8.1

For the half-wave phase-controlled rectifier of Fig. 8.4, assume the *on*-state voltage drop of the SCR is V_{SCR}.

(a) Sketch the waveforms for v_o and i_o.

(b) Derive the expression for the average output voltage, V_o.

(c) Calculate V_o, pf, THD, and $V_{o,\text{rms}}$ using the following values: $V_s = 25$ V, $R = 10$ Ω, $V_{\text{SCR}} = 1.5$ V, and $\alpha = \pi/5$.

ANSWER   16 V, 0.76, 95%, 13.5 V

EXERCISE 8.2

For the half-wave phase-controlled rectifier of Fig. 8.7(a), use $v_s = 78\sin 377t$, $L = 10$ mH, $R = 5$ Ω, to design α for $V_o = 16$ V, obtain t_1 and t_2 in Fig. 8.7 (b).

ANSWER   37°, 1.2 ms, 17.9 ms

Consider the case where a free-wheeling diode is added to Fig. 8.7(a) as shown in Fig. 8.8(a). The presence of the diode results in a continuous load current, and the output voltage is always positive. The corresponding waveforms are shown in Fig. 8.8(b).

Notice that the principal operation of this circuit is similar to that of the half-wave diode rectifier with free-wheeling diode discussed in Chapter 3, with the exception that the intervals in which the two circuit equations are applied are shifted by α.

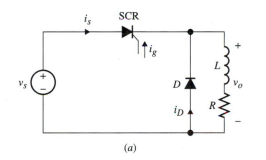

(a)

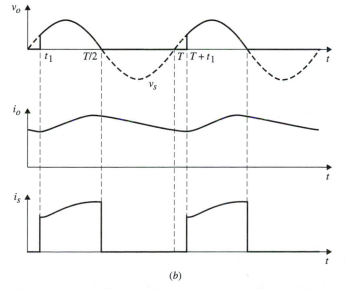

(b)

Figure 8.8 (a) Half-wave phase-controlled rectifier with free-wheeling diode. (b) Typical waveforms.

For the time interval in which the free-wheeling diode D is conducting, the circuit equation is given by

$$L\frac{di_o}{dt} + Ri_o = 0 \qquad \pi \le \omega t \le 2\pi + \alpha \tag{8.11}$$

and for the time interval in which the SCR is conducting and D is reverse biased,

$$L\frac{di_o}{dt} + Ri_o = V_s \sin \omega t \qquad \alpha \le \omega t \le \pi \tag{8.12}$$

By applying the proper boundary conditions at $\omega t = \alpha$ and $\omega t = \pi$, an expression for i_o can be derived (see Problem 8.3). It is interesting to note that since v_o does not go negative, the average output is no longer a function of the load inductance. It is given by

$$V_o = \frac{1}{T}\int_{t_1}^{T/2} V_s \sin \omega t\ dt = \frac{V_s}{2\pi}(1 + \cos \alpha) \tag{8.13}$$

Note that this circuit does not support a negative output voltage (no inversion).

EXAMPLE 8.1

Draw the waveforms for i_s and v_o in Fig. 8.8(a). Assume $L/R \gg T/2$.

SOLUTION   The equivalent circuit of Fig. 8.8(a) is shown in Fig. 8.9(a), and the relevant waveforms are shown in Fig. 8.9(b). The control characteristic curve is the same as the one given in Fig. 8.6.

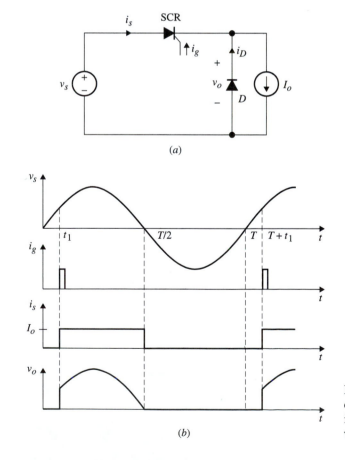

Figure 8.9 (a) Equivalent circuit for Fig. 8.8(a) with infinitely large L. (b) Relevant waveforms.

8.3   FULL-WAVE PHASE-CONTROLLED RECTIFIERS

8.3.1   Resistive Load

A phase-controlled full-wave bridge rectifier is shown in Fig. 8.10. This circuit is more practical than the phase-controlled half-wave rectifier shown in Fig. 8.4.

Like the uncontrolled full-wave bridge rectifier, the diagonal pairs SCR_1-SCR_2 and SCR_3-SCR_4 conduct during the positive and negative half-cycles, respectively. Assume the gating signal pairs are both delayed by α in their respective half-cycles as shown in Fig. 8.11.

The output average value is calculated from the following integral:

$$V_o = \frac{2}{T}\int_{t_1}^{T/2} V_s \sin \omega t \; dt$$

$$= \frac{V_s}{\pi}(\cos \alpha + 1) \tag{8.14}$$

From Eq. (8.14), it is clear that the control characteristic curve of the average output voltage versus α for this full-wave topology is doubled compared to the half-wave topology given in Fig. 8.6.

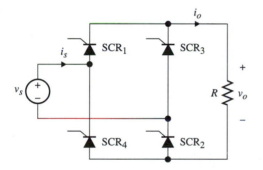

Figure 8.10 Full-bridge phase-controlled rectifier.

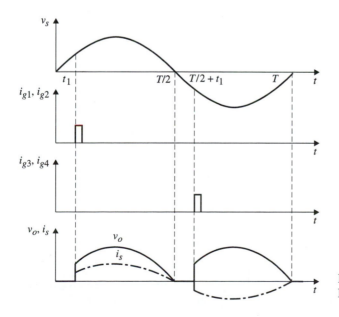

Figure 8.11 Waveforms for Fig. 8.10.

EXAMPLE 8.2

Determine the input power factor and the THD for the full-bridge topology of Fig. 8.10.

SOLUTION   The rms value of v_s is $V_s/(\sqrt{2})$, and the rms value for i_s, calculated from the waveform of Fig. 8.11, is

$$
\begin{aligned}
I_{s,\,rms} &= \sqrt{\frac{1}{T/2}\int_{t_1}^{T/2}\left(\frac{V_s}{R}\sin\omega t\right)^2 dt} \\
&= \frac{V_s}{\sqrt{2}R}\sqrt{1-\frac{\alpha}{\pi}+\frac{\sin 2\alpha}{2\pi}}
\end{aligned}
\tag{8.15}
$$

The average input power can be calculated from the following relation:

$$
\begin{aligned}
P_{in} &= \frac{1}{T/2}\frac{1}{R}\int_{t_1}^{T/2}(V_s\sin\omega t)^2 dt \\
&= \frac{V_s^2}{2R}\left[\left(1-\frac{\alpha}{\pi}\right)+\frac{\sin 2\alpha}{2\pi}\right]
\end{aligned}
\tag{8.16}
$$

From the definition of the power factor, and using Eqs. (8.15) and (8.16), we can obtain the following relation:

$$
\begin{aligned}
\text{pf} &= \frac{P_{ave}}{I_{s,\,rms}V_{s,\,rms}} \\
&= \sqrt{\left(1-\frac{\alpha}{\pi}\right)+\frac{\sin 2\alpha}{2\pi}}
\end{aligned}
\tag{8.17}
$$

A plot of pf versus α is given in Fig. 8.12.
    The THD can be found as

$$
\text{THD} = \sqrt{\frac{2\alpha+2\cos(2\alpha)-1}{2\pi-2\alpha-2\cos(2\alpha)+1}}
$$

EXERCISE 8.3

Calculate the input power factor for $\alpha = 30°$ and $150°$ using Fig. 8.10.

ANSWER   0.75, 0.36

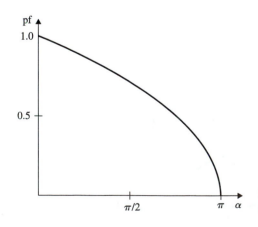

Figure 8.12 Power factor versus the delay angle for the circuit of Fig. 8.10.

EXERCISE 8.4

Consider the waveform for $i_s(t)$ given in Fig. 8.11.

(a) Derive the expression of the fundamental component, $i_{s1}(t)$, for the input current i_s.

(b) Calculate the THD for $\alpha = 30°$ and $\alpha = 150°$.

ANSWER   10.2%, 18.9%

8.3.2 Inductive Load

The full-wave phase-controlled rectifier under an inductive load is shown in Fig. 8.13. Let us first consider the waveforms for i_o and v_o and assume that the time constant L/R is not large compared to $T/2$. Again, the thyristor pairs SCR$_1$-SCR$_2$ and SCR$_3$-SCR$_4$ are triggered diagonally, each delayed by α in the respective half-cycles. It can be shown that, as far as the load current is concerned, there are two modes of operation: discontinuous and continuous conduction modes. If we assume that in steady-state operation, the inductor current value reaches zero before $t = T/2 + t_1$, then the rectifier is known to operate in the discontinuous conduction mode (dcm). For dcm to occur, the load current must reach zero before the opposite pair of SCRs are fired. However, if the inductor current (load) is not allowed to reach zero, i.e., the second diagonal pair of SCRs are triggered while the current is flowing, then since the current in the inductor does not change instantaneously, the current commutates instantaneously in the other thyristors. Under this case, the rectifier is known to operate in the continuous conduction mode (ccm). Figures 8.14 and 8.15 show the waveforms for v_o and i_o under dcm and ccm, respectively.

EXAMPLE 8.3

Derive the expression for $i_o(t)$ under both the continuous and discontinuous conduction modes, and determine the condition at which the boundary between ccm and dcm occurs.

SOLUTION   Since the dcm is a special case of the ccm, let us first consider the continuous conduction mode case. As shown in the previous chapter, the total solution for $i_o(t)$ is given by

$$i_o(t) = Ae^{-(t-t_1)/\tau} + \frac{V_s}{|Z|}\sin(\omega t - \theta) \tag{8.18}$$

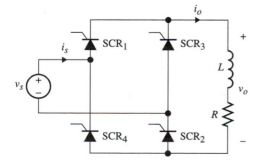

Figure 8.13 Full-wave phase-controlled rectifier with inductive load.

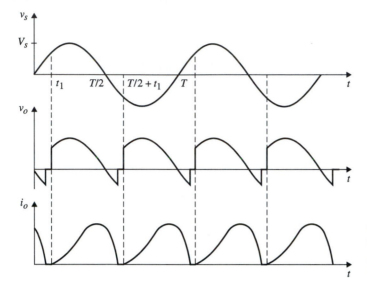

Figure 8.14 Waveforms for Fig. 8.13 under discontinuous conduction mode.

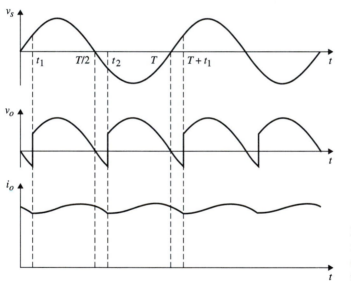

Figure 8.15 Waveforms for Fig. 8.13 under continuous conduction mode.

where A is a constant and must be determined from the initial condition, and the other parameters are given by

$$\tau = L/R$$

$$|Z| = \sqrt{(\omega L)^2 + R^2}$$

$$\theta = \tan^{-1}\frac{\omega L}{R}$$

If we let the initial condition of $i_o(t)$ at $t = t_1$ be I_{o1}, from Eq. (8.18) we have

$$i_o(t_1) = I_{o1} = A + \frac{V_s}{Z}\sin(\alpha - \theta)$$

Solving for A, we obtain

$$A = I_{o1} - \frac{V_s}{|Z|}\sin(\alpha - \theta)$$

Therefore, the expression for the load current is given by

$$i_o(t) = \left[I_{o1} - \frac{V_s}{|Z|}\sin(\alpha - \theta)\right]e^{(t-t_1)/\tau} + \frac{V_s}{|Z|}\sin(\omega t - \theta) \tag{8.19}$$

In steady-state operation, $i_o(t_1)$ must equal $i_o(T/2 + t_1)$; hence, evaluating Eq. (8.19) at $t = T/2 + t_1$ and solving for I_{o1}, we obtain

$$I_{o1} = \frac{-\dfrac{V_s}{|Z|}(e^{-\pi/\omega\tau} - 1)\sin(\alpha - \theta)}{1 - e^{-\pi/\omega\tau}} \tag{8.20}$$

This initial condition determines the mode of operation as follows:

$$I_{o1} : \begin{cases} < 0 & \text{for} \quad \alpha > \theta \quad \text{dcm} \\ = 0 & \text{for} \quad \alpha = \theta \quad \text{boundary condition} \\ > 0 & \text{for} \quad \alpha < \theta \quad \text{ccm} \end{cases}$$

If we assume that the load time constant L/R is much larger than the half-period of the applied voltage, then we can model the load current by a constant current source, I_o. In this case, the waveform of the current i_s is a squarewave of magnitude $\pm I_o$, shifted by the triggering angle $\alpha = \omega t_1$ as shown in Fig. 8.16.

The average output voltages of the full-bridge phase-controlled circuits with the waveforms of Figs. 8.14, 8.15, and 8.16 are the same and given by

$$\begin{aligned} V_o &= \frac{1}{T/2}\int_{t_1}^{T/2 + t_1} V_s \sin\omega t \; dt \\ &= \frac{2V_s}{\pi}\cos\alpha \end{aligned} \tag{8.21}$$

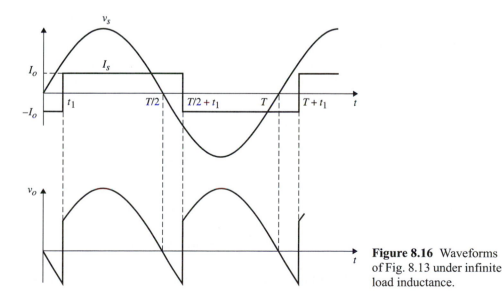

Figure 8.16 Waveforms of Fig. 8.13 under infinite load inductance.

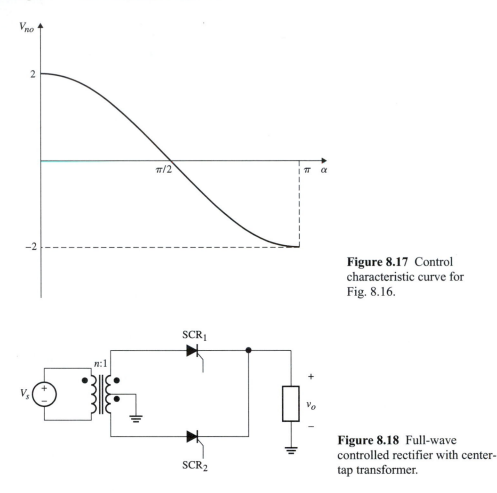

Figure 8.17 Control characteristic curve for Fig. 8.16.

Figure 8.18 Full-wave controlled rectifier with center-tap transformer.

The control characteristic curve for the normalized average output voltage $V_{no} = \pi V_o / V_s$ is shown in Fig. 8.17. Notice that when $\pi/2 \le \alpha \le \pi$, this circuit produces a negative average output voltage. Since the average load current is always positive, the direction of the power flow in this range of α is from the dc output side to the ac input side. Consequently, as stated before, the circuit is known to operate in the inversion mode, whereas for $0 \le \alpha \le \pi/2$ the circuit operates in the rectification mode.

If the application requires the use of an isolated transformer, a center-tap full-wave controlled rectifier can be obtained using two SCRs as shown in Fig. 8.18. All waveforms are similar to the waveforms under resistive and inductive loads.

EXAMPLE 8.4

Derive the power factor for the phase-controlled full-wave bridge rectifier under a constant load current as shown in Fig. 8.19(a).

SOLUTION   The waveform for $i_s(t)$ is redrawn in Fig. 8.19(b). From the Fourier series, we have

$$i_s(t) = I_{dc} + \sum_{n=1}^{\infty} a_n \cos n\omega t + b_n \sin(n\omega t)$$

From the symmetry of $i_s(t)$, the dc average value is zero and only its odd harmonics exist.

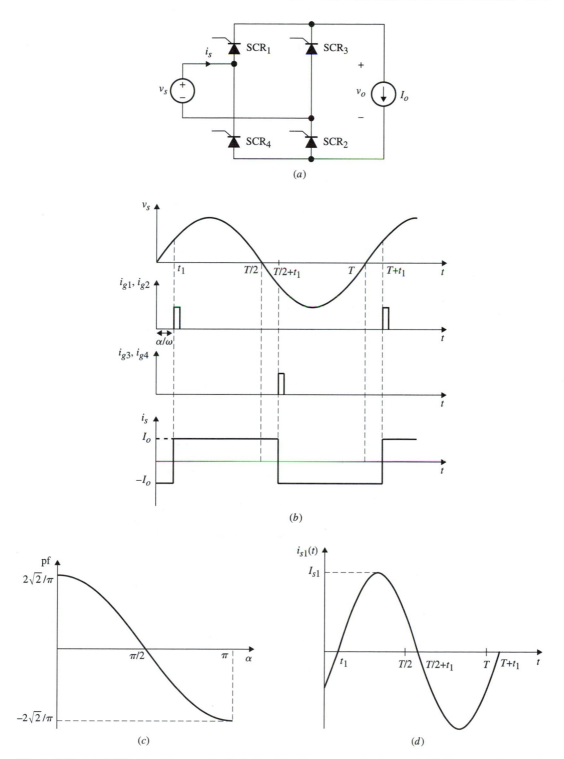

Figure 8.19 (a) Full-bridge phase-controlled circuit under constant load current. (b) Source voltage and current waveforms. (c) Power factor. (d) Fundamental component of $i_s(t)$.

The fundamental component is given by

$$i_{s1}(t) = I_{s1}\sin(\omega t + \theta)$$

where

$$I_{s1} = \sqrt{a_1^2 + b_1^2}$$

$$\theta = \tan^{-1}\frac{a}{b}$$

The coefficients a_1 and b_1 are evaluated from the following relations:

$$a_1 = \frac{2}{T}\left[\int_{t_1}^{T/2+t_1} I_o\cos(\omega t)\ dt + \int_{T/2+t_1}^{T+t_1} -I_o\cos(\omega t)\ dt\right]$$

$$= -\frac{4I_o}{\pi}\sin\alpha$$

$$b_1 = \frac{2}{T}\left[\int_{t_1}^{T/2+t_1} I_o\sin(\omega t)\ dt + \int_{T/2+t_1}^{T+t_1} -I_o\sin(\omega t)\ dt\right]$$

$$= -\frac{4I_o}{\pi}\cos\alpha$$

Hence, the peak value of the fundamental component is given by

$$I_{s1} = \frac{4I_o}{\pi}$$

and its rms value is

$$I_{s1,rms} = \frac{4I_o}{\sqrt{2}\,\pi}$$

The rms value of $i_s(t)$ is I_o.

From the preceding expression, we obtain the power factor:

$$pf = \frac{4}{\sqrt{2}\,\pi}\cos\theta$$

where the displacement angle is given by

$$\theta = -\tan^{-1}\left(\frac{\sin\alpha}{\cos\alpha}\right) = -\tan^{-1}(\tan\alpha) = -\alpha$$

The power factor is expressed in terms of the delay angle, α, as

$$pf = \frac{4}{\sqrt{2}\,\pi}\cos\alpha$$

which is sketched in Fig. 8.19(c). The fundamental component is shown in Fig. 8.19(d).

EXAMPLE 8.5

Consider the full-wave phase-controlled rectifier given in Fig. 8.20(a), with two SCRs and a flyback diode D_3.

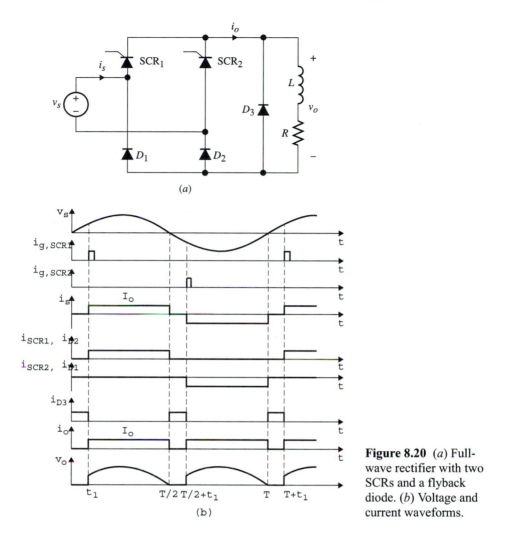

Figure 8.20 (*a*) Full-wave rectifier with two SCRs and a flyback diode. (*b*) Voltage and current waveforms.

(a) Assume $L/R \gg T/2$. Draw the waveforms for i_s, i_{SCR1}, i_{SCR2}, i_{D1}, i_{D2}, i_{D3}, i_o, and v_o for $\alpha = 30°$.

(b) Show that the fundamental component of $i_s(t)$ is given by

$$i_{s1}(t) = I_{s1} \sin\left(\omega t - \frac{\alpha}{2}\right)$$

where

$$I_{s1} = \frac{2\sqrt{2}I_o}{\pi}\sqrt{1 + \cos 2\alpha}$$

I_o is the load current, assumed constant, and $\alpha = \omega t_1$.

(c) Derive the expression for the power factor and THD.

SOLUTION

(a) To sketch the waveforms, we must understand the basic circuit operation. Since we assume $L/R \gg T/2$, i_o is considered constant as I_o. Let the firing times for SCR$_1$ and SCR$_2$ occur at t_1 and $T/2 + t_1$, respectively, as shown in Fig. 8.20(*b*). Before SCR$_1$ is gated at $t = t_1$, the load current is flowing in the flyback diode D_3, since SCR$_2$ is off because of the negative source voltage. At $t = t_1$, SCR$_1$ and D_2 turn on. At $t = T/2$, the negative source voltage turns D_3 on, and SCR$_1$

and D_2 cease to conduct. At $t = T/2 + t_1$, SCR_2 is gated, turning it on and causing D_3 to turn off and D_1 to turn on. This mode of operation continues until $t = T$, when SCR_2 and D_1 turn off, causing D_3 to turn on again. The cycle repeats at $t = T = t_1$, when SCR_1 is gated again.

(b) The fundamental component of $i_s(t)$ is given by

$$i_{s1}(t) = I_{s1} \sin(\omega t + \theta)$$

The parameters I_{s1} and θ are obtained from

$$I_{s1} = \sqrt{a_1^2 + b_1^2}$$

$$\theta = \tan^{-1}\frac{a_1}{b_1}$$

where

$$a_1 = \frac{2}{T}\int_0^T i_s(t)\cos\omega t \; dt$$

$$= \frac{2}{T}\left[\int_{t_1}^{T/2} I_o\cos\omega t \; dt - \int_{T/2+t_1}^{T} I_o\cos\omega t \; dt\right]$$

$$= \frac{-2I_o}{\pi}\sin\alpha$$

and

$$b_1 = \frac{2}{T}\int_0^T i_s(t)\sin\omega t \; dt$$

$$= \frac{2}{T}\left[\int_{t_1}^{T/2} I_o\sin\omega t \; dt - \int_{T/2+t_1}^{T} I_o\sin\omega t \; dt\right]$$

$$= \frac{2I_o}{\pi}(1 + \cos\alpha)$$

Hence, I_{s1} is given by

$$I_{s1} = \frac{2\sqrt{2}I_o}{\pi}\sqrt{1 + \sin 2\alpha}$$

(c) The rms expression is given by

$$I_{s,\text{rms}}^2 = \frac{1}{2\pi}\left[\int_\alpha^\pi I_o^2\, d\omega t + \int_{\pi+\alpha}^{2\pi} I_o^2\, d\omega t\right]$$

$$= \frac{1}{2\pi}[I_o^2(\pi - \alpha) + I_o^2(2\pi - (\pi + \alpha))]$$

$$= \frac{I_o^2(-2\alpha + 2\pi)}{2\pi}$$

$$= I_o^2\left(1 - \frac{\alpha}{\pi}\right)$$

Hence, the power factor is expressed as

$$\text{pf} = \frac{2\sqrt{1 + \cos\alpha}}{\pi}\frac{\cos\dfrac{\alpha}{2}}{\sqrt{1 - \dfrac{\alpha}{\pi}}}$$

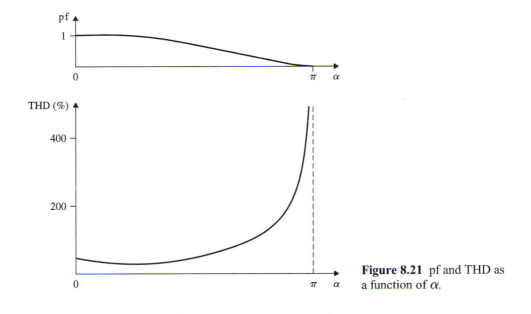

Figure 8.21 pf and THD as a function of α.

and the total harmonic distortion is given by

$$\text{THD} = \sqrt{\frac{\pi(\pi - \alpha)}{4(1 + \cos \alpha)} - 1}$$

The pf and THD plots are given in Fig. 8.21.

It is possible to design a full-wave phase-controlled converter that can provide bidirectional flow of the output voltage and the output current. The system of Fig. 8.22(a) can operate in the four quadrants as shown in Fig. 8.22(b).

The preceding general characteristics can be achieved by using dual full-wave phase-controlled converters connected in parallel, with their SCR connections opposite to each other. Two other configurations for full-wave bridge converters are known as series and parallel converters, shown in Fig. 8.23 and Fig. 8.24, respectively.

Figure 8.25 shows a phase-controlled full-wave circuit with a dc voltage source in the load side. Again, in the analysis we assume that the time constant L/R is very large compared to the period of the applied voltage. Hence, we can model the inductor load current as a constant I_o. The waveforms of the output voltage, v_o, and the source current, i_s, are the same as those of Fig. 8.19(a). In the steady state, the average output current is given by

$$I_o = \frac{V_o}{R} - \frac{V_{dc}}{R} = 2\frac{V_s}{\pi R}\cos \alpha - \frac{V_{dc}}{R} \tag{8.22}$$

EXERCISE 8.5

Consider the phase-controlled full-wave rectifier shown in Fig. 8.25 with $v_s = 110 \sin 2\pi(60)t$, $L = 38$ mH, $R = 25\ \Omega$, and $V_{dc} = 12$ V. Assume the firing angle is $45°$ and the load current is constant. Find:

(a) The average output voltage

(b) The average output current

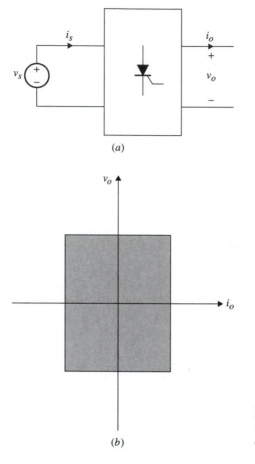

(a)

(b)

Figure 8.22 (*a*) Block diagram representation for a single-phase controlled rectifier. (*b*) Four-quadrant operation.

(c) The input and output average powers

(d) The power factor

ANSWER   49.5 V, 1.5 A, 74.26 W, 0.64

EXAMPLE 8.6

Determine the range of the delay angle α in Fig. 8.25 so that the input power factor is at least 0.75 and the minimum output power delivered to a 5 Ω load resistor is 1.2 kW. Assume $v_s = 240 \sin 377t$, $L = 125$ mH, and $V_{dc} = 20$ V .

SOLUTION   The time constant is $L/R = 25$ ms , which is larger than $T/2$. Thus, the ripple current is considered very small. For an output power equal to 1.2 kW, the load current I_o equals 2.83 A from Eq. (8.22), and $\alpha = 50.37°$. For a power factor of 0.75, the conduction angle must be smaller than 33.6°. As a result, the range of α is $0° \le \alpha \le 33.6°$. The power delivered to the load at $\alpha = 0°$ and $\alpha = 0°$ is 3.53 kW and 2.304 kW, respectively.

8.4   EFFECT OF AC-SIDE INDUCTANCE

In this section, we turn our attention to the analysis of the SCR rectifier and inverter circuits by including the ac-side inductance L_s. As with the diode rectifier circuits, here both the half-wave and the full-wave configurations will be considered.

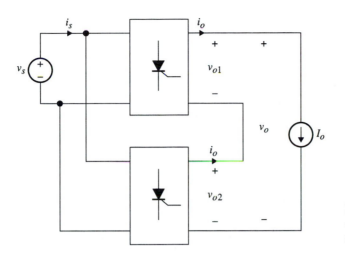

Figure 8.23 Series-connected full-wave SCR circuits.

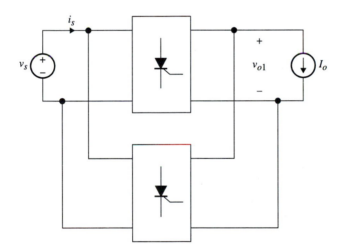

Figure 8.24 Parallel-connected full-wave SCR circuits.

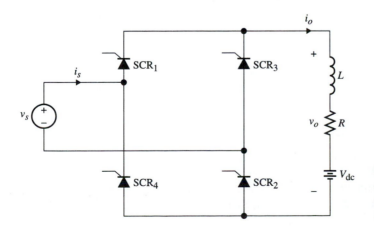

Figure 8.25 Phase-controlled full-wave circuit with load voltage source.

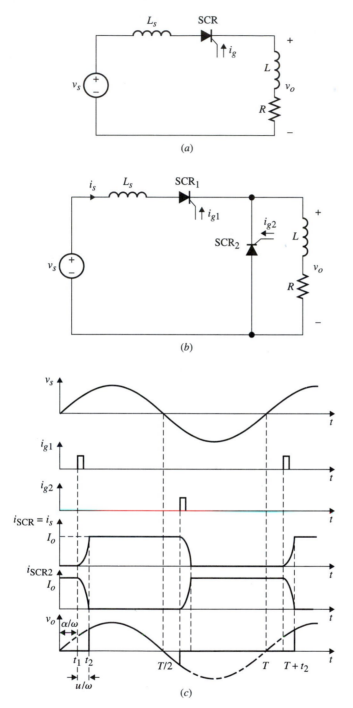

Figure 8.26 Half-wave SCR rectifier with ac-side reactance: (*a*) without flyback SCR, (*b*) with flyback SCR$_2$. (*c*) Voltage and current waveforms.

8.4.1 Half-Wave Circuits

Figure 8.26(*a*) shows the half-wave phase-controlled circuit including L_s with an inductive load. Without a flyback SCR across R-L, the circuit behaves like the half-wave rectifier under an inductive load, discussed in Section 8.2. Figure 8.26(*b*) shows the same circuit including a flyback SCR$_2$.

The analysis of this circuit is easily carried out by assuming $(L/R) \gg T$. Figure 8.26(c) shows the waveforms for i_s, i_{SCR1}, and v_o under a constant load current.

Notice that at $t = T/2$, SCR_2 cannot be turned on because its gate signal is not available until time $t = t_1 + T/2$. The operation of this circuit is similar to that of the diode circuit with an ac-side inductance discussed in the previous chapter. At $t = t_1 + T/2$, when SCR_2 is fired, the current in SCR_1 cannot change to zero because of L_s; hence, both SCR_1 and SCR_2 remain on for a period of $(t_2 - t_1) = \mu/\omega$ shown in Fig. 8.26(c).

$$V_o = \frac{1}{T} \int_{t_2}^{T/2 + t_1} V_s \sin \omega t \, dt \tag{8.23}$$

During this interval, the current in SCR_1 decreases to zero and the current in SCR_2 increases to I_o at the same rate. The average output voltage is given by evaluating Eq. (8.23) and using $\alpha = \omega t_1$ and $u = \omega(t_2 - t_1)$, leading to the following equation for V_o:

$$V_o = \frac{V_s}{2\pi} (\cos \alpha + \cos(\alpha + u)) \tag{8.24}$$

The load current in terms of the commutation angle u and the firing angle α is obtained by integrating Eq. (8.25) from $t = t_1$ to t.

$$L_s \frac{di_s}{dt} = v_s(t) \tag{8.25}$$

Equation (8.25) is obtained from Fig. 8.26(a) when both SCR_1 and SCR_2 are conducting. From Eq. (8.25), we obtain $i_s(t)$ using the initial condition $i_s(t_1) = 0$, to yield

$$\begin{aligned} i_s(t) &= \frac{-V_s}{L_s \omega} (\cos \omega t - \cos \omega t_1) \\ &= \frac{V_s}{L_s \omega} (\cos \alpha - \cos \omega t) \end{aligned} \tag{8.26}$$

Evaluating Eq. (8.26) at $t = t_2$, where $i_s(t_2) = I_o$, we obtain

$$I_o = \frac{V_s}{\omega L_s} [\cos \alpha - \cos(u + \alpha)] \tag{8.27}$$

Substitute the above equation into Eq. (8.24), to give

$$V_o = \frac{V_s}{\pi} \left(\cos \alpha - \frac{\omega L_s I_o}{2 V_s} \right) \tag{8.28}$$

In terms of the normalized output voltage and the normalized output current, Eq. (8.28) may be written as follows:

$$V_{no} = \frac{1}{\pi} \left(\cos \alpha - \frac{I_{no}}{2} \right) \tag{8.29}$$

The characteristic curve for I_{no} as a function of V_{no} under different firing angles is shown in Fig. 8.27.

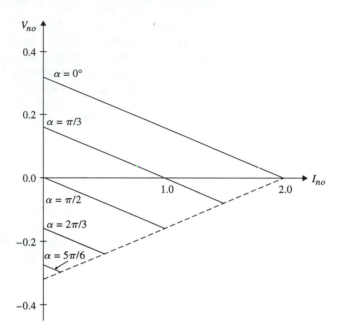

Figure 8.27 Characteristic curves for V_{no} versus I_{no} under different values of α.

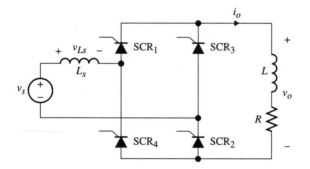

Figure 8.28 Controlled full-bridge rectifier with ac-side inductance.

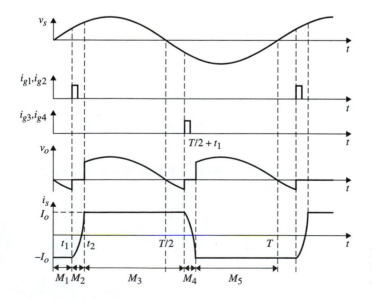

Figure 8.29 Output voltage and source current waveforms.

8.4.2 Full-Wave Bridge Circuits

Figure 8.28 shows the controlled full-wave bridge rectifier with an ac-side inductance. Assuming constant I_o, the waveforms for v_s, i_s, and v_o are shown in Fig. 8.29.

The basic circuit operation is similar to that of the full-wave diode rectifier with an ac-side inductance. We first assume that SCR$_1$ and SCR$_2$ are off during the negative cycle of $v_s(t)$ for $t \leq t_1$. The voltage across L_s is zero; hence, the entire source voltage appears across the output as shown in Fig. 8.30(a), which is represented as mode 1 (M_1).

In this mode we have, $v_{Ls} = 0$, $i_s = -I_o$, and $v_o = -v_s$. This mode continues while SCR$_1$ and SCR$_2$ remain off, until $t = t_1$, when they are triggered on. At this time, SCR$_3$ and SCR$_4$ remain on to maintain continuity of the inductor current. The resultant circuit is shown in Fig. 8.30(b) and designated as mode 2.

The inductor current, i_s, is obtained from the following relation:

$$v_{Ls} = L_s \frac{di_s}{dt}$$

$$= v_s(t)$$

(8.30)

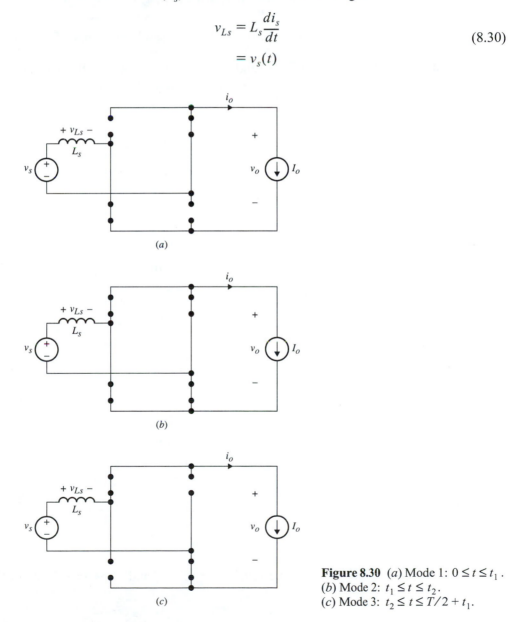

(a)

(b)

(c)

Figure 8.30 (a) Mode 1: $0 \leq t \leq t_1$.
(b) Mode 2: $t_1 \leq t \leq t_2$.
(c) Mode 3: $t_2 \leq t \leq T/2 + t_1$.

Using the initial condition $i_s(t_1) = -I_o$ in Eq. (8.30), we obtain

$$i_s(t) = \int_{t_1}^{t} \frac{V_s}{L_s} \sin \omega t \; dt - I_o$$

$$= \frac{V_s}{L_s \omega}(\cos \alpha - \cos \omega t) - I_o \tag{8.31}$$

At $t = t_2$, the currents in SCR_3 and SCR_4 reach zero and the circuit enters mode 3, as shown in Fig. 8.30(c). In this mode we have $i_s = +I_o$, $v_{Ls} = 0$, and $v_o = v_s$.

Mode 4 starts at $t = T/2 + t_1$, when SCR_3 and SCR_4 are turned on, resulting in the same equivalent circuit of mode 2 with $i_s(t)$ given by

$$i_s(t) = \frac{-V_s}{L_s \omega}(\cos \alpha + \cos \omega t) + I_o \tag{8.32}$$

At $t = T/2 + t_2$, the current $i_s(t)$ reaches $-I_o$, resulting in mode 5, where SCR_3 and SCR_4 are on, and SCR_1 and SCR_2 are off, which is similar to mode 1. The cycle repeats at $t = T + t_1$, when SCR_1 and SCR_2 are triggered. To obtain an expression for the commutation angle, u, we evaluate Eq. (8.32) at $t = T/2 + t_2$ with $i_s(T/2 + t_2) = -I_o$; hence, we have

$$-I_o = I_o - \frac{V_s}{L_s \omega}[\cos \omega(T/2 + t_2) + \cos \omega t_1]$$

Using $u = \omega(t_2 - t_1)$, the commutation angle may be given by

$$u = \cos^{-1}\left(\cos \alpha - \frac{2 I_o \omega L_s}{V_s}\right) - \alpha \tag{8.33}$$

In terms of the normalized load current, Eq. (8.33) gives

$$u = \cos^{-1}(\cos \alpha - 2 I_{no}) - \alpha \tag{8.34}$$

The average output voltage is

$$V_o = \frac{2}{T}\int_{t_2}^{T/2+t_1} v_s(t) \; dt$$

$$= \frac{2 V_s}{2 \pi}(\cos \omega t_1 + \cos \omega t_2) \tag{8.35}$$

$$= \frac{V_s}{\pi}(\cos \alpha + \cos(\alpha + u))$$

Substituting for $\cos(\alpha + u) = \cos \alpha - 2 L_s I_o / V_s$, we get

$$V_o = \frac{2 V_s}{\pi}\left(\cos \alpha - \frac{\omega L_s I_o}{V_s}\right) \tag{8.36}$$

In terms of normalized values, Eq. (8.36) becomes

$$V_{no} = \frac{2}{\pi}(\cos \alpha - I_{no}) \tag{8.37}$$

Figure 8.31 shows the regulation curve for I_{no}, v_s, and V_{no} under different firing angles.

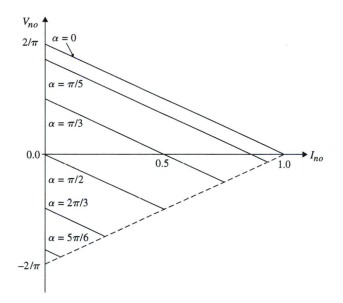

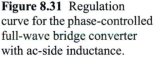

Figure 8.31 Regulation curve for the phase-controlled full-wave bridge converter with ac-side inductance.

EXERCISE 8.6

Consider the circuit given in Fig. 8.28. Assume $v_s = 110 \sin 2\pi(60)t$, $\alpha = 60°$, $L_s = 20$ mH, and $I_o = 10$ A. Determine the average output voltage V_o and the approximate rms input current. (*Hint*: Assume current commutation is linear.)

ANSWER   −13 V, 8.15 A

EXERCISE 8.7

Consider the full-wave circuit of Fig. 8.28 with $L_s = 5$ mH, $V_s = 110$ V, $\alpha = 60°$, and $\omega = 377$ rad/s. Use Fig. 8.31 to determine:

(a) The average output voltage for $I_o = 10$ A

(b) The range of α if the local current changes from 10 A to 20 A while V_o remains the same as in part (*a*)

(c) The range of α if we assume the input voltage V_s changes by ±20% of its nominal value of 110 V while the output voltage remains constant as given in part (*a*)

ANSWER   23 V, $47.8° < \alpha < 60°$, $51.3° < \alpha < 65.4°$

8.5   THREE-PHASE CONTROLLED CONVERTERS

8.5.1   Half-Wave Converters

Figure 8.32(*a*) shows a phase-controlled half-wave rectifier circuit under a resistive load. Its current and voltage waveforms are shown in Fig. 8.32(*b*). Figure 8.32(*c*) shows the waveforms under a highly inductive load.

If we assume all SCRs are triggered simultaneously at $\alpha = \omega t_1$, then it is apparent from the circuit that only one positive source current can flow at any given time. This type of arrangement resembles the operation of an OR gate in digital systems. In practice, each SCR is triggered at α within its cycle; i.e., SCR_1, SCR_2, and SCR_3 are triggered at α, $2\pi/3 + \alpha$, and $4\pi/3 + \alpha$, respectively, as shown in Fig. 8.32(*b*). Figure 8.32(*b*) and

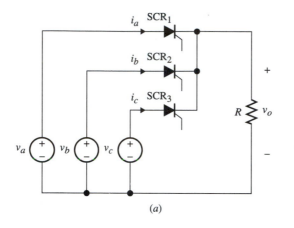

(a)

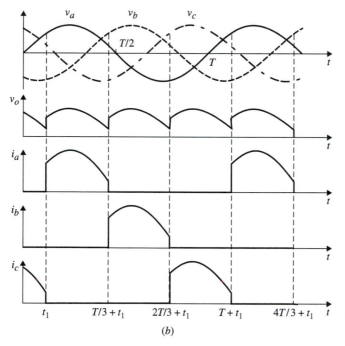

Figure 8.32 (a) Three-phase controlled half-wave rectifier. (b) Output voltage and line current waveforms under resistive load.

(b)

(c) shows relevant current and voltage waveforms under a purely resistive load, and a highly inductive load, respectively. Here we have assumed the three-phase voltages are position sequence with their line to natural voltages given as $v_a = V_s \sin \omega t$, $v_b = V_s \sin(\omega t - 120°)$, and $v_c = V_s \sin(\omega t - 240°)$.

The average output voltage is given by

$$V_o = \frac{3}{T} \int_{t_1}^{T/3+t_1} V_s \sin \omega t \; dt$$

$$= \frac{-3V_s}{T\omega}[\cos \omega(T/3 + t_1) - \cos \omega t_1]$$

(8.38)

$$= \frac{3V_s}{2\pi}[\cos \alpha - \cos \omega(T/3 + t_1)]$$

where $\alpha = \omega t_1$.

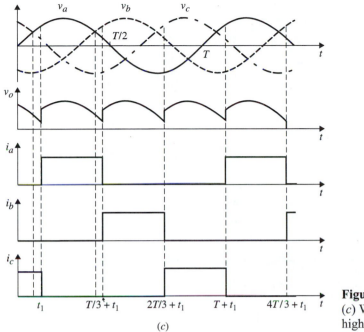

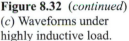

Figure 8.32 (*continued*) (*c*) Waveforms under highly inductive load.

8.5.2   Full-Wave Converters

Figure 8.33(*a*) shows the phase-controlled full-wave three-phase converter under an inductive-resistive load with a Y-connected three-phase voltage source.

The waveforms for $L = 0$ and $L = \infty$ are shown in Fig. 8.33(*b*) and (*c*), respectively, using $\alpha = 15°$. To illustrate how the full-wave bridge circuit works, we will give a detailed mode-by-mode operation for $\alpha = 15°$. If we select a positive set of voltages as in the half-wave case, then we must obtain a line-to-line voltage sequence for the converter operation.

The line-to-line voltage set is v_{ab}, v_{bc}, and v_{ca}, given in Chapter 7 and whose waveforms are redrawn in Fig. 8.34(*a*), showing mode 1 to mode 6 for $\alpha = 0°$ (SRCs act like diodes) with the corresponding conducting SCRs. Figure 8.34(*b*) shows the firing sequence for $\alpha = 0°, 30°, 60°, 90°, 120°, 150°$, and $180°$. For $\alpha > 90°$, the voltage starts to become negative and the converter operates in the *inversion mode*. For simplicity, these waveforms are plotted against angle ωt. As shown in Fig. 8.34(*b*), the most positive rectified voltage in mode 1 (M_1) is $|v_{bc}|$. Hence, the conducting SCRs are SCR_3 and SCR_5, and the equivalent circuit is shown in Fig. 8.35(*a*). In this mode, $v_o = -v_{bc}$, $i_a = 0$, $i_b = -i_c = -I_o$ and the SCR voltages are: $v_{SCR1} = -v_{ca}$, $v_{SCR2} = v_{bc}$, $v_{SCR4} = v_{bc}$, $v_{SCR6} = -v_{ab}$. It is clear from these voltages that all voltages across the other SCRs are negative.

At $\omega t = 30° + \alpha$, v_{ab} becomes the most positive and SCR_1 and SCR_5 are triggered to start mode 2. At $60°$ later, as shown in Fig. 8.35, it can be shown that all the SCR voltages are negative in this mode also. Note from Fig. 8.35 that during each voltage phase, two gating signals are applied during the positive and negative cycles. For example, during the positive cycle of $v_{bc}(t)$, SCR_2 and SCR_6 are triggered with $\alpha = 45°$ at $\omega t = \pi/2 + \pi/4$, and SCR_3 and SCR_5 are triggered in the negative cycle at $\omega t = 3\pi/2 + \pi/4$.

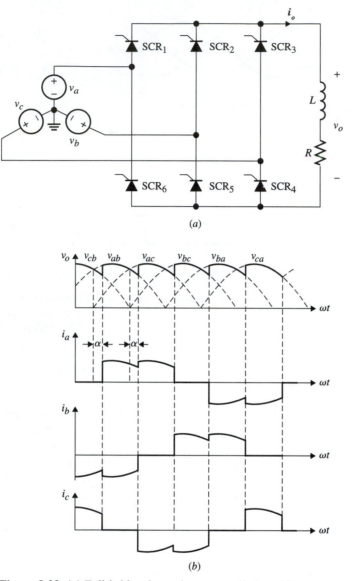

(a)

(b)

Figure 8.33 (a) Full-bridge three-phase controlled rectifier. (b) Output voltage and current waveforms under purely resistive load.

The output voltages for $\alpha = 0°$, $\alpha = 30°$, $\alpha = 60°$, and $\alpha = 90°$ are shown in Fig. 8.36(a), and the output voltages for $\alpha = 120°$, $\alpha = 150°$, and $\alpha = 180°$ are shown in Fig. 8.36(b). For clarity, Fig. 8.36 shows the absolute voltage values for v_{ab}, v_{bc}, and v_{ca}.

The average output voltage can be evaluated over any of the given six pulse intervals. Let us select the second pulse between t_1 and t_2 indicated in the phase voltages of Fig. 8.36(a). The average output voltage is given by

$$V_o = \frac{6}{T} \int_{t_1}^{t_2} v_{ab} \, dt$$

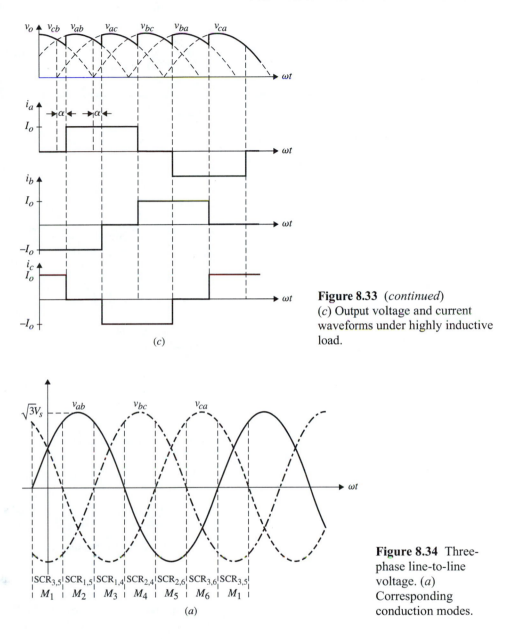

Figure 8.33 (*continued*) (*c*) Output voltage and current waveforms under highly inductive load.

Figure 8.34 Three-phase line-to-line voltage. (*a*) Corresponding conduction modes.

Substituting for $v_{ab} = \sqrt{3}\,V_s \sin(\omega t + 30°)$ and $\omega t_1 = 30° + \alpha$ and $\omega t_2 = 90° + \alpha$, we have

$$V_o = \frac{3}{\pi} \int_{\pi/6+\alpha}^{\pi/2+\alpha} \sqrt{3}\,V_s \sin\left(\omega t + \frac{\pi}{6}\right) d\omega t$$

$$= \frac{-3\sqrt{3}}{\pi} V_s \left[\cos\left(\alpha + \frac{2\pi}{3}\right)\cos\left(\alpha + \frac{\pi}{6}\right)\right] \qquad (8.39)$$

$$= \frac{3\sqrt{3}\,V_s}{\pi}\cos\alpha$$

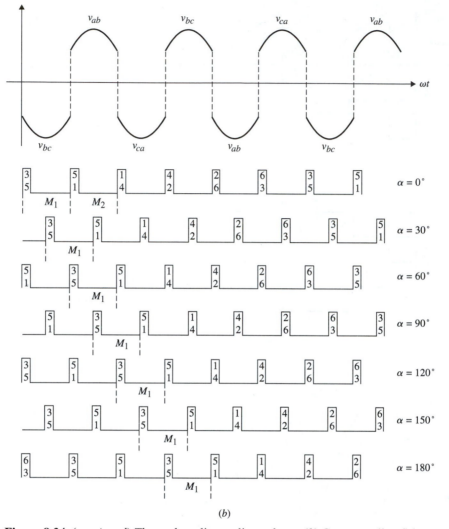

Figure 8.34 (*continued*) Three-phase line-to-line voltage. (*b*) Corresponding firing angles.

The natural conduction is at $\omega t = \pi/6$ if all SCRs are triggered continuously. In this case modes 1 to 6 are the same as in the previous chapter. In this section we are investigating the effect of delaying the trigger by α with respect to the angle of natural conduction. In Fig. 8.36, the natural conduction angle in M_1 is $-\pi/6$ when SCR_3 and SCR_5 are acting like diodes. Now let us consider M_1 by delaying the triggering of SCR_3 and SCR_5 by α. Prior to $\omega t = \alpha$, v_{ca} is the most positive; hence, SCR_3 and SCR_6 are conducting.

For illustration purposes, consider the case for $\alpha = 60°$ We notice that for $\alpha > 60°$, the output voltage starts becoming negative. For $\alpha = 90°$, the positive and negative areas are equal, resulting in a zero average output voltage.

EXERCISE 8.8

Show that the average output power for Fig. 8.35 with $L = \infty$ is given by Eq. (8.40).

$$P = \frac{3\sqrt{3}}{2\pi}V_s I_o \cos\alpha \qquad (8.40)$$

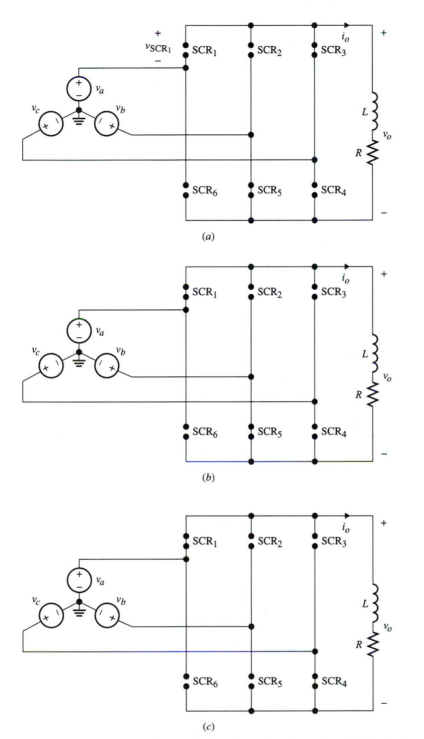

Figure 8.35 Equivalent circuits for the six modes of operation. (*a*) Mode 1. (*b*) Mode 2. (*c*) Mode 3.

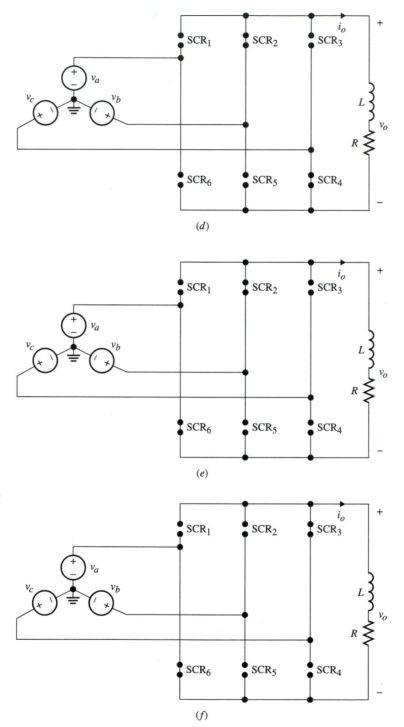

Figure 8.35 (*continued*) Equivalent circuits for the six modes of operation. (*d*) Mode 4. (*e*) Mode 5. (*f*) Mode 6.

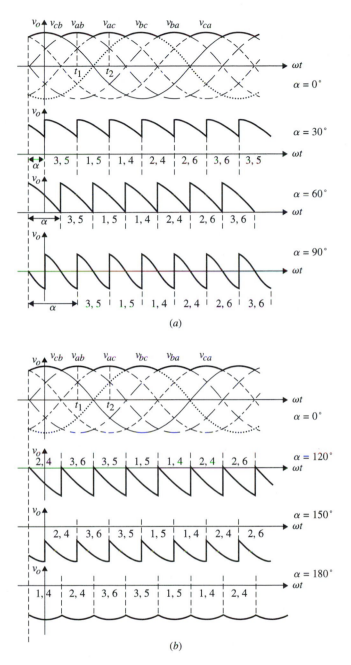

Figure 8.36 Output voltages for firing angles. (*a*) $\alpha = 0°$, $\alpha = 30°$, $\alpha = 60°$, and $\alpha = 90°$. (*b*) $\alpha = 120°$, $\alpha = 150°$, and $\alpha = 180°$

PROBLEMS

Single-Phase Half-Wave Rectifiers

8.1 Derive the power factor and THD expressions for the half-wave phase-controlled rectifier given in Eqs. (8.5) and (8.6).

8.2 Calculate the average power delivered to a resistive load in a half-wave phase-controlled rectifier with $v_s = 200 \sin 377t$, $\alpha = 30°$, and $R_L = 25\ \Omega$.

8.3 Derive the load current equation, i_o, for the inductive-load phase-controlled rectifier with a freewheeling diode in Fig. 8.8(*a*).

8.4 Determine the delay angle α such that the total power to be delivered to a 20 Ω load resistance with $\omega L = 20\ \Omega$ in Fig. 8.8(*a*) equals 40 W. Assume $v_s = 100 \sin 377t$.

Single-Phase Full-Wave Rectifiers

8.5 An alternative way to make a full-wave bridge rectifier is to use a center-tap transformer as shown in Fig. P8.5.

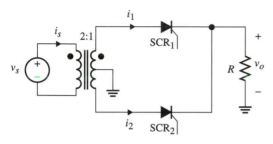

Figure P8.5

(a) Sketch the waveforms for i_s, i_1, i_2, and v_o and determine the power factor.

(b) Determine α for a power factor equal to 0.95. Assume $v_s = 100 \sin 377t$ and $\alpha = 40°$.

8.6 (a) Determine the input power factor for the inductive-load phase-controlled full-wave rectifier shown in Fig. P8.6. Assume $v_s = 100 \sin 377t$, $\alpha = 60°$, and the load current is constant.

(b) Determine α for a power factor equal to 0.75.

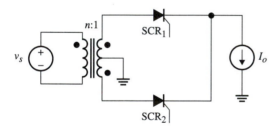

Figure P8.6

8.7 Figure P8.7 shows a full-wave phase-controlled rectifier with an inductive load and a free-wheeling diode.

(a) Sketch the waveform for v_o and give the expression for the average output voltage.

(b) Show that the load current is given by the following two equations:

$$i_o(t) = I_{o1} e^{-(t-t_1)/\tau} + \frac{V_s}{|Z|} \sin[\omega(t - t_1)]$$

$$t_1 \leq t \leq \frac{T}{2}$$

$$i_o(t) = I_{o2} e^{-(t-T/2)/\tau} \quad \frac{T}{2} \leq t \leq T - t_1$$

where

$$|Z| = \sqrt{(\omega L)^2 + R^2}, \ \alpha = \omega t_1, \ \theta = \tan^{-1}\frac{\omega L}{R},$$

$$\tau = L/R$$

$$I_{o1} = e^{-(\pi + \alpha/\tau\omega)} + e^{\alpha/\tau\omega} \sin(\alpha + \theta)$$

$$I_{o2} = e^{-(\pi/\tau\omega)} + e^{\alpha/\tau\omega} \sin(\alpha + \theta)$$

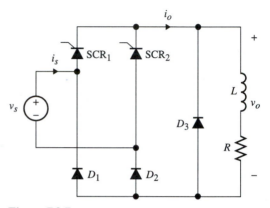

Figure P8.7

8.8 Derive the THD for the full-bridge phase-controlled rectifier of Fig. 8.13.

8.9 Show that the average power delivered to the load in Fig. 8.25 is zero when

$$\alpha = \cos^{-1}\frac{V_{dc}\,\pi}{2V_s}$$

8.10 Sketch the waveforms for i_s, i_o, and v_o, and determine the average output voltage for the circuit of Fig. P8.10, under the conditions (a) $C = 0$ and (b) $C = \infty$, respectively. Assume the devices are triggered in a similar way to those of Fig. 8.11.

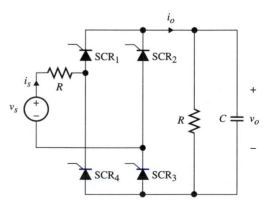

Figure P8.10

8.11 Consider a power electronic circuit that generates the waveforms i_s and v_s shown in Fig. P8.11. Determine the THD.

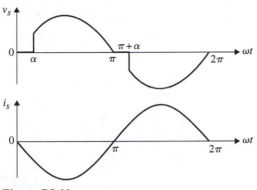

Figure P8.11

8.12 Sketch the waveforms for i_s and v_o for Fig. P8.12. Assume $L/R \gg T$, $v_s = V_s \sin \omega t$, and $V_{dc} < V_s$.

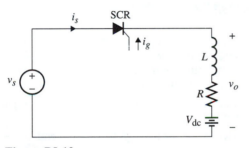

Figure P8.12

8.13 Repeat Exercise 8.5 for $\alpha = 150°$.

8.14 Sketch the waveforms for v_o, i_{SCR1}, and i_{SCR2} using the circuit shown in Fig. P8.15 and obtain the expression for the fundamental component of $i_s(t)$. Assume v_s is a squarewave of amplitude $\pm V_s$.

8.15 Calculate the input pf for $I_o = 5$ A and $\alpha = 60°$, and the current THD for Problem 8.14.

8.16 Consider the full-wave phase-controlled circuit shown in Fig. P8.16. Calculate: (a) the average output voltage V_o, (b) the average output current I_o, (c) the rms values of $i_s(t)$, and (d) the input power factor. Assume $L/R \gg T/2$ and $v_s = 110 \sin 377t$ and $\alpha = 45°$.

8.17 Repeat Problem 8.16 for $\alpha = 150°$.

8.18 Figure P8.18 shows a half-wave phase-controlled converter with a flyback SCR_2. Assume $v_s = 110 \sin 377t$.

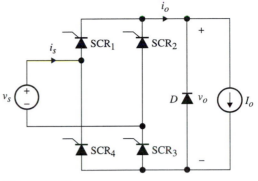

Figure P8.15

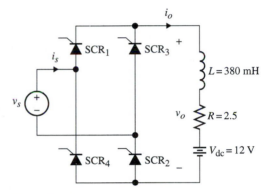

Figure P8.16

(a) Calculate V_o, u, and the power delivered to the load for $\alpha = 60°$.

(b) If the load resistance is reduced by 50%, find the new value of α for maintaining the same load current.

(c) Repeat part (b) by assuming L_s has increased by 20%.

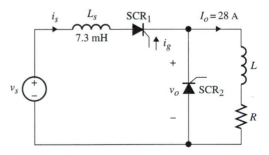

Figure P8.18

8.19 Calculate the rms value for $i_s(t)$ given in Problem 8.18 by assuming that the commutation period is linear.

8.20 Calculate the average load current in the circuit shown in Fig. P8.20. Assume $\alpha = 60°$ and $v_s = 110 \sin 377t$.

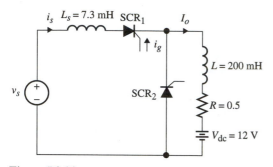

Figure P8.20

8.21 Draw the waveforms for i_s, i_{D1}, i_{s1}, and v_o in Fig. P8.21 under the following load conditions: (a) $L = 0$, (b) $L = \infty$. Label your axes and show which device(s) conduct for each time interval. Assume $v_s = V_s \sin \omega t$ and the firing angle is $30°$.

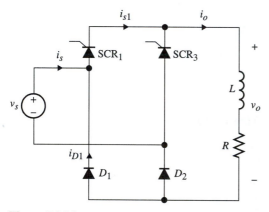

Figure P8.21

8.22 Consider the SCR circuit in Fig. P8.22 with $\alpha = 30°$. Assume v_{s1} and v_{s2} are $180°$ out of phase.

(a) Sketch the waveforms for i_{s1}, i_{s2}, and v_o for two periods.

(b) Derive the expressions for i_{s1}, i_{s2}, and v_o.

(c) Find the commutation interval during which both SCRs are conducting.

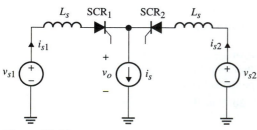

Figure P8.22

D8.23 Consider the half-controlled diode/SCR full-bridge circuit shown in Fig. P8.23. Assume the input is given by $v_s(t) = V_s \sin \omega t$. Assume the firing angle is $\alpha = 30°$.

(a) Sketch v_o, i_{D1}, i_{D2}, i_{SCR1}, i_{SCR2}, and i_s.

(b) Derive the expression for v_o.

(c) Design for α so that the average output voltage is 48 V. Assume $v_s(t) = 100 \sin 377t$ V.

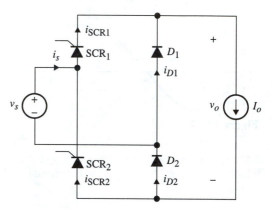

Figure P8.23

D8.24 Consider the controlled half-wave rectifier of Fig. 8.26(a) with $L_s = 7.3$ mH, $\alpha = 60°$, which supplies an average load current of 20 A. Assume $v_s = 110 \sin (2\pi 60t)$.

(a) Use Fig. 8.27 to determine the average output voltage.

(b) Determine the average power delivered to the load.

(c) Find the commutation angle.

(d) Design for the inductor L in the load side; $V_o = 8.75$ V and $R = 0.44$ Ω.

(e) If the load resistance is decreased by 50%, what is the new α needed to maintain the same average output voltage?

(f) If the source voltage is increased by 20%, what is the new α needed to maintain the same average load?

Three-Phase Converters

8.25 The three-phase half-wave controlled rectifier with a commutating diode across the load is shown in Fig. P8.25. Assume that the voltage drops across the thyristor are negligible and the supply voltage is 100 V rms. For a firing delay angle of 45°, sketch the waveform of v_o and determine its average value.

8.26 (a) Sketch the waveforms for v_o, i_{SCR3}, and i_o for the three-phase controlled rectifier shown in Fig. P8.26 for $\alpha = 30°$ and $\alpha = 60°$.

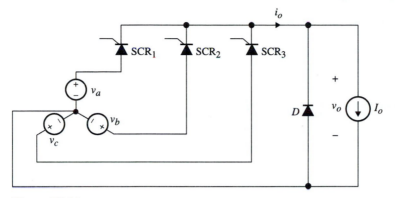

Figure P8.25

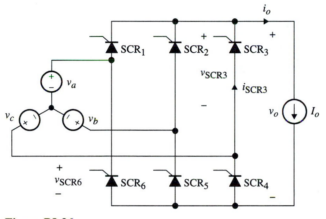

Figure P8.26

(b) Derive the expression for the average output voltage.

(c) Sketch the waveform of the voltage across SCR_6 and SCR_3. Assume $v_a = V_s \sin \omega t$, $v_b =$ $V_s \sin(\omega t - 2\pi/3)$, and $v_c = V_s \sin(\omega t - 4\pi/3)$.

8.27 Determine the average output voltage for Problem 8.26 using $V_s = 110$ V.

Chapter 9

dc-ac Inverters

INTRODUCTION

In this chapter, we will consider power electronic circuits that produce variable-frequency ac output voltages from dc sources. As discussed in Chapter 3, depending on whether the source is dc or ac, power electronic circuits with ac output voltages are referred to as *dc-ac inverters* or *ac-ac cycloconverters*. In converting ac to ac, if the output voltage frequency is different from the source frequency, the converter is called an *ac voltage controller*. Traditionally, dc-to-ac inverters (also known as *static inverters*) use fixed dc sources to produce symmetrical ac output voltages at fixed or variable frequency or magnitude. The output ac voltage system can be of the single-phase or three-phase type at frequencies of 50 Hz, 60 Hz, and 400 Hz with a voltage magnitude range of 110–380 VAC. Inverter circuits are used to deliver power from a dc source to a passive or active ac load employing conventional SCRs or gate-driven semiconductor devices such as GTOs, IGBTs, and MOSFETs. Due to their increased switching speed and power capabilities coupled with complex control techniques, today's inverters can operate in wide ranges of regulated output voltage and frequency with reduced harmonics. Medium- and high-power half-bridge and full-bridge switching devices using MOSFETs and IGBTs are available as packages.

Dc-to-ac inverters are used in applications where the only source available is a fixed dc source and the system requires an ac load such as in uninterruptible power supply (UPS). Applications where dc-ac inverters are used include aircraft power supplies, variable-speed ac motor drives, and lagging or leading var generation. For example, an inverter used to provide necessary changes in the frequency of the ac output is used to regulate the speed of an induction motor and is also used in a UPS system to produce a fixed ac frequency output when the main power grid system is out.

9.1 BASIC BLOCK DIAGRAM OF DC-AC INVERTER

Figure 9.1 shows a typical block diagram of a power electronic circuit utilizing a dc-to-ac inverter with input and output filters used to smooth the output ac signal.

The feedback circuit is used to sense the output voltage and compare it with a sinusoidal reference signal as shown in Fig. 9.1. Depending on the arrangement of the power switches and their types, various control techniques are normally adopted in industry. The control objective is to produce a controllable ac output from an uncontrollable dc voltage source. Even though the desired output voltage waveform is purely sinusoidal, practical inverters are not purely sinusoidal but include significant high-frequency harmonics. This is why inverters normally employ a high-frequency switching technique to reduce such harmonics. Two alternative control methods of inversions will be discussed in this chapter: uniform pulse width modulation (PWM) and sinusoidal PWM. Depending on the output power level, both techniques find their way into practice. Sinusoidal PWM is widely used in motor drive applications with high-frequency operation.

The front end of the power electronic circuit in Fig. 9.1 is the line ac-to-dc converter discussed in Chapters 7 and 8. Unlike the line ac-to-dc converter circuits in which diodes and SCRs are commutated naturally, in dc-to-ac inverters forced commutation is used to turn on and off the power switches. The negative portion of the source voltage is used naturally to turn off the diode and/or the SRC in line ac-dc converters. As a result, the ac-dc converter circuit has one ac frequency—the line frequency—and the device relies on line commutation for switching. In dc-to-ac inverters, the

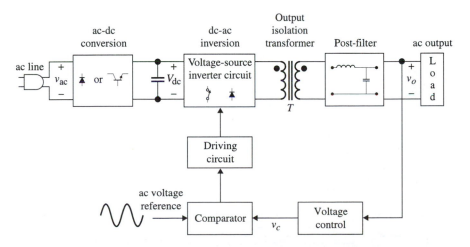

Figure 9.1 Block diagram of a typical power electronic circuit with dc-to-ac inverter.

ac frequency is not necessarily the line frequency. Hence, if an SCR is used, we must devise a means to force it to turn off in order to produce the desired ac output. This feature, the variable-frequency output voltage, causes these circuits to play a very important role in all kinds of industrial applications, such as controlling the speed of motors. Variable-speed drives are used to control the speed of electric vehicles, pumps, rollers, and conveyors. Therefore, inverters circuits require more elaborate control signals to shape the ac voltage.

The load in Fig. 9.1 is broadly classified as either passive or active. If the load consists of impedance only (i.e., is passive), its time domain response is determined by the nature of the load and cannot be controlled externally. If the load consists of a source (i.e., is active), its time domain information can be controlled externally, as in the case of machine loads.

One final note should be made about power switches. The simplest inverter is one that employs a switching device that can be gate-controlled to interrupt current flow, resulting in a naturally commutating or self-commutated inverter. These switching devices are the gate-driven types such as GTOs, IGBTs, MOS-FETs, and power BJTs. When SCRs are used, the converter requires an additional external circuit to force the SCR to turn off. The reason for the mandatory forced commutation in such inverters is that the dc input voltage across the SCR devices causes them to be in the forward conduction state. The higher the power rating of the gate-controlled devices, the less SCRs are used in the design of inverters.

Voltage- and Current-Source Inverters

Since a practical source can provide either a constant voltage or a constant current, broadly speaking, inverters are divided into voltage-source inverters (VSIs) and current-source inverters (CSIs). The dc source in a VSI is a fixed voltage such as a battery, fuel cell, solar cell, dc generator, or rectified dc source. In a CSI, the dc source is a nearly constant current source. The nature of the source, whether it is a dc current source or a dc voltage source, makes the power inverter clearly distinguishable and its practical application more defined. Block diagrams for a voltage-source and a current-source inverter are shown in Fig. 9.2(a) and (b), respectively.

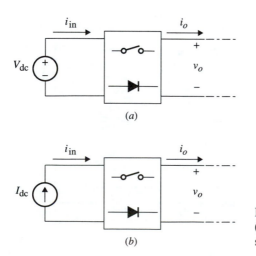

Figure 9.2 Block diagram representations for (a) voltage source inverter and (b) current-source inverter.

In the voltage-source inverter (VSI), the output voltage, v_o, is a function of the inverter operation; the load current, i_o, is a function of the nature of the load; and the dc input, V_{dc}, is a constant input voltage. In the current-source inverter, the output voltage is a function of the inverter operation; the load current, i_o, is a function of the nature of the load; and the source, I_{dc} is a constant input current.

Inverter Configurations

Figure 9.3(*a*), (*b*), and (*c*) shows, respectively, three possible single-phase inverter arrangements: biphase, half-bridge, and full-bridge. The biphase inverter, also known as a push-pull inverter, is drawn in two different ways in Fig. 9.3(*a*).

Output Voltage Control

If the output voltage is controlled by varying the dc source voltage, this can be accomplished by either controlling the dc input by using a front dc-dc converter, as shown in Fig. 9.4(*a*), or by using an ac-dc phase control converter, as shown in Fig. 9.4(*b*). In many applications, varying the input dc voltage is not possible or is costly.

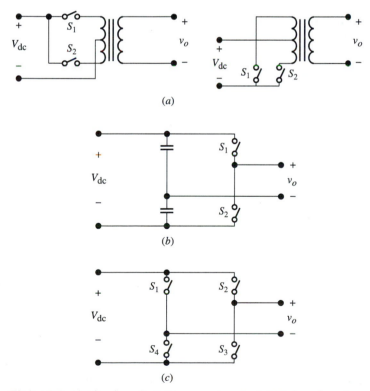

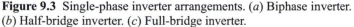

Figure 9.3 Single-phase inverter arrangements. (*a*) Biphase inverter. (*b*) Half-bridge inverter. (*c*) Full-bridge inverter.

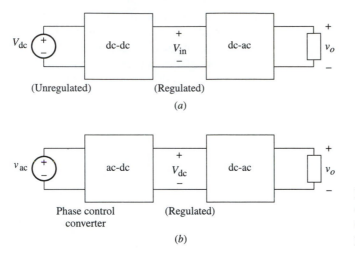

Figure 9.4 Controlling the dc input using (*a*) dc-dc converter or (*b*) ac-dc phase-controlled rectifier.

9.2   BASIC HALF-BRIDGE INVERTER CIRCUITS

9.2.1   Resistive Load

To illustrate the basic concept of a dc-to-ac inverter circuit, we consider a half-bridge voltage-source inverter circuit under a resistive load as shown in Fig. 9.5(*a*). The switching waveforms for S_1, S_2, and the resultant output voltage are shown in Fig. 9.5(*b*).

The circuit operation is very simple. S_1 and S_2 are switched on and off alternatively at a 50% duty cycle, as shown in the switching waveform in Fig. 9.5(*b*). This shows that the circuit generates a square ac voltage waveform across the load from a

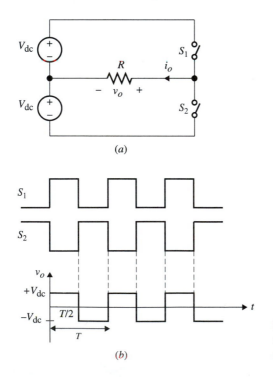

Figure 9.5 (*a*) Half-bridge inverter under resistive load. (*b*) Switching and output voltage waveforms.

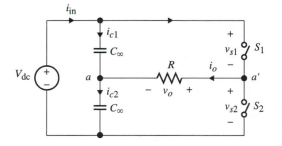

Figure 9.6 Half-bridge inverter circuit with large splitting capacitors.

constant dc source. The voltages V_{dc} and $-V_{dc}$ are applied across R when S_1 is on and S_2 is off, and when S_2 is on and S_1 is off, respectively. One observation to be made here is that the frequency of the output voltage is equal to $1/T$ and is determined by the switching frequency. This is true as long as S_1 and S_2 are switched complementarily. Moreover, the rms value of the output voltage is simply V_{dc}. Hence, to control the rms value of the output voltage, we must control the rectified V_{dc} voltage source. Another observation is that the load power factor is unity since we have a purely resistive load. This is rarely encountered in practical applications.

Finally, we should note that in practice the circuit does not require two equal dc voltage sources as shown in Fig. 9.5(a). Instead, large splitting capacitors are used to produce two equal dc voltage sources as shown in Fig. 9.6. The two capacitors, C_∞, are equal and very large, so that RC_∞ is much larger than the half switching period. This will guarantee that the midpoint, a, between the capacitors has a fixed potential at one-half of the supply voltage V_{dc}. The current from the source, V_{dc}, equals one-half of the load current, i_o. In steady state, the average capacitor currents are zero; hence, capacitors C_∞ are used to block the dc component of i_o.

The limitation of this half-bridge configuration is that varying the switching sequence cannot control the output voltage.

EXAMPLE 9.1

Sketch the current and voltage waveforms for i_{in}, i_o, v_{s1}, and v_{s2}, for the circuit shown in Fig. 9.6 for $\theta = 0$ and $\theta \neq 0$ by using the switching waveforms for S_1 and S_2 shown in Fig. 9.7(a). Determine the average output voltage in terms of V_{dc} and θ when the inverter operates in the steady state.

SOLUTION   Let us consider mode 1, when S_1 is on and S_2 is off. Then the output voltage and current equations are given by

$$v_o = \frac{V_{dc}}{2}$$

$$i_o = \frac{v_o}{R} = \frac{V_{dc}}{2R}$$

Since we assumed that the capacitors are equal, the load current, i_o, splits equally, i.e.,

$$i_{c1} = -\frac{1}{2}i_o = -\frac{V_{dc}}{4R}$$

$$i_{c2} = i_{in} = -i_{c1} = \frac{V_{dc}}{4R}$$

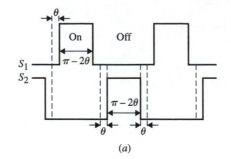

(a)

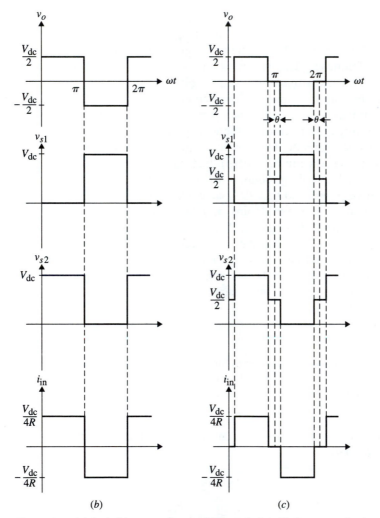

(b)   (c)

Figure 9.7 (a) Switching waveforms for Example 9.1. (b) Current and voltage waveforms for $\theta = 0$. (c) Current and voltage waveforms for $\theta \neq 0$.

The voltages across the switches are

$$v_{s1} = 0$$

$$v_{s2} = v_o + \frac{V_{dc}}{2} = V_{dc}$$

Mode 2 starts when S_1 and S_2 are off during the short interval θ.

$$v_o = 0$$

$$i_o = 0$$

$$i_{c1} = i_{c2} = i_{in} = 0$$

$$v_{s1} = v_{s2} = \frac{V_{dc}}{2}$$

Mode 3 starts when S_2 is on and S_1 is off, which yields the following equations:

$$v_o = -\frac{V_{dc}}{2}$$

$$i_o = \frac{v_o}{R} = -\frac{V_{dc}}{2R}$$

$$i_{c1} = \frac{-i_o}{2} = +\frac{V_{dc}}{4R}$$

$$i_{c2} = -i_{c1} = \frac{-V_{dc}}{4R}$$

$$i_{in} = i_{c1} = +\frac{V_{dc}}{4R}$$

$$v_{s1} = \frac{V_{dc}}{2} - v_o = V_{dc}$$

$$v_{s2} = 0$$

Mode 4 is similar to mode 2 since both switches are open.
The average output voltage is given by

$$v_{o,\text{rms}} = \sqrt{\frac{1}{T}\int_0^T v_o^2(t)\, dt}$$

$$v_{o,\text{rms}} = \sqrt{\frac{1}{2\pi}\left[\int_\theta^{\pi-\theta}\left(\frac{V_{dc}}{2}\right)^2 d\theta + \int_{\pi+\theta}^{2\pi-\theta}\left(\frac{-V_{dc}}{2}\right)^2 d\theta \right]} \tag{9.1}$$

$$v_{o,\text{rms}} = \frac{V_{dc}}{2}\sqrt{\left(1 - \frac{2\theta}{\pi}\right)}$$

Notice that when $\theta = 0$, the rms value of the output is $V_{dc}/2$, as expected. Notice also that in a practical situation under an inductive load, the two switches are not allowed to switch off simultaneously. Figure 9.7(b) and (c) shows the currents and voltages for $\theta = 0$ and $\theta \neq 0$, respectively. Since the waveform is symmetric, the average output voltage will always be zero.

9.2.2 Inductive-Resistive Load

Figure 9.8(a) shows a half-bridge inverter under an inductive-resistive load, with its equivalent circuit and the output waveforms shown in Fig. 9.8(b) and (c), respectively.

With S_1 and S_2 switched complementarily, each at a 50% duty cycle at a switching frequency f, then the load between terminals a and a' is excited by a square voltage

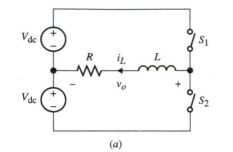

(a)

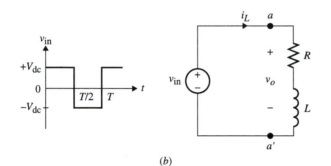

(b)

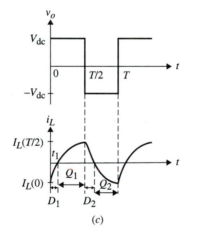

(c)

Figure 9.8 (a) Half-bridge inverter with inductive-resistive load. (b) Equivalent circuit. (c) Steady-state waveforms.

waveform $v_{in}(t)$ of amplitudes $+V_{dc}$ and $-V_{dc}$ as shown in Fig. 9.8(b); i.e., $v_{in}(t)$ is defined as follows:

$$v_{in}(t) = \begin{cases} +V_{dc} & 0 \le t < T/2 \\ -V_{dc} & T/2 \le t < T \end{cases} \tag{9.2}$$

The switches are implemented using a conventional SCR (requiring an external forced commutation circuit) or fully controlled power switching devices such as IG-BTs, GTOs, BTJs, or MOSFETs. Notice that from the direction of the load current i_L, these switches must be bidirectional. An example is the half-bridge inverter circuit shown in Fig. 9.9 with S_1 and S_2 implemented by MOSFETs.

Assume the inverter operates in the steady state and its inductor current waveform is shown in Fig. 9.8(c). For $0 \le t < t_1$, the inductor current is negative, which means that while S_1 is on the current actually flows in the reverse direction, i.e., in the

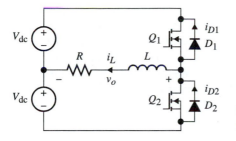

Figure 9.9 MOSFET implementation for S_1 and S_2 in the half-bridge inverter of Fig. 9.8(a).

body (flyback) diode of the bidirectional switch S_1. At $t = t_1$, the current flows through transistor Q_1, as shown in Fig. 9.8(c). At $t = T/2$, when S_2 is turned on, since the current direction is positive, the flyback diode, D_2, turns on until $t = T/2 + t_1$, when Q_2 starts conducting.

In steady state, the following conditions must hold:

$$i_L(0) = -i_L(T/2)$$

$$i_L(0) = i_L(T)$$

During the first interval $(0 \leq t < T/2)$, when S_1 is on and S_2 is off, $v_{in}(t) = +V_{dc}$, resulting in the following equation for $i_L(t)$:

$$L\frac{di_L}{dt} + Ri_L = V_{dc} \tag{9.3}$$

If the inductor initial value equals $I_L(0)$, the solution for $i_L(t)$ is given by

$$i_L(t) = +\left(I_L(0) \mp \frac{V_{dc}}{R}\right)e^{-t/\tau} + \frac{V_{dc}}{R} \tag{9.4}$$

where $\tau = L/R$. Since $I_L(T/2) = -I_L(0)$, the initial condition at $t = 0$ is constant and given by

$$I_L(0) = -\frac{V_{dc}}{R}\frac{1 - e^{-T/2\tau}}{1 + e^{-T/2\tau}} \tag{9.5}$$

The second half-cycle for $t > T/2$ produces the following expression for $i_L(t)$ with the initial condition at $t = T/2$ equaling $-I_L(0)$:

$$i_L(t) = \left(I_L(0) + \frac{V_{dc}}{R}\right)e^{-(t - T/2)/\tau} - \frac{V_{dc}}{R} \tag{9.6}$$

This expression equals $-i_L(t)$ of Eq. (9.4), which is valid for the interval $(T/2 \leq t < T)$.

The exact expression for the average power delivered to the load can be obtained from the following relation:

$$P_{o,ave} = \frac{1}{T}\int_0^T i_L(t)v_o(t)\, dt$$

$$= \frac{2V_{dc}}{T}\int_0^{T/2}\left[\left(I_L(0) \mp \frac{V_{dc}}{R}\right)e^{-t/\tau} + \frac{V_{dc}}{R}\right] dt \tag{9.7}$$

where $I_L(0)$ is given by Eq. (9.5).

To determine the device's ratings, we must find the average and rms current values through the switching devices and flyback diodes. It is clear that when i_L changes polarity at $t = t_1$ during the first interval, the load current changes polarity and, hence, commutates from D_1 to Q_1. Similarly, at $t = t_1 + T/2$, the current commutation goes from D_2 to Q_2 as shown in Fig. 9.8(c). The time at which $i_L(t)$ becomes zero, $t = t_1$, is obtained by setting $i_L(t)$ in Eq. (9.4) to zero at $t = t_1$, to yield

$$t_1 = \tau \ln \frac{2}{1 + e^{-T/2\tau}} \tag{9.8}$$

Obtaining the value of t_1 allows us to find the average and rms values for the switching devices and flyback diodes.

Average Transistor and Diode Currents

To help us obtain quantitatively the expressions for the diode and transistor currents, we represent the load voltage and current by their fundamental components as shown in Fig. 9.10(a).

Let the fundamental components of $v_o(t)$ and $i_L(t)$ be given by

$$v_{o1}(t) = V_{o1} \sin \omega t \tag{9.9}$$

$$i_{L1}(t) = I_{o1} \sin(\omega t + \theta) \tag{9.10}$$

where $V_{o1} = 2V_{dc}/\pi$, and I_{o1} and θ are the peak and the phase angle of $i_{L1}(t)$ as shown in Fig. 9.10(a), which are given by

$$I_{o1} = \frac{2V_{dc}}{\pi|Z|}, \quad \theta = \tan^{-1}\left(\frac{\omega L}{R}\right) \tag{9.11}$$

where $|Z| = \sqrt{R^2 + (\omega L)^2}$. Using the fundamental component expression, the rms values for each of the diode and transistor currents are given by

$$
\begin{aligned}
I_{D,\mathrm{rms}} &= \sqrt{\frac{1}{2\pi} \int_0^\theta i_{L1}^2(t)\, d\omega t} \\
&= \sqrt{\frac{1}{2\pi} \int_0^\theta I_{o1}^2 \sin^2 \omega t\, d\omega t} \\
&= \sqrt{\frac{I_{o1}^2}{4\pi}\left(-\frac{\sin 2\theta}{2} + \theta\right)} = \frac{I_{o1}}{2}\sqrt{\frac{\theta}{\pi} - \frac{\sin\theta\cos\theta}{2\pi}}
\end{aligned}
\tag{9.12}
$$

$$
\begin{aligned}
I_{Q,\mathrm{rms}} &= \sqrt{\frac{1}{2\pi} \int_0^\pi I_{o1}^2 \sin^2 \omega t\, d\omega t} \\
&= \sqrt{\frac{I_{o1}^2}{4\pi}(\pi + \cos\theta\sin\theta - \theta)} = \frac{I_{o1}}{2}\sqrt{1 - \frac{\theta}{\pi} + \frac{\sin\theta\cos\theta}{\pi}}
\end{aligned}
\tag{9.13}
$$

and the average values for each of the diode and transistor currents are given by

$$
\begin{aligned}
I_{D,\mathrm{ave}} &= \frac{1}{2\pi} \int_0^\theta I_{o1} \sin \omega t\, d\omega t \\
&= \frac{I_{o1}}{2\pi}(1 - \cos\theta)
\end{aligned}
\tag{9.14}
$$

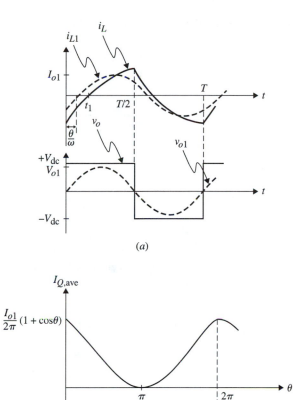

(a)

(b)

Figure 9.10 (a) Output voltage and current waveforms. (b) Average transistor and diode current waveforms as a function of θ.

$$I_{Q,\text{ave}} = \frac{I_{o1}}{2\pi}(1 + \cos\theta) \qquad (9.15)$$

Notice that under a resistive load, the lagging phase angle is zero and the entire load current flows through Q_1 and Q_2 with D_1 and D_2 always reverse biased. This is clear from the zero average diode current obtained from the preceding equation. Under a purely inductive load, the phase angle is 90° and the load current flows through the diode and transistors for an equal time, resulting in equal average and rms current values for the diode and transistor. Figure 9.10(b) shows the average transistor and diode current values as functions of the phase angle.

The average output power delivered to the load is approximated by

$$P_{o,\text{ave}} = V_{o1,\text{rms}} I_{o1,\text{rms}} \cos\theta \tag{9.16}$$

where

$$V_{o1,\text{rms}} = \frac{V_{o1}}{\sqrt{2}} = \frac{\sqrt{2}\, V_{dc}}{\pi}$$

$$I_{o1,\text{rms}} = \frac{I_{o1}}{\sqrt{2}}$$

Equation (9.16) yields the following expression for $P_{o,\text{ave}}$:

$$P_{o,\text{ave}} = \frac{2 V_{dc}^2}{\pi^2 |Z|} \cos\theta \tag{9.17}$$

The ripple voltage in the traditional dc-to-ac converter is defined by

$$V_{o,\text{ripple}} = \sqrt{V_{o,\text{rms}}^2 - V_{o,\text{ave}}^2} \tag{9.18}$$

Similarly, the input ripple current expression in the dc-to-ac inverter is defined by

$$I_{\text{in,ripple}} = \sqrt{I_{o,\text{rms}}^2 - I_{o,\text{ave}}^2} \tag{9.19}$$

$$I_{\text{in,ripple}} \approx I_{o1,\text{rms}} = \frac{\sqrt{2}\, V_{dc}}{\pi |Z|} \tag{9.20}$$

EXAMPLE 9.2

Consider the half-bridge inverter of Fig. 9.9 with the following circuit components: $V_{dc} = 408$ V, $R = 8\ \Omega$, $f = 400$ Hz, and $L = 40$ mH.

(a) Derive the exact expression for $i_L(t)$.

(b) Derive the expression for the fundamental component of $i_L(t)$.

(c) Determine the average diode and transistor currents.

(d) Determine the average power delivered to the load.

SOLUTION   (a) The exact solution for $i_L(t)$ was derived before, and is given again in Eq. (9.21).

$$i_L(t) = \begin{cases} -\left(I_L(0) + \dfrac{V_{dc}}{R}\right) e^{-t/\tau} + \dfrac{V_{dc}}{R} & 0 \le t \le T/2 \\[2ex] +\left(I_L(0) + \dfrac{V_{dc}}{R}\right) e^{-(t-T/2)/\tau} - \dfrac{V_{dc}}{R} & T/2 \le t \le T \end{cases} \tag{9.21}$$

Since $i_L(T/2) = -I_L(0)$, we have

$$I_L(0) = -\frac{V_{dc}}{R} \frac{1 - e^{-T/2\tau}}{1 + e^{-T/2\tau}} \quad \text{where } \tau = \frac{L}{R} \tag{9.22}$$

(b) To calculate the fundamental component of $i_L(t)$, we first determine the a_1 and b_1 coefficients:

$$I'_{L1}(t) = \frac{2}{T}\left[\int_0^{T/2} i_L(t)\cos\omega t\ dt + \int_{T/2}^{T} i_L(t)\cos\omega t\ dt\right] \tag{9.23a}$$

$$I''_{L1}(t) = \frac{2}{T}\left[\int_0^{T/2} i_L(t)\sin\omega t\ dt + \int_{T/2}^{T} i_L(t)\sin\omega t\ dt\right] \tag{9.23b}$$

The fundamental component of $i_L(t)$ is given by

$$i_{L1}(t) = I'_{L1}(t)\cos\omega t + I''_{L1}(t)\sin\omega t \tag{9.24}$$

$$i_{L1}(t) = -0.57\cos(2513.3t + 57.8°)\ \text{A}$$

The rms value of $i_L(t)$ is

$$I_{L,\text{rms}} = \sqrt{\frac{1}{T}\left[\int_0^{T/2} i_L^2(t)\ dt + \int_{T/2}^{T} i_L^2(t)\ dt\right]}$$

$$I_{L,\text{rms}} = 64.1\ \text{A}$$

(c) The average diode current is given by Eq. (9.14) as

$$I_{D,\text{ave}} = \frac{I_{o1}}{2\pi}(1 - \cos\theta)$$

with $\theta = \tan^{-1}(\omega L/R) = 85.45°$ and

$$I_{o1} = \frac{2V_{\text{dc}}}{\pi|Z|} = \frac{2\times408}{\pi\times100.85} = 2.58\ \text{A}$$

$$|Z| = \sqrt{R^2 + (\omega L)^2} = 100.85\ \Omega$$

If we substitute these values, then $I_{D,\text{ave}}$ turns out to be

$$I_{D,\text{ave}} = \frac{2.58}{2\pi}(1 - \cos 85.45°) = 0.38\ \text{A}$$

Similarly, $I_{Q,\text{ave}}$ is equal to

$$I_{Q,\text{ave}} = \frac{I_{o1}}{2\pi}(1 + \cos\theta) = 0.44\ \text{A}$$

(d) The average power is given by Eq. (9.17):

$$P_{\text{ave}} = V_{o1,\text{rms}}I_{o1,\text{rms}}\cos\theta$$

$$= \frac{2V_{\text{dc}}^2}{\pi^2|Z|}\cos\theta$$

$$P_{\text{ave}} = \frac{2(408)^2}{\pi^2(100.85)}\cos(85.45) = 26.53\ \text{W}$$

EXERCISE 9.1

Figure E9.1 shows a half-bridge inverter under a purely inductive load with $V_{\text{dc}} = 24$ V, $L = 10$ mH, and $f = 60$ Hz. Assume Q_1 and Q_2 are operating at a 50% duty cycle and the circuit has reached the steady-state operation.

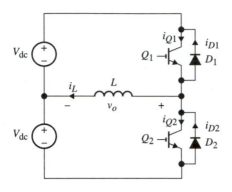

Figure E9.1 Half-bridge inverter with purely inductive load.

(a) Sketch the waveforms for i_L, i_{Q1}, i_{D1}, i_{Q2}, i_{D2}, and v_o.

(b) Calculate the rms value of $i_L(t)$.

(c) Determine the average output power delivered to the load.

ANSWER $I_{L,\,\text{rms}} = 11.54$ A , $P_{\text{ave}} = 0$ W

EXERCISE 9.2

Consider the single-phase inverter of Fig. 9.8(*a*) with an inductive-resistive load that delivers 400 W to a 60 Hz load from a dc source of 420 V. Assume the output voltage and current are represented by their fundamental components with a lagging power factor of 0.6. Determine the average and rms current values for the transistors and diodes.

ANSWER $I_{D,\text{rms}} = 0.81$ A, $I_{D,\text{ave}} = 0.45$ A, $I_{Q,\text{rms}} = 2.31$ A, $I_{Q,\text{ave}} = 1.80$ A

EXAMPLE 9.3

Draw the output voltage and v_{s1} waveforms for the center-tap biphase inverter shown in Fig. 9.11. Assume S_1 and S_2 are bidirectional switches and are switched at a 50% duty cycle. It is used in a low-input-voltage application to reduce losses, since the current only flows half-period in a section of the transformer (the transformer is not fully utilized). The two modes of operations are shown in Fig. 9.11(*b*) and (*c*); the waveforms are shown in Fig. 9.11(*d*).

SOLUTION   The equivalent circuit for mode 1, when switch S_1 is on, is shown in Fig. 9.11(*b*). The output voltage is obtained as follows:

$$\frac{v_o}{V_{dc}} = \frac{n_2}{n_1}$$

$$v_o = \frac{n_2}{n_1} V_{dc}$$

Figure 9.11(*c*) shows the equivalent circuit for mode 2, with v_o given by

$$\frac{v_o}{V_{dc}} = -\frac{n_2}{n_1}$$

$$v_o = -\frac{n_2}{n_1} V_{dc}$$

The waveforms for v_o are shown in Fig. 9.11(*d*).

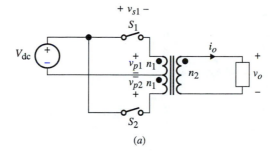

(a)

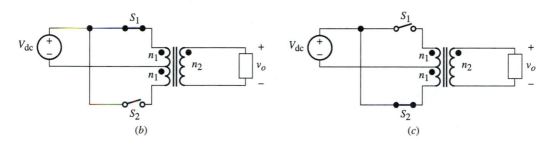

(b) (c)

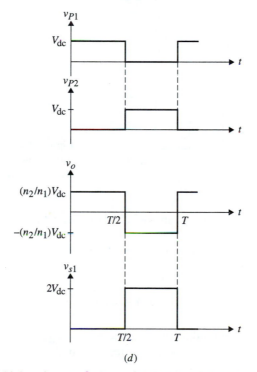

(d)

Figure 9.11 (a) Center-tap biphase inverter for Example 9.3. (b) Mode 1. (c) Mode 2. (d) Voltage waveforms.

9.3   FULL-BRIDGE INVERTERS

Figure 9.12 shows the full-bridge circuit configuration for a voltage-source inverter under resistive load.

Depending on the switching sequence of S_1, S_2, S_3, and S_4, the output voltage can be controlled either by varying V_{dc} only or by controlling the phase shift between the

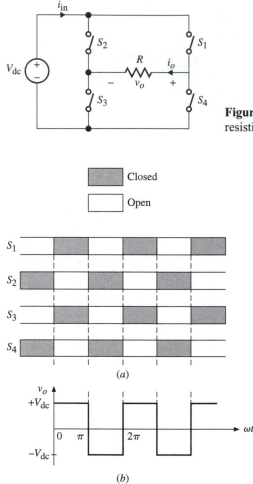

Figure 9.12 Full-bridge inverter under a purely resistive load.

Figure 9.13 (*a*) Switching sequence for full-bridge voltage-source inverter at 50% duty cycle. (*b*) Output voltage waveform.

switches. If S_1-S_3 and S_2-S_4 are switched on and off at a 50% duty cycle as shown in Fig. 9.13(*a*), the output voltage, shown in Fig. 9.13(*b*), is a symmetrical squarewave whose fundamental rms value is controlled by varying only V_{dc}. The fundamental value of $v_o(t)$ is given by

$$v_{o1}(t) = \frac{4V_{dc}}{\pi} \sin \omega t \tag{9.25}$$

The rms value of the fundamental component is $V_{dc}/\sqrt{2}\,\pi$. The use of SCR rectifier circuits or dc-dc switch-mode converters normally accomplishes varying of the dc voltage source.

Another method to control the rms value of the output voltage is to use the switching sequence shown in Fig. 9.14(*a*). This switching sequence is obtained by shifting the timing sequence of S_1 and S_4 in Fig. 9.13(*a*) to the left by a phase angle α, and S_2 and S_3 to the right by the same angle.

The fundamental component of the output voltage, $v_{o1}(t)$ is given by

$$v_{o1}(t) = V_{dc}\sqrt{\frac{1}{2} - \frac{\alpha}{\pi}} \sin(\omega t - \theta) \tag{9.26}$$

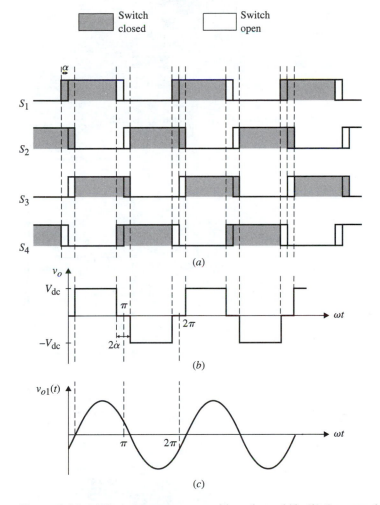

Figure 9.14 (*a*) Switching sequence with α phase shift. (*b*) Output voltage. (*c*) Fundamental component for $v_o(t)$.

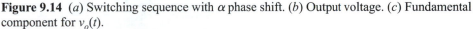

and the rms value is given by $V_{dc}\sqrt{\frac{1}{4} - \alpha/(2\pi)}$. It is clear from this relation that the rms value or the peak of the fundamental component can be controlled by the amount of the phase shift between the switching signals. We notice that, unlike the switching sequence given in Fig. 9.13(*b*), which has two states for the output voltage—$+V_{dc}$ and $-V_{dc}$—the switching sequence in Fig. 9.14(*a*) gives three states for the output voltage, $+V_{dc}$, 0, and $-V_{dc}$; such inverters are known as *tri-state inverters*. Since the load is resistive, the four switches can be implemented by SCRs with a unidirectional current flow. This is because the load current can reverse direction instantaneously as the voltage v_o reverses its direction. However, under an inductive load, for the circuit to work using SCRs, a diode must be added in parallel with each SCR as shown in Fig. 9.15.

It can be seen from the switching sequence of Fig. 9.14(*a*) that in the steady state, there are four modes of operation, as shown in Fig. 9.16. The switch implementation for S_1, S_2, S_3, and S_4 is quite simple since there is no need for bidirectional current flow. The average power delivered to the load is $(V_{o,\text{rms}}^2/R)$, where $V_{o,\text{rms}}$ is equal to V_{dc} or $V_{dc}\sqrt{\frac{1}{4} - \alpha/2\pi}$, depending on whether the switching sequence of Fig. 9.13(*a*) or Fig. 9.14(*a*), respectively, is used.

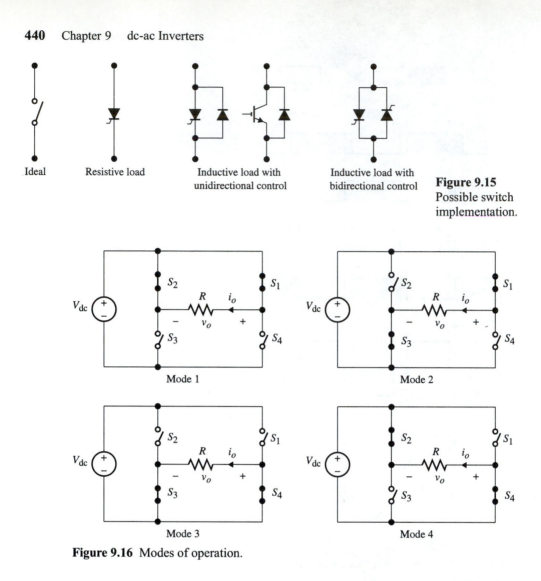

Ideal   Resistive load   Inductive load with   Inductive load with
                                   unidirectional control   bidirectional control

Figure 9.15
Possible switch
implementation.

Mode 1   Mode 2

Mode 3   Mode 4

Figure 9.16 Modes of operation.

EXAMPLE 9.4

Consider the resistive-load full-bridge voltage-source inverter shown in Fig. 9.12 with the following circuit parameters: $V_{dc} = 150$ V, $R = 12\ \Omega$, and $f_s = 60$ Hz. Sketch the waveforms
for v_o and i_{in} and determine the average power delivered to the load for the two switching sequences shown in Fig. 9.13(a) and Fig. 9.14(a), with $\alpha = 10°$.

SOLUTION   For the switch sequence shown in Fig. 9.13(a), v_o and i_o are symmetric and given by

$$v_o = \begin{cases} +V_{dc} & 0 \leq t < T/2 \\ -V_{dc} & T/2 \leq t < T \end{cases}$$

$$i_o = \begin{cases} +\dfrac{V_{dc}}{R} & 0 \leq t < T/2 \\ -\dfrac{V_{dc}}{R} & T/2 \leq t < T \end{cases}$$

The average output power is given by

$$P_{o,\text{ave}} = \frac{1}{T}\int_0^T v_o i_o \, dt$$

For the switch sequence shown in Fig. 9.14(a), the average output power is given by

$$P_{o,\text{ave}} = \frac{V_{o,\text{rms}}^2}{R}$$

where the rms value is expressed as

$$V_{o,\text{rms}} = V_{dc}\sqrt{\frac{1}{4} - \frac{\alpha}{2\pi}}$$

The resultant average output power for $\alpha = 10°$ is given by

$$P_{o,\text{ave}} = \frac{V_{dc}^2\left(1 - \dfrac{2\alpha}{\pi}\right)}{R}$$

$$= 1666.67 \text{ W}$$

As stated earlier, practical loads do not consist of a simple resistor with a unity power factor, but rather have some sort of an inductance. Figure 9.17(a) shows a full-bridge inverter under an inductive-resistive load. If the switches are operating at a 50% duty cycle with a two-state output, then the current and voltage waveforms are as shown in Fig. 9.17(b). The analysis of this inverter is similar to that for the half-bridge voltage-source inverter discussed earlier.

To obtain an expression for the average power absorbed by the load, we revert to using the Fourier series as our analysis technique. The rms values for v_o and i_o of Fig. 9.17(b) are based on a 50% squarewave output given by

$$I_{o,\text{rms}}^2 = I_{1,\text{rms}}^2 + I_{2,\text{rms}}^2 + \cdots + I_{n,\text{rms}}^2 \tag{9.27a}$$

$$V_{o,\text{rms}}^2 = V_{dc} \tag{9.27b}$$

where $I_{n,\text{rms}} = I_n/\sqrt{2}$ and I_n is the peak current of the nth harmonic of $i_o(t)$.

Since the output voltage is a squarewave with a 50% duty cycle, its Fourier series may be expressed as follows:

$$v_o(t) = \frac{4V_{dc}}{\pi}\left(\sin \omega t + \frac{\sin 3\omega t}{3} + \frac{\sin 5\omega t}{5} + \cdots + \frac{\sin n\omega t}{n}\right) \tag{9.28}$$

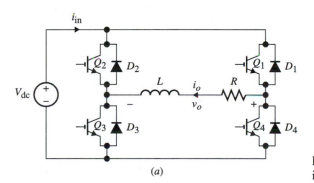

(a)

Figure 9.17 (a) Full-bridge inverter under R-L load.

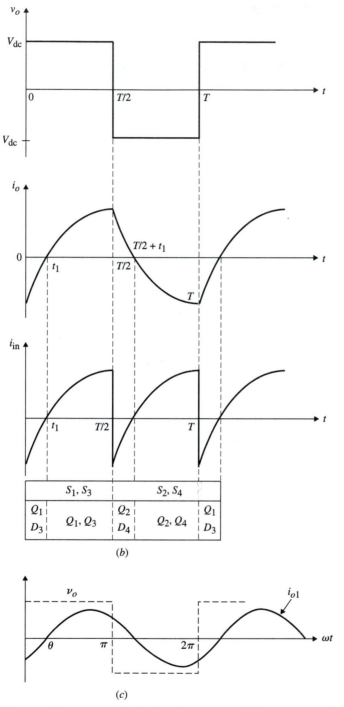

(b)

(c)

Figure 9.17 (*continued*) (*b*) Waveforms under 50% duty cycle. (*c*) Fundamental component of the inductor current.

Therefore, $i_o(t)$ is given by

$$i_o(t) = \frac{4V_{dc}}{\pi}\left[\frac{\sin(\omega t - \theta_1)}{\sqrt{R^2 + (\omega L)^2}} + \frac{\sin(3\omega t - \theta_2)}{3\sqrt{R^2 + (3\omega L)^2}} + \frac{\sin(5\omega t - \theta_5)}{5\sqrt{R^2 + (5\omega L)^2}} + \cdots\right.$$

$$\left. + \frac{\sin(n\omega t - \theta_n)}{n\sqrt{R^2 + (n\omega L)^2}}\right] \qquad (9.29)$$

and the rms value for the nth current component is given by

$$I_{n,\mathrm{rms}} = \frac{2\sqrt{2}\,V_{dc}}{n\pi|Z_n|}$$

where

$$|Z_n| = \sqrt{R^2 + (n\omega L)^2}$$

If we assume that the major part of the average output power is delivered at the fundamental frequency, then $P_{o,\mathrm{ave}}$ is given by

$$P_{o,\mathrm{ave}} = \frac{8V_{dc}^2}{\pi^2\sqrt{R^2 + (\omega L)^2}}\cos\theta_1 \qquad (9.30)$$

9.3.1   Approximate Analysis

An approximate solution for the load current can be obtained by assuming that $L/R \gg T/2$. This will allow us to represent the load current by its first harmonic. As illustrated in the previous example, Fig. 9.17(c) shows the inductor current represented by its fundamental component. The load current may be approximated by

$$i_{o1} \approx I_{o1}\sin(\omega t - \theta) \qquad (9.31)$$

where

$$\theta = \tan^{-1}\frac{\omega L}{R}$$

$$I_{o1} = \frac{V_{o1}}{\sqrt{(\omega L)^2 + R^2}}$$

$$V_{o1} = \frac{4V_{dc}}{\pi}$$

The average power delivered to the load is given by

$$P_{\mathrm{ave}} = I_{o1,\mathrm{rms}}V_{o1,\mathrm{rms}}\cos\theta$$

$$= \frac{I_{o1}V_{o1}}{2}\cos\theta$$

$$P_{\mathrm{ave}} = \frac{8V_{dc}^2 R}{\pi^2\sqrt{(\omega L)^2 + R^2}}\cos\theta \qquad (9.32)$$

where $I_{o1,\mathrm{rms}}$ is the rms of the fundamental component of the inductor current.

Including the contribution of higher harmonics to the power delivered to the load leads to more complex equations for the power factor and total harmonic distortion.

EXERCISE 9.3

Consider the full-bridge voltage-source inverter under an R-L load of Fig. 9.17(a) with $V_{dc} = 220$ V, $L = 6$ mH, $R = 16\ \Omega$, and $f_s = 50$ Hz. Calculate the average power delivered to the load up to the seventh harmonic.

ANSWER 2.73 W

9.3.2 Generalized Analysis

As in the half-bridge configuration, the preceding analysis under 50% duty cycle control with no α overlap does not allow output control. Figure 9.18(a) shows a typical output voltage under α control that is produced using the switching sequence of Fig. 9.14(a). The equivalent circuit for the single-phase bridge inverter is shown in Fig. 9.18(b).

Let us assume we have a generalized impedance load, Z_{load}, as shown in Fig. 9.19. The Fourier analysis representation for $v_o(t)$ is given by

$$v_o(t) = \sum_{n=1, 3, 5, \dots}^{\infty} V_n \sin n\omega t \tag{9.33}$$

where

$$V_n = \frac{4 V_{dc}}{n\pi} \cos n\alpha \tag{9.34}$$

and $i_o(t)$ is obtained from

$$i_o(t) = \sum_{n=1,3,5,\dots}^{\infty} I_n \sin(n\omega t - \theta_n) \tag{9.35}$$

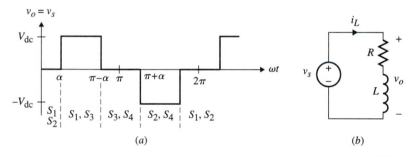

Figure 9.18 (a) Output voltage using switching sequence given in Fig. 9.14(a). (b) Equivalent circuit for the full-bridge inverter.

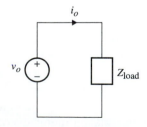

Figure 9.19 Generalized load representation.

where

$$I_n = \frac{V_n}{|Z_n|} = \frac{4V_{dc}}{n\pi|Z_n|}\cos n\alpha \qquad (9.36a)$$

$$\theta_n = \angle Z_n \qquad (9.36b)$$

where $|Z_n|$ is the magnitude of the nth harmonic impedance, and $\angle Z_n$ is the angle of the nth harmonic impedance. The overall rms output voltage is given by

$$V_{o,\text{rms}} = V_{dc}\sqrt{1 - \frac{2\alpha}{\pi}} \qquad (9.37)$$

In terms of the nth harmonic, the rms value of the voltage for each harmonic is given by

$$V_{on,\text{rms}} = \frac{4V_{dc}}{n\sqrt{2}\,\pi}\sqrt{\sum_{n=1,3,5,\dots}^{\infty}\cos^2 n\alpha} \qquad (9.38)$$

and the rms value of the current for each harmonic is given by

$$I_{on,\text{rms}} = \sqrt{\sum_{n=1,3,5,\dots}^{\infty} I_{n,\text{rms}}^2} \qquad (9.39)$$

The rms for the nth harmonic is given by

$$I_{n,\text{rms}} = \frac{4V_{dc}}{\sqrt{2}\,n\pi|Z_n|}\cos n\alpha \qquad (9.40)$$

and the rms of the fundamental output current is given by

$$I_{o1,\text{rms}} = \frac{4V_{dc}}{\sqrt{2}\,\pi|Z_1|}\cos\alpha \qquad (9.41)$$

where

$$|Z_1| = \sqrt{R^2 + (\omega_0 L)^2}$$

and ω_0 is the fundamental frequency.

The total average power delivered to the load resistance is given by

$$P_{o,\text{ave}} = \frac{1}{T}\int_0^T i_o v_o \, dt \qquad (9.42)$$

The voltage and current THDs are given in Eqs. (9.43a) and (9.43b), respectively.

$$\text{THD}_v = \sqrt{\left(\frac{V_{o,\text{rms}}}{V_{o1,\text{rms}}}\right)^2 - 1} \qquad (9.43a)$$

$$= \frac{\pi}{2\cos\alpha}\sqrt{\frac{1}{2} - \frac{\alpha}{\pi}}$$

$$\text{THD}_i = \sqrt{\left(\frac{I_{o,\text{rms}}}{I_{o1,\text{rms}}}\right)^2 - 1} \qquad (9.43b)$$

$$= \sqrt{\left(\sum_{n=1,3,5,\dots}^{\infty} \frac{|Z_1|}{n|Z_n|}\frac{\cos\alpha}{\cos n\alpha}\right)^2 - 1}$$

For the nth harmonic, the average power of Eq. (9.42) is given by

$$P_{on,\text{ave}} = I_{on,\text{rms}} V_{on,\text{rms}} \cos\theta_n$$

So the total average power is given by

$$P_{o,\text{ave}} = \sum_{n=1,3,5,\dots}^{\infty} I_{on,\text{rms}} V_{on,\text{rms}} \cos\theta_n \qquad (9.44)$$

EXAMPLE 9.5

Consider the full-bridge inverter whose equivalent circuit is represented in Fig. 9.19 with the four different loads shown in Fig. 9.20 with $R = 8\ \Omega$, $L = 30\ \mu\text{H}$, $C = 147\ \text{mF}$, $f_o = 60\ \text{Hz}$, $V_{\text{dc}} = 120\ \text{V}$, and $\alpha = \pi/6$.

(a) Determine the rms for i_o and v_o for the first, third, fifth, and seventh harmonics.

(b) Determine the total average power delivered to the load for each of the above harmonics.

(c) Determine the output current and voltage total harmonic distortion.

SOLUTION   **(a)** To determine the rms for i_o and v_o for the first, third, fifth, and seventh harmonics, we use the nth harmonic, given by

$$v_{n,o} = \frac{4V_{\text{dc}}}{n\pi}\cos n\alpha\ \sin n\omega t \qquad (9.45)$$

$$I_{n,\text{rms}} = \frac{4V_{\text{dc}}}{\sqrt{2}n\pi|Z_n|}\cos n\alpha$$

(i) For the resistive load:

$$V_{n,\text{rms}} = \frac{2\sqrt{2}V_{\text{dc}}}{n\pi}\cos n\alpha \qquad\qquad I_{n,\text{rms}} = \frac{V_{n,\text{rms}}}{|Z_n|},\ |Z_n| = R = 8\ \Omega,\ \theta_n = 0,\ \alpha = \pi/6.$$

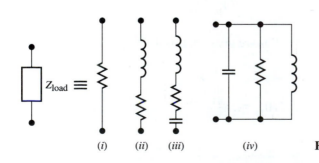

(i)   (ii)   (iii)   (iv)   **Figure 9.20** Various types of loads.

$$V_{1,\text{rms}} = \frac{2\sqrt{2} \cdot 120}{\pi} \cos\left(\frac{\pi}{6}\right) = 93.56 \text{ V} \qquad I_{1,\text{rms}} = \frac{V_{1,\text{rms}}}{|Z_1|} = \frac{V_{1,\text{rms}}}{R} = 11.69 \text{ A}$$

$$V_{3,\text{rms}} = \frac{2\sqrt{2} \cdot 120}{3\pi} \cos\left(\frac{\pi}{2}\right) = 0 \text{ V} \qquad I_{3,\text{rms}} = \frac{V_{3,\text{rms}}}{|Z_3|} = \frac{V_{3,\text{rms}}}{R} = 0 \text{ A}$$

$$V_{5,\text{rms}} = \frac{2\sqrt{2} \cdot 120}{5\pi} \cos\left(\frac{5\pi}{6}\right) = -18.71 \text{ V} \qquad I_{5,\text{rms}} = \frac{V_{5,\text{rms}}}{|Z_5|} = \frac{V_{5,\text{rms}}}{R} = -2.34 \text{ A}$$

$$V_{7,\text{rms}} = \frac{2\sqrt{2} \cdot 120}{7\pi} \cos\left(\frac{7\pi}{6}\right) = -13.37 \text{ V} \qquad I_{7,\text{rms}} = \frac{V_{7,\text{rms}}}{|Z_7|} = \frac{V_{7,\text{rms}}}{R} = -1.67 \text{ A}$$

Using the first four harmonics, the approximate rms values for v_o and i_o are given by

$$V_{o,\text{rms}} = \sqrt{V_{1,\text{rms}}^2 + V_{3,\text{rms}}^2 + V_{5,\text{rms}}^2 + V_{7,\text{rms}}^2} = 96.34 \text{ V}$$

$$I_{o,\text{rms}} = \sqrt{I_{1,\text{rms}}^2 + I_{3,\text{rms}}^2 + I_{5,\text{rms}}^2 + I_{7,\text{rms}}^2} = 12.04 \text{ A}$$

The exact rms values are given by

$$V_{o,\text{rms}} = \sqrt{\frac{2}{2\pi} \int_\alpha^\pi V_{\text{dc}}^2 \, d\omega t} = \sqrt{\frac{5}{6}} V_{\text{dc}} = 109.59 \text{ V}$$

$$I_{o,\text{rms}} = \frac{V_{o,\text{rms}}}{|Z_n|} = 13.69 \text{ A}$$

(ii) For the series RL load: The rms value for the output voltage is the same as above, since the output voltage is independent of the load.

$$I_{n,\text{rms}} = \frac{V_{n,\text{rms}}}{|Z_n|}, \qquad |Z_n| = \sqrt{R^2 + (n\omega L)^2}, \qquad \theta_n = \tan^{-1}\left(\frac{n\omega L}{R}\right)$$

$$I_{1,\text{rms}} = \frac{V_{1,\text{rms}}}{\sqrt{R^2 + (\omega L)^2}} = 6.75 \text{ A}$$

$$I_{3,\text{rms}} = \frac{V_{3,\text{rms}}}{\sqrt{R^2 + (3\omega L)^2}} = 0 \text{ A}$$

$$I_{5,\text{rms}} = \frac{V_{5,\text{rms}}}{\sqrt{R^2 + (5\omega L)^2}} = -0.33 \text{ A}$$

$$I_{7,\text{rms}} = \frac{V_{7,\text{rms}}}{\sqrt{R^2 + (7\omega L)^2}} = -0.168 \text{ A}$$

(iii) For the series RLC load: The rms value for the output voltage is the same as in *(i)*, since the output voltage is independent of the load.

$$|Z_n| = \sqrt{R^2 + \left(n\omega L - \frac{1}{n\omega C}\right)^2}, \qquad \theta_n = \tan^{-1}\left(\frac{n\omega L - \dfrac{1}{n\omega C}}{R}\right)$$

$$I_{1,rms} = \frac{V_{1,rms}}{\sqrt{R^2 + \left(\omega L - \dfrac{1}{\omega C}\right)^2}} = 8.95 \text{ A}$$

$$I_{3,rms} = \frac{V_{3,rms}}{\sqrt{R^2 + \left(3\omega L - \dfrac{1}{3\omega C}\right)^2}} = 0 \text{ A}$$

$$I_{5,rms} = \frac{V_{5,rms}}{\sqrt{R^2 + \left(5\omega L - \dfrac{1}{5\omega C}\right)^2}} = -0.35 \text{ A}$$

$$I_{7,rms} = \frac{V_{7,rms}}{\sqrt{R^2 + \left(7\omega L - \dfrac{1}{7\omega C}\right)^2}} = -0.17 \text{ A}$$

(iv) For the parallel RLC load: The rms value for the output voltage is the same as the one shown above, since the output voltage is independent of the load.

$$|Z_n| = \frac{1}{\sqrt{\left(\dfrac{1}{R}\right)^2 + \left(\dfrac{n^2\omega^2 LC - 1}{n\omega L}\right)^2}}, \qquad \theta_n = -\tan^{-1}\left(\frac{n^2\omega^2 LRC - R}{n\omega L}\right)$$

$$I_{1,rms} = \frac{V_{1,rms}}{1 \Big/ \sqrt{\left(\dfrac{1}{R}\right)^2 + \left(\dfrac{1^2\omega^2 LC - 1}{\omega L}\right)^2}} = 12.1 \text{ A}$$

$$I_{3,rms} = \frac{V_{3,rms}}{1 \Big/ \sqrt{\left(\dfrac{1}{R}\right)^2 + \left(\dfrac{3^2\omega^2 LC - 1}{3\omega L}\right)^2}} = 0 \text{ A}$$

$$I_{5,rms} = \frac{V_{5,rms}}{1 \Big/ \sqrt{\left(\dfrac{1}{R}\right)^2 + \left(\dfrac{5^2\omega^2 LC - 1}{5\omega L}\right)^2}} = -5.39 \text{ A}$$

$$I_{7,rms} = \frac{V_{7,rms}}{1 \Big/ \sqrt{\left(\dfrac{1}{R}\right)^2 + \left(\dfrac{7^2\omega^2 LC - 1}{7\omega L}\right)^2}} = -5.29 \text{ A}$$

(b) Power calculations are obtained from the following equation:

$$P_{on,ave} = I_{on,rms} V_{on,rms} \cos \theta_n$$

(i) For the resistive load:

$$\cos \theta_n = 1 \quad \text{for } n = 1, 3, 5, \ldots$$

$$P_{o1,ave} = (93.56)(11.69) = 1094.65 \text{ W}$$

$$P_{o3,ave} = 0 \text{ W}$$

$$P_{o5,\text{ave}} = (-18.71)(-2.34) = 43.78 \text{ W}$$

$$P_{o7,\text{ave}} = (-13.37)(-1.67) = 22.33 \text{ W}$$

$$P_o = \sum_{n = 1, 3, 5, \ldots}^{\infty} I_{on,\text{rms}} V_{on,\text{rms}} \cos \theta_n = 1160.76 \text{ W}$$

(ii) For the series RL load:

$$P_{o1,\text{ave}} = (93.56)(6.75)\cos(54.73°) = 364.67 \text{ W}$$

$$P_{o3,\text{ave}} = 0 \text{ W}$$

$$P_{o5,\text{ave}} = (-18.71)(-0.33)\cos(81.95°) = 0.865 \text{ W}$$

$$P_{o7,\text{ave}} = (-13.37)(-0.168)\cos(84.23°) = 0.226 \text{ W}$$

$$P_{o,\text{ave}} = \sum_{n = 1, 3, 5, \ldots}^{\infty} I_{on,\text{rms}} V_{on,\text{rms}} \cos \theta_n = 365.75 \text{ W}$$

(iii) For the series RLC load:

$$P_{o1,\text{ave}} = (93.56)(9.95)\cos(-40.093°) = 640.58 \text{ W}$$

$$P_{o3,\text{ave}} = 0 \text{ W}$$

$$P_{o5,\text{ave}} = (-18.71)(-0.35)\cos(81.36°) = 0.98 \text{ W}$$

$$P_{o7,\text{ave}} = (-13.37)(-0.17)\cos(84.037°) = 0.24 \text{ W}$$

$$P_{o,\text{ave}} = \sum_{n = 1, 3, 5, \ldots}^{\infty} I_{on,\text{rms}} V_{on,\text{rms}} \cos \theta_n = 641.8 \text{ W}$$

(iv) For the parallel RLC load:

$$P_{o1,\text{ave}} = (93.56)(12.1)\cos(-14.76°) = 1094.2 \text{ W}$$

$$P_{o3,\text{ave}} = 0 \text{ W}$$

$$P_{o5,\text{ave}} = (-18.17)(-5.39)\cos(-64.274°) = 43.78 \text{ W}$$

$$P_{o7,\text{ave}} = (-13.37)(-5.29)\cos(-71.579°) = 22.35 \text{ W}$$

$$P_{o,\text{ave}} = \sum_{n = 1, 3, 5, \ldots}^{\infty} I_{on,\text{rms}} V_{on,\text{rms}} \cos \theta_n = 1160.3 \text{ W}$$

(c) The total harmonic distortion can be obtained from Eq. (9.43):

$$\text{THD}_i = \sqrt{\left(\frac{I_{o,\text{rms}}}{I_{o1,\text{rms}}}\right)^2 - 1} = 0.31$$

EXERCISE 9.4

Consider the inductive-resistive–load inverter shown in Fig. 9.17(*a*) with the following parameters: $\alpha = \pi/3$, $L = 100$ mH, $R = 16$ Ω, $V_{\text{dc}} = 250$ V, and $f_o = 60$ Hz. Determine the fundamental current and voltage components.

ANSWER $i_{o1} = \dfrac{12.2}{\pi} \sin(\omega t - 67°)$ A

The general analysis presented in this section is based on a passive load Z_L that produces a fixed load phase shift $\angle Z_L = \theta$; therefore, the output voltage control can be achieved only by varying α. However, when the load contains a voltage source, as in motor and utility grid applications, then it is an active load that allows both the phase shift θ and the switching displacement α to be used as control variables.

EXAMPLE 9.6

Consider the active load in a bridge inverter that consists of an R-L load and an ac sinusoidal voltage source as shown in Fig. 9.21 with the load voltage v_o as shown. Obtain the expression for the fundamental load current and the average power delivered to the load for the following circuit parameters: $v_{ac} = 100 \sin(2\pi 100t - 30°)$, $V_{dc} = 180$ V, $L = 42$ mH, $R = 0.5\ \Omega$, $\alpha = 15°$, and $f_o = 60$ Hz.

SOLUTION   To obtain the expression for the fundamental load current, we can use the following equations (the arrow indicates phasor notation):

$$\overrightarrow{I_{o1}} = \frac{\overrightarrow{V_{o1}} - \overrightarrow{V_{ac}}}{R + j\omega L}$$

We have

$$V_{o1} = \frac{4V_{dc}}{\pi}\cos\alpha = \frac{4 \times 180}{\pi}\cos(15°) = 221.37 \text{ V}$$

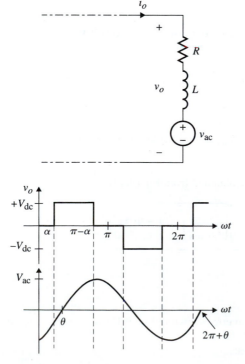

Figure 9.21 Inverter example with an active load.

Hence,

$$\overrightarrow{V_{o1}} = 221.37\angle 0° \text{ V}$$

$$\overrightarrow{V_{ac}} = 100\angle 30° \text{ V}$$

$$\overrightarrow{I_{o1}} = \frac{221.37\angle 0° \text{ V} - 100\angle 30°}{(0.5 + j2\pi(60))(42 \times 10^{-3})} = 9.08\angle -110.25° \text{ A}$$

The total power delivered to the load is given by the following equation:

$$\overrightarrow{P_T} = \frac{1}{2}\overrightarrow{I_{o1}}\overrightarrow{V_{o1}}$$

$$\overrightarrow{P_T} = \frac{1}{2}(221.37\angle 0°)(9.08\angle -110.25°) = 1005.02\angle -110.25° \text{ W}$$

9.4   HARMONIC REDUCTION

Harmonic reduction includes the elimination and cancellation of certain harmonics of the output voltage. Reducing the harmonic content of the ac output is one of the most difficult challenges of dc-ac inverter design. If the inverter drives an electromechanical load (ac motor), the harmonics can excite a mechanical resonance, causing the load to emit acoustic noise.

Unlike the case in dc-to-dc converters, harmonics in the output waveforms are very significant. This is why the output filters perform different functions and their design is quite different in terms of complexity and size. In designing output filters in dc-to-dc converters, the objective is to limit the output voltage ripple to a certain desired percentage of the average output voltage. In other words, this is a passive filter approach that is limited by the physical size of inductors and capacitors. In dc-to-dc inverters, the reduction or cancellation of the output harmonics is done actively by controlling the switching technique of the inverter. Compared with ac-to-dc conversion, harmonic filtering in dc-to-ac is harder since it will affect the attenuation and/or the phase shift of the fundamental component.

The harmonics that are present in the inverter's output voltage are high for many practical applications, especially when the output voltage needs to be near sinusoidal. To help produce a near sinusoidal output, electronic low-pass filters are normally added at the output to remove third and higher harmonics. The design of such filters tends to be challenging when the third and fifth harmonics close to the fundamental are high. In most cases, effective filters require large size and high numbers of filtering capacitive and inductive components, resulting in bulkiness. By controlling the width or the number of pulses, certain harmonic content can be removed without the need for complex harmonic filtering circuits.

It is possible to cancel certain harmonics by simply selecting the duration of the pulse in the half-cycle of the output voltage. Recall that the peak component of the nth harmonic for Fig. 9.18(a) is given by

$$V_n = \frac{4V_{dc}}{n\pi}\cos n\alpha \tag{9.46}$$

For the third harmonic, we have

$$V_3 = \frac{4V_{dc}}{3\pi}\cos 3\alpha \tag{9.47}$$

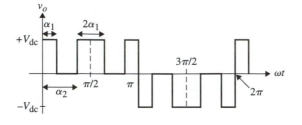

Figure 9.22 Two-angle shift control of inverter output voltage.

To cancel the third harmonic, we set $\alpha = \pi/6$; this results in the cancellation of all harmonics of the order of $3n$.

Consider the case of a two-pulse output of Fig. 9.22 with angles of α_1 and α_2 as shown. It can be shown that the nth harmonic is given by

$$V_{on} = \frac{4V_{dc}}{n\pi}(1 - \cos n\alpha_1 + \cos n\alpha_2) \tag{9.48}$$

For example, to eliminate the third and fifth harmonics, we make $V_{on} = 0$ at two frequencies as follows:

$$1 - \cos 3\alpha_1 + \cos 3\alpha_2 = 0$$

$$1 - \cos 5\alpha_1 + \cos 5\alpha_2 = 0$$

Solve for α_1 and α_2 to obtain $\alpha_1 = 17.8°$ and $\alpha_2 = 38°$.

EXAMPLE 9.7

Consider two cascaded push-pull inverters as shown in Fig. 9.23(a); the switching waveforms for $S_1 - S_4$ are shown in Fig. 9.23(b). Sketch the waveforms for the outputs v_{o1}, v_{o2}, and v_o.

SOLUTION   The waveforms for the output voltage are shown in Fig. 9.23(b).
The fundamental component of v_o is given by

$$v_{o1}(t) = \frac{8V_{dc}}{\pi}\cos\alpha\sin\omega t$$

where $\alpha = \omega t_1$. It is clear that the magnitude of the fundamental component of v_{o1}, has been reduced by $\cos\alpha$ compared to the case when no phase is present ($\alpha = 0$).

Harmonic Analysis

The harmonics content present in the output of the inverter could be significant. Depending on the application, reducing the effect of these harmonics can be very important. Recall that the nth component for a squarewave output ($\alpha = 0$) is $4V_{dc}/n\pi$. For $\alpha \neq 0$, the expression for $v_o(t)$ is given by

$$v_o(t) = \sum_{n = 1, 3, 5, \ldots} V_n \sin\omega t \tag{9.49}$$

where

$$V_n = \frac{4V_{dc}}{n\pi}\cos n\alpha$$

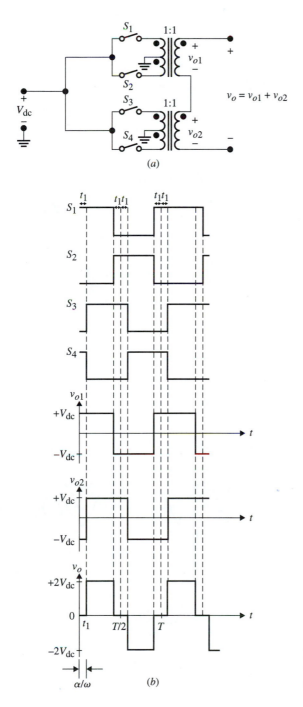

Figure 9.23 (*a*) Two push-pull inverters for Example 9.7. (*b*) Typical switching waveforms.

The fundamental output voltage component is given by

$$v_{o1} = \frac{4V_{dc}}{\pi} \cos \alpha \sin \omega t \qquad (9.50)$$

Figure 9.24(*a*), (*b*), and (*c*) shows the resultant output voltage where harmonics up to the ninth are included for $\alpha = 0$, $\alpha = \pi/6$, and $\alpha = \pi/3$, respectively. The sketch of the magnitude of the harmonics of Fig. 9.24, as a function of α, is shown in Fig. 9.25.

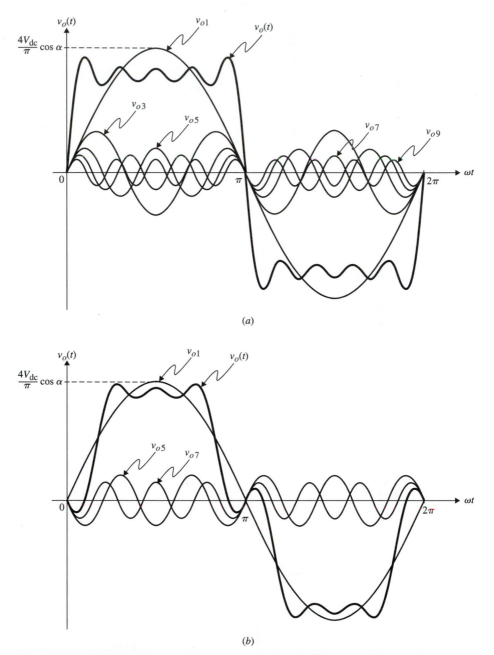

Figure 9.24 The first nine output harmonics: (a) $\alpha = 0$. (b) $\alpha = \pi/6$.

It is clear from the earlier equation that the magnitude of the harmonics is inversely proportional to the magnitude of n and α. Under a wide range of variation of α, the filtering of the harmonics might not be an easy task.

The total harmonic distortion for $v_o(t)$ is given by

$$\text{THD}_V = \sqrt{\left(\frac{V_{o,\text{rms}}}{V_{o1,\text{rms}}}\right)^2 - 1} \tag{9.51}$$

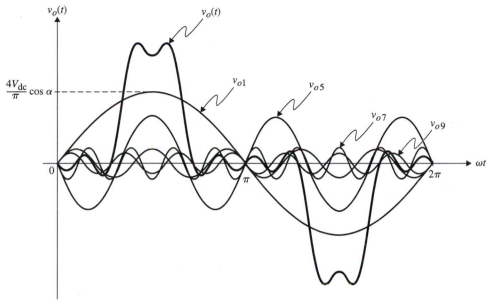

(c)

Figure 9.24 (*continued*) (c) $\alpha = \pi/3$.

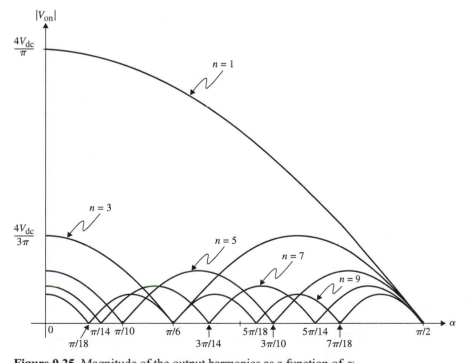

Figure 9.25 Magnitude of the output harmonics as a function of α.

$$V_{o,\,\mathrm{rms}}^2 = V_{o1,\,\mathrm{rms}}^2 + V_{o2,\,\mathrm{rms}}^2 + \cdots + V_{on,\,\mathrm{rms}}^2$$

$$= \sum_{n=1,\,3,\,5,\,\ldots} V_{on,\mathrm{rms}}^2 \tag{9.52}$$

where

$$V_{on,\,\mathrm{rms}} = \frac{V_{on}}{\sqrt{2}} \tag{9.53a}$$

$$V_{on} = \frac{4V_{\mathrm{dc}}}{n\pi} \cos n\alpha \tag{9.53b}$$

Substitute Eqs. (9.52) and (9.53) into Eq. (9.51) to yield

$$\mathrm{THD}_V = \sqrt{\left(\dfrac{\displaystyle\sum_{n=1,3,5,\ldots} V_{on,\,\mathrm{rms}}^2}{V_{o1,\,\mathrm{rms}}^2}\right) - 1}$$

$$\mathrm{THD}_V = \sqrt{\left(\dfrac{\displaystyle\sum_{n=3,5,7,\ldots} V_{on,\,\mathrm{rms}}^2}{V_{o1,\,\mathrm{rms}}^2}\right)} \tag{9.54}$$

$$\mathrm{THD}_V = \frac{1}{\cos\alpha}\sqrt{\sum_{n=3,5,\ldots}\frac{1}{n^2}\cos^2 n\alpha}$$

The plots of the THD for $\alpha = 0$ to $\alpha = \pi/2$ are shown in Fig. 9.26.

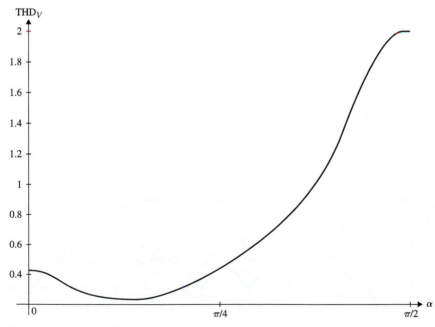

Figure 9.26 Plots of THD as a function of α.

EXERCISE 9.5

Show that the ratio of the harmonic magnitude with respect to its fundamental component for Fig. E9.5 is given by

$$\frac{V_{on}}{V_{o1}} = \frac{\sin(n\theta/2)}{n\sin(\theta/2)}$$

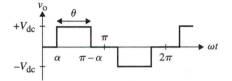

Figure E9.5 Inverter voltage with α control.

EXERCISE 9.6

Determine the value of $\alpha > 0$ that produces the largest THD in Eq. (9.54) when only the third harmonics are included.

ANSWER $\alpha = \pi/2$

9.5   PULSE-WIDTH MODULATION

Figure 9.27 shows the simplified block diagram representation for a single-phase switching-mode inverter. The output $v_o'(t)$ shows possible types of output waveforms that can be produced depending on the pulse-width modulation (PWM) control technique employed.

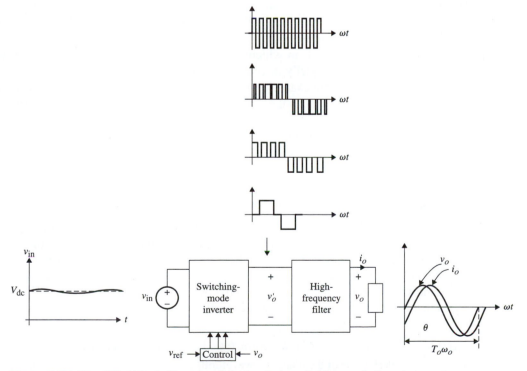

Figure 9.27 Simplified block diagram of single-phase switching-mode inverter.

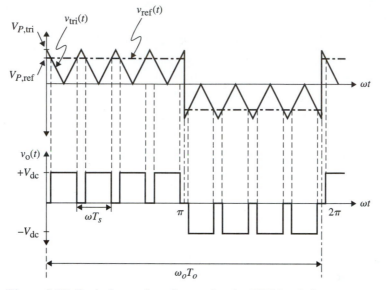

Figure 9.28 Typical waveform for equal-pulse PWM technique.

In the single-phase inverter, one or more pulses in a given half-cycle are used to control the output voltage. Since varying the width of these pulses within the half-cycle carries out the control, the process is appropriately known as pulse-width modulation (PWM). By modulating the width of several pulses per half-cycle, a more efficient method of controlling the output voltage of the inverter is obtained. Using the PWM process, we can extract a low-frequency signal from a train of high-frequency squarewaves.

Generally speaking, the PWM method may be divided into two classes, depending on the modulation technique:

1. Nonsinusoidal PWM, in which all pulses have the same width and are normally modulated equally to control the output voltage, as shown in the four-pulse-per-half-cycle example in Fig. 9.28. The widths of these pulses are adjusted equally to control the output voltage. Eliminating a selected number of series harmonics as discussed before requires a very complex control technique to generate the required switching sequence.

2. Sinusoidal PWM, which allows the pulse width to be modulated sinusoidally; i.e., the width of each pulse is proportional to the instantaneous value of a reference sinusoid whose frequency equals that of the fundamental components as shown in the six-pulse-per-half-cycle example in Fig. 9.29.

Normally the pulses in the nonsinusoidal PWM method are arranged to produce an odd-function output voltage that is symmetrical around $T_o/2$. This results in the cancellation of all the cosine terms and the even harmonics of the sine terms.

We notice that the reference voltage $v_{ref}(t)$ is square and sinusoidal for the equal-pulse and sinusoidal PWM methods, respectively. It is important that we first define some terms that pertain to the square and sinusoidal PWM inverters that will be discussed.

$v_{tri}(t)$:	Repetitive triangular waveform (also known as a carrier signal)
$V_{P,tri}$:	Peak value of the triangular waveform

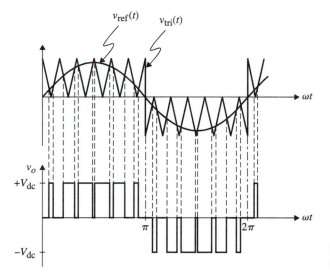

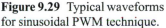

Figure 9.29 Typical waveforms for sinusoidal PWM technique.

T_s, f_s :	Period and frequency of the triangular waveform (f_s is also known as the carrier or switching frequency)
v_{ref} :	Reference signal, which can be either a square or a sinusoidal waveform (also known as a control signal)
$V_{P, ref}$:	Peak value of the reference signal
T_o, f_o :	Desired inverter output period and output frequency, which are equal to the period and frequency of the reference or control signal
m_a :	Inverter amplitude modulation index
m_f :	Inverter frequency modulation index
k :	Number of pulses per half-cycle

The amplitude and frequency modulation indices are defined as follows:

$$m_a = \frac{V_{P,ref}}{V_{P,tri}} \tag{9.55}$$

$$m_f = \frac{f_s}{f_o} \tag{9.56}$$

Next we discuss the two well-known PWM techniques.

9.5.1   Equal-Pulse (Uniform) PWM

The equal-pulse PWM technique, known also as single-pulse PWM control, is very old and less popular nowadays. The technique is very simple and requires simple control since all generated pulses have equal widths. Generating the equal and multiple pulses is achieved by comparing a square reference voltage waveform $v_{ref}(t)$ to a triangular control (carrier) voltage waveform, $v_{cont}(t)$, as shown in Fig. 9.28. The op-amp produces a triggering signal every time the carrier signal goes below or above the reference signal, as shown in the figure. It is clear that the frequency of the reference voltage waveform determines the frequency of the output voltage, and the frequency of the control signal determines the number of equal pulses in each half-cycle.

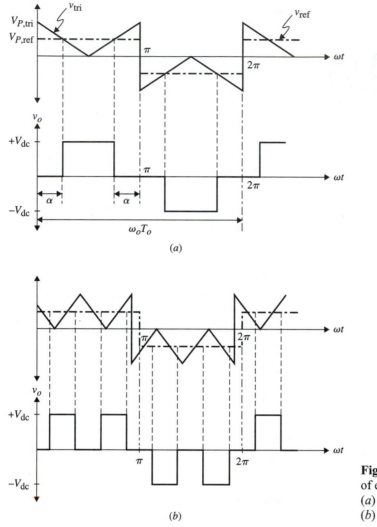

Figure 9.30 Examples of equal pulses.
(*a*) One-pulse output.
(*b*) Two-pulse output.

Figure 9.30(*a*), (*b*), (*c*), and (*d*) shows examples of one-, two-, three-, and seven-pulse outputs, respectively. As the magnitude of the reference signal increases, the pulse width increases as well, reaching its maximum value of π when the magnitude of the reference signal becomes equal to the peak of the modulating signal.

It can be shown the α can be expressed in terms of $V_{P,\text{tri}}$ and $V_{P,\text{ref}}$ as follows:

$$\alpha = -\frac{\pi}{2}\left(\frac{V_{P,\text{ref}}}{V_{P,\text{tri}}} - 1\right) = \frac{\pi}{2}(1 - m_a) \qquad (9.57)$$

Notice that the frequency of the control signal, f_{tri}, is twice the frequency of the reference signal, f_{ref}. It is clear from these waveforms that the number of pulses, k, is equal to the number of periods of the control signal per half-period of the reference signal; i.e., k is the number of switching periods, T_s, in the $T_o/2$ period, which can be expressed as

$$k = \frac{1}{2}\frac{f_s}{f_o} \qquad (9.58)$$

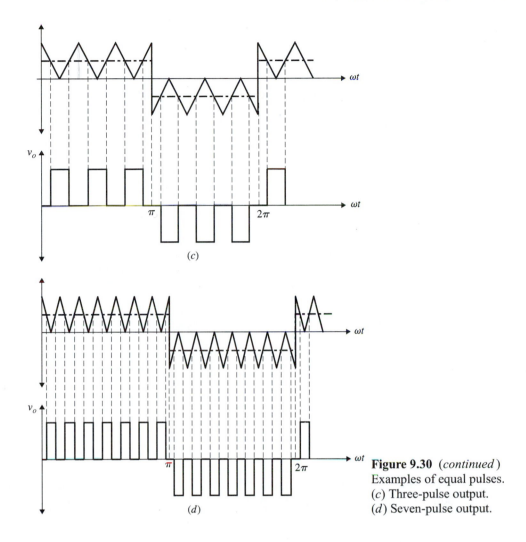

Figure 9.30 (*continued*) Examples of equal pulses. (*c*) Three-pulse output. (*d*) Seven-pulse output.

In terms of the frequency modulation index, m_f, k may be expressed by the following relation:

$$k = \frac{1}{2}m_f \tag{9.59}$$

Notice that $k = 1$ is a special case where $f_s/f_o = 2$.

We should point out that the output frequency is equal to the frequency of the reference signal, (i.e., $f_o = f_{ref}$) and the switching frequency is equal to the frequency of the carrier or triangle signal ($f_s = f_{tri}$).

For example, in Fig. 9.30(*d*), $k = 7$ and $m_f = 14$. The maximum width of each pulse occurs when $m_a = 1$ and is given by

$$t_{width,max} = \frac{T_o}{2k} \tag{9.60}$$

The maximum conduction angle width of each pulse is given by

$$\theta_{width,max} = \omega_o t_{width,max} = \frac{\pi}{k} \tag{9.61}$$

Derive Eq. (9.57).

Next we will derive a general expression for the ith pulse width in a given k-pulse output in terms of i, k, and m_a. Referring to Fig. 9.31, which shows a k-pulse inverter output, the start of the ith pulse is given by

$$t_i = (i-1)T_s + \frac{T_s}{2}\left(1 - \frac{V_{P,\text{ref}}}{V_{P,\text{tri}}}\right) \tag{9.62}$$

Substituting for $f_s = 1/T_s$ from Eq. (9.58) into Eq. (9.62), t_i becomes

$$t_i = \frac{T_o}{2k}(i-1) + \frac{T_o}{4k}(1 - m_a) \tag{9.63}$$

In terms of the starting angle, $\theta_i = \omega_o t_i$, of the ith pulse, Eq. (9.63) may be written as follows:

$$\theta_i = \frac{\pi}{k}\left[i - \frac{m_a}{2} - \frac{1}{2}\right] \tag{9.64}$$

For example, for the two-pulse waveform, $k = 2$ and $m = 0.5$, the angles at which the pulses start are given by

$$\theta_1 = \frac{\pi}{2}\left[1 - \frac{1}{4} - \frac{1}{2}\right] = \frac{\pi}{8} \qquad \theta_3 = \frac{\pi}{2}\left[3 - \frac{1}{4} - \frac{1}{2}\right] = \frac{9\pi}{8}$$

$$\theta_2 = \frac{\pi}{2}\left[2 - \frac{1}{4} - \frac{1}{2}\right] = \frac{5\pi}{8} \qquad \theta_4 = \frac{\pi}{2}\left[4 - \frac{1}{4} - \frac{1}{2}\right] = \frac{13\pi}{8}$$

Notice that because of the symmetry, $\theta_3 = \theta_1 + \pi$ and $\theta_4 = \theta_2 + \pi$. It can be shown that, in general, the width of each pulse is given by

$$\theta_{\text{width}} = \pi\frac{m_a}{k} \tag{9.65}$$

For $k = 2$ and $m = 0.5$, $\theta_{\text{width}} = \pi/4$, as expected.

The Output Voltage

It can be shown that the average output voltage over a period of T_s is given by

$$V_{o,\text{ave}} = m_a V_{\text{dc}} \qquad 0 < m_a \le 1 \tag{9.66}$$

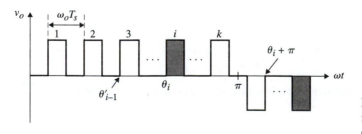

Figure 9.31 k-pulse inverter output in a half-cycle.

The rms value for the ith pulse is given by

$$V_{o,\text{rms}} = \sqrt{\frac{1}{2\pi}\int_{\theta_i}^{\theta_1 + \theta_{\text{width}}} 2kV_{\text{dc}}^2 \, d\omega t}$$

$$= V_{\text{dc}}\sqrt{\frac{k}{\pi}\theta_{\text{width}}}$$

(9.67)

Recall that from Eq. (9.65) we have $\theta_{\text{width}} = \pi(m_a/k)$; hence, Eq. (9.67) becomes

$$V_{o,\text{rms}} = V_{\text{dc}}\sqrt{\frac{k}{\pi}\frac{\pi m_a}{k}} = V_{\text{dc}}\sqrt{m_a}$$

(9.68)

The rms of the output voltage is a function of the modulation index.

Harmonics of k Equal Pulses

Let us determine the harmonic components for the ith pulse shown in Fig. 9.32. Since it is an odd function, only the odd harmonics exist in the b_n coefficients. The harmonics of the output voltage due to the ith pulse acting alone is given by

$$V_{on,i} = \frac{V_{\text{dc}}}{\pi}\left[\int_{\theta_i}^{\theta_i + \theta_{\text{width}}} \cos n\omega t \, d\omega t - \int_{\theta_i + \pi}^{\theta_i + \theta_{\text{width}} + \pi} \cos n\omega t \, d\omega t\right]$$

$$= \frac{2V_{\text{dc}}}{n\pi}\sin\left(\frac{n\theta_{\text{width}}}{2}\right)\left[\sin\left(n\left(\theta_i + \frac{\theta_{\text{width}}}{2}\right)\right) - \sin\left(n\left(\pi + \theta_i + \frac{\theta_{\text{width}}}{2}\right)\right)\right]$$

(9.69)

For the total k pulses, V_{on} is given by

$$V_{on} = \sum_{i=1}^{k}\frac{2V_{\text{dc}}}{n\pi}\sin\left(\frac{n\theta_{\text{width}}}{2}\right)\left[\sin\left(n\left(\theta_i + \frac{\theta_{\text{width}}}{2}\right)\right) - \sin\left(n\left(\pi + \theta_i + \frac{\theta_{\text{width}}}{2}\right)\right)\right]$$

$$\text{for } n = 1, 3, 5, \ldots$$

(9.70)

In terms of k, m_a, and θ_{width}, using Eqs. (9.64) and (9.65), θ_i can be expressed as

$$\theta_i = \frac{\pi}{k}\left(i - \frac{1}{2}\right) - \frac{\theta_{\text{width}}}{2} \quad \text{where } i = 1, 2, \ldots, k$$

(9.71)

where

$$\theta_{\text{width}} = \frac{\pi}{k}m_a$$

(9.72)

It can be shown that the nth harmonic component for the k-pulse output voltage may be expressed as follows:

$$V_{on} = \frac{4V_{\text{dc}}}{n\pi}\sin\left(\frac{m_a\pi}{2k}n\right)\sum_{i=1}^{k}\sin\left(\frac{\pi}{k}(i - 1/2)n\right)$$

(9.73)

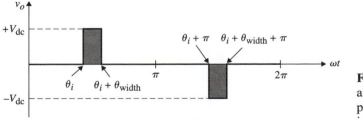

Figure 9.32 Symmetrical representation of ith pulse for a given inverter output.

For illustration purposes, let us consider the example with $m = 0.5$ and the first nine harmonics, and then evaluate the total harmonic distortion as a function of the number of pulses.

To verify the above equation for a single pulse, we calculate the harmonic components for $k = 1$, $i = 1$, and $\theta_i = \alpha$, to yield

$$V_n = \frac{4V_{dc}}{n\pi} \sin\left(\frac{n\pi}{2}m_a\right) \sum_{i=1}^{1} \sin\left(\frac{\pi}{k}(i - 1/2)n\right) \tag{9.74}$$

Substituting for m_a from Eq. (9.72), Eq. (9.74) becomes

$$V_n = \left(\frac{4V_{dc}}{n\pi}\right) \sin\left(n\frac{\theta_{width}}{2}\right) \quad \text{for odd } n \tag{9.75}$$

Equation (9.75) represents the harmonic components of the output voltage for $k = 1$ as a function of the pulse width θ_{width}. This equation is similar to what we derived previously as a function of α.

EXAMPLE 9.8

For uniform PWM with a value of $k = 5$ and a modulation index $m_a = 0.2$, calculate the output harmonic components up to the fifteenth harmonic.

SOLUTION   Using Eq. (9.73), Table 9.1 shows the values of the first 15 harmonics. Figure 9.33 shows the plot for the harmonic contents of Table 9.1. Figure 9.34 shows the harmonic ratios with respect to the fundamentals for $m = 0.2$ and $k = 1$ to $k = 7$ pulses per half-cycle.

From Table 9.1 for $m_a = 0.2$, we observe that as the number of pulses increases per half-cycle, the magnitude of the lower harmonics (third, fifth, seventh) decreases with respect to the fundamental component. Furthermore, there is an increase in magnitude for the higher-order harmonics with respect to the fundamental; however, such higher-order harmonics produce a negligible ripple that can be easily filtered out. Still, the ratio of the harmonic to the fundamental is relatively unchanged as the number of pulses increases within a half-cycle. The THD for this example is 223%! The higher the modulation index, the lower the THD.

EXERCISE 9.8

Consider two equal pulses placed at $\omega t = \theta_i$ and $\omega t = \theta_i + \pi$, respectively, each with a width of θ_{width} as shown in Fig. 9.32. Show that the nth harmonic component is given by

$$v_{on} = V_n \sin n\omega t$$

Table 9.1 Normalized Harmonics for $k = 5$ and $m_a = 0.2$

Magnitude harmonic coefficient	V_n/V_{dc}
V_1	0.258715
V_3	0.098301
V_5	0.078691
V_7	0.095728
V_9	0.245304
V_{11}	–0.238761
V_{13}	–0.088251
V_{15}	–0.068671

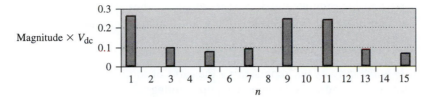

Figure 9.33 Harmonic contents for $k = 5$ and $m_a = 0.2$.

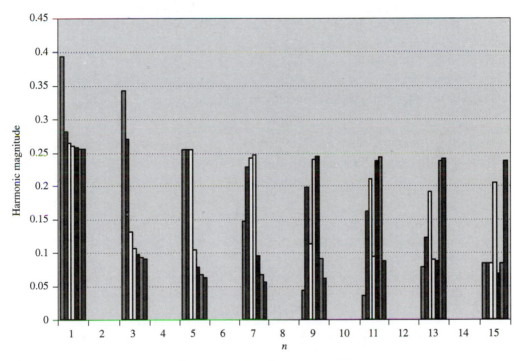

Figure 9.34 Harmonic contents for $k = 1$ to $k = 7$, and $m_a = 0.2$.

where

$$V_n = \frac{2V_{dc}}{n\pi} \sin\left(\frac{n\theta_{\text{width}}}{2}\right)\left[\sin n\left(\theta_i + \frac{\theta_{\text{width}}}{2}\right) - \sin n\left(\pi + \theta_i + \frac{\theta_{\text{width}}}{2}\right)\right]$$

9.5.2 Sinusoidal PWM

Basic Concept

To illustrate the process of sinusoidal PWM, we refer to the simplified buck converter shown in Fig. 9.35. Recall that in PWM dc-dc converters, the duty cycle is modulated

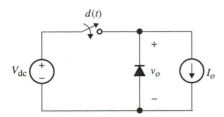

Figure 9.35 Simplified buck converter.

between 0 and 1 to regulate the dc output voltage. In the steady-state, the duty cycle in PWM switch-mode converters is relatively constant and does not vary with time:

$$V_o = DV_{dc} \tag{9.76}$$

where D is the duty cycle, representing the ratio of the *on* time of the switch to the switching period, and V_o is the average output voltage.

If the duty cycle, $d(t)$, varies or is modulated according to a certain time function, with a modulating frequency, f_o, then it is possible to shape the output voltage waveform, v_o, in such a way that its average value over the modulating period synthesizes a sinusoidal waveform. For example, if the duty cycle is defined according to the function

$$d(t) = D_{dc} + D_{max} \sin \omega_o t \tag{9.77}$$

where

$$D_{dc} = \text{dc duty cycle when no modulation exists}$$
$$D_{max} = \text{maximum modulation constant}$$
$$\omega_o = \text{frequency of modulation}$$

then the output voltage, v_o, is given by

$$
\begin{aligned}
v_o &= d(t)V_{dc} \\
&= V_{dc}D_{dc} + V_{dc}D_{max} \sin \omega_o t
\end{aligned}
\tag{9.78}
$$

For a buck converter, since the output voltage cannot be negative, then $D_{max} \le D_{dc}$, as shown in Fig. 9.36.

As an example, if $D_{dc} = 0.5$, $D_{max} = 0.8D_{dc}$, and $f_o = f_s/12$, the duty cycle is given by

$$d(t) = 0.5 + 0.4 \sin(2\pi f_o t_i)$$

where $t_i = 0, 1, 2, 3, \ldots, 12$.

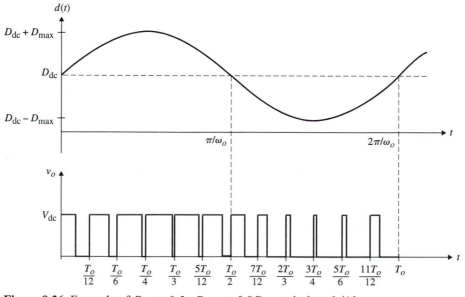

Figure 9.36 Example of $D_{dc} = 0.5$, $D_{max} = 0.8D_{dc}$, and $f_o = f_s/12$.

Table 9.2   Example of 12-Pulse Sinusoidal PWM Signal

Pulse	Time, t_i	$d(t) = 0.5 + 0.4\sin(2\pi f_o t_i)$
1	0	0.50
2	$T_o/12$	0.70
3	$T_o/6$	0.85
4	$T_o/4$	0.90
5	$T_o/3$	0.85
6	$5T_o/12$	0.70
7	$T_o/2$	0.50
8	$7T_o/12$	0.30
9	$2T_o/3$	0.15
10	$9T_o/12$	0.10
11	$5T_o/6$	0.15
12	$11T_o/12$	0.30

Since the switching frequency is 12 times faster than the modulating frequency, f_o, then $d(t)$ is sampled 12 times between $0 \le t < T_o$. This is illustrated in Table 9.2 for $0 \le t < T_o/2$ and six pulses in a half-cycle.

The preceding will now be applied to sinusoidal PWM inverters.

PWM Modulation Function

Since the waveform is symmetrical around the $T_o/2$ point, we define $m(t)$ only in the first half-cycle. In the single-pulse inverter with a two-state output of Fig. 9.13(*b*), the modulation function is simply unity.

$$v_o = m(t)V_{dc} \quad m(t) = 1 \quad 0 \le t < T/2 \tag{9.79}$$

In the single-pulse inverter with a tri-state output of Fig. 9.18(*a*), the modulation function in a half-cycle is given by

$$v_o = m(t)V_{dc} \quad m(t) = \begin{cases} 0 & 0 \le t < \alpha/\omega \\ +1 & \alpha/\omega \le t < T_o/2 - \alpha/\omega \\ 0 & T_o/2 - \alpha/\omega \le t < T_o/2 \end{cases} \tag{9.80}$$

In a k-pulse inverter with a constant duration, the modulation function is given by

$$v_o = m(t)V_{dc} \quad m(t) = \begin{cases} +1 & 0 \le t < d(t) \\ 0 & d(t) \le t < T_s \end{cases} \tag{9.81}$$

In the sinusoidal PWM technique, the pulse durations are adjusted to change slowly to follow the sinusoidal function. The modulation function is normally limited between 0 and 1. Since it is desired to have an output voltage with zero dc, the modulation function must be symmetrical around zero.

For the sinusoid PWM waveforms, we define as follows:

$$m(t) = M_{\max} \cos \omega_o t \tag{9.82}$$

where w_o is the modulation frequency, which must be less than the switching frequency; i.e., $\omega_o \ll \omega_s$. $M_{\max}$ is the gain of the modulation function, which varies between 0 and 1. Normally the switching frequency is in the range of a few hertz to a 100 kHz while the modulation frequency is less than 500 Hz. As stated before, using the PWM process, we can extract a low-frequency signal from a train of high-frequency squarewaves. The higher the switching frequency of the squarewave output with respect to the desired low-frequency output signal, the more the output waveform approximates a sinusoidal.

A modulation gain of unity represents the largest possible output voltage, whereas a modulation gain of zero means the output waveform frequency equals the switching frequency, and the output voltage is the smallest.

Switching Schemes

Depending on the switching sequence, the output voltage in PWM inverters can be either bipolar or unipolar. Figure 9.37 shows a bipolar output voltage in a PWM inverter. When the reference sinusoidal signal is larger or smaller than the triangular wave, the output equals $+V_{dc}$ or $-V_{dc}$, respectively. Both the half-bridge and full-bridge configurations are used in practice to generate a PWM output voltage waveform.

In bipolar voltage switching, m_f is an odd number that is the same as the switching frequency, f_s. The output frequency, f_o, in unipolar voltage switching is twice the frequency in bipolar voltage switching (m_f is doubled).

If both positive and negative sinusoidal control signals are available, then the switching sequence will produce a unipolar output waveform as shown in Fig. 9.38. The output waveforms for v_{o1} and v_{o2} are shown in Fig. 9.38(a) and (b), respectively, and Fig. 9.38(c) shows $v_o = v_{o1} - v_{o2}$. Here the triangular signal is chosen to be a sawtooth function.

Signal Generation

Advanced digital and analog techniques exist in today's inverters to generate the driving signals that produce sinusoidal PWM. For illustration purposes, Fig. 9.39 shows a comparator that compares a triangular signal to a sinusoidal reference signal.

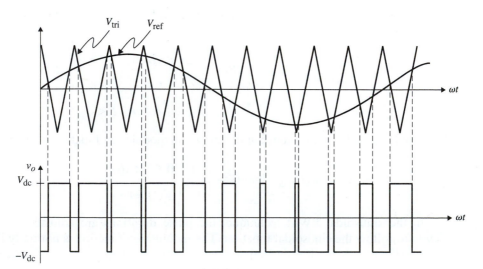

Figure 9.37 Example of a bipolar PWM output waveform.

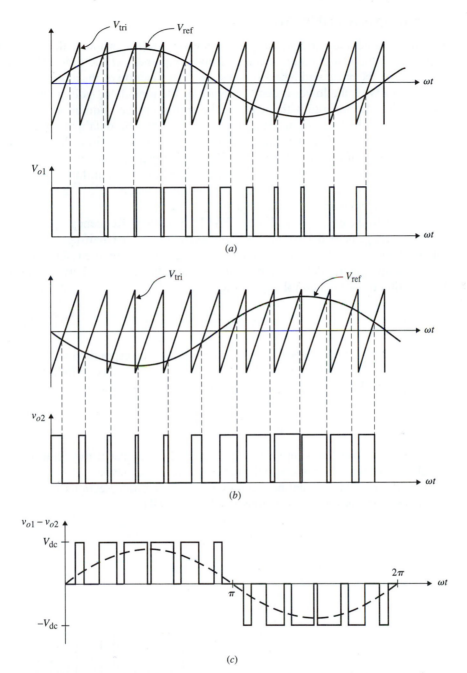

Figure 9.38 Unipolar PWM output. (a) A positive sinusoidal reference to produce v_{o1}. (b) Positive sinusoidal reference to produce v_{o2}. (c) The differential output $v_o = v_{o1} - v_{o2}$.

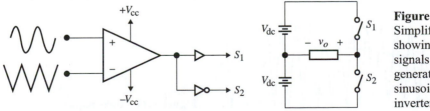

Figure 9.39 Simplified circuit showing how signals are generated in sinusoidal PWM inverters.

Analysis of Sinusoidal PWM

As stated before, the width of each pulse is varied in proportion to the instantaneous integrated value of the required fundamental component at the time of its event. In other words, the pulse width becomes a sinusoidal function of the angular position. In sinusoidal PWM, the lower-order harmonics of the modulated voltage waveform are highly reduced compared with the use of the uniform pulse-width modulation.

The output voltage signal in sinusoidal PWM can be obtained by comparing a control signal, v_{cont}, against a sinusoidal reference signal, v_{ref}, at the desired frequency. At the first half of the output period, the output voltage takes a positive value $(+V_{dc})$ whenever the reference signal is greater than the control signal. In the second half of the output period, the output voltage takes a negative value $(-V_{dc})$ whenever the reference signal is less than the control signal.

Similar to the case of equal-pulse PWM, the control frequency f_{cont} (equal to the switching frequency) determines the number of pulses per half-cycle of the output voltage signal. Also, the output frequency f_o is determined by the reference frequency f_{ref}. The amplitude modulation index, m_a, is defined as the ratio between the sinusoidal magnitude and the control signal magnitude.

$$m_a = \frac{V_{P,ref}}{V_{P,cont}} \tag{9.83}$$

The duration of the pulses is proportional to the corresponding value of the sine-wave at the corresponding position. Then the ratio of any pulse duration to its corresponding time duration is constant, as shown in Figs. 9.40 and 9.42.

$$\frac{\beta_1}{y_1} = \frac{\beta_2}{y_2} = \frac{\beta_3}{y_3} \Rightarrow \frac{\beta_i}{y_i} = \text{constant} \tag{9.84}$$

The most important simplifying assumption here is that if the control frequency signal is very high with respect to the reference frequency signal, $(m_f \gg 1)$, then the value of the reference signal between two consecutive intersections with the control signal is almost constant. This is illustrated in Fig. 9.41.

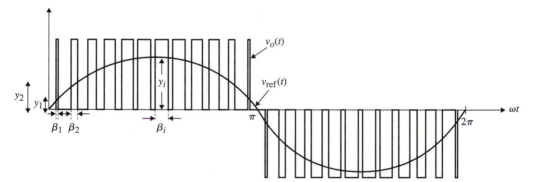

Figure 9.40 PWM figure illustrating the constant ratio between the width and height of a given pulse.

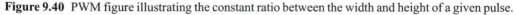

Figure 9.41 High-frequency sinusoidal PWM.

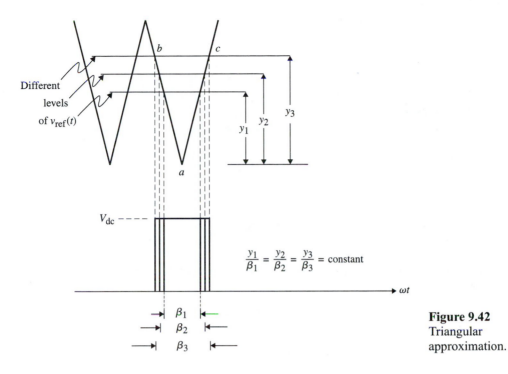

$$\frac{y_1}{\beta_1} = \frac{y_2}{\beta_2} = \frac{y_3}{\beta_3} = \text{constant}$$

Figure 9.42
Triangular
approximation.

The proportional variation of each pulse width with respect to the corresponding sinewave amplitude can be seen by applying a triangular relationship, as shown in Fig. 9.42.

Output Voltage Harmonics

The calculation of the sinusoidal PWM output voltage is the same as that of the uniform PWM output voltage. However, for sinusoidal PWM, the width of each pulse varies according to its position. The expression for the output voltage is obtained using a Fourier series transformation for v_o, given by

$$v_o(t) = \sum_{n = 1, 2, \ldots}^{\infty} (a_n \cos n\omega t + b_n \sin n\omega t) \tag{9.85}$$

Since the inverter output voltage is an odd function, only odd harmonics exist.

The calculation of the output voltage harmonic components can be done using a single pair of pulses as shown in Fig. 9.43.

$$V_{n, i} = \frac{1}{\pi} \int_0^{2\pi} v_o(\omega t) \sin(n\omega t) \, d(\omega t)$$

$$V_{n, i} = \frac{1}{\pi} \left[\int_{\theta_i}^{\theta_i + \theta_{wi}} V_{dc} \sin(n\omega t) \, d(\omega t) + \int_{\pi + \theta_i}^{\pi + \theta_i + \theta_{wi}} -V_{dc} \sin(n\omega t) \, d(\omega t) \right] \tag{9.86}$$

Using the trigonometric relationship

$$\cos x - \cos y = -\left(2 \sin \frac{x+y}{2} \right) \left(\sin \frac{x-y}{2} \right)$$

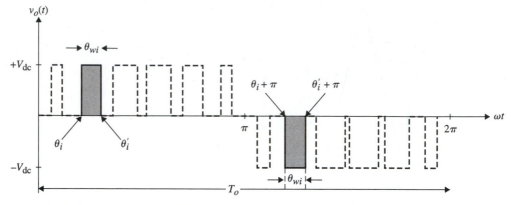

Figure 9.43 Single sinusoidal PWM pair of pulses.

it can be shown that the harmonic component for a single pair of pulses is given by

$$V_{n,i} = \left(\frac{2V_{dc}}{n\pi}\right)\sin\left(n\frac{\theta_{wi}}{2}\right)\left[\sin n\left(\theta_i + \frac{\theta_{wi}}{2}\right) - \sin n\left(\theta_i + \frac{\theta_{wi}}{2} + \pi\right)\right] \tag{9.87}$$

Adding the contribution from all other pulses, the ith component of v_o is given by

$$V_n = \sum_{i=1}^{k}\left(\frac{2V_{dc}}{n\pi}\right)\sin\left(n\frac{\theta_{wi}}{2}\right)\left\{\sin n\left(\theta_i + \frac{\theta_{wi}}{2}\right) - \sin n\left(\theta_i + \frac{\theta_{wi}}{2} + \pi\right)\right\} \tag{9.88}$$

where θ_i is the starting angle of the ith pulse and θ_{wi} is the pulse width at the corresponding angular position. Next, we estimate the width of each pulse, θ_{wi}, for each ith pulse.

Approximating the Pulse Width, θ_{wi}

Assume each pulse is located at the discrete value of θ_i, which represents the first intersection for the generation of the ith pulse. Then the approximated mathematical relation for the width is found using a geometrical relation as shown in Fig. 9.44.

From the geometry of the triangle ABC in Fig. 9.44, we have

$$h_y = V_{P,\text{tri}}, \qquad h_x = V_{P,\text{ref}}\sin\theta_i, \qquad y = \frac{T_o}{2k}\omega_o = \frac{\pi}{k}$$

The approximated width of the ith pulse, $\theta_{wi,\text{app}}$, is given by

$$\theta_{wi,\text{app}} = x \tag{9.89}$$

Since $\alpha = \beta$, then

$$\frac{h_x}{x} = \frac{h_y}{y} \tag{9.90}$$

Substituting for h_y, h_x, y, and x in Eq. (9.90) and using $m_a = V_{P,\text{ref}}/V_{P,\text{tri}}$, Eq. (9.89) becomes

$$\theta_{wi,\text{app}} = \left(\frac{\pi}{k}\right)m_a\sin\theta_i \tag{9.91}$$

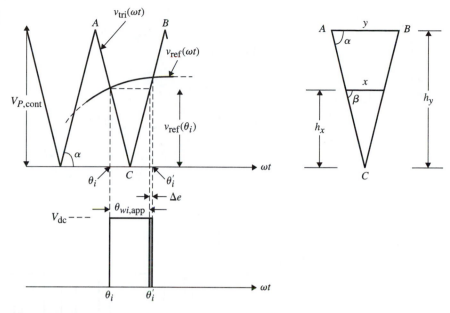

Figure 9.44 Approximated pulse width, i.e., $\Delta e \approx 0$.

It can be seen that the exact width is produced for the consecutive intersections of the control signal and the sinusoidal signal at positions θ_i and θ'_i. Therefore, an error is originated when the signal is considered constant and the value of the width is found using the approximated procedure. To reduce the width approximation error, it is necessary to increase the carrier frequency. Increasing the index frequency m_f produces a considerable reduction in the difference of the reference signal values evaluated at θ_i and θ'_i respectively.

Exact Expression for θ_{wi}

For the ith pulse with an angle θ_i, it can be easily shown that the point of intersection between the control signal and the ωt axis is $((2i-1)\pi/2k, 0)$, as shown in Fig. 9.45.

The general expression for $v_{cont}(\omega t)$ having all lines of negative slope is given by

$$v_{cont}(\omega t) = -\frac{m_f}{\pi} V_{P,tri}\left(\omega t - (2i-1)\frac{\pi}{m_f}\right) \quad i = 0, 1, 2, \ldots, k \tag{9.92}$$

The expression for the reference signal is given by

$$v_{ref}(\omega t) = V_{P,ref}\sin \omega t \tag{9.93}$$

Evaluating Eqs. (9.92) and (9.93) at $\omega t = \theta_i$, $v_{cont}(\theta_i) = v_{ref}(\theta_i)$, yields

$$-\frac{m_f}{\pi} V_{P,tri}\left(\theta_i - (2i-1)\frac{\pi}{m_f}\right) = V_{P,ref}\sin \theta_i \tag{9.94}$$

Equation (9.94) can be rewritten as

$$m_a \sin \theta_i = -\frac{m_f}{\pi}\theta_i + (2i-1) \tag{9.95}$$

The value of θ_i can be found by solving Eq. (9.95) numerically.

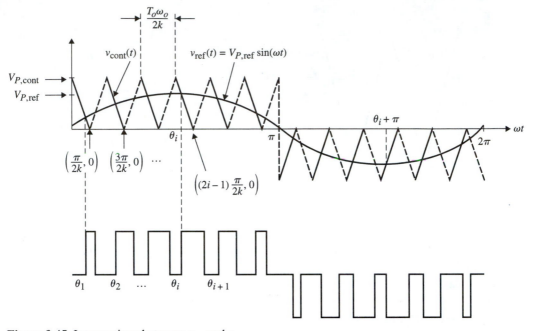

Figure 9.45 Intersections between v_{ref} and v_{cont}.

Consider the cases for different numbers of pulses per half-cycle shown in Fig. 9.46. The exact width for each pulse is found by adding the approximated width and the corresponding error Δe in each interval. As shown in Fig. 9.46, for the two-pulse case we have $\Delta e_1 = \Delta e_2$ by symmetry, as well as for the cases of three and four pulses.

Generalizing for any symmetric pair of pulses, the calculation for any width can be done as shown in Fig. 9.47. From the triangle ABC

$$\tan \alpha = \frac{v_{\text{ref}}(\theta'_i) - v_{\text{ref}}(\theta_i)}{\Delta e_i} \tag{9.96}$$

From the triangle DEF

$$\tan \alpha = \frac{V_{P,\text{tri}}}{\left(\dfrac{T_o \omega_o}{4k}\right)} = \frac{V_{P,\text{tri}}}{\dfrac{1}{2}\dfrac{\pi}{k}} = \frac{2k}{\pi} V_{P,\text{tri}} \tag{9.97}$$

Equating Eqs. (9.96) and (9.97), Δe_i may be expressed by

$$\Delta e_i = \left(\frac{\pi}{2k}\right)\left(\frac{v_{\text{ref}}(\theta'_i) - v_{\text{ref}}(\theta_i)}{V_{P,\text{tri}}}\right) \tag{9.98}$$

Substituting for $v_{\text{ref}}(\omega t) = V_{P,\text{ref}} \sin(\omega t)$, $k = \frac{1}{2}m_f$, and $m_a = V_{P,\text{ref}}/V_{P,\text{cont}}$ in Eq. (9.98), Δe_i becomes

$$\Delta e_i = \frac{\pi m_a}{2k}(\sin \theta'_i - \sin \theta_i) \tag{9.99}$$

From the supplementary angle relation $\sin x = \sin(\pi - x)$, Eq. (9.99) becomes

$$\Delta e_i = \left(\frac{\pi m_a}{2k}\right)(\sin \theta_{k+1-i} - \sin \theta_i)$$

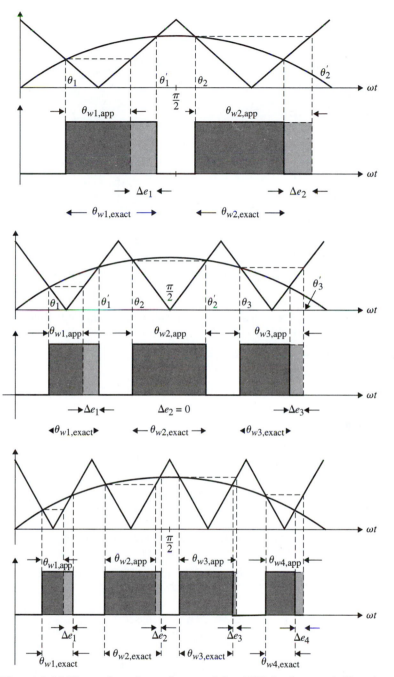

Figure 9.46 Illustration of two, three, and four PWM pulses per half-cycle.

and since $\theta_{wi} = \theta_{wi,\text{app}} \pm \Delta e_i$, the general expression for the exact width is given by

$$\theta_{wi} = \frac{\pi m_a}{k}\left(\sin\theta_i \pm \frac{1}{2}\left(\sin\theta_i - \sin\theta_{k+i-1}\right)\right) \qquad (9.100)$$

An increase in the control frequency causes a decrease in the value of Δe and $\theta_{wi,\text{app}}$. Therefore, when the control frequency tends to infinity, the width of Δe tends to zero. It can be shown that Δe decreases at a higher rate than does $\theta_{wi,\text{app}}$.

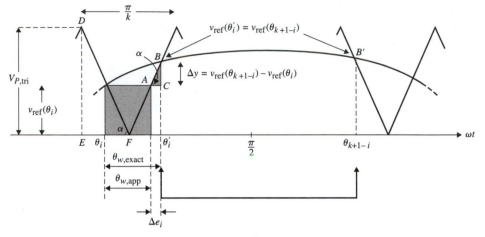

Figure 9.47 Symmetric pair of pulses about $\pi/2$.

EXAMPLE 9.9

Consider a full-bridge inverter with an $R - L$ load with a sinusoidal PWM output voltage with a reference frequency f_o of 60 Hz, $V_{dc} = 280$ V, $m_a = 0.6$, and $m_f = 24$.

(a) Find the carrier frequency (triangular wave).

(b) Find the number of pulses per half-cycle.

(c) Find the angles of intersection and the pulse widths in a half-cycle.

(d) Find the harmonic components.

(e) Find the total harmonic distortion.

SOLUTION

(a) $f_s = m_f \times f_o = 24 \times 60$ Hz $= 1.44$ kHz

(b) $k = \dfrac{24}{2} = 12$ pulses

(c) The exact angles were numerically calculated using Mathcad as shown in the following table.

Starting angle θ_i (degrees)	Arriving angle θ_i (degrees)	Pulse width θ_{wi} (degrees)
6.955	8.137	1.182
20.895	24.364	3.469
34.926	40.419	5.493
49.099	56.242	7.143
63.474	71.773	8.299
78.093	86.994	8.901
93.006	101.903	8.897
108.226	116.526	8.300
123.759	130.901	7.106
139.582	145.076	5.494
155.644	159.104	3.460
171.863	173.045	1.182

(d) Harmonic components, in volts:

$$V_1 = 167.931 \qquad V_{11} = -0.019 \qquad V_{21} = 19.909$$
$$V_3 = -0.048 \qquad V_{13} = 0.085 \qquad V_{23} = 103.541$$
$$V_5 = 0.127 \qquad V_{15} = -0.021 \qquad V_{25} = -103.74$$
$$V_7 = 0.03 \qquad V_{17} = -0.071 \qquad V_{27} = -19.736$$
$$V_9 = -0.118 \qquad V_{19} = 1.058 \qquad V_{29} = -0.898$$

These values are plotted in Fig. 9.48.

Using the approximated width, we obtain the following, using MATLAB:

$$V_1 = 167.742 \qquad V_{11} = 0.02 \qquad V_{21} = 16.071$$
$$V_3 = 0.769 \qquad V_{13} = -0.021 \qquad V_{23} = 108.129$$
$$V_5 = 6.917 \times 10^{-3} \qquad V_{15} = 8.579 \times 10^{-3} \qquad V_{25} = -99.036$$
$$V_7 = 7.869 \times 10^{-3} \qquad V_{17} = -4.769 \times 10^{-3} \qquad V_{27} = -23.464$$
$$V_9 = -6.94 \times 10^{-3} \qquad V_{19} = 0.564 \qquad V_{29} = -2.089$$

These results are plotted in Fig. 9.49.

(e) The total harmonic distortion is approximated by

$$\text{THD} \approx \sqrt{\frac{1}{V_1^2}(V_{21}^2 + V_{23}^2 + V_{25}^2 + V_{27}^2 + V_{29}^2)} = 0.89$$

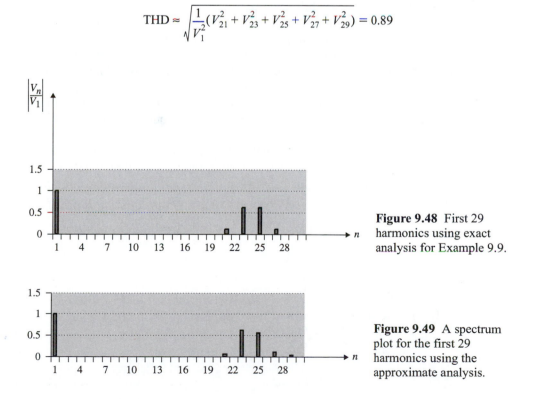

Figure 9.48 First 29 harmonics using exact analysis for Example 9.9.

Figure 9.49 A spectrum plot for the first 29 harmonics using the approximate analysis.

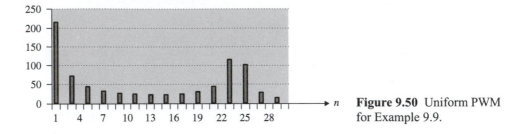

Figure 9.50 Uniform PWM for Example 9.9.

If we compare the uniform PWM approach with the sinusoidal PWM approach, we can notice a considerable reduction in the harmonic components. The next graph, Fig. 9.50, shows a plot of the normalized magnitude of the harmonic components using uniform pulse-width modulation with a modulation index of 0.6 for 12 pulses per half-period.

As we know, the inverter will produce a rectangular pulse waveform. To transform the rectangular waveform generated by the switching process to an output waveform that resembles a sinusoid wave, we need to use a harmonic filter. A low-pass filter can be used due to the high attenuation property without affecting the fundamental harmonic.

However, a perfect low-pass filter is impractical to realize, and the lower harmonics that are close to the fundamental will affect our desired output. Therefore, we need to maximally reduce the lower-order harmonics so the filter will allow the first harmonic to pass in our variable-frequency output.

9.6   THREE-PHASE INVERTERS

Consider the three-phase full-bridge dc-ac inverter shown in Fig. 9.51. To obtain a set of balanced line-to-line output voltages, the switching sequence of switches S_1-S_6 should produce a sequence of pulses whose summation at any given time is zero. As a result, it can be shown that in a one-pulse phase voltage, the conduction angle is $\pi/3$. The switch numbering follows the sequence of switching. The bidirectional switch implementation of S_1-S_6 allows an inductive-load current flow.

Figure 9.52(a) and (b) shows two switching sequences for S_1-S_6 with each switch conducting for π and $2\pi/3$, respectively. Both sequences produce a similar output voltage. To avoid shorting the voltage source V_{dc}, the switching sequence of S_1-S_6 must ensure that the S_1-S_4, S_3-S_6, and S_5-S_2 pairs are not switched on at the same time.

Figure 9.53(a) shows the circuit configuration for a three-phase inverter with a wye load and splitting input capacitors. The inverters generate three-phase output voltages. The switches are switched in such a way that the voltages v_b and v_c are shifted by $2\pi/3$. Figure 9.53(b) shows the three-phase output voltages v_a, v_b, and v_c. Figure 9.53(c) shows the three-phase line-to-line voltages v_{ab}, v_{bc}, and v_{ca}.

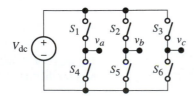

Figure 9.51 Three-phase inverter.

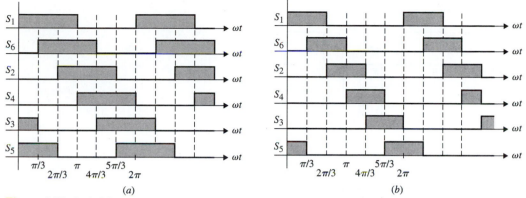

Figure 9.52 Switching sequence. (*a*) Conduction equals π. (*b*) Conduction equal $2\pi/3$.

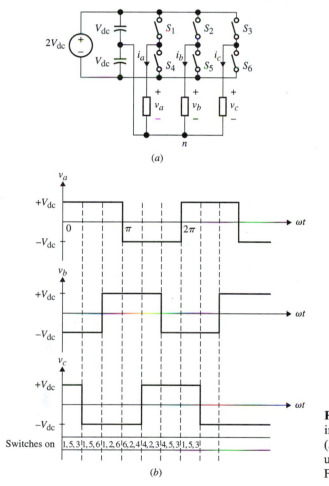

Figure 9.53 (*a*) Three-phase inverter with wye load. (*b*) Phase voltage waveforms using switching sequence of Fig. 9.52(*a*).

Figure 9.54 shows another line-to-line voltage using the switching sequence of Fig. 9.52(*a*). Three-phase inverters are also implemented by employing three single-phase inverters as shown in Fig. 9.55. The transformers' primary windings must be isolated from one another, and the transformers' secondary windings may be connected in a delta or wye configuration.

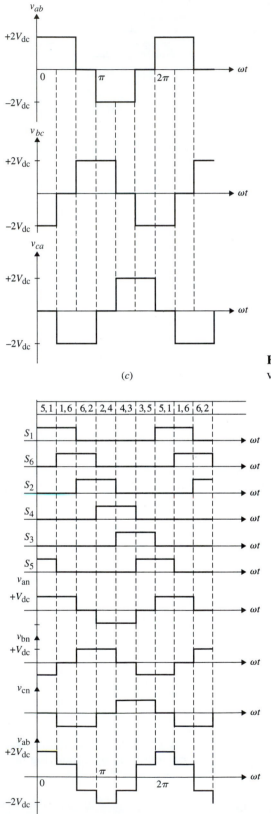

Figure 9.53 (*continued*) (*c*) Line-to-line voltages v_{ab}, v_{bc}, and v_{ca}.

(*c*)

Figure 9.54 Switching sequence for three-phase inverter to produce v_{ab}.

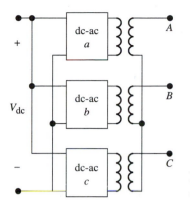

Figure 9.55 Three-phase inverter using three single-phase dc-ac inverters.

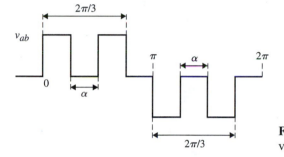

Figure 9.56 α control in line-to-line phase voltage.

From Fourier analysis, v_{ab} is given by

$$v_{ab}(t) = \sum_{n=1,3,5,\ldots}^{\infty} \frac{4V_{dc}}{n\pi} \cos\frac{n\pi}{6} \sin n\left(\omega t + \frac{\pi}{6}\right)$$

$$v_{bc}(t) = \sum_{n=1,3,5,\ldots}^{\infty} \frac{4V_{dc}}{n\pi} \cos\frac{n\pi}{6} \sin n\left(\omega t - \frac{\pi}{2}\right) \qquad (9.101)$$

$$v_{ab}(t) = \sum_{n=1,3,5,\ldots}^{\infty} \frac{4V_{dc}}{n\pi} \cos\frac{n\pi}{6} \sin n\left(\omega t - \frac{7\pi}{6}\right)$$

Notice all triple harmonics ($n = 3, 6, 9, 12, \ldots$) are zero.

Similarly, the output voltage control can be implemented using three-phase inverters. Figure 9.56 shows one possible three-phase output under α control.

EXAMPLE 9.10

Consider the three-phase inverter circuit shown in Fig. 9.57 under an inductive-resistive load. Sketch the voltage waveform for the phase voltages v_{an}, v_{bn}, and v_{cn} and line-to-line voltages v_{ab}, v_{bc}, and v_{ca} using the switching sequence shown in Fig. 9.58(a).

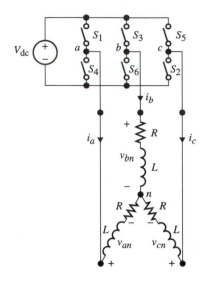

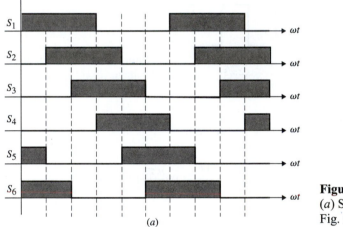

Figure 9.57 Three-phase inverter under an inductive-resistive load.

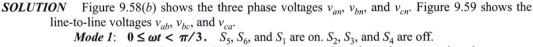

Figure 9.58
(a) Switching sequence for Fig. 9.57.

(a)

SOLUTION   Figure 9.58(*b*) shows the three phase voltages v_{an}, v_{bn}, and v_{cn}. Figure 9.59 shows the line-to-line voltages v_{ab}, v_{bc}, and v_{ca}.

    Mode 1: $0 \le \omega t < \pi/3$. S_5, S_6, and S_1 are on. S_2, S_3, and S_4 are off.

It can be shown that since the loads are the same, the pulse voltages are given by

$$v_{an} = \frac{1}{3}V_{dc}, \qquad v_{bn} = -\frac{2}{3}V_{dc}, \qquad v_{cn} = \frac{1}{3}V_{dc}$$

The equivalent circuit for this mode is shown in Fig. 9.60(*a*).

    The line-to-line voltages are given by

$$v_{ab} = V_{dc}, \qquad v_{bc} = -V_{dc}, \qquad v_{ca} = 0$$

    Mode 2: $\pi/3 \le \omega t < 2\pi/3$. S_1, S_2, and S_6 are on. S_3, S_4, and S_5 are off. The equivalent circuit for this mode is shown in Fig. 9.60(*b*).

    The line-to-line voltages are given by

$$v_{ab} = V_{dc}, \qquad v_{bc} = 0, \qquad v_{ca} = -V_{dc}$$

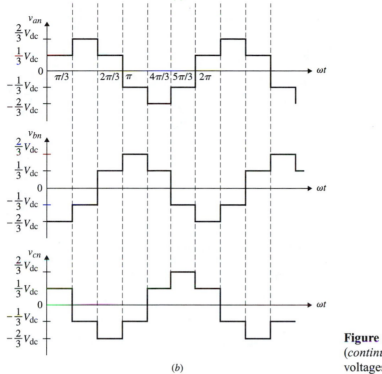

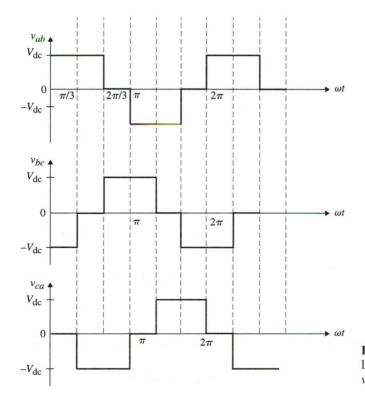

(b)

Figure 9.58
(*continued*) (*b*) Phase
voltages v_{an}, v_{bn}, and v_{cn}.

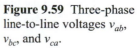

Figure 9.59 Three-phase
line-to-line voltages v_{ab},
v_{bc}, and v_{ca}.

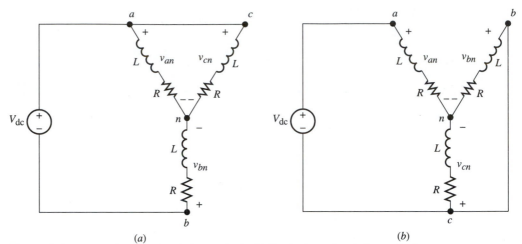

(a)   (b)

Figure 9.60 (a) Equivalent circuit for mode 1. (b) Equivalent circuit for mode 2.

and the phase voltages are given by

$$v_{an} = \frac{2V_{dc}}{3}, \quad v_{bn} = -\frac{V_{dc}}{3}, \quad v_{cn} = \frac{-V_{dc}}{3}$$

It can be shown that the remaining modes will produce the waveforms in Fig. 9.58(b) and Fig. 9.59.

EXERCISE 9.9

Sketch the load currents i_{ab}, i_{bc}, and i_{ca} for the three-phase inverter shown in Fig. E9.9. Assume the device conduction sequence is as given in Fig. 9.58.

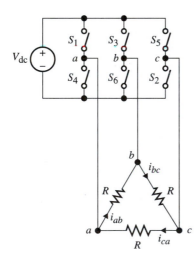

Figure E9.9 Three-phase inverter for Exercise 9.9.

EXERCISE 9.10

Show that the fundamental components for the load currents in Exercise 9.9 are given by

$$i_{ab,1} = \frac{4V_{dc}}{R\pi} \cos\frac{\pi}{6} \sin(\omega t + \pi/6)$$

$$i_{bc,1} = \frac{4V_{dc}}{R\pi}\cos\frac{\pi}{6}\sin(\omega t - \pi/2)$$

$$i_{ca,1} = \frac{4V_{dc}}{R\pi}\cos\frac{\pi}{6}\sin(\omega t - 7\pi/6)$$

9.7   CURRENT-SOURCE INVERTERS

In the current-source inverter, the dc input is a current source regardless of the input voltage variation. Practically, the dc current source is implemented using a large dc inductor in series with the dc voltage source as shown in Fig. 9.61. The dc supply is of a high impedance (because of the high input inductance). Since the dc inductance (L_{dc}) is large, I_{dc} is nearly constant.

The output current waveform is determined by the circuit's topology and switching sequence, while the output voltage waveform is determined by the nature of the load. Loads with low impedance to the harmonics are normally used. Figure 9.62 shows a tall-bridge current-source inverter. The switches S_1-S_4 are implemented using GTOs and diodes; the SCR implementation is difficult in the current-source inverter since the turn-off time is hard to set.

As with the voltage-source inverter, depending on the switching sequence of S_1, S_2, S_3, and S_4, two possible output currents may be obtained, as shown in Fig. 9.63 (*a*) and (*b*). Figure 9.63(*a*) uses a 50% duty cycle, and the output current and its rms value are controlled by varying the value of I_{dc}. In Fig. 9.63(*b*) the duty cycle is less than

Figure 9.61 Practical dc current source implementation.

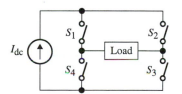

Figure 9.62 Full-bridge current-source inverter.

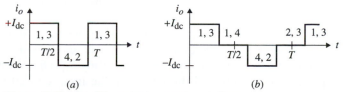

Figure 9.63 Possible output current waveforms.

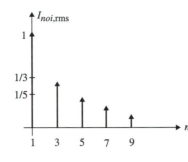

Figure 9.64 Frequency spectra for the normalized rms current harmonics.

50%. There are three output states: $+I_{dc}$, 0, and $-I_{dc}$. In both cases, the output voltage waveforms will depend on the nature of the load.

From Fourier analysis, the load current is given by

$$i_o(t) = \frac{4I_{dc}}{\pi}\left[\sin \omega t + \frac{\sin 3\omega t}{3} + \cdots + \frac{\sin n\omega t}{n}\right] \quad n = 1, 3, 5, 7, \ldots \quad (9.102)$$

The fundamental component of $i_o(t)$ is given by

$$i_{o1}(t) = \frac{4I_{dc}}{\pi}\sin \omega t \quad (9.103)$$

The peak and rms values of the fundamental component are

$$i_{o1,peak} = \frac{4I_{dc}}{\pi} \quad (9.104a)$$

$$i_{o1,rms} = \frac{2I_{dc}\sqrt{2}}{\pi} \quad (9.104b)$$

Let the normalized rms of the ith harmonic be given by

$$I_{no,rms} = \frac{I_{oi,rms}}{I_{o1,rms}} \quad i = 1, 3, 5, \ldots \quad (9.105)$$

Harmonics	1	3	5	7	9
$I_{no,rms}$	1	1/3	1/5	1/7	1/9

The frequency spectra for the normalized rms current harmonics are shown in Fig. 9.64.

Consider the ideal current-source inverter shown in Fig. 9.65(a). Assume $v_o = V_o \sin(\omega t + \theta)$ and S_1-S_4 are switched according to the sequence shown in Fig. 9.65(b), with a 50% switching duty cycle with no α control. By shifting S_1 to the left by α and S_3 to the right by α, we allow an interval during which the input current source is shorted, as shown in Fig. 9.65(c).

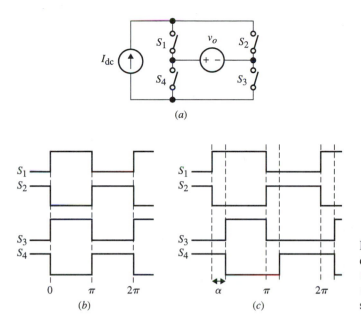

Figure 9.65 (a) Ideal current-source inverter ($\alpha = 0$). (b) 50% switch ($\alpha = 0$). (c) Less than 50% switch ($\alpha \neq 0$).

PROBLEMS

Half-Bridge and Full-Bridge Inverters

9.1 Derive Eq. (9.5), which gives the initial inductor current in a half-bridge inverter under a resistive-inductive load.

9.2 Repeat Exercise 9.2 by including the first, third, and fifth harmonics.

9.3 Consider the half-bridge inverter and its driven waveforms shown in Fig. P9.3 with the following inverter parameters: $V_{dc} = 185$ V, $R = 10\ \Omega$, and $f = 60$ Hz.

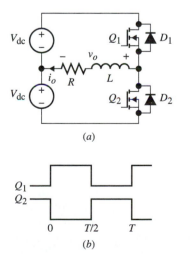

(a)

(b)

Figure P9.3

(a) Determine the average diode current for $L = 1$ mH, 5 mH, 10 mH, and 20 mH.

(b) As the load's time constant increases with respect to $T/2$, what is the effect on the load's harmonics?

9.4 Consider the half-bridge inverter of Fig. 9.9 with the following circuit components: $V_{dc} = 408$ V, $R = 6\ \Omega$, and $f = 60$ Hz, and $L = 20$ mH.

(a) Derive the exact expression for i_L.

(b) Derive the expression for the fundamental component of $i_L(t)$.

(c) Determine the average diode and transistor currents.

(d) Determine the average power delivered to the load.

9.5 Consider the full-bridge inverter shown in Fig. P9.5(a) with the given parameters. Assume the switching sequence of S_1-S_4 produces the output voltage waveform shown in Fig. P9.5(b).

(a) Determine the expression for $i_o(t)$ and sketch it.

(b) With S_1-S_4 replaced by a parallel combination of switch and diode, find the peak and average switch and diode currents.

(c) Find the average power delivered to the load.

(d) Determine the fundamental and third harmonic peak voltage. What is the percentage of the third harmonic with respect to the fundamental?

(e) Repeat part (d) for $L = 50$ mH, 100 mH, and 200 mH.

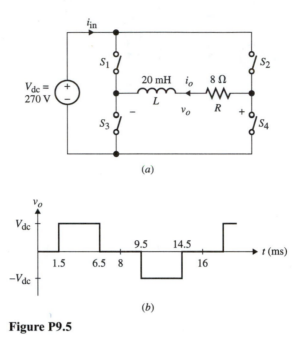

(a)

(b)

Figure P9.5

9.6 Consider the ideal dc-to-ac inverter shown in Fig. P9.6. Assume S_1-S_4 and S_2-S_3 are switched alternatively at a 50% duty cycle with a switch period of T. Let the output current be given by $i_o(t) = I_P \sin \omega t$, where $\omega = 2\pi/T$.

(a) Derive the expression for the instantaneous power delivered to the load.

(b) Determine the average power delivered to the load.

(c) Discuss the requirement for a practical switch implementation for S_1-S_4.

(d) Repeat part (c) for $i_o(t) = I_P \sin(\omega t - \theta)$.

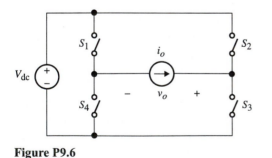

Figure P9.6

9.7 Consider the full-bridge inverter circuit with an RLC load shown in Fig. P9.7(a). Use $L = 42$ mH, $R = 18$ Ω, $C = 900$ μF, and $V_{dc} = 270$ V. Assume the switch sequence of

S_1-S_4 produces the output voltage waveform shown in Fig. P9.7(b). Determine the peak fundamental component for the load current $i_o(t)$.

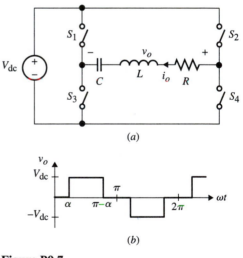

(a)

(b)

Figure P9.7

9.8 *(Approximation analysis for an RL load)* Consider the half-bridge inverter of Fig. 9.8(b) with an *RL* load. Assume the load's time constant, L/R, is longer than half of the switching period, i.e., $L/R \gg T/2$, so that the third and higher harmonics are neglected. Show that the average

power delivered to the load is given by

$$P_{ave} = |S|\cos\theta$$

where $|S|$ is the magnitude of the reactive power, which is given by

$$|S| = \frac{8V_{dc}^2}{\pi^2|Z|}\cos^2\alpha$$

$$\theta = \tan^{-1}(\omega L/R)$$

and

$$|Z| = \sqrt{R^2 + (\omega L)^2}$$

9.9 Consider the switching RL circuit shown in Fig. P9.9(a) with v_s given in Fig. P9.9(b).

(a) Derive the expression for $i_L(t)$ in terms of V_{dc} and τ_n, where $\tau_n = \tau/T = L/RT$.

(b) Give the expressions for $i_L(0)$, $i_L(t_1)$, and $i_L(T/2)$ in terms of V_{dc} and τ_n.

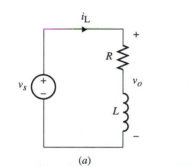

(a)

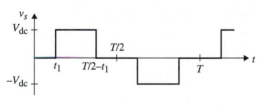

(b)

Figure P9.9

Harmonics

9.10 Show that the harmonics for the generalized squarewave with a pulse width equal to $\pi - 2\alpha$ and a delay angle α as shown in Fig. P9.10 is given by

$$v_o(t) = V_1\sin(\omega t) + V_3\sin(3\omega t) +$$

$$V_5\sin 5(\omega t) + \cdots + V_n\sin(n\omega t)$$

where

$$V_n = \frac{-4}{n\pi}\cos n\alpha$$

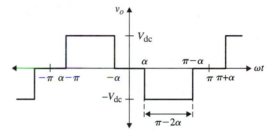

Figure P9.10

9.11 Consider the output voltage of an inverter shown in Fig. P9.11 that uses α control to eliminate two harmonics from the output.

(a) For $\beta_1 = \pi/4$, $\beta_2 = \pi/3$, and $\beta_3 = 3\pi/4$ find the fundamental, third, and fifth harmonics.

(b) Redesign the problem for α_1, α_2, and α_3 to eliminate the third and fifth harmonics.

9.12 Consider an inverter circuit that produces the stepped output waveform shown in Fig. P9.12. It can be shown that by properly selecting α_1 and α_2, the third and fifth harmonics can be eliminated from the output voltage. Determine α_1 and α_2 to accomplish this.

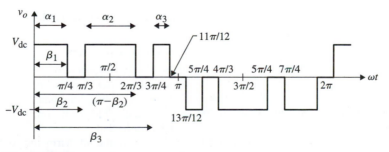

Figure P9.11

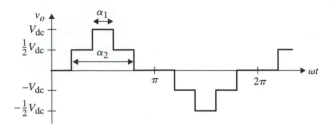

Figure P9.12

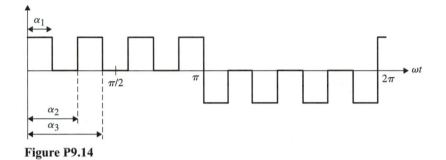

Figure P9.14

9.13 Determine the THD for the stepped waveform shown in Problem 9.12 by using only up to the ninth harmonic.

9.14 In a certain output voltage for an inverter, it is desired to eliminate the third, fifth, and seventh harmonics by controlling the switching sequence of the power devices. The following equation was obtained for the nth output voltage harmonic of Fig. P9.14.

$$V_{on} = \frac{4 V_{dc}}{n\pi}(1 - \cos n\alpha_1 + \cos n\alpha_2 - \cos n\alpha_3)$$

Determine α_1, α_2, and α_3 for a zero value of third, fifth, and seventh harmonics. (*Hint:* Use a numerical solution method or an iterative technique.)

Uniform Pulse-Width Modulation

9.15 Derive the relation for θ_i given by Eq. (9.64),

$$\theta_i = \frac{\pi}{k}\left[i - \frac{m_a}{2} - \frac{1}{2}\right]$$

for the ith pulse in a k-pulse output.

9.16 Derive the expression given in Eq. (9.65), which shows that the width of the pulse in a uniform PWM output is constant and is given by $\theta_{\text{width}} = \pi m_a / k$.

9.17 Determine the rms value for a uniform PWM inverter output with $V_{dc} = 260$ V and $m_a = 0.8$.

Determine the width of each pulse if the output has 24 pulses per cycle.

9.18 Derive Eq. (9.75).

9.19 Repeat Example 9.8 for $k = 10$ and $m_a = 0.4$. How does the THD compare to the THD obtained in that example?

Sinusoidal Pulse-Width Modulation

9.20 Sketch the output voltage for the buck converter whose output is modulated according to Eq. (9.78). Assume $D_{dc} = D_{\text{max}} = 0.5$, $V_{dc} = 1$ V, and $f_o = f_s / 10$.

9.21 Derive Eq. (9.88).

9.22 Derive the closed-form expression for θ_i given in Eq. (9.95).

9.23 Determine the approximate width of the fourteenth pulse in a sinusoidal PWM inverter output of Fig. 9.44 with $k = 18$ and $m_a = 0.5$.

9.24 Repeat Example 9.9 for $m_a = 1.0$.

Three-Phase Inverters

9.25 The waveform shown in Fig. P9.25 represents one possible way to implement a line-to-line output voltage in a three-phase inverter control. By varying the angle α, the rms value of the output can be controlled.

Show that the nth harmonic component for the line-to-line voltage is given by

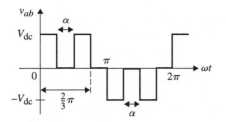

Figure P9.25

$$V_{ab,n} = \frac{4}{n\pi} V_{in}\left(\sin n\frac{\pi}{3} - \sin \frac{n\alpha}{2} \right)$$

$$\left(\cos n\left(\omega t - \frac{\pi}{3} - \frac{\alpha}{2} \right) \right)$$

and show that the total harmonic distortion is given by

$$\text{THD}_v^2 = \frac{\pi(2\pi/3 - \alpha)}{8(\sin(\pi/3) - \sin(\alpha/2))^2} - 1$$

9.26 Derive the harmonic components for v_{ab}, v_{bc}, and v_{ca} in Fig. 9.53(c).

9.27 One way to reduce the harmonics of a line-to-line voltage in a three-phase inverter is to produce an output waveform like the one shown in Fig. 9.54. Determine the nth harmonic component of v_{ab}.

9.28 Derive the harmonic component for v_{ab} of Fig. P9.28. (*Hint*: See Appendix B.)

9.29 The equivalent circuit for a line-to-line voltage in a three-phase inverter is given in Fig. P9.29. **(a)** Show that in the steady state the inductor current at $\omega t = 0$ is given by

$$I_L(0) = -\frac{V_{dc}(1 + e^{-1/6\tau_n} - e^{-1/3\tau_n} - e^{-1/2\tau_n})}{3R \qquad 1 + e^{-1/2\tau_n}}$$

where

$$\tau_n = \frac{\tau}{T} = \frac{L}{RT}, \qquad T = \frac{2\pi}{\omega}$$

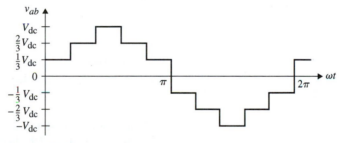

Figure P9.28

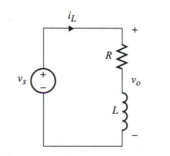

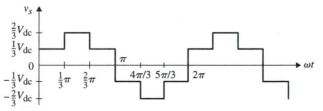

Figure P9.29

(b) Show that the steady-state inductor current value at $\omega t = \pi/3$ is given as a function of $I_L(0)$ as follows:

$$I_L(\pi/3) = \left(I_L(0) - \frac{V_{dc}}{3R}\right)e^{-1/6\tau_n} + \frac{V_{dc}}{3R}$$

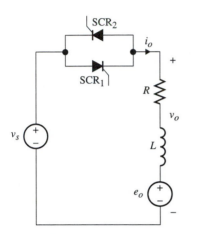

Figure P9.30

(c) Determine the time at which the flyback diode of S_1 stops conducting, assuming $L = 12$ mH, $R = 18\ \Omega$, $f = 400$ Hz, and $V_{dc} = 260$ V.

General Problems

9.30 Consider the single-phase ac controller shown in Fig. P9.30, with the induction motor load modeled by a resistor, inductor, and emf ac source.
Assume the emf voltage, $e_o(t)$, is given by

$$e_o(t) = V_o \sin(\omega t + \theta_o)$$

and

$$v_s(t) = V_s \sin(\omega t)$$

Here we assume that the source frequency and the emf voltage frequency are equal. Also assume SCR_1 and SCR_2 are triggered continuously at $\alpha_1 = 30°$ and $\alpha_2 = 330°$, respectively. Use $R = 0.85\ \Omega$, $L = 10.2$ mH, $V_s = 120$ V, $\omega = 2\pi(60)$, and $V_o = 85$ V.
(a) Sketch the waveform of $i_o(t)$ for the full cycle.
(b) Derive the expression of $i_o(t)$ for the first half-cycle.

9.31 Figure P9.31(a) shows a single-phase power electronic circuit with an $R-L$ load known as an *ac controller.* These controllers are widely used in single- or three-phase arrangements for various residential and commercial applications such as heating, motor speed control, and power factor correction. A possible switch implementation is shown in Fig. P9.31(b) with back-to-back SCRs to produce bidirectional current flow and bidirectional voltage blocking.
Discuss the operation of the circuit that produces the output waveform shown in Fig. P9.31(c).

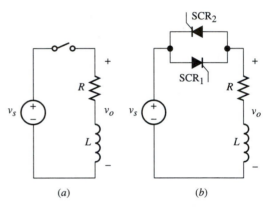

(a)   (b)

Figure P9.31 (a) ac controller with inductive-resistive load. (b) SCR implementation.

9.32 (a) Show that if the exact expression for $i_L(t)$ is used in Fig. 9.9, then the fundamental component is given by

$$i_L(t) = I_{L1} \sin(\omega t + \theta)$$

where

$$I_{L1} = \frac{4V_{dc}}{R\pi}\sqrt{\frac{4\pi^2\tau_n^2(e^{-1/\tau_n} - e^{-1/2\tau_n}) + 4\pi^2\tau_n^2 + 1}{4\pi^2\tau_n^2 + 1}}$$

$$\theta = \tan^{-1}\left[2\pi\tau_n e^{-1/2\tau_n} - \frac{1}{2\pi\tau_n} - 2\pi\tau_n\right]e^{1/(2\tau_n)}$$

(b) Show that the simplified expression when $L/R \gg T$ is given by

$$I_{L1} \approx \frac{4V_{dc}}{\pi}\frac{1}{\sqrt{R^2 + (\omega L)^2}}$$

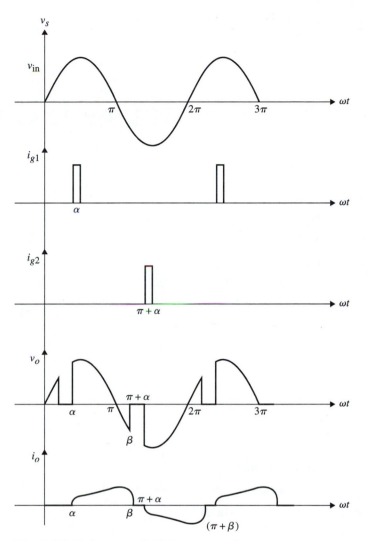

Figure P9.31 (*continued*) (*c*) Output waveform.

Appendix A

Introduction to Magnetic Circuits

INTRODUCTION

Modern power electronic systems are rarely designed without inductors or transformers. Because of the importance of these components, understanding magnetic and magnetic circuits is necessary for the successful design of power electronic systems. Moreover, if not included as discrete components, inductors exist as parasitic components, especially when the power systems are operated at high frequencies. Normally, the magnetic components are the most expensive and difficult to design in power systems.

Inductors are used to accomplish one or more of the following functions:

1. Filter switch waveforms, at both the input and output sides.

2. Form resonant circuits along with capacitors in order to create sinusoidal waveforms for various applications.

3. Limit the rate of change of load currents in switching circuits.

4. Limit transients at power-up of electric systems.

A transformer may be considered as two inductors coupled through a shared magnetic circuit with common flux. Because of this common flux, it is possible to change the ac electric energy at a given voltage level to another voltage level (same form of energy) with high efficiency. Transformers, whether operated at line or higher frequencies, are used to accomplish one or more of the following functions:

1. Step up or step down the voltage to service various needs and applications.

2. Provide isolation between power systems to reduce various EMI problems and for safety considerations.

3. Provide phase shift in multiphase systems to generate systems with three or more phases.

4. Provide a means to store energy to be utilized at later times, especially in high-frequency applications.

5. Provide a coupling mechanism between the gate or base drive circuits of high-power semiconductor switching devices and the power circuits in various power electronic systems.

6. Provide sensing for voltage and/or current in various control feedback systems.

Because of the wide variation of parameters that characterize magnetic circuits (inductors and transformers), it is extremely difficult to standardize the size, values, and ratings of inductors and transformers. As a result, stockpiling such components is highly impractical and normally not done. Among the parameters that characterize magnetic components are current, voltage, and power ratings; energy storage and dissipation; frequency of operation; magnetizing and leakage inductances; turn ratio; fabrication difficulties; size; weight; and cost. An attempt to minimize one or more components will be done at the expense of one or more other parameters. As a result, designing magnetic circuits is a very tricky and challenging engineering problem, and experience plays a very important role.

A.1   TYPES OF MAGNETIC MATERIALS

Just as the types of conductors in electric circuits are identified based on their ability to conduct electric current, the types of magnetic materials are identified based on their degree of receptivity to the magnetic field. A good conductor allows the current to flow freely, and a good magnetic material allows the magnetic field to pass through freely. Even though the two materials experience totally different physical phenomena in the process, many of the concepts applied to electric circuits can be also identified with magnetic circuits. This analogy will be discussed in more detail in Section A.4.

Depending on the degree of magnetization in the presence of an electric field, magnetic materials are divided into three major groups:

1. *Diamagnetics*: These materials experience extremely weak magnetization in the presence of magnetic fields, i.e., they exclude magnetic fields. This property is similar to the property of insulators in electric circuits.

2. *Paramagnetics*: These materials experience slight magnetization. This property is similar to the conductive property of semiconductor materials in electric circuits.

3. *Ferromagnetics*: These materials experience strong magnetization in the presence of magnetic fields. This property is similar to the property of a good conductor material.

A single parameter that characterizes the property of magnetic material, therefore, its type, is known as the permeability, μ, which is expressed in the following constitutive relation:

$$\mathbf{B} = \mu \mathbf{H} \tag{A.1}$$

where

μ is the permeability of the magnetic material

$\mathbf{B}$ is the magnetic flux density[1]

$\mathbf{H}$ is the magnetic field intensity

Normally, the value of μ is identified relative to the permeability of the vacuum, μ_o, as follows:

$$\mu = \mu_r \mu_o \tag{A.2}$$

where

μ_o is the permeability of the vacuum

μ_r is the *relative permeability* of the magnetic material

Depending on the magnetic material, the value of μ_r can vary from a small fraction of μ_o to thousands of μ_o. The following is a fairly accurate classification of magnetic materials in terms of μ_r:

$\mu_r < 1$ for diamagnetic materials

$\mu_r = 1$ for the vacuum

$\mu_r > 1$ for paramagnetic materials

$\mu_r \gg 1$ for ferromagnetic materials

The value of μ_o is equal to $4\pi \times 10^{-7}$ henrys/meter. For many applications, normally it is desired that a large magnetic field be produced with the smallest possible current in the coil. As a result, materials with high μ are desired. This also means that a preponderant portion of the produced flux is confined to the magnetic material.

Ferromagnetic materials are normally obtained from iron with certain alloys such as cobalt (Co), tungsten (Tn), nickel (Ni), aluminum (Al), silicon (Si), manganese (Mn), and zinc (Zn) alloys. Alloys are selected based on several factors, such as maximum saturation flux density, $\mathbf{B}_{sat}$; operating frequency of the magnetic circuit; cost; and resistivity.

Both the conductivity and the frequency of operation affect the total energy losses in the magnetic material, known as *core losses*. For example, the iron-silicon alloy with low silicon content has relatively high losses and high $\mathbf{B}_{sat}$, whereas a high–silicon content material is more expensive and has reduced core losses. As a result, the latter material is used mostly in high-efficiency line-frequency applications, and the former material is used when cost is the determining factor. For high-frequency applications, high-permeability magnetic material is used, such as iron with Ni alloy, with relatively low $\mathbf{B}_{sat}$. However, for higher saturation flux density, Co alloys are used. Compared with the Si alloys, Co and Ni alloys are more expensive.

In power electronic systems, operating frequencies in the several megahertz range are commonplace. A special ceramic known as ferrite, made from various combinations of iron with Mn, Ni, and Zn alloys, has been very popular in the design of magnetic circuits in power electronics. For example Mn-Zn and Ni-Zn ferrites are used in power electronic applications with frequencies in the range of a few to tens of megahertz, respectively.

---

[1]Boldface letters indicate vector quantities.

A.2 MAGNETIC FIELDS

As you may recall from the study of electromagnetic fields, Maxwell's equations form the basis of the fundamental theory of electromagnetism, governing the relationship between the strength of electric (**E**) and magnetic (**H**) fields, and the electric (**D**) and magnetic (**B**) flux densities. In this section, we focus on only two of Maxwell's equations as they relate to magnetism. Maxwell's equations (A.3)[2] and (A.4) governing the magnetostatic field **B** and **H** are given by

$$\nabla \times \mathbf{H} = \mathbf{J} \tag{A.3}$$

$$\nabla \cdot \mathbf{B} = 0 \tag{A.4}$$

where **J** is the current density in amperes per square meter, and the symbol ∇ represents the partial-differential operator; $\nabla \times \mathbf{H}$ is called the *curl of* **H**, and $\nabla \cdot \mathbf{B}$ is called *the divergence of* **B**. Equation (A.3) is known as Ampere's law, and Eq. (A.4) is known as Gauss's magnetic law. Notice that as charge is the source of the electrostatic field, current is the source of the magnetostatic field.

Since our interest in this appendix is to derive magnetic fields resulting from a current-carrying wire wound around a core of magnetic material, Maxwell's equations in this form are not useful. It is normally more convenient to use different forms of Maxwell's equations to solve for the magnetic field under these circumstances. It can be shown, by using Stokes's theorem, that Eq. (A.3) may be expressed as follows:

$$\oint_C \mathbf{H} \cdot dl = \int_S \mathbf{J} \cdot dS \tag{A.5}$$

The left-hand side gives the integration of the magnetic field along a closed contour C, and the right-hand side gives the total current density, **J** ($\mathbf{J} = \mathbf{J}_1 + \mathbf{J}_2 + \cdots + \mathbf{J}_n$), flowing through the surface area, S, enclosed by the contour C as shown in Fig. A.1. It is possible to simplify Ampere's law further by assuming the current is confined to a

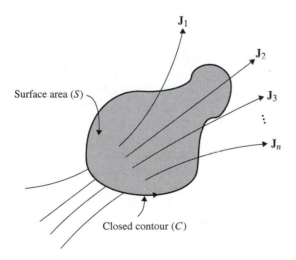

Figure A.1 Current density ($\mathbf{J} = \mathbf{J}_1 + \mathbf{J}_2 + \mathbf{J}_3 + \cdots + \mathbf{J}_n$) passing through area S enclosed by contour C subject to the integral of Eq. (A.5).

[2]Maxwell's equation (A.3) is based on the electrostatic field; i.e., the electric field does not vary with time. In time-varying magnetic fields, an additional term, known as *displacement current*, is used in this equation, to form $\nabla \times \mathbf{H} = \mathbf{J} + \partial \mathbf{D}/\partial t$. Here, the displacement current will be assumed negligible since its contribution is significant only at extremely high frequencies.

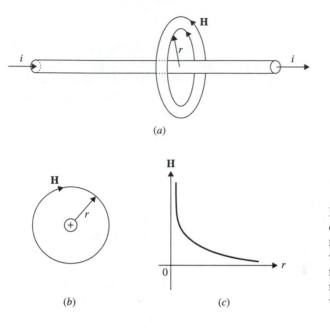

(a)

(b)   (c)

Figure A.2 Infinitely long wire conducting current i. (a) The magnetic field surrounding the wire. (b) The direction of the magnetic field at a distance of radius r. (c) The magnetic field versus distance from the wire.

wire rather than being distributed over a region in space. Consequently, the right-hand side gives i, where i is the total current flowing in the wire. Moreover, if we assume there are N wires each carrying the same current and encircled by the integral loop, then Ampere's law can be expressed as follows:

$$\oint_C \mathbf{H}dl = Ni \tag{A.6}$$

As an example, consider solving the above integral for an infinitely long wire carrying a current i as shown on Fig. A.2(a). The direction of the magnetic field is subject to the right-hand rule (the thumb is in the direction of the current and the fingers are in the direction of the magnetic field). The notations $\oplus$ and $\odot$ indicate that the current is flowing into and out of the paper, respectively.

To evaluate the integral, we apply Ampere's law to a circle with radius r and centered in the origin, as shown in Fig. A.2(b). A plot of the magnetic field versus the radius is shown in Fig. A.2(c). It is clear that the magnetic field due to the current decreases in the radial direction away from the conducting wire. The magnetic field lines are called *flux lines* or simply the *flux* of the magnetic field, denoted by ϕ, and the amount of the flux or the density of the flux is defined as follows:

$$\mathbf{B} = \frac{\phi}{A} \tag{A.7}$$

where

ϕ is the flux in webers (Wb)

$\mathbf{B}$ is the flux density in Wb/m² or teslas (T)

A is the area through which the flux flows

A.2.1 Toroidal Structure

One of the important applications of magnetics to the field of power electronics is when a wire is wound around a doughnut-shaped magnetic material to form what is

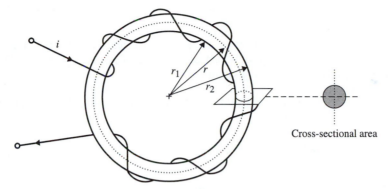

Cross-sectional area

Figure A.3 Toroidal structure created by winding wire around doughnut-shaped magnetic material.

known as *toroid*, as shown in Fig. A.3. This is not only because a large class of inductors and transformers are manufactured with this structure, but also because the analysis of such a structure can be easily applied to other magnetics with various arrangements. We will apply Ampere's law to the magnetic structure and solve for its magnetic field.

When solving for **H**, we will assume that the permeability of the magnetic material is much higher than that of air; hence, the entire magnetic field is assumed to be confined within the toroid core. Also, it is accurate to assume that the flux density is constant throughout the structure. Notice that **H** can be determined by taking the integral contour at the inner radius r_1 or the outer radius r_2. It can be easily shown that if the radial dimension $(r_2 - r_1)$ is very small compared to r_1, then **H** can be assumed constant and to give a good approximation when evaluated over the mean radius $(r_1 + r_2)/2$. Applying Ampere's law to a line integral with radius r and the circle loop shown in the figure, we obtain

$$\oint_C \mathbf{H} \cdot dl = Ni$$

$$\mathbf{H} \int_0^{2\pi} d\theta = Ni$$

Solving for **H**, we have

$$\mathbf{H} = \frac{Ni}{2\pi r}$$

$$= \frac{Ni}{l}$$

where r and l are the mean radius and length of the toroid, respectively.

The product Ni is normally referred to as the *magnetomotive force* (mmf), which is the magnetic field potential difference tending to force flux around the toroid. Moreover, by using the constitutive relation $\mathbf{B} = \mu\mathbf{H}$ and the definition of **B** given in Eq. (A.7), the flux in the magnetic material is given by

$$\phi = \frac{NiA\mu}{l} \tag{A.8}$$

where A is the cross-sectional area of the toroid.

EXAMPLE A.1

Consider a rectangular cross-sectional area of the toroidal magnetic core shown in Fig. A.4 with $r_1 = 3.5$ cm, $r_2 = 4$ cm, and a height of 1 cm. Assume $\mu = 2000\mu_o$ and $\mathbf{B}_{sat} = 0.3$ T. Determine:

(a) Maximum core flux before entering saturation

(b) The mmf required to produce this flux

(c) Number of turns that must be wound if the coil must carry 10 mA and still avoid saturation

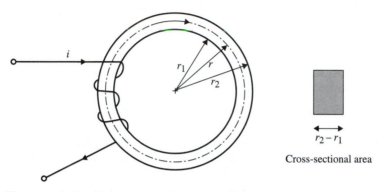

Figure A.4 Toroidal structure of Example A.1.

SOLUTION   (a) The cross-sectional area of the core is given by

$$A = (r_2 - r_1)h$$

$$= 50 \times 10^{-6} \text{ m}^2$$

The maximum flux is given by

$$\phi = \mathbf{B}A$$

$$= 15 \times 10^{-6} \text{ Wb}$$

(b) To obtain the mmf, we must determine the mean length of the flux path, which is given by

$$l = 2\pi \frac{r_1 + r_2}{2}$$

$$= 23.56 \times 10^{-2} \text{ m}$$

The mmf is given by

$$\text{mmf} = \frac{\phi l}{\mu A}$$

$$= \frac{(15 \times 10^{-6})(23.56 \times 10^{-2})}{(2000 \times 4\pi \times 10^{-7})(50 \times 10^{-6})}$$

$$= 28.125 \text{ A-turn}$$

(c) From the value of mmf, the number of turns, N, for $i = 10$ mA is 4.

A.3 MAGNETIC CIRCUITS

For many practical applications in power electronics, the design of special coil and core configurations are performed to produce various magnetic fields. As stated earlier, it is normally desirable to produce the maximum magnetic field with minimum magnetizing current in the coil. Because of the direct proportionality between current and magnetic field density, an attempt to increase the magnetic field intensity will result in an increase in the coil current. However, because of the relation $\mathbf{B} = \mu\mathbf{H}$, it is possible to increase the flux density by using material with high permeability compared to the air permeability. As discussed in the previous section, these materials are called *ferromagnetics*, from which almost all cores are made. Consequently, this choice is based on the assumption that most of the flux is confined to the magnetic material, with no flux flow in the surrounding medium, air or otherwise. This is because the flux will flow in a high μ path rather than in the low μ of the air. This phenomenon is analogous to the way in which current flows in wires, which have a conductivity much higher than the conductivity of the air. This observation, and others to be made later on in this section, have led to new approaches to analyze magnetic circuits. It has been customary to make this analogy between "electric circuits" and "magnetic circuits" in order to simplify the analysis and design of magnetic circuits, since electric circuits are well understood by electrical engineers.

Magnetic circuit analysis has proven useful in the design of inductors, transformers, and other special magnetic devices. In electric circuits, based on Kirchhoff's voltage and current laws (KVL and KCL), all branch currents and nodal voltages are normally determined. In magnetic circuits, the following questions arise: How much current is needed to magnetize or demagnetize a given core? What is the resultant flux? When do we obtain maximum flux?

The analysis of magnetic circuits will be based on the following two Maxwell's equations:

$$\oint_C \mathbf{H} \cdot dl = Ni \qquad \text{(Ampere's law)} \qquad \text{(A.9)}$$

$$\oint_C \mathbf{B} \cdot dS = 0 \qquad \text{(Gauss's magnetic law)} \qquad \text{(A.10)}$$

Equation (A.9) suggests that the total magnetic field in a closed path of length l must equal the total applied magnetomotive force. In other words, we may consider the right-hand side of this equation as a current source and the left-hand side as the resultant or induced mmf. This is analogous to state that the source voltage around a closed loop equals the total resultant voltage (drop) in the same loop. This is like dealing with KVL for mmf, rather than for emf.

Similarly, Eq. (A.10) suggests that the sum of the total magnetic flux, ϕ, into a closed region in space is equal to zero. This is analogous to KCL for ϕ rather than for i. In other words, we may state that the sum of all magnetic flux into a small region must be zero. It is obvious that the mechanism of the flow of flux in magnetic circuits is similar to the way current flows in electric circuits.

To further illustrate the similarities between magnetic and electric circuits, consider the equation for the flux derived in the previous section, namely:

$$\phi = \frac{Ni}{l} A \mu \qquad \text{(A.11)}$$

If we let a parameter known as reluctance $\Re$ be defined as

$$\Re = \frac{1}{\mu A} \qquad \text{(A.12)}$$

then the flux may be expressed as

$$\phi = \frac{Ni}{\Re} \qquad \text{(A.13)}$$

or

$$\begin{aligned} \text{mmf} &= Ni \\ &= \Re\phi \end{aligned}$$

(A.14)

Equation (A.14) reminds us of the way we define the resistivity for a conductor of length l, cross-sectional area A, and conductivity σ:

$$R = \frac{l}{\sigma A}$$

(A.15)

Moreover, Eq. (A.14) is similar to Ohm's law relating voltage drop to the resistor and current. Just as the resistance opposes the current flow in the conductor, the reluctance opposes the flow of the flux in the magnetic circuit.

The flux linkage is normally defined by

$$\lambda = N\phi$$

(A.16)

and the inductance is defined as

$$L = \frac{\lambda}{i}$$

(A.17)

In terms of magnetic circuit geometry, L can be written as

$$L = \frac{\mu A N^2}{l}$$

(A.18)

It is also customary to express the flux using the following relation:

$$\phi = \mathscr{P}Ni$$

(A.19)

where $\mathscr{P}$ is known as the *permeance*. In terms of the reluctance, the permeance is given by

$$\mathscr{P} = 1/\Re$$

(A.20)

Table A.1 summarizes the analogy between the electric and magnetic circuit parameters.

Table A.1   Analogy between Magnetic and Electric Circuits

Electric circuits	Magnetic Circuits
Electromotive force (emf) (volts)	Magnetomotive force (mmf) (Ni)
Current (I)	Magnetic flux (ϕ)
Voltage drop (volts)	Magneto volts (Hl)
Resistance, R	Reluctance, $\Re$
Current density, $J = I/A$	Flux density, $B = \phi/A$
KVL for emf	KVL for mmf
KCL for currents	KCL for flux
Conductance, $G = 1/R$	Permeance, $\mathscr{P} = 1/\Re$
Conductivity, σ	Permeability, μ
Conductors	Ferromagnetics
Insulators	Diamagnetics

EXAMPLE A.2

Consider a rectangular core with a winding N that turns as shown in Fig. A.5. Assume the cross-sectional areas of the side and top/bottom segments of the core are A_1 and A_2, respectively. Develop the electric circuit equivalence and determine the reluctance, flux, and inductance of this configuration.

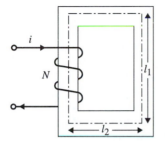

Figure A.5 Magnetic structure for Example A.2.

SOLUTION   Since the permeability of the core is much higher than that of the air, we assume that the entire flux produced by Ni is confined to the core. There are four segments (legs) in this core: the side segments, of the same reluctance $\Re_1$, and the top/bottom segments, also of the same reluctance $\Re_2$. The coil and the excitation current are represented as an electromotive source, as shown in Fig. A.6.

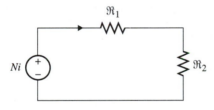

Figure A.6 Equivalent electric circuit of Fig. A.5.

The reluctances $\Re_1$ and $\Re_2$ are given by

$$\Re_1 = \frac{2l_1}{\mu A_1}$$

$$\Re_2 = \frac{2l_2}{\mu A_2}$$

where l_1 and l_2 are the mean lengths of the core as shown in Fig. A.5, and μ is the permeability of the core.

The flux produced in the winding is given by

$$\phi = \frac{Ni}{\Re_T}$$

where $\Re_T = \Re_1 + \Re_2$, and the inductance is given by

$$L = \frac{N^2}{\Re_T}$$

EXAMPLE A.3

Figure A.7 shows a three-legged magnetic structure made of material with $\mu = 3000\mu_o$. Winding of N turns is placed at the center leg and an air gap of width l_g is made in the right leg with the dimensions shown. Develop an equivalent electric circuit and determine the flux and the inductance of the winding.

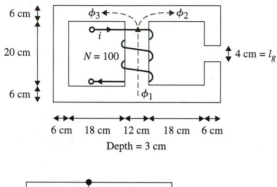

6 cm · 18 cm · 12 cm · 18 cm · 6 cm
Depth = 3 cm

Figure A.7 Magnetic structure for Example A.3.

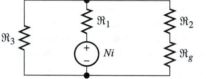

Figure A.8 Equivalent electric circuit of Fig. A.7.

SOLUTION   As in Example A.2, it will be assumed $\mu \gg \mu_o$ and the cross-sectional area of the air gap is the same as the cross-sectional area of the core. (Neglect fringing effects, to be discussed later.) The equivalent electric circuit is shown in Fig. A.8, where

$\Re_1$ = Reluctance of the center leg
$\Re_2$ = Reluctance of the right-leg core segment
$\Re_3$ = Reluctance of the left-leg core segment
$\Re_g$ = Reluctance of the free space (air)

From Fig. A.7, we obtain the mean length for these reluctances:

$$l_1 = 26 \text{ cm}$$

$$l_g = 4 \text{ cm}$$

$$l_2 = 2(27) + 22 = 76 \text{ cm}$$

$$l_3 = 2(27) + 26 = 80 \text{ cm}$$

The cross-sectional areas of the center leg, A_1, and the remaining segments, A_2 (including the air), are given by

$$A_1 = 36 \text{ cm}^2$$

$$A_2 = 18 \text{ cm}^2$$

The reluctances are calculated as follows:

$$\Re_1 = \frac{l_1}{\mu A_1}$$

$$= 6.89 \times 10^5$$

$$\mathfrak{R}_g = \frac{l_g}{\mu_o A_2}$$

$$= 1.77 \times 10^7$$

$$\mathfrak{R}_2 = \frac{l_2}{\mu A_2}$$

$$= 11.2 \times 10^4$$

$$\mathfrak{R}_3 = \frac{l_3}{\mu A_2}$$

$$= 11.78 \times 10^4$$

The total flux in the center leg is calculated from

$$\phi_1 = \frac{Ni}{\mathfrak{R}_T}$$

$$= 0.62 \text{ mWb}$$

where $\mathfrak{R}_T = (\mathfrak{R}_2 + \mathfrak{R}_g) \parallel \mathfrak{R}_3 + \mathfrak{R}_1 = 8.06 \times 10^5$.
The flux in $\mathfrak{R}_2$ and $\mathfrak{R}_3$ is given by

$$\phi_2 = \phi_1 \frac{\mathfrak{R}_3}{\mathfrak{R}_2 + \mathfrak{R}_3 + \mathfrak{R}_g}$$

$$= 4.07 \text{ mWb}$$

$$\phi_3 = \phi_1 \frac{\mathfrak{R}_2 + \mathfrak{R}_g}{\mathfrak{R}_2 + \mathfrak{R}_3 + \mathfrak{R}_g}$$

$$= 0.616 \text{ mWb}$$

The inductance of the winding is given by

$$L = N \frac{\phi_1}{i}$$

$$= \frac{N^2}{\mathfrak{R}_T}$$

$$= 12.41 \text{ mH}$$

As illustrated in the examples, the calculation of flux and inductance is straight-forward and quite simple. Even though these calculations are approximate, they often give satisfactory results. Some sources of inaccuracy in calculating the flux in the core can be summarized as follows:

1. Leakage flux into the low-permeability surrounding air is not exactly zero. This is because the conductivity of a current-carrying wire in an electric circuit is higher than the conductivity of surrounding air by 12 to 15 orders of magnitude, whereas the conductivity difference between magnetic materials and the surrounding medium is around 2 to 3 orders of magnitude. This is a fundamental difference between magnetic and electric circuits.

2. In air-gapped cores, the cross-sectional area of the air is not quite that of the core. This is due to the additional flux that exists outside of the volume directly between the core segments, which results in an increase in the

effective cross-sectional area of the gap. This phenomenon is known as the *fringing effect*, and the flux is called *fringing flux*.

3. Due to the nonlinear nature of ferromagnetic material, its permeability is not always constant under any current excitation. As a result, reluctances and, consequently, inductances exhibit a nonlinear relation between the mmf and flux. This is another fundamental difference between magnetic and electric circuits.

4. To simplify the evaluation of the closed-loop integral to obtain the magnetic field density, it is assumed that the flux path is an *average* or mean length. Even though the difference is not significant, it contributes to the error in the calculations.

5. The magnetic field density and intensity are not uniform throughout the cross-sectional area of the core. Unlike the cross-sectional area of the conductivity of an electric wire, which is very small when used to calculate the resistivity of the conductive material, the cross-sectional area in a magnetic circuit is not necessarily small. Hence, magnetic fields are not really uniform.

A.4   THE MAGNETIZING CURVE

Thus far, based on the constitutive relation $\mathbf{B} = \mu \mathbf{H}$, it has been assumed that, regardless of the size of the applied mmf, the permeability of the magnetic material is constant. This assumption is only true in the free space, where μ_o is constant. The permeability is highly nonlinear for iron and ferromagnetic materials, as will be illustrated in this section.

To understand the behavior of magnetic material, we first consider the relationship between the flux ϕ and magnetizing force Ni. Consider the structure made of a ferromagnetic material shown in Fig. A.9(a), with a variable mmf source. We start at point A in Fig. A.9(b) with the material that has never been magnetized, i.e., $\phi = i = 0$. Now we apply magnetizing force Ni to the coil; this will result in a proportional increase in

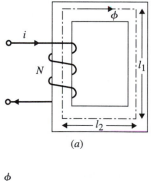

(a)

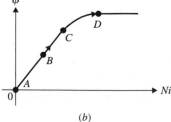

(b)

Figure A.9 Flux versus mmf in ferromagnetic material. (a) Magnetic structure excited by mmf force. (b) Typical ϕ vs. Ni plot for ferromagnetic material.

the magnetic flux to point B as shown on the curve. Notice that at this point, a small increase in mmf produces a large increase in the flux, ϕ. If we continue to increase the current, a point will be reached at which any additional increase in the mmf produces a relatively small increase in the flux, as shown at point C. At point D, any further increase in the current produces no change in the flux. At this point and beyond, the magnetic material is said to be in *saturation*. This curve is known as a *magnetizing curve*. The shape of the linear region and the magnitude of the saturation vary from one type of ferromagnetic material to another.

Another interesting property of a magnetic material is known as *hysteresis*. This phenomenon is illustrated in Fig. A.10, where the loop 1234561 is known as the *hysteresis loop*. To illustrate how this loop is created, let us assume that the applied magnetic force, **H**, is produced by a sinusoidally varying excitation current. First we assume that the material is magnetized and has reached its saturation point 1. Then when the current starts decreasing, the magnetizing current starts following a different path. At point 2, even though the magnetic force is zero, a magnetic flux remains in the material; B_R known as *residual flux* or *remnant magnetization*. This is how permanent magnets are made. At point 3, the material experiences a negative magnetic force that forces the flux to zero, i.e., demagnetization. The value that causes the material to demagnetize is known as *coercive mmf*, $\mathbf{H}_C$. The material reaches negative saturation at point 4. As **H** starts increasing again, the curve follows a different path until it reaches positive saturation once again at point 1. Points 5 and 6 correspond to remnant magnetization and coercive mmf, respectively. One important feature of the **B-H** curve is that **B** is not a single-valued function of **H**. In other words, the amount of flux at any given point in the **B-H** curve depends not only on the value of **H** at that point, but also on the previous value of **B** or on the history of **B**. As a result, ferromagnetic materials can be used as memory devices.

The physical explanation for the behavior of magnetic material can be clearly understood by studying the quantum mechanical forces between the atoms in the presence of a magnetic field. Only a brief qualitative description of this behavior is given in this section.

All magnetic materials, whether diamagnetic, paramagnetic, or ferromagnetic, possess net *magnetic moment* due to both the orbital and spinning motions of the electrons in their corresponding atoms. In the absence of an externally applied magnetic field, the degree to which the net magnetic moment exists varies between materials. In diamagnetic materials, the net magnetic moment is zero. In the presence of an external magnetic field, a net magnetic moment is created due to the combined interaction between the external magnetic field and spinning and orbital electrons, which results

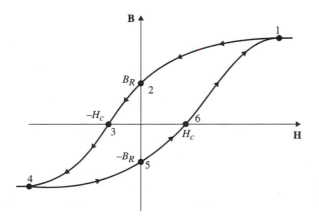

Figure A.10 B-H magnetizing curve.

in perturbation in their angular velocities. This effect of magnetization is very small in this type of material and too weak to be considered for any practical application. Moreover, diamagnetic material exhibits no permanent magnetization.

Unlike the case for diamagnetics, the magnetic moments due to orbiting and spinning electrons in paramagnetic material do not cancel completely. When an external magnetic field is applied, the alignment of magnetic moments in the direction of the magnetic field takes place. Consequently, an increase in the net flux density is obtained. Still, the resultant magnetic flux is quite small and the relative permeability of such material is close to 1.

Unlike diamagnetics and paramagnetics, ferromagnetic material shows a high degree of susceptibility to an externally applied magnetic field. In ferromagnetic materials, it has been experimentally shown that due to the strong forces exerted by the neighboring magnetic dipoles, which are produced from the spinning electrons in the atoms, all the magnetic dipoles are aligned in parallel, resulting in magnetized small regions known as *domains*. These domains range in linear dimension between 10^{-6} and 10^{-4} meters, and each contains about 10^{16} atoms.

Figure A.11 depicts a simplified symbolic domain model, with each domain containing a number of magnetic dipoles aligned in parallel to produce a magnetized domain, with the domains magnetized in different directions. The domains are clearly separated by regions of hundreds of atoms. Because of the random nature of the direction of magnetization, the net magnetization of the ferromagnetic material is zero.

Under an externally applied magnetic field, some domains align themselves in the direction of the applied field, increasing the magnetic flux density. The stronger the applied field, the more domains are aligned and the larger the magnetic flux density gets. Depending on the type of the ferromagnetic material, there exists a certain magnetic force beyond which the process of alignment is not reversible. In other words, once the external field is removed, the magnetized domains do not return to their original position. This can be seen in the **B-H** curve by noting that B_R does not become zero once the external field is removed. This phenomenon can be easily explained if we recognize that if energy is needed to orient or rotate the magnetic dipoles in the direction of the externally applied field, energy must also be applied to rotate these dipoles to their original random positions. This is why a *coercive force*, H_C, must be applied to demagnetize the material. Typical values of H_C for permanent magnets range from one hundred to thousands of amperes per meter. Another way to demagnetize a ferromagnetic material is to

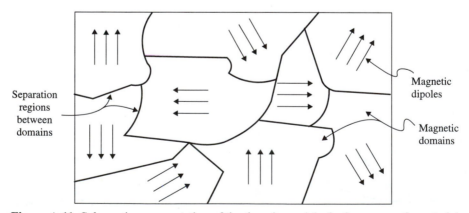

Figure A.11 Schematic representation of the domain model of a ferromagnetic material.

raise its temperature to a level that causes the domains to disorganize and return to their random orientation. The temperature at which this occurs is known as the *Curie temperature, T_c*. Typical values of T_c range from 200 to 1000°C. Another important point to note is that once all domains are aligned in one direction, the ferromagnetic material becomes saturated and starts behaving like paramagnetic material, i.e., magnetically transparent. At this point, its permeability decreases to μ_o.

A.5 INDUCTORS

Inductance, whether self- or mutual, results from the application of a third Maxwell's equation known as Faraday's law, which is given by

$$\nabla \times \mathbf{E} = \frac{-\partial \mathbf{B}}{\partial t} \quad \text{(Faraday's law)} \tag{A.21}$$

By integrating Eq. (A.21) over a specified surface area, applying Stokes's theorem to convert a surface integral to a line integral around a path that bounds this surface, and substituting for $\mathbf{B} = \phi/A$, Eq. (A.21) can be rewritten as follows:

$$\oint_C \mathbf{E} \cdot dl = -\frac{d\phi}{dt} \tag{A.22}$$

If we assume the conducting coil has N turns, then the total flux bound by the surface integral through which the magnetic field exists is given by $N\phi$, which is known as the *flux linkage*, as defined in Eq. (A.16). Recognizing the left-hand side of Eq. (A.22) as an induced voltage (we also assume the electric field exists only in the conducting wire), we may write Eq. (A.22) as

$$e = -N\frac{d\phi}{dt}$$
$$= -\frac{d\lambda}{dt} \tag{A.23}$$

The minus sign is expression of Lenz's law, which indicates that the induced voltage opposes the increase in the flux flow.

EXAMPLE A.4

Consider the toroidal structures shown in Fig. A.12.

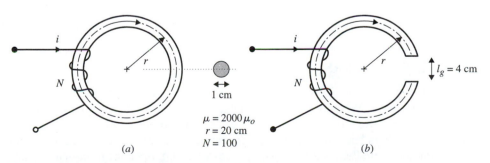

Figure A.12 Magnetic structures for Example A.4.

(**a**) Determine the inductances for both the ungapped and gapped magnetic structures made from the same material shown in Fig. A.12(*a*) and (*b*), respectively.
(**b**) If the permeability of the magnetic material is increased by 15%, determine the inductances of part (*a*).

SOLUTION   (**a**) The cross-sectional area is given by

$$A = \pi r^2$$

$$= 0.7854 \times 10^{-4}$$

The total reluctance of the ungapped core is given by

$$\mathfrak{R}_T = \frac{2\pi r}{\mu A}$$

$$= 159154.6 \text{ A-turn/Wb}$$

The reluctance of the gapped core is given by

$$\mathfrak{R}_{T,\,gap} = \frac{2\pi r - l_g}{\mu A} + \frac{l_g}{\mu_o A}$$

$$= 4.0524 \times 10^8 \text{ A-turns/Wb}$$

Hence, the inductances of the gapped and ungapped cores are given by

$$L_{gap} = \frac{N^2}{\mathfrak{R}_{T,\,gap}}$$

$$= 24.68 \ \mu H$$

$$L = \frac{N^2}{\mathfrak{R}_T}$$

$$= 62.8 \text{ mH}$$

(**b**) If μ is increased by 15%, then the new reluctances for $\mu = 23000\mu_o$ are

$$\mathfrak{R}_T = 1.384 \times 10^5 \quad \text{A-turn/Wb}$$

$$\mathfrak{R}_{T,\,gap} = 3.524 \times 10^8 \quad \text{A-turn/Wb}$$

resulting in the following new gapped and ungapped core inductances:

$$L_{gap} = 28.38 \ \mu H$$

$$L = 72.2 \text{ mH}$$

It is important to notice that an increase of 15% in μ results in an increase of 15% and 0.25% in the inductances of the ungapped and gapped core structures, respectively.

A.6 TRANSFORMERS

The typical structure of a transformer with two windings on a common core is shown in Fig. A.13. When one of the windings with N_1 turns (primary) is excited by an external source, an induced voltage is created in the other winding with N_2 turns (secondary). This is because both windings share the same magnetic core and, consequently, link the same flux. Depending on the power rating, operating frequencies, and intended

Figure A.13 Magnetic structure of a two-winding transformer.

application, various types and shapes of core structures are used. Line-frequency transformers (50 Hz, 60 Hz, or 400 Hz) are normally used for isolation and stepping down/up the primary voltage. Depending on whether they are used for distribution, transmission, or household applications, the power rating varies widely. In power electronics applications, high-frequency transformers are very common, whose functions are mainly to transfer voltage, provide isolation, and store energy. Special types of transformers are used in instrumentation for sensing purposes. For example, *potential* and *current* transformers are used to sample a high primary voltage and a high line current, respectively. Regardless of their application, transformers generally have four relevant ratings: voltage, current, apparent power, and frequency. The apparent power (the product of the rms current and the rms voltage) sets the limit on the maximum I^2R loss in the transformer windings in order to limit heating of the transformer coils, which could damage the insulation or drastically shorten the transformer's life. Transformer voltage and frequency ratings serve two purposes: to limit the core losses and to prevent the transformer from saturation, as explained later in this section. Depending on the number of windings on a shared magnetic circuit, transformers can be single- or multiple-phase types. Because of their widespread applications, we will study only the single-phase and three-phase transformers.

A.6.1 Ideal Transformers

The operation of an ideal transformer is easy to understand because it's assumed that both coil and core losses are negligible and the two windings (primary and secondary) are perfectly coupled; i.e., they are linked with the same flux. Since the majority of the applications of a transformer are for voltage stepping and voltage isolation, its terminal voltages are important parameters. Because of this, additional expressions are needed to relate the voltage with the magnetic field and flux. The key relation in understanding the transformer operation is Faraday's law, which is given by

$$\oint_C \mathbf{E} \cdot dl = -\frac{d\phi}{dt} \qquad \text{(Faraday's law)} \qquad (A.24)$$

The left-hand side of Eq. (A.24) is an expression of the induced voltage or emf resulting from the presence of the magnetic flux, ϕ. In other words, Faraday's law states that the induced voltage across a conducting wire is proportional to the rate of change of the flux through the wire with respect to time. If we assume the conducting wire has N turns, then Eq. (A.24) can be rewritten in terms of the total induced voltage as follows:

$$v(t) = -N\frac{d\phi}{dt} \qquad (A.25)$$

Applying Eq. (A.24) to the transformer shown in Fig. A.13, we obtain

$$v_1(t) = -N_1 \frac{d\phi_1}{dt} \tag{A.26}$$

$$v_2(t) = -N_2 \frac{d\phi_2}{dt} \tag{A.27}$$

where ϕ_1 and ϕ_2 are the fluxes in the primary and secondary windings, respectively. Since perfect coupling is assumed, ϕ_1 and ϕ_2 must be the same; consequently, the terminal voltage relation for the transformer is given by

$$\frac{v_1}{v_2} = \frac{N_1}{N_2} \tag{A.28}$$

Since we also assume no losses, the total input and output powers must be equal; hence, the following terminal current relation can be obtained:

$$\frac{i_1}{i_2} = \frac{N_2}{N_1} \tag{A.29}$$

As stated earlier, one function of the transformer is impedance transformation. If an impedance Z_L is connected across the secondary winding, then it can be easily shown that the input impedance seen in the primary winding is given by

$$Z_{\text{in}} = Z_L \left(\frac{N_1}{N_2} \right)^2 \tag{A.30}$$

This equation is useful when the entire circuit is to be reflected from one side of the winding to the other side. This approach could simplify the analysis of complex circuits.

To identify the direction of the windings, whether clockwise or counterclockwise, a "dot" convention has been adopted. If the positive reference direction of either voltage is applied to one winding, then the dotted end of the other winding is positive. For example, in Fig. A.14(a) and (b) the windings for the primaries are wound in the counterclockwise direction and the secondaries are wound in the counterclockwise and clockwise directions, respectively.

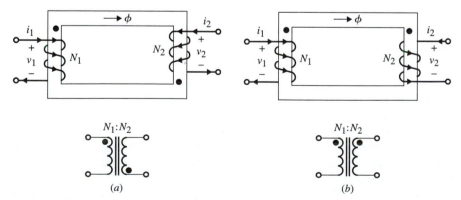

(a)   (b)

Figure A.14 Transformer "dot" notation. (a) The primary and secondary are wound in the counterclockwise direction. (b) The secondary is wound in the clockwise direction.

EXAMPLE A.5

Consider an RLC load connected to the secondary side of a transformer in a converter circuit as shown in Fig. A.15(a). Draw an equivalent circuit by reflecting the entire load circuit to the primary side.

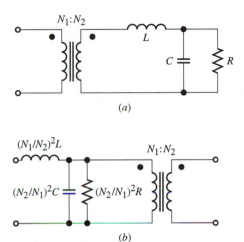

Figure A.15 (a) Transformer circuit with LRC used. (b) Equivalent circuit of (a) with the entire secondary side reflected to the primary side.

SOLUTION   The resultant configuration is shown in Fig. A.15(b). The terminal voltage and current are scaled by N_1/N_2 and N_2/N_1, respectively.

A.6.2   Nonideal Transformers

The practical transformer is far from ideal due to the following nonidealities that are always present in the magnetic structure.

Nonperfect Coupling

Practical transformers are not perfectly coupled. This is because the flux linkage between the two windings is not the same; some flux leaks into the surrounding medium. As stated earlier, unlike in electric circuits, the air permeability is only a few orders of magnitude smaller than the core permeability, resulting in a small flux leaving the magnetic path and returning through the air as shown in Fig. A.16. This leaked flux is the origin of the

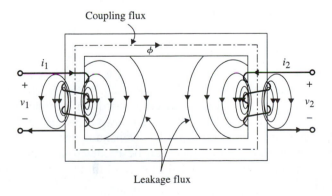

Figure A.16 Transformer magnetic structure illustrating coupling and leakage flux.

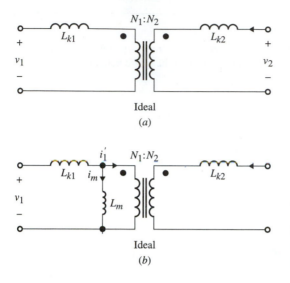

Figure A.17 Transformer model including (*a*) primary and secondary leakage inductances, and (*b*) magnetizing inductance reflected in the primary side.

imperfection in coupling. To simplify the analysis and design of transformer circuits, it is always useful to come up with models for any nonidealities. The leakage flux in the primary and secondary windings is normally modeled by two series inductive elements known as *leakage inductances*, L_{k1} and L_{k2}, respectively, as shown in Fig. A.17.

It is clear from Fig. A.17(*a*) that because of the presence of L_{k1} and L_{k2}, the terminal voltage relation given in Eq. (A.28) is no longer valid. Voltage drops are present across these inductances, resulting in energy loss and, therefore, less efficiency. The need to understand leakage effects and their origins is not due to concern for efficiency. In fact, when high-permeability cores are used with good coupling, it is common to design transformers with efficiency above 90%. The major concern is with the adverse effects these leakage fields have on the overall transformer performance. This concern is more serious in high-frequency power electronics applications, where the voltage drops across these leakage inductances become very large, causing regulation problems. An additional drawback in such applications is when the leakage inductors form resonating circuits with the distributed capacitances of the transformer windings and other diode and transistor junction capacitors, causing high voltage and current peaks that could damage the power switching devices and other circuit components.

Modeling the leakage effects in transformers as linear, lumped inductor elements in series with both windings with assumed constant values for a large range of operating frequencies is very accurate since the major portion of the closed leakage flux is through the air with constant permeability.

Finite Permeability

The permeability of the core is not infinite. Therefore, in order to produce a flux in the secondary winding, the primary winding must be excited by Ni. Even when the secondary winding is open-circuited, the primary current needed to magnetically couple the secondary winding will not be zero. This current is known as *magnetizing current* and flows through the *magnetizing inductance, L_m*. Depending on the winding at which L_m is measured, magnetizing inductance can be modeled at both windings. Normally, the magnetizing inductance is modeled as a shunt finite element in the primary side as shown in Fig. A.17.

The magnetizing current, i_m, shown in Fig. A.17(b) is the current required to create the flux in the primary winding. Theoretically speaking, if μ of the core is infinite, then it is possible to produce the flux in the primary winding, consequently coupling the secondary winding without magnetizing current. This represents the case for an ideal transformer with infinite L_m. Note that under dc excitation, the input current to the transformer, i_1', is zero since the magnetizing inductance becomes short-circuited.

Core Losses

Nonideal transformers experience two types of core losses: hysteresis and eddy current losses, to be discussed in Section A.6.6. Because of these losses, additional current to i_m will flow in the primary winding, even when the secondary winding is open-circuited. This current is known as *core current*, i_{core}. Figure A.18 shows a modified equivalent transformer model including a shunt resistor, R_{core}, across L_m to model the core losses.

Copper Loss

Due to the current conduction in both winding coils of the transformer, usually additional equivalent resistors, R_{Cu1} and R_{Cu2}, are placed in series with L_{k1} and L_{k2} to model the copper losses in the primary and secondary windings, respectively, as shown in Fig. A.19. A more simplified model, known as a transformer π model, can be obtained by reflecting the secondary-side circuit to the primary side as shown in Fig. A.20. Furthermore, if we assume that the magnetizing current is smaller than the terminal current, i_1, it is possible to further simplify the transformer model of Fig. A.20 as shown in Fig. A.21. The reflected equivalent resistance and inductance are given by

$$R_{eq} = R_{Cu1} + \left(\frac{N_1}{N_2}\right)^2 R_{Cu2}$$

$$L_{k,eq} = L_{k1} + \left(\frac{N_1}{N_2}\right)^2 L_{k2}$$

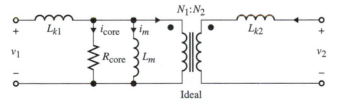

Figure A.18 Transformer model including the core losses and the leakage and magnetizing inductances.

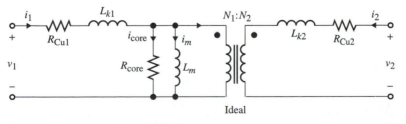

Figure A.19 Transformer model including the core losses, leakage and magnetizing inductances, and copper losses.

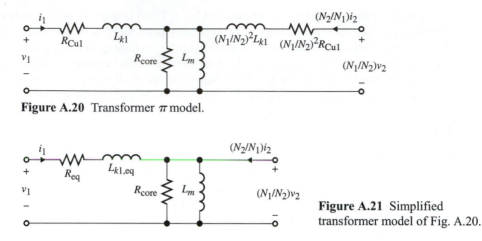

Figure A.20 Transformer π model.

Figure A.21 Simplified transformer model of Fig. A.20.

We should point out that the parasitic capacitance has not been included in the model since its effect becomes dominant only at extremely high frequencies.

A.6.3 Transformer Equations

In the previous section, we described the model of a nonideal transformer using its leakage and magnetizing inductances as well as its coil and core losses. However, a more useful way to represent a nonideal transformer is by describing its terminal voltages and currents as linear time-invariant coupled-circuit differential equations in terms of self- and mutual inductances.

Consider a two-winding transformer as shown in Fig. A.22 including the effect of the primary and secondary leakage flux $N_1 i_1$ and $N_2 i_2$, respectively. The mutual fluxes ϕ_{m1} and ϕ_{m2} produced by $N_1 i_1$ and $N_2 i_2$ are also shown in the figure. The total fluxes produced at the primary and secondary windings are given in Eqs. (A.31) and (A.32), respectively.

$$\phi_1 = \phi_{l1} + \phi_{m1} + \phi_{m2} \tag{A.31}$$

$$\phi_2 = \phi_{l2} + \phi_{m1} + \phi_{m2} \tag{A.32}$$

Applying Faraday's law to Eqs. (A.31) and (A.32), we obtain

$$v_1 = N_1 \frac{d\phi_1}{dt}$$
$$= N_1 \frac{d}{dt}(\phi_{l1} + \phi_{m1}) + N_1 \frac{d\phi_{m2}}{dt} \tag{A.33}$$

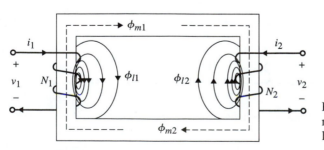

Figure A.22 Transformer magnetic core with mutual and leakage fluxes.

$$v_1 = N_2 \frac{d\phi_2}{dt}$$

$$= N_2 \frac{d}{dt}(\phi_{l2} + \phi_{m2}) + N_2 \frac{d\phi_{m1}}{dt}$$

(A.34)

The self-inductances of the primary and secondary are defined as follows:

$$L_1 = \frac{N_1(\phi_{l1} + \phi_{m1})}{i_1}$$

(A.35)

$$L_2 = \frac{N_2(\phi_{l2} + \phi_{m2})}{i_2}$$

(A.36)

The mutual inductances at the primary and secondary windings are defined as follows:

$$L_{21} = \frac{N_1 \phi_{m2}}{i_2}$$

(A.37)

$$L_{12} = N_2 \frac{\phi_{m1}}{i_1}$$

(A.38)

If the magnetic structure is reciprocal, then $L_{12} = L_{21} = M$; hence, Eqs. (A.37) and (A.38) give

$$L_1 = L_{k1} + \frac{N_1}{N_2} M$$

(A.39)

$$L_2 = L_{k2} + \frac{N_2}{N_1} M$$

(A.40)

From Eqs. (A.33), (A.34), (A.39), and (A.40), the transformer equations in terms of terminal voltages and currents are given by

$$v_1 = L_1 \frac{di_1}{dt} + M \frac{di_2}{dt}$$

(A.41)

$$v_2 = M \frac{di_1}{dt} + L_2 \frac{di_2}{dt}$$

(A.42)

A.6.4 Leakage Inductances

Calculating the leakage inductance of a given magnetic geometry for an inductor or transformer is not a straightforward problem. Detailed dimensions of the windings and their physical location in the magnetic structure must be known; only then can an estimate of the leakage inductance be obtained. To minimize the transformer leakage inductances, normally the two windings are placed one on top of the other to enhance the flux linkage. It can be shown that if the number of turns, winding thickness (build-up), and insulation thickness (space between windings) are decreased and the winding length is increased, a smaller leakage inductance can be obtained (see Problem A.21). Furthermore, it is possible to experimentally determine the total leakage inductance of a transformer by

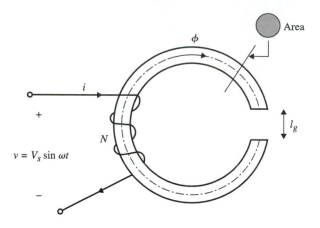

Figure A.23 Toroidal inductor with air gap.

performing two measurements known as open- and short-circuit tests, as illustrated in Problem A.15.

A.6.5 Saturation

As stated earlier, when the magnetic material saturates, a large change in the excitation current in the primary of the transformer results in no change in the coupling flux. The core becomes magnetically transparent. In limited applications, such as magnetic amplifiers and fluorescent lamp ballasts, the saturation property of the magnetic material is essential. Otherwise, saturation is avoided in most applications. Voltage, current, and frequency may cause the core to saturate as illustrated in the air-gapped toroidal inductor shown in Fig. A.23.

The inductance of Fig. A.23 can be shown to be independent of the permeability of the material if we assume $l_g \gg l\mu_o/\mu$. Since at saturation the permeability of the core reduces to that of the air, this relation no longer holds. Consequently, the inductance becomes a function of both permeability and material type, and therefore unstable. The question normally asked is, when does the core saturate? It is obvious that the core saturates when its magnetic density, **B**, exceeds B_{sat} of the material. Since the external terminals of any transformer and inductor are voltages and currents, it is more useful to address the question of saturation in terms of these parameters. If we assume the excitation voltage in Fig. A.23 is a sinusoid given by $v_s(t) = V_s \sin \omega t$, then it can be shown that the following relation must hold to avoid saturation:

$$\frac{V_s}{N} \le 2\pi B_{sat} A f \tag{A.43}$$

where A is the core cross-sectional area, V_s is the maximum voltage, and f is the frequency of the applied signal. The left-hand side of Eq. (A.43) gives the maximum volts per turn allowed to prevent saturation. If the peak voltage is fixed, it is possible to set a limit on the minimum frequency necessary to avoid saturation, which is given by

$$f_{min} = \frac{V_s}{2\pi N A B_{sat}} \tag{A.44}$$

A.6.6 Magnetic Losses

Two types of losses exist in magnetic structures: *core losses* and *copper losses*. The copper losses exist in the winding coils due to the current conduction and are given by

$$P_{Cu} = I_{rms}^2 R_{Cu} \tag{A.45}$$

The core losses are attributed to both hysteresis losses and eddy current losses. The former are the result of unrecovered energy spent in the constant rotation or realignment of magnetic domains within the magnetic material. The larger the area enclosed by the hysteresis loop, the higher the loss. The loss due to the hysteresis loop is proportional to the frequency and magnetic field:

$$P_{hys} \propto f B_R^n \quad 2 \geq n \geq 1.5 \tag{A.46}$$

The energy lost due to eddy currents is proportional to the square of the frequency and the square of the magnetic field:

$$P_{eddy} \propto f^2 B^2 \tag{A.47}$$

The larger the size of eddy current loops, the higher the losses within the core. To break these loops into smaller ones, it is customary to use small stripes on lamination cores.

A.6.7 Core Material and Types

The choice of the appropriate type and shape of core material has become an extremely important consideration in designing inductors and transformers for today's power circuits. This is in large part due to a substantial increase in the operating frequencies of power converters, which have reached 5–10 MHz. This push for higher and higher frequencies is mainly driven by the ever-increasing demand for smaller and lighter power converters. At these frequencies, smaller magnetic components can be used; also, the effect of the leakage inductances, self-capacitance of the winding, and inter-winding capacitance could become more significant. As a result, the choices of magnetic material, core geometry, conducting wires, and fabrication techniques are challenging considerations. The following four types of material are currently in wide use in the design of magnetic circuits.

1. *Air-core*. This type of core is used when magnetic field distortion due to the nonlinearity of the magnetic structure is not desirable. Air-core structures produce stable inductor and transformer values, since μ_o is constant. The winding is normally wound around a prefabricated form to sustain it.

2. *Laminated iron*. This material is used to minimize the effect of eddy current losses. This is achieved by coating lamination with some sort of electric insulation. Iron lamination cores are used mostly in line- and low-frequency applications. Lamination thickness and number, as well as the size of the air gap, are some design parameters.

3. *Powdered iron*. Depending on the core geometry, this type is normally used for frequencies up to 1 MHz and in high-Q inductors. The core losses and permeability are normally controlled by the size and composition of magnetic particles.

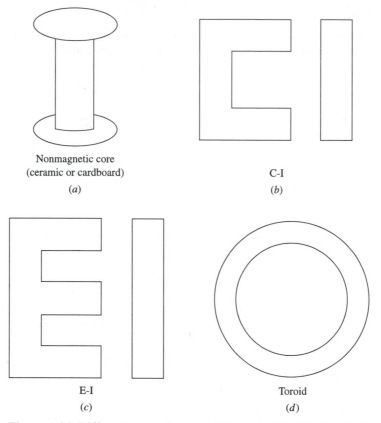

Figure A.24 Different types of cores. (*a*) Nonmagnetic. (*b*), (*c*), (*d*) Magnetic.

4. *Ferrite*. Because it has high permeability and high resistivity to eddy currents, ferrite core material is used in high-frequency applications up to 100 MHz. However, its B_{sat} is much lower than that of laminated or powdered iron core materials. As a result, ferrite cores are limited to lower-current applications. They also have a relatively low Curie temperature. Nevertheless, the ferrite core type is the most popular in the design of high-frequency power electronics.

Figure A.24 shows different shapes of cores widely used in today's power electronics applications. Generally speaking, the shape and geometry of magnetic cores are dictated by various electrical and electromechanical requirements. Examples of core arrangements are pot, toroid, E-I, and C-I.

A.7 THREE-PHASE TRANSFORMERS

Three-phase systems are the most popular ac electrical systems in the world, whether generated, transmitted, or distributed. This is in large part due to the economic advantages in the distribution and generation of three-phase systems using rotating machines. Moreover, unlike the power in single-phase systems, the total instantaneous power in a three-phase system is constant. This interesting feature of three-phase

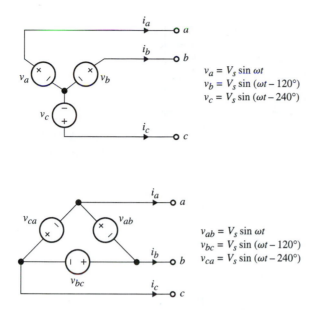

$$v_a = V_s \sin \omega t$$
$$v_b = V_s \sin (\omega t - 120°)$$
$$v_c = V_s \sin (\omega t - 240°)$$

Figure A.25 Y-connected three-phase system.

$$v_{ab} = V_s \sin \omega t$$
$$v_{bc} = V_s \sin (\omega t - 120°)$$
$$v_{ca} = V_s \sin (\omega t - 240°)$$

Figure A.26 Δ-connected three-phase system.

systems makes them attractive for power rotating machinery since it causes less vibration compared to a single-phase system, whose instantaneous power frequency is twice the line frequency.

Because of their importance in the implementation of three-phase systems, three-phase transformers are important devices in modern power electronic systems. Hence, a brief discussion of three-phase transformers is presented in this section.

Depending on whether the voltage source has a common reference point, known as the *neutral point*, voltages in three-phase systems can be represented as *wye* (Y) or *delta* (Δ) connections. In both system configurations, the voltage sources are equal in amplitude and frequency and have a 120° phase shift between each other. The Y-connected three-phase system is shown in Fig. A.25, where currents i_a, i_b, and i_c are line currents and voltages v_{ab}, v_{bc}, and v_{ca} are line voltages. It is apparent from Fig. A.25 that the line currents are the same as the phase currents, and it can be easily shown that the amplitudes of the line voltages are increased by $\sqrt{3}$ and phase-shifted by 30° from the phase voltages, v_a, v_b, and v_c. Figure A.26 shows the alternative way of connecting three-phase voltages, the Δ connection. Notice the line voltages and the phase voltages are the same, whereas the line currents i_a, i_b, and i_c have magnitudes $\sqrt{3}$ times the magnitude of the phase currents i_{ba}, i_{cb}, and i_{ac} and are shifted by 30°.

One easy and straightforward way to make a three-phase transformer is to take three single-phase transformers and connect them in various Δ and Y configurations. An alternative way is to use a common core to create a set of three windings as shown in Fig. A.27. For economy and to increase efficiency and power density, this approach is more common these days. Like the voltage source connections in a three-phase system, each of the windings in the three-phase transformer can also be connected in Δ and Y configurations. As a result, there are four possible three-phase transformer connections: Δ-Δ, Δ-Y, Y-Δ, and Y-Y. Examples of a Y-Δ connection created from three single-phase transformers and a common core structure are shown in Fig. A.28(*a*) and (*b*), respectively, with the equivalent transformer schematic shown in Fig. A.28(*c*).

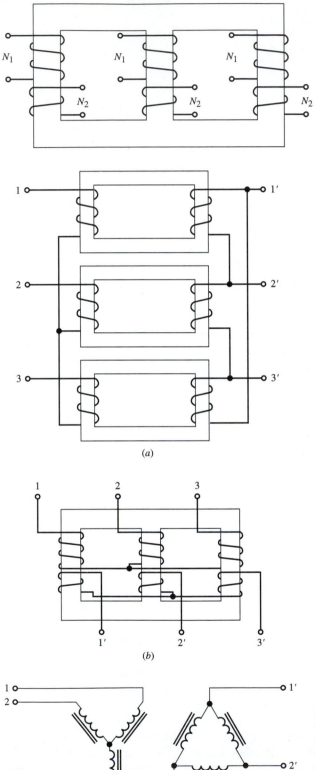

Figure A.27 Three-set of windings on a common core.

(a)

(b)

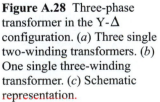

(c)

Figure A.28 Three-phase transformer in the Y-Δ configuration. (a) Three single two-winding transformers. (b) One single three-winding transformer. (c) Schematic representation.

PROBLEMS

A.1 Consider a coaxial line consisting of a solid inner and outer conductors of thickness a and $(c-b)$, respectively, each carrying current I, as shown in Fig. PA.1. Show that the magnetic field is expressed in the following function:

$$\mathbf{H} = \begin{cases} \dfrac{Ir}{2\pi a^2} & 0 < r \le a \\[2mm] \dfrac{I}{2\pi r} & a < r \le b \\[2mm] \dfrac{I}{2\pi r}\dfrac{(c^2 - r^2)}{(c^2 - b^2)} & b < r \le c \\[2mm] 0 & c < t \le \infty \end{cases}$$

Assume the currents are uniformly distributed and the current into the solid inner conductor equals the current in the outer shell.

A.2 By using the magnetic circuit approach, determine the inductances of the magnetic structures shown in Fig. PA.2. Assume $\mu = 2000\,\mu_o$ and the depth of the core cross-section is 4 cm.

A.3 Repeat Problem A.2 for Figure PA.2(a) and (d) by assuming the permeability of the core is decreased by 20%. Determine the percentage change in the new inductance values.

A.4 Consider Fig. PA.2(e) with $i = 4$ A flowing in the 250-turn coil. Determine the flux, flux density, and magnetic field intensity in each leg segment.

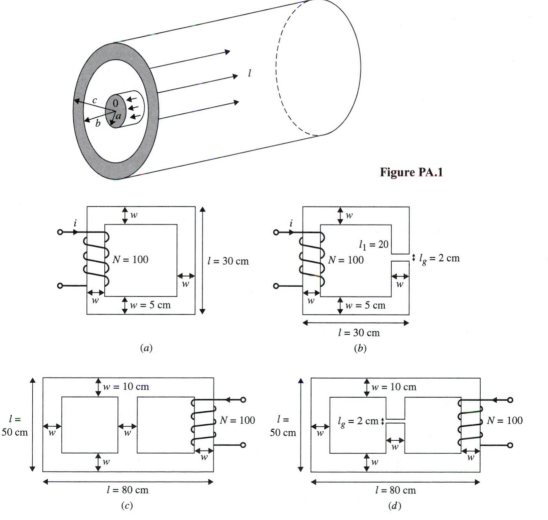

Figure PA.1

Figure PA.2

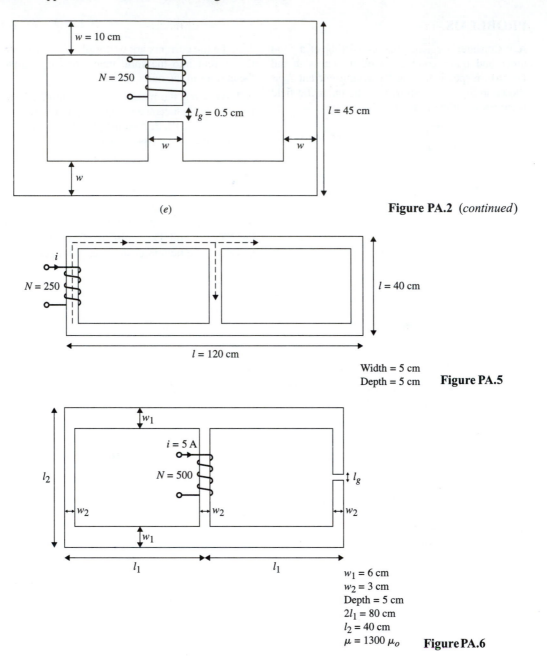

(e)   **Figure PA.2** (*continued*)

Width = 5 cm
Depth = 5 cm   **Figure PA.5**

$w_1 = 6$ cm
$w_2 = 3$ cm
Depth = 5 cm
$2l_1 = 80$ cm
$l_2 = 40$ cm
$\mu = 1300\,\mu_o$ **Figure PA.6**

A.5 Consider the magnetic structure shown in Fig. PA.5 with $\mu = 3000\,\mu_o$. Determine the coil current that will produce 0.4 T in the center leg.

A.6 It is desired to produce a flux of 1.5 mWb in the air gap of the core shown in Fig. PA.6. Determine the air gap that will establish this flux.

A.7 Determine the inductance of a coil wound in air with the dimensions shown in Fig. PA.7 for $l_g = 0$, 0.5, and 1 cm.

A.8 Consider a toroidal coil structure with N turns wound around a circular cross-sectional core as

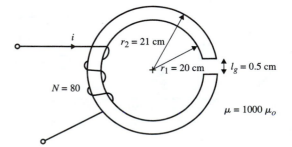

Figure PA.7

shown in Fig. PA.8(*a*). Assume the resistance of the copper coil is modeled by R_{Cu}, resulting in a possible equivalent circuit as shown in Fig. PA.8(*b*).

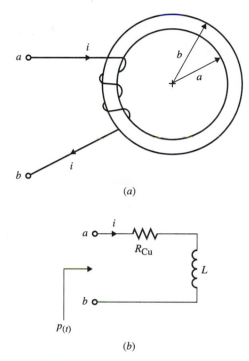

(*a*)

(*b*)

Figure PA.8

(**a**) If the excitation current is sinusoidal, $i(t) = I_p \sin \omega t$, show that the instantaneous power at the terminals *ab* is given by

$$p(t) = R_{Cu}I_p^2 \sin^2 \omega t + \frac{N^2 \mu A I_p^2 \omega}{\pi(b+a)} \cos \omega t \sin \omega t$$

(**b**) Show that the average input power between the terminals *ab* is given by

$$P_{ave} = R_{Cu}\frac{I_p^2}{2}$$

(**c**) Show that the total instant energy stored in the magnetic field is given by

$$W = \frac{\pi^2(b-a)^2(b+a)}{8\mu}$$

A.9 Consider the air-gapped toroidal core with the dimensions shown in Fig. PA.9.
(**a**) Determine the inductance L.
(**b**) Repeat part (*a*) for $\mu = 500\mu_o$.
(**c**) Determine the minimum length of the air gap, l_g, so that the inductances obtained in parts (*a*) and (*b*) are within $\pm 2\%$.

A.10 Figure PA.10 shows a square core with air gap, l_g. Assume the applied voltage is sinusoidal with peak voltage 3000 V and frequency 60 Hz, and $\mu = 4500\mu_o$.
(**a**) Determine the maximum flux ϕ.
(**b**) If $B_{sat} = 2.2$ T, determine the minimum frequency that will keep the core out of saturation.

A.11 The fringing effect and leakage flux could affect the value of the inductance. Consider the magnetic circuit with air gap given in Fig. PA.11.
(**a**) Calculate L with leakage flux and fringing effect negligible.
(**b**) Repeat part (*a*) by assuming that the air-gapped effective area is increased by 5%.
(**c**) Repeat part (*b*) by further assuming that 5% of the produced flux leaks to the surrounding air.
(**d**) What is the percentage change of the calculated inductances of parts (*b*) and (*c*) with respect to part (*a*)?

A.12 A group of electromechanical devices known as actuators consist of fixed and movable sections of ferromagnetic material. When excited by injecting current, *i*, the resultant magnetic field magnetizes the movable part and consequently lifts a mechanical load. Figure PA.12 shows one possible ferromagnetic structure for an actuator. Assume $B_{sat} = 1.2$ T and $\mu = 2000\mu_o$, and ignore leakage flux.

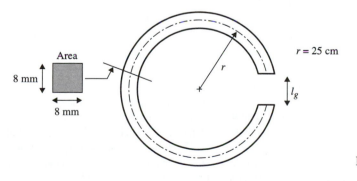

Figure PA.9

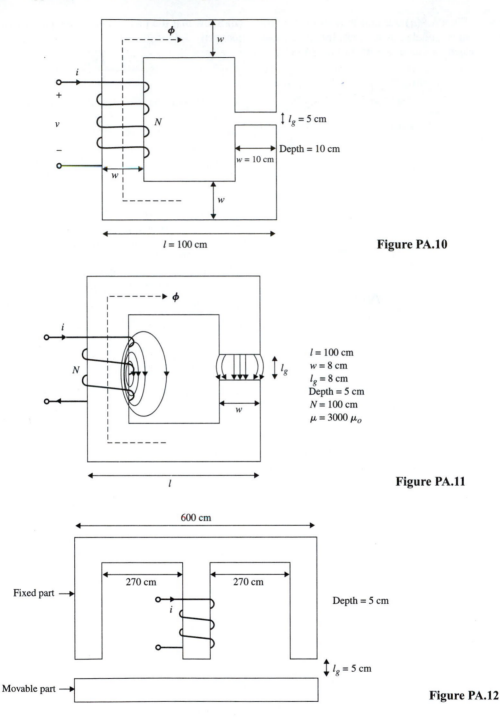

Figure PA.10

Figure PA.11

Figure PA.12

(a) Determine the maximum air-gap flux produced in the center leg.

(b) Determine the current i that would produce a flux density equal to $0.5B_{sat}$ in each of the two side air gaps.

A.13 Determine the maximum air gap, l_g, needed to establish $0.8B_{sat}$ in the center leg for $i = 10$ A in Problem A.12.

A.14 Consider a transformer with a turn ratio N_1/N_2 = 200 and rated at 240 kVA (44 kV/220 V) used to step down a 60 Hz voltage in a distributed system.

(a) Determine the rated primary and secondary currents.

(b) Determine the load impedance seen between the primary terminals when the load is fully loaded.

A.15 **(a)** A 60 Hz, 110 VA rms voltage is applied to the primary of a transformer with the secondary winding left open-circuited. The following two measurements are obtained:

(*i*) The primary current is found to be 5 A rms.

(*ii*) The average (real) power between the primary terminals is found to be 80 W.

This test is known as the open-circuit test. Use the transformer equivalent circuit shown in Fig. A.20 to determine the core resistance R_{core} and the magnetizing inductance L_m. (*Hint:* Assume the voltage drops across R_{Cu1} and L_{k1} are negligible.)

(b) With the same applied voltage as in part (*a*), the output voltage is shorted and the short-circuit current is measured and found to be 150 mA rms, and new measurements are taken as follows:

(*i*) The primary current is found to be 40 mA rms.

(*ii*) The average power at the primary terminal is found to be 25 W.

This test is known as the *short-circuit test*. Determine the total copper resistance, R_{Cu}, given by

$$R_{Cu} = R_{Cu1} + \left(\frac{N_1}{N_2}\right)^2 R_{Cu2}$$

(*Hint:* Assume the current in the magnetizing circuit, R_{core} and L_m, is negligible.)

A.16 Because of the series impedance in the primary side of the transformer, the output voltage will change under various load conditions, even if the input voltage remains constant. The degree to which the output voltage change between low-load and full-load conditions is known as *load regulation*, LR, defined as

$$LR = \left| \frac{v_{o,l} - v_{o,f}}{v_{o,f}} \right| 100\%$$

where $v_{o,l}$ and $v_{o,f}$ are the low- and full-load output voltages, respectively. This parameter is very important to the design of transformers.

(a) Using the simplified transformer equivalent circuit given in Fig. PA.16 with load impedance Z_l, show that LR is given by

$$LR = \frac{Z_s Z_m (Z_{l,l} - Z_{l,f})}{(Z_m Z_{l,l} + Z_s Z_m + Z_s Z_{l,l}) Z_{l,f}}$$

where $Z_{l,f}$ and $Z_{l,l}$ indicate the load impedance at full- and low-load conditions, respectively.

(b) If the low-load condition occurs at $Z_{l,l} = \infty$ and assuming $R_{core} = \infty$ and $R_{Cu} = 0 \ \Omega$, show that LR is given by

$$LR = \frac{\omega \left(L_{k1} + \left(\frac{N_1}{N_2}\right)^2 L_{k2} \right)}{|Z_{l,f}|}$$

(c) If a 250 kHz input voltage source with $V_s = 200$ V is applied to the primary winding, determine LR for the following values: $R_{eq} = 0.2 \ \Omega$, $L_{eq} = 125 \ \mu H$, $R_{core} = 125 \ \Omega$, $L_m = 450 \ \mu H$, $Z_{l,l} = 57 \angle 6° \ \Omega$, and $Z_{l,l} = 0.5 \angle 78° \ \Omega$.

A.17 It is customary to represent transformers by their primary and secondary coupling coefficients k_1 and k_2, respectively. If k_1 and k_2 are defined as

$$k_1 = \frac{L_{m1}}{L_{m1} + L_{k1}}$$

$$k_2 = \frac{L_{m2}}{L_{m2} + L_{k2}}$$

$$M = \sqrt{k_1 k_2}$$

show that the transformer model given in Fig. A.20 can be presented as shown in Fig. PA.17.

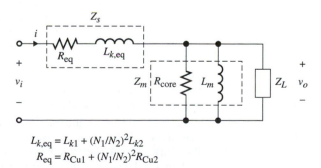

$$L_{k,eq} = L_{k1} + (N_1/N_2)^2 L_{k2}$$
$$R_{eq} = R_{Cu1} + (N_1/N_2)^2 R_{Cu2}$$

Figure PA.16

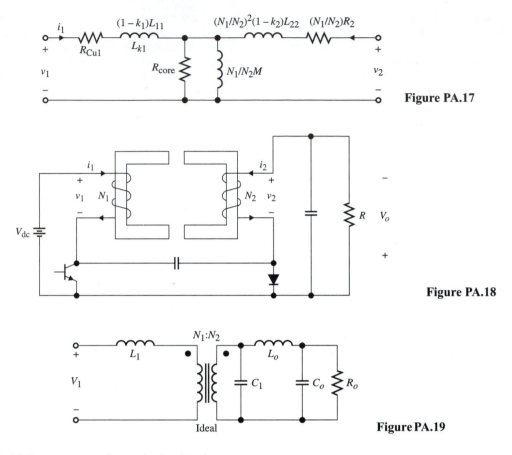

Figure PA.17

Figure PA.18

Figure PA.19

A.18 In some power electronic circuits where more than one inductor is used, it is desirable to implement two or more inductors on a single magnetic structure. Such inductors are known as *coupled inductors*. The Cuk converter shown in Fig. PA.18 is one example of a coupled inductor circuit.

If it is assumed v_1 and v_2 are identical ac voltages and the leakage inductances of the N_1 and N_2 windings are modeled by $\Re_{l,1}$ and $\Re_{l,2}$ respectively, show that:

(a) The current $i_1(t) = 0$ when

$$\frac{N_2}{N_1} = \frac{\Re_{l2}}{\Re_{l1} + \Re_m}$$

(b) The current $i_2(t) = 0$ when

$$\frac{N_1}{N_2} = \frac{\Re_{l1}}{\Re_{l2} + \Re_m}$$

where $\Re_m$ is the total reluctance of the magnetic circuit, including the reluctance of the air.

A.19 Derive the equivalent circuit for Fig. PA.19 by reflecting the entire impedance to the primary side. Assume ideal transformers.

A.20 Determine the inductance of a coil wound in the air as shown in Fig. PA.20.

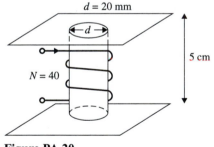

Figure PA.20

A.21 To reduce the transformer leakage inductances, normally the primary and secondary windings are placed one on top of the other as shown in Fig. PA.21(a). By assuming that the magnetic field is distributed as shown in Fig. PA.21(b), derive an expression for the leakage inductance in terms of the given parameters. Assume the cross-sectional area of the core is A and the number of turns is N.

A.22 It is common to find transformers with more than two windings on a single core. Figure PA.22(a) and (b) shows a transformer structure with two and three windings respectively. Assume that the cross-sectional area of the center leg is A_1 and that of the remaining parts is A_2.

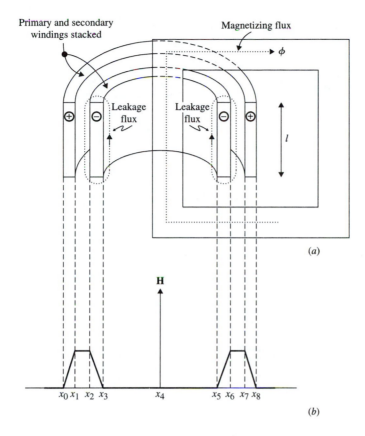

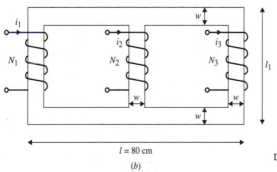

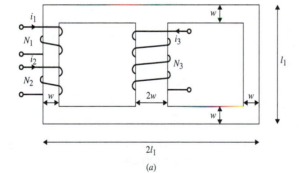

Figure PA.21

Figure PA.22

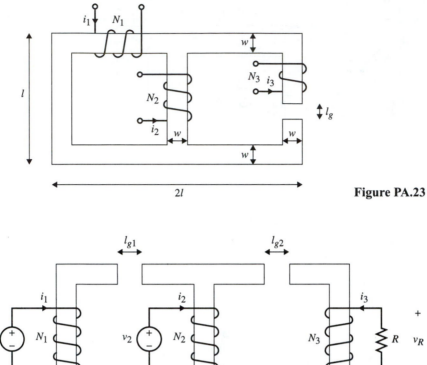

Figure PA.23

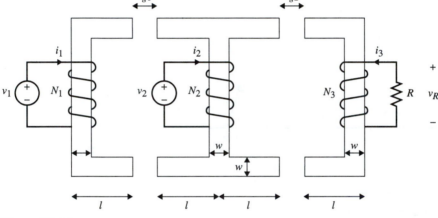

Figure PA.24

(a) Develop the magnetic circuit equivalents of these structures

(b) Determine the flux in the center leg in terms of the shown parameters.

A.23 Derive the expression for the flux, ϕ, in the center leg of the magnetic circuit of Fig. PA.23. Assume uniform cross-sectional area A and width W. The flux will be expressed in terms of N_1, N_2, N_3, i_1, i_2, i_3, l_g, l, w, A, and μ.

A.24 Determine the voltage, v_R, in terms of the given parameters in the three-winding magnetic circuit shown in Fig. PA.24. Assume the cross sectional area is A and the permeability is μ.

A.25 **(a)** Determine the flux in the center leg of the three-winding magnetic structure shown in

Fig. PA.22(b). Assume $N_1 i_1$, $N_2 i_2$, and $N_3 i_3$ are 5, 10, and 8 A-turn, respectively, and $\mu = 3000 \, \mu_o$.

(b) Determine the inductance for winding N_1 with $i_2 = i_3 = 0$.

A.26 Show how a set of three single-phase transformers can be connected to create a three-phase configuration for each of the following connections:

(a) Y-Δ

(b) Δ-Y

(c) Δ-Δ

A.27 Show that in a balanced Y-connected three-phase system, the line currents and the line voltages are also balanced three-phase systems.

Appendix B

Fourier Series for Common Waveforms

INTRODUCTION

B.1   SQUARE WAVEFORMS

B.2   SINUSOIDAL WAVEFORMS

INTRODUCTION

This appendix gives the Fourier components for waveforms commonly encountered in power electronic circuits. This appendix does not attempt to repeat the detailed treatment given in Chapter 3, but rather is meant to provide the reader with basic techniques to obtain Fourier components for selected square and sinusoidal waveforms without the need to rely on tedious and complex mathematics.

B.1   SQUARE WAVEFORMS

Let us first consider waveforms that have the general shape of a squarewave. Figure B.1 shows a common squarewave function with amplitudes $\pm V$.

Since $f_1(\omega t)$ is odd and half-wave symmetric, only odd harmonics for the sine components are present. The b_n coefficients are given by

$$b_n = \frac{1}{\pi}\left[\int_0^{\pi} V \sin(n\omega t)\, d(\omega t) + \int_{\pi}^{2\pi} (-V) \sin(n\omega t)\, d(\omega t)\right]$$

$$= \frac{V}{\pi}\left\{\left[-\frac{1}{n}\cos(n\omega t)\right]\Big|_0^{\pi} + \left[\frac{1}{n}\cos(n\omega t)\right]\Big|_{\pi}^{2\pi}\right\}$$

$$= \frac{V}{\pi}[-\cos(n\pi) + \cos(0) + \cos(2n\pi) - \cos(n\pi)]$$

$$= \frac{2V}{n\pi}[1 - \cos(n\pi)]$$

Hence, $f_1(\omega t)$ may be expressed as

$$f_1(\omega t) = \sum_{n=1,3,5,\ldots}^{\infty} \frac{4V}{n\pi}\sin n\omega t \qquad (B.1)$$

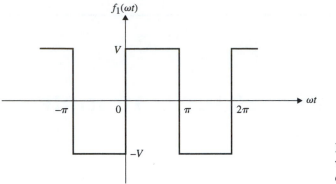

Figure B.1 Square waveform with no dc component.

Figure B.2(*a*) and (*b*) shows forms of Fig. B.1 shifted by $\pi/2$ and $\pi/3$ to the left, respectively. The wave of Fig. B.2(*a*) is even and is half-wave symmetric with only the odd harmonics of the cosine terms present as follows:

$$a_n = \frac{2}{\pi}\left[\int_{-\pi/2}^{\pi/2} V \cos(n\omega t)\, d(\omega t)\right]$$

$$= \frac{2V}{n\pi}\left[\sin(n\omega t)\big|_{-\pi/2}^{\pi/2}\right]$$

$$= \frac{2V}{n\pi}\left[\sin\left(\frac{n\pi}{2}\right) + \sin\left(\frac{n\pi}{2}\right)\right]$$

$$= \frac{4V}{n\pi}\sin\left(\frac{n\pi}{2}\right) \quad \text{for } n = 1, 3, 5, \ldots$$

Hence, the harmonics are given by

$$f_2(\omega t) = \sum_{n=1,3,5,\ldots}^{\infty} \frac{4V}{n\pi}\sin\frac{n\pi}{2}\cos n\omega t \tag{B.2}$$

It can be shown that shifting $f_1(\omega t)$ by $\pi/2$ will produce the same result shown in Eq. (B.2). This is illustrated as follows:

$$f_2(\omega t) = f_1(\omega t + \pi/2)$$

Therefore, $f_2(\omega t)$ is given by

$$f_2(\omega t) = \frac{4V}{\pi}\sum_{n=1,3,5,\ldots}^{\infty} \frac{1}{n}\sin\left[n\omega\left(t + \frac{\pi}{2}\right)\right]$$

$$= \frac{4V}{\pi}\sum_{n=1,3,5,\ldots}^{\infty} \frac{1}{n}\left[\cos(n\omega t)\sin\left(\frac{n\pi}{2}\right) + \sin(n\omega t)\cos\left(\frac{n\pi}{2}\right)\right]$$

$$= \frac{4V}{\pi}\sum_{n=1,3,5,\ldots}^{\infty} \frac{1}{n}\cos(n\omega t)\sin\left(\frac{n\pi}{2}\right)$$

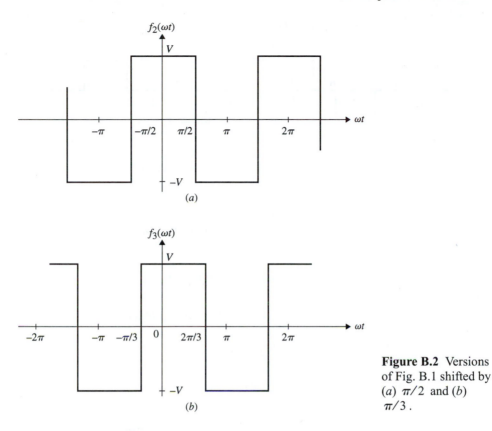

Figure B.2 Versions of Fig. B.1 shifted by (a) $\pi/2$ and (b) $\pi/3$.

$$f_2(\omega t) = \sum_{n=1,3,5,\ldots}^{\infty} \frac{4V}{n\pi} \sin\frac{n\pi}{2} \cos n\omega t \tag{B.3}$$

Similarly, the Fourier series for $f_3(\omega t)$ of Fig. B.2(b) is given by

$$f_3(\omega t) = f_1\left(\omega t + \frac{\pi}{3}\right)$$

$$= \sum_{n=1,3,5,\ldots}^{\infty} \frac{4V}{n\pi} \sin\ n\left(\omega t + \frac{\pi}{3}\right) \tag{B.4}$$

We should point out that many of the Fourier components can be obtained simply by manipulating the waveforms of Figs. B.1 and B.2.

Three-Level Square Waveform

Next we consider a square waveform with symmetric dead angle, α, as shown in Fig. B.3. The waveform of Fig. B.3 is odd and half-wave symmetric with only odd harmonics of the sine terms as follows:

$$b_n = \frac{2}{\pi}\left[\int_0^\pi f(\omega t) \sin(n\omega t)\ d(\omega t)\right]$$

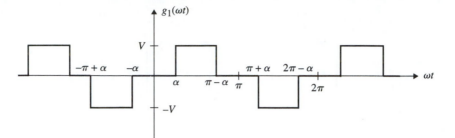

Figure B.3 Squarewave function with symmetric dead angle, α.

$$= \frac{2}{\pi}\left[\int_{\alpha}^{\pi-\alpha} V \sin(n\omega t)\, d(\omega t)\right]$$

$$= \frac{2V}{n\pi}\left[-\cos(n\omega t)\big|_{\alpha}^{\pi-\alpha}\right]$$

$$= \frac{2V}{n\pi}[-\cos n(\pi-\alpha)+\cos(n\alpha)]$$

$$= \frac{4V}{n\pi}\cos n\alpha$$

The harmonic expression for $g_1(\omega t)$ is given by

$$g_1(\omega t) = \sum_{n=1,3,5,\ldots}^{\infty} \frac{4V}{n\pi}\cos n\alpha \sin n\omega t \tag{B.5}$$

Figure B.4(*a*) and (*b*) shows versions of Fig. B.3 shifted by α to the right and left, respectively.

The function $g_2(\omega t)$ is obtained by shifting $g_1(\omega t)$ by α as follows:

$$g_2(\omega t) = g_1(\omega t - \alpha)$$

$$= \sum_{n=1,3,5,\ldots}^{\infty} \frac{4V}{n\pi}\cos n\alpha \sin n(\omega t - \alpha) \tag{B.6}$$

Similarly, $g_3(\omega t)$ is expressed as follows:

$$g_3(\omega t) = g_1(\omega t + \alpha)$$

$$= \sum_{n=1,3,5,\ldots}^{\infty} \frac{4V}{n\pi}\cos n\alpha \sin n(\omega t + \alpha) \tag{B.7}$$

Multi-Pulse Waveforms

Let us consider multi-level pulse square waveforms normally encountered in three-phase circuits. One way to analyze such waveforms is to consider the addition of several square waveforms with different dc levels.

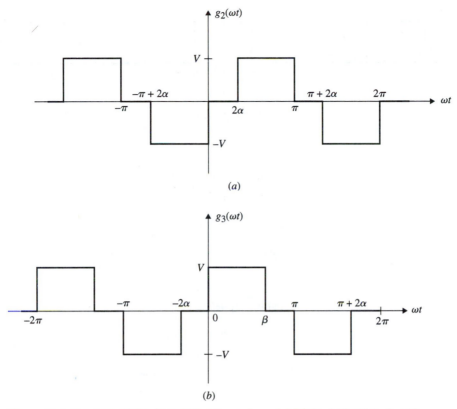

Figure B.4 Versions of Fig. B.3 shifted by angle α. (*a*) Right shift. (*b*) Left shift.

Figure B.5 is the same as Fig. B.2 but includes a dc component. The function $h_1(\omega t)$ is expressed as follows:

$$h_1(\omega t) = \frac{1}{2}f_2(\omega t) + \frac{V}{2}$$

where the dc value of $h_1(\omega t)$ is $V/2$. As a result, Eq. (B.2) yields the following Fourier series for $h_1(\omega t)$:

$$h_1(\omega t) = \frac{V}{2} + \sum_{n=1,3,5,\dots}^{\infty} \frac{2V}{n\pi} \sin\frac{n\pi}{2} \cos n\omega t \qquad \text{(B.8)}$$

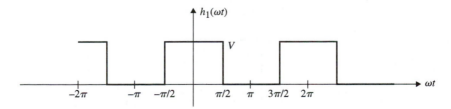

Figure B.5 Single-pulse waveform with width π.

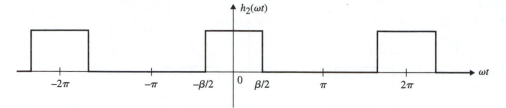

Figure B.6 Single-pulse waveform with pulse width β.

Figure B.6 shows a pulse square waveform with a general pulse width of β. The waveform of $h_2(\omega t)$ has the following coefficients:

$$a_o = \frac{1}{2\pi}\int_{-\beta/2}^{\beta/2} V\, d(\omega t) = \frac{\beta V}{2\pi}$$

$$a_n = \frac{1}{\pi}\int_{-\beta/2}^{\beta/2} V\cos n\omega t\, d(\omega t) = \frac{2V}{n\pi}\sin(n\beta)$$

Therefore, the function $h_2(\omega t)$ is expressed as

$$h_2(\omega t) = a_o + a_n \cos n\omega t$$

$$= \frac{\beta V}{2\pi} + \sum_{n=1,3,5,\dots}^{\infty} \frac{2V}{n\pi}\sin n\beta \cos n\omega t \tag{B.9}$$

Figure B.7 is obtained by shifting Fig. B.6 to the right by $\beta/2$. Its Fourier components are given as follows:

$$h_3(\omega t) = h_2\left(\omega t - \frac{\beta}{2}\right)$$

$$= \frac{\beta V}{2\pi} + \sum_{n=1,3,5,\dots}^{\infty} \frac{2V}{n\pi}\sin n\beta \cos n\left(\omega t - \frac{\beta}{2}\right) \tag{B.10}$$

Next we consider multi-pulse square waveforms such as the two-pulse waveform shown in Fig. B.8(a). It can be shown that its harmonics may be obtained by adding $h_5(\omega t)$ and $h_6(\omega t)$ of Fig. B.8(b) and (c), respectively. Figure B.8(d) is the same as Fig. B.8(a) but shifted to the left by $\alpha/2$.

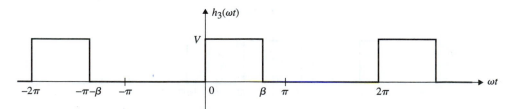

Figure B.7 Version of Fig. B.6 shifted to the right by $\beta/2$.

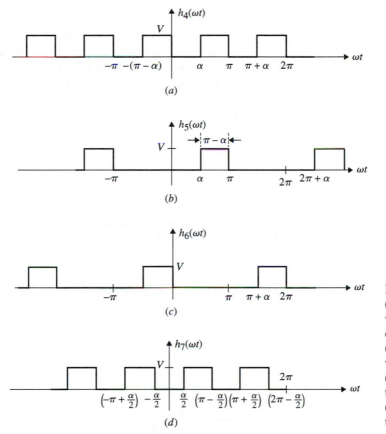

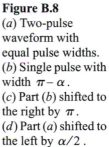

Figure B.8
(a) Two-pulse waveform with equal pulse widths.
(b) Single pulse with width $\pi - \alpha$.
(c) Part (b) shifted to the right by π.
(d) Part (a) shifted to the left by $\alpha/2$.

Figure B.8(d) shows the waveform $h_7(\omega t)$ obtained by shifting $h_4(\omega t)$ by $\alpha/2$ to the left. Fourier series for $h_4(\omega t)$ and $h_7(\omega t)$ are given by

$$h_4(\omega t) = h_5(\omega t) + h_6(\omega t)$$

$$= \frac{V}{\pi}(\pi - \alpha) + \sum_{n=2,4,6,\dots}^{\infty} \frac{-4V}{n\pi} \sin\left(\frac{n\alpha}{2}\right) \cos n\left(\omega t - \frac{\alpha}{2}\right) \qquad (B.11)$$

where

$$h_5(\omega t) = V\left(\frac{\pi - \alpha}{2\pi}\right) + \frac{2V}{\pi} \sum_{n=1,3,5,\dots} \frac{1}{n} \sin\, n\left(\frac{\pi - \alpha}{2}\right) \cos n\left[\omega t - \left(\frac{\pi}{2} + \frac{\alpha}{2}\right)\right]$$

$$h_6(\omega t) = V\left(\frac{\pi - \alpha}{2\pi}\right) + \frac{2V}{\pi} \sum_{n=1,3,5,\dots} \frac{1}{n} \sin\, n\left(\frac{\pi - \alpha}{2}\right) \cos n\left[\omega t - \left(\frac{3\pi}{2} + \frac{\alpha}{2}\right)\right]$$

and

$$h_7(\omega t) = h_4\left(\omega t + \frac{\alpha}{2}\right)$$

$$= \frac{V}{\pi}(\pi - \alpha) + \sum_{n=2,4,6,\dots}^{\infty} \frac{-4V}{n\pi} \sin\left(\frac{n\alpha}{2}\right) \cos n\omega t \qquad (B.12)$$

Two- and Three-step Waveforms

Figure B.9(a) is a two-step square waveform whose harmonic contents are difficult to obtain by applying the original Fourier series equations. Instead, we simply add two previously analyzed waveforms shown again in Figs. B.9(b) and (c).

The harmonics of $f(\omega t)$ are obtained from the following relation:

$$f(\omega t) = f_1(\omega t) + g_1(\omega t)$$

$$= \frac{4V}{\pi} \left[\sum_{n=1,3,5,\dots}^{\infty} \frac{1}{n} \sin n\omega t \left(1 + \cos \frac{n\pi}{3} \right) \right]$$

$$= \frac{4V}{\pi} \sum_{n=1,3,5,\dots}^{\infty} \frac{1}{n} \left(1 + \cos \frac{n\pi}{3} \right) \sin n\omega t$$

$$= \sum_{n=1,3,5,\dots}^{\infty} \frac{4V}{n\pi} \left(1 + \cos \frac{n\pi}{3} \right) \sin n\omega t \qquad (B.13)$$

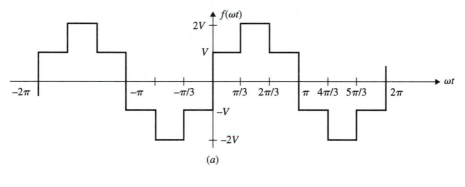

(a)

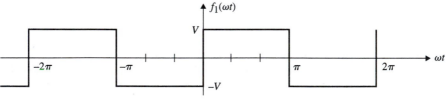

(b)

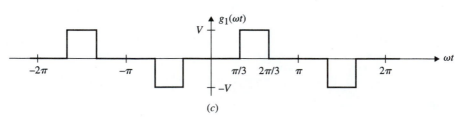

(c)

Figure B.9 (a) Two-step waveform obtained by adding parts (b) and (c).

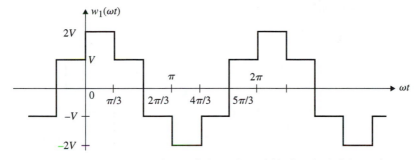

Figure B.10 Two-step waveform of Fig. B.9(a) shifted to the left by $\pi/3$.

where $f_1(\omega t)$ and $g_1(\omega t)$ are given by

$$f_1(\omega t) = \frac{4V}{\pi} \sum_{n=1,3,5,\dots}^{\infty} \frac{1}{n} \sin n\omega t$$

$$g_1(\omega t) = \frac{4V}{\pi} \sum_{n=1,3,5,\dots}^{\infty} \frac{1}{n} \cos\frac{n\pi}{3} \sin n\omega t$$

Figure B.10 is obtained by shifting $f(\omega t)$ of Fig. B.9(a) to the left by $\pi/3$, giving the following Fourier series:

$$
\begin{aligned}
w_1(\omega t) &= f\left(\omega t + \frac{\pi}{3}\right) \\
&= \sum_{n=1,3,5,\dots}^{\infty} \frac{4V}{n\pi}\left(1 + \cos\frac{n\pi}{3}\right) \sin n\left(\omega t + \frac{\pi}{3}\right)
\end{aligned}
\tag{B.14}
$$

Finally, we consider the three-step waveform shown in Fig. B.11(a), obtained by adding Fig. B.11(b), (c), and (d).

The waveform $w_2(\omega t)$ of Fig. B.11(a) is obtained by adding the harmonics of Fig. B.11(b), (c), and (d), which results in the following harmonic series:

$$w_2(\omega t) = \sum_{n=1,3,5,\dots}^{\infty} \frac{4V}{3\pi}\frac{1}{n}\left(1 + \cos\frac{n\pi}{3} + \cos\frac{2n\pi}{5}\right) \sin n\omega t \tag{B.15}$$

B.2   SINUSOIDAL WAVEFORMS

In this section, we will consider sinusoidal waveforms frequently encountered in power electronic circuits. These include half- and full-wave diode and SCR sinusoidal waveforms.

B.2.1   Diode Sinusoidal Waveforms

Figure B.12 shows a half-wave rectifier waveform, with its Fourier components given in Eq. (B.16).

The waveform is an even function, with its harmonics containing only cosine terms and a dc value, as follows:

$$a_o = \frac{1}{2\pi}\int_{-\pi/2}^{\pi/2} V\cos(\omega t)\, d(\omega t) = \frac{V}{\pi}$$

$$a_n = \frac{1}{\pi}\int_{-\pi/2}^{\pi/2} V\cos(\omega t)\cos n(\omega t)\, d(\omega t) = \frac{2V}{\pi(1-n^2)}\cos\frac{n\pi}{2}$$

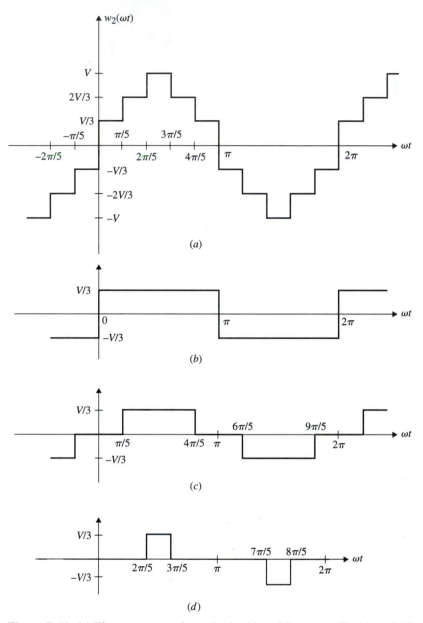

Figure B.11 (a) Three-step waveform obtained by adding parts (b), (c), and (d).

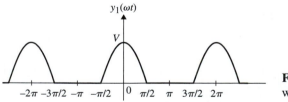

Figure B.12 Half-wave sinusoidal waveform.

The Fourier series for Fig. B.12 is given by

$$y_1(\omega t) = \frac{V}{\pi} + \frac{V}{2}\cos\omega t + \sum_{n=2,4,6,...}^{\infty} \frac{2V}{\pi(1-n^2)}\cos\frac{n\pi}{2}\cos n\omega t \qquad \text{(B.16)}$$

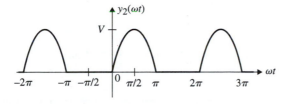

Figure B.13 Shifted half-wave sinusoidal waveform.

Figure B.13 is the same as Fig. B.12 but shifted to the right by $\pi/2$. Hence, $y_2(\omega t) = y_1(\omega t - \pi/2)$, and its harmonic components are given by

$$y_2(\omega t) = \frac{V}{\pi} + \frac{V}{2}\sin \omega t + \sum_{n=2,4,6,\ldots}^{\infty} \frac{2V}{\pi(1-n^2)}\cos\frac{n\pi}{2}\cos n\left(\omega t - \frac{\pi}{2}\right) \qquad (B.17)$$

Another way to obtain the Fourier series for Fig. B.12 is to use the product of two known waveforms. The product of $\sin(\omega t)$ and $h_3(\omega t)$ for $\beta = \pi$, redrawn in Fig. B.14, will produce the waveform $y_2(\omega t)$.

The harmonic function of Fig. B.14(a) is expressed as

$$y_2(\omega t) = \sin(\omega t) \cdot h_3(\omega t)$$

where $h_3(\omega t)$ is obtained from Eq. (B.10) by setting $\beta = \pi$, to yield

$$h_3(\omega t) = \frac{V}{2} + \sum_{n=1,3,5,\ldots}^{\infty} \frac{2V}{n\pi}\sin(n\pi)\cos n\left(\omega t - \frac{\pi}{2}\right)$$

$$h_3(\omega t) = \frac{V}{2} + \frac{2V}{\pi}\left[\cos\left(\omega t - \frac{\pi}{2}\right) - \frac{1}{3}\cos 3\left(\omega t - \frac{\pi}{2}\right) + \frac{1}{5}\cos 5\left(\omega t - \frac{\pi}{2}\right)\cdots\right]$$

$$= \frac{V}{2} + \frac{V}{\pi}\left[\sin \omega t + \frac{1}{3}\sin 3\omega t + \frac{1}{5}\sin 5\omega t\cdots\right]$$

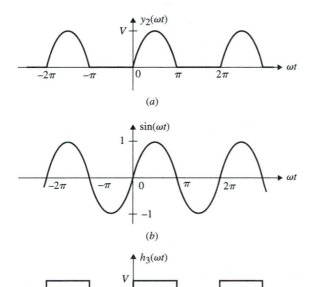

Figure B.14 (a) Sinusoidal waveform obtained by the product of parts (b) and (c).

Hence,

$$y_2(\omega) = \sin \omega t \cdot h_3(\omega t)$$

$$= \frac{V}{\pi}\left[1 + \frac{\pi}{2}\sin(\omega t) - \frac{2}{3}\cos 2\omega t - \frac{2}{15}\cos 4\omega t - \frac{2}{35}\cos 6\omega t \cdots\right] \qquad \text{(B.18)}$$

which is similar to Eq. (B.17).

Similarly, the harmonics of Fig. B.15(a) are obtained by multiplying the waveforms of Fig. B.15(b) and (c).

Another common waveform is the full-wave rectifier waveform shown in Fig. B.16(a). The harmonic coefficients of Fig. B.16(a) can be written as follows:

$$a_o = \frac{1}{2\pi}\int_0^{2\pi} y_3(\omega t)\, d(\omega t) = \frac{1}{\pi}\int_0^{\pi} V \sin(\omega t)\, d(\omega t) = \frac{2V}{\pi}$$

$$a_n = \frac{2}{\pi}\int_0^{T/2} V \sin \omega t \, \cos n\omega t \, d(\omega t) = \frac{4V}{\pi}\int_0^{\pi/2} \sin \omega t \, \cos n\omega t \, d(\omega t)$$

$$= \frac{2V}{\pi}\left[\int_0^{\pi/2}\sin(1+n)\omega t \, d(\omega t) + \int_0^{\pi/2}\sin(1-n)\omega t \, d(\omega t)\right]$$

$$= -\frac{2V}{\pi}\left[\frac{\cos(1+n)\omega t}{1+n}\Big|_0^{\pi/2} + \frac{\cos(1-n)\omega t}{1-n}\Big|_0^{\pi/2}\right]$$

$$= -\frac{2V}{\pi}\left[\frac{\cos(1+n)\frac{\pi}{2}}{1+n} - \frac{1}{1+n} + \frac{\cos(1-n)\frac{\pi}{2}}{1-n} - \frac{1}{1-n}\right]$$

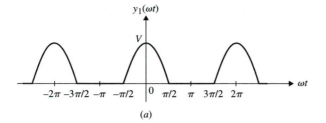

(a)

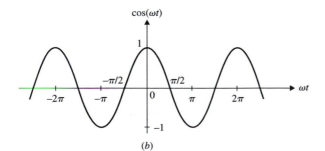

(b)

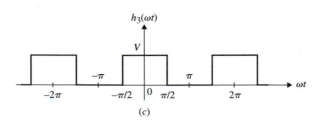

(c)

Figure B.15 (*a*) Waveform obtained as the product of parts (*b*) and (*c*).

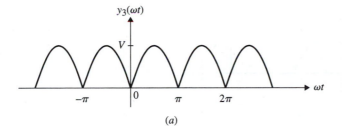

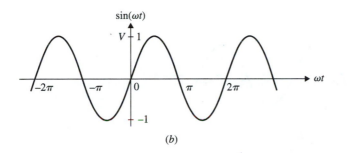

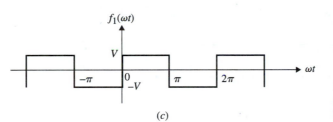

Figure B.16 (*a*) Full-wave sinusoidal waveform obtained by multiplying parts (*b*) and (*c*).

$$a_n = \frac{2V}{\pi}\left[\frac{1}{1+n} + \frac{1}{1-n}\right] = \frac{4V}{(1-n^2)\pi}$$

Hence, the harmonics of $y_3(\omega t)$ are given by

$$y_3(\omega t) = \frac{2V}{\pi} + \sum_{n=2,4,6,\dots}^{\infty} \frac{4V}{(1-n^2)\pi}\cos n\omega t \qquad (B.19)$$

It can be shown that $y_3(\omega t)$ of Fig. B.16(*a*) may be obtained by multiplying the waveforms of Fig. B.16(*b*) and (*c*).

B.2.2   SCR Sinusoidal Waveforms

Figure B.17(*a*) shows a typical half-wave SCR voltage waveform. Its harmonic contents were studied in Chapter 8. The same harmonic relationship can be obtained by multiplying the waveforms $\sin(\omega t)$ and $g_1(\omega t)$ shown in Fig. B.17(*b*) and (*c*), respectively. Therefore, $x_1(\omega t)$ is given by

$$x_1(\omega t) = \sin(\omega t)\cdot g_1(\omega t)$$

$$= V\sin\omega t\left[\frac{(\pi-\alpha)}{2\pi} + \sum_{n=1,3,5,\dots}^{\infty}\frac{2}{n\pi}\sin n\left(\frac{\pi-\alpha}{2}\right)\cos n\left(\omega t - \left(\frac{\pi}{2}+\frac{\alpha}{2}\right)\right)\right] \qquad (B.20)$$

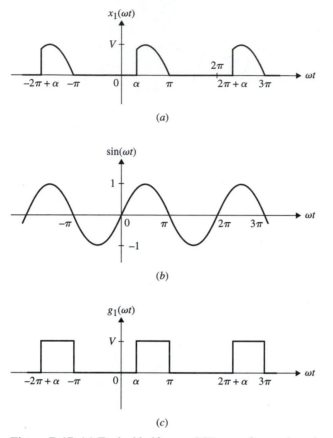

Figure B.17 (*a*) Typical half-wave SCR waveform, whose harmonics are obtained by multiplying parts (*b*) and (*c*).

Figure B.18(*a*) shows a typical voltage waveform produced by a bidirectional SCR circuit or triac. Its harmonic contents are obtained by multiplying $\sin(\omega t)$ and $h_4(\omega t)$ of Fig. B.18(*b*) and (*c*).

The Fourier series for $x_2(\omega t)$ is given by

$$x_2(\omega t) = \sin(\omega t) \cdot h_4(\omega t)$$

$$= V \sin \omega t \left[\frac{(\pi - \alpha)}{\pi} - \sum_{n=2,4,6,\ldots}^{\infty} \frac{4}{n\pi} \sin \frac{n\alpha}{2} \cos n\left(\omega t - \frac{\alpha}{2}\right)\right] \qquad \text{(B.21)}$$

Finally, Fig. B.19(*a*) shows a full-wave rectified SCR voltage waveform whose Fourier series can be obtained by multiplying Fig. B.19(*b*) and (*c*). The harmonics for Fig. B.19(*a*) are given by

$$x_3(\omega t) = \sin(\omega t) \cdot g_2(\omega t)$$

$$= V \sin \omega t \left[\sum_{n=1,3,5,\ldots}^{\infty} \frac{4}{n\pi} \cos \frac{n\alpha}{2} \sin n\left(\omega t - \frac{\alpha}{2}\right)\right] \qquad \text{(B.22)}$$

Table B.1 is a summary of common waveforms with their harmonic spectrum plots.

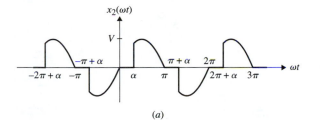

(a)

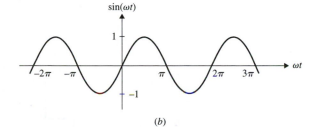

(b)

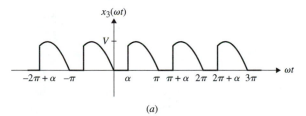

(c)

Figure B.18 (a) Typical bidirectional SCR voltage circuit waveform obtained by multiplying parts (b) and (c).

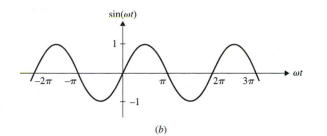

(a)

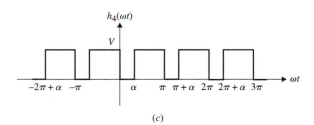

(b)

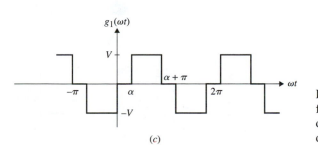

(c)

Figure B.19 (a) Typical rectified full-wave SCR voltage waveform obtained by multiplying waveforms of parts (b) and (c).

Table B.1 Summary of Common Waveforms with Their Harmonic Spectrum Plots

Waveform	Fourier series	$\text{THD} = \sqrt{\dfrac{1}{k_{\text{dist}}^2} - 1}$	Harmonic spectrum (normalized to fundamental component)
	$f_1(\omega t) = \displaystyle\sum_{n=1,3,5,\dots}^{\infty} \frac{4V}{n\pi}\sin n\omega t$	0.438	
	$g_3(\omega t) = \displaystyle\sum_{n=1,3,5,\dots}^{\infty} \frac{4V}{n\pi}\cos n\alpha \, \sin n\omega t$	0.262 $(\alpha = \pi/6)$	
	$h_1(\omega t) = \dfrac{V}{2} + \displaystyle\sum_{n=1,3,5,\dots}^{\infty} \frac{2V}{n\pi}\sin\frac{n\pi}{2}\cos n\omega t$	1.194	

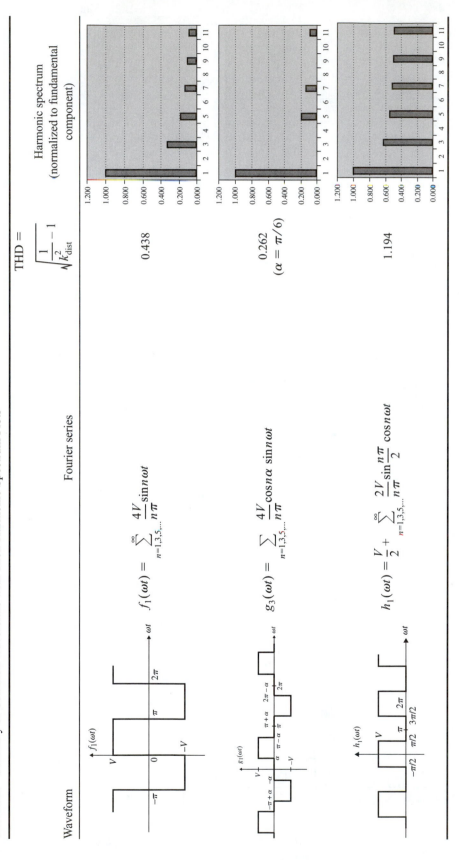

Table B.1 *(continued)*

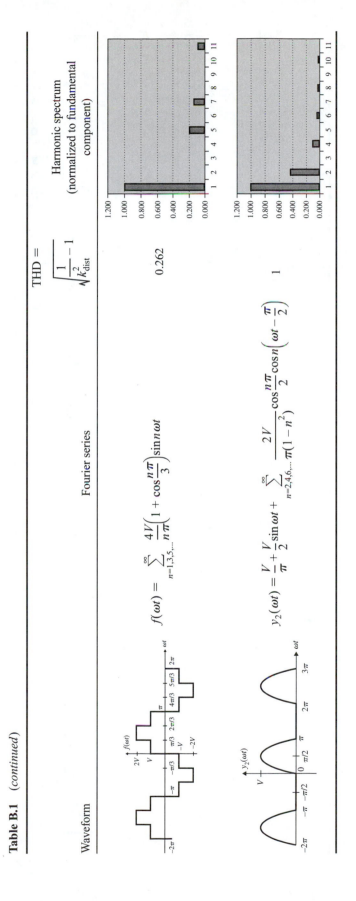

Waveform	Fourier series	THD = $\sqrt{\dfrac{1}{k_{\text{dist}}^2} - 1}$	Harmonic spectrum (normalized to fundamental component)
	$f(\omega t) = \displaystyle\sum_{n=1,3,5,\dots}^{\infty} \frac{4V}{n\pi}\left(1 + \cos\frac{n\pi}{3}\right)\sin n\omega t$	0.262	
	$y_2(\omega t) = \dfrac{V}{\pi} + \dfrac{V}{2}\sin\omega t + \displaystyle\sum_{n=2,4,6,\dots}^{\infty} \frac{2V}{\pi(1-n^2)}\cos\frac{n\pi}{2}\cos n\left(\omega t - \frac{\pi}{2}\right)$	1	

Table B.1 *(continued)*

Waveform	Fourier series	THD = $\sqrt{\dfrac{1}{k_{dist}^2} - 1}$	Harmonic spectrum (normalized to fundamental component)

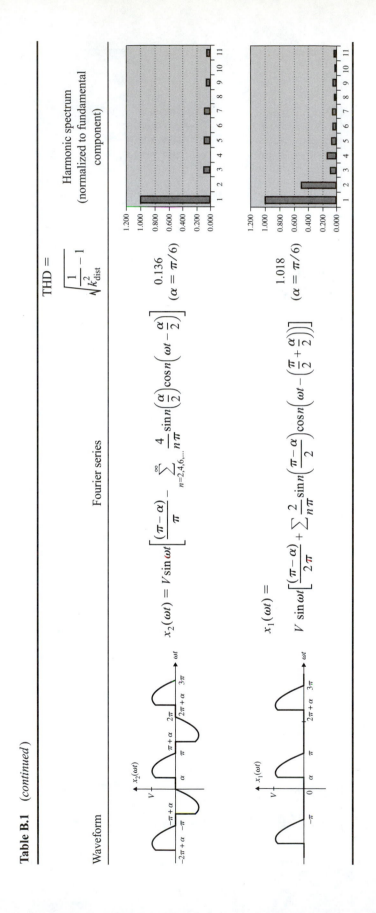

$$x_2(\omega t) = V\sin\omega t\left[\frac{(\pi-\alpha)}{\pi} - \sum_{n=2,4,6,\ldots}^{\infty}\frac{4}{n\pi}\sin n\left(\frac{\alpha}{2}\right)\cos n\left(\omega t - \frac{\alpha}{2}\right)\right]$$

$$0.136$$
$$(\alpha = \pi/6)$$

$$x_1(\omega t) =$$
$$V\sin\omega t\left[\frac{(\pi-\alpha)}{2\pi} + \sum\frac{2}{n\pi}\sin n\left(\frac{\pi-\alpha}{2}\right)\cos n\left(\omega t - \left(\frac{\pi}{2} + \frac{\alpha}{2}\right)\right)\right]$$

$$1.018$$
$$(\alpha = \pi/6)$$

Appendix C

Useful Functions

C.1 TRIGONOMETRIC IDENTITES

ωt	$-\theta$	$\theta \mp \dfrac{\pi}{2}$	$\theta \mp \pi$	$\theta \mp \dfrac{3\pi}{2}$	$\theta \mp 2\pi$
$\cos(\omega t)$	$\cos\theta$	$\pm\sin\theta$	$-\cos\theta$	$\mp\sin\theta$	$\cos\theta$
$\sin(\omega t)$	$-\sin\theta$	$\cos\theta$	$\pm\sin\theta$	$\cos\theta$	$\mp\sin\theta$

$$\sin(\theta_1 \pm \theta_2) = \sin\theta_1\cos\theta_2 \pm \cos\theta_1\sin\theta_2$$

$$\cos(\theta_1 \pm \theta_2) = \cos\theta_1\cos\theta_2 \mp \sin\theta_1\sin\theta_2$$

$$\tan(\theta_1 \pm \theta_2) = \frac{\tan\theta_1 \pm \tan\theta_2}{1 \mp \tan\theta_1\tan\theta_2}$$

$$\sin 2\theta = 2\sin\theta\cos\theta$$

$$\cos 2\theta = 1 - 2\sin^2\theta = 2\cos^2\theta - 1 = \cos^2\theta - \sin^2\theta$$

$$\tan 2\theta = \frac{2\tan\theta}{1 - \tan^2\theta}$$

$$\sin\theta_1 + \sin\theta_2 = 2\sin\frac{\theta_1 + \theta_2}{2}\cos\frac{\theta_1 - \theta_2}{2}$$

$$\sin\theta_1 - \sin\theta_2 = 2\cos\frac{\theta_1 + \theta_2}{2}\sin\frac{\theta_1 - \theta_2}{2}$$

$$\cos\theta_1 + \cos\theta_2 = 2\cos\frac{\theta_1 + \theta_2}{2}\cos\frac{\theta_1 - \theta_2}{2}$$

$$\cos\theta_1 - \cos\theta_2 = 2\sin\frac{\theta_1 + \theta_2}{2}\sin\frac{\theta_2 - \theta_1}{2}$$

$$\sin\theta_1\sin\theta_2 = \frac{1}{2}[\cos(\theta_1 - \theta_2) - \cos(\theta_1 + \theta_2)]$$

$$\cos\theta_1\cos\theta_2 = \frac{1}{2}[\cos(\theta_1 - \theta_2) + \cos(\theta_1 + \theta_2)]$$

$$\sin\theta_1\cos\theta_2 = \frac{1}{2}[\sin(\theta_1 - \theta_2) + \sin(\theta_1 + \theta_2)]$$

549

$$\sin^2\theta = \frac{1}{2}(1 - \cos 2\theta)$$

$$\cos^2\theta = \frac{1}{2}(1 + \cos 2\theta)$$

$$a\cos\theta + b\sin\theta = \left[\sqrt{a^2+b^2}\right]\cos\left[\theta + \tan^{-1}\left(\frac{-b}{a}\right)\right]$$

$$= \left[\sqrt{a^2+b^2}\right]\sin\left[\theta + \tan^{-1}\left(\frac{a}{b}\right)\right]$$

C.2 SOME LAPLACE TRANSFORMATIONS

$f(t)[t > 0^-]$	$F(s)$
$\delta(t)$	1
1	$\dfrac{1}{s}$
t	$\dfrac{1}{s^2}$
$e^{-\alpha t}$	$\dfrac{1}{(s+\alpha)}$
$\sin\omega t$	$\dfrac{\omega}{(s^2+\omega^2)}$
$\cos\omega t$	$\dfrac{s}{(s^2+\omega^2)}$
$f'(t)$	$sF(s) - f(0)$
$f''(t)$	$s^2 F(s) - sf(0) - f'(0)$
$te^{-\alpha t}$	$\dfrac{1}{(s+\alpha)^2}$
$e^{-\alpha t}\sin\omega t$	$\dfrac{\omega}{(s+\alpha)^2+\omega^2}$
$e^{-\alpha t}\cos\omega t$	$\dfrac{s+\alpha}{(s+\alpha)^2+\omega^2}$

C.3 DERIVATIVES AND INTEGRALS

Given $f_1(\omega t)$ and $f_2(\omega t)$ with ω and α constants:

$$\frac{d}{dt}(\alpha f_1) = \alpha\frac{df_1}{dt}$$

$$\frac{d}{dt}(f_1 \cdot f_2) = f_1 \frac{df_2}{dt} + f_2 \frac{df_1}{dt}$$

$$\frac{d}{dt}(\alpha f_1^n) = n\alpha f_1^{n-1} \frac{df_1}{dt}$$

$$\frac{d}{dt}(\sin \omega t) = \omega \cos \omega t$$

$$\frac{d}{dt}(\cos \omega t) = -\omega \sin \omega t$$

$$\int \alpha \, dt = \alpha t + C$$

$$\int f_1^n(t) \, df_1 = \frac{f_1^{n+1}(t)}{n+1} + C, \quad n \neq -1$$

$$\int e^{\alpha t} \, dt = \frac{1}{\alpha} e^{\alpha t} + C$$

$$\int x e^{\alpha t} \, dt = \frac{e^{\alpha t}}{\alpha^2}(\alpha t - 1) + C$$

$$\int \sin \omega t \, dt = -\frac{1}{\omega} \cos \omega t + C$$

$$\int \cos \omega t \, dt = \frac{1}{\omega} \sin \omega t + C$$

$$\int \sin^2 \omega t \, dt = \frac{t}{2} - \frac{\sin 2\omega t}{4\omega} + C$$

$$\int \cos^2 \omega t \, dt = \frac{t}{2} + \frac{\sin 2\omega t}{4\omega} + C$$

$$\int \sin \omega_1 t \, \sin \omega_2 t \, dt = \frac{\sin(\omega_1 - \omega_2)t}{2(\omega_1 - \omega_2)} - \frac{\sin(\omega_1 + \omega_2)t}{2(\omega_1 + \omega_2)} + C, \quad \text{for } \omega_1^2 \neq \omega_2^2$$

$$\int \sin \omega_1 t \, \cos \omega_2 t \, dt = -\frac{\cos(\omega_1 - \omega_2)t}{2(\omega_1 - \omega_2)} - \frac{\cos(\omega_1 + \omega_2)t}{2(\omega_1 + \omega_2)} + C, \quad \text{for } \omega_1^2 \neq \omega_2^2$$

$$\int \cos \omega_1 t \, \cos \omega_2 t \, dt = \frac{\sin(\omega_1 - \omega_2)t}{2(\omega_1 - \omega_2)} + \frac{\sin(\omega_1 + \omega_2)t}{2(\omega_1 + \omega_2)} + C, \quad \text{for } \omega_1^2 \neq \omega_2^2$$

$$\int t \sin \omega t \, dt = \frac{1}{\omega^2}(\sin \omega t - \omega t \cos \omega t) + C$$

$$\int t \cos \omega t \, dt = \frac{1}{\omega^2}(\cos \omega t + \omega t \sin \omega t) + C$$

$$\int e^{\alpha t} \sin \omega t \ dt = \frac{e^{\alpha t}}{\alpha^2 + \omega^2}(\alpha \sin \omega t - \omega \cos \omega t) + C$$

$$\int e^{\alpha t} \cos \omega t \ dt = \frac{e^{\alpha t}}{\alpha^2 + \omega^2}(\alpha \cos \omega t + \omega \sin \omega t) + C$$

C.4 DEFINITE INTEGRALS

For n and m integers:

$$\int_0^{2\pi} \sin nt \ dt = 0$$

$$\int_0^{2\pi} \cos nt \ dt = 0$$

$$\int_0^{\pi} \sin^2 nt \ dt = \int_0^{\pi} \cos^2 nt \ dt = \frac{\pi}{2}$$

$$\int_0^{\pi} \sin nt \ \cos mt \ dt = \begin{cases} 0, & m + n \text{ even} \\ \dfrac{2n}{n^2 - m^2}, & m + n \text{ odd} \end{cases}$$

$$\int_0^{\pi} \sin nt \ \sin mt \ dt = \int_0^{\pi} \cos nt \ \cos mt \ dt = 0; \quad m \neq n$$

Bibliography

TEXTBOOKS

Introduction to Power Electronics

Ang, S. *Power Switching Converters*. Marcel Dekker, 1995.

Bird, B., K. King, and D. Pedder. *An Introduction to Power Electronics*, 2nd ed. John Wiley, 1993.

Bradly, D. A. *Power Electronics*, 2nd ed. Chapman & Hall, 1995.

Dewan, S. B., and A. Straughen. *Power Semiconductor Circuits*. John Wiley, 1975.

Erickson, R. *Fundamentals of Power Electronics*. Chapman & Hall, 1997.

Fisher, M. *Power Electronics*. PWS-KENT, 1991.

Hart, D. *Introduction to Power Electronics*. Prentice-Hall, 1997.

Heumann, Klemens. *Basic Principles of Power Electronics*. Springer-Verlag, 1986.

Hoft, Richard. *Semiconductor Power Electronics*. Van Nostrand Reinhold, 1986.

Kassakian, J., M. Schlecht, and G. Vergese. *Principles of Power Electronics*. Addison-Wesley, 1991.

Krein, Philip. *Elements of Power Electronics*. Oxford University Press, 1998.

Lander, Gyril. *Power Electronics*, 2nd ed. McGraw-Hill, 1987.

Mohan, N., T. Undeland, and W. Robbins. *Power Electronics: Converters, Applications, and Design*, 3rd ed. John Wiley, 2002.

Motto, J. W. *Introduction to Solid State Power Electronics*. Westinghouse, 1977.

Ohno, Eiichi. *Introduction to Power Electronics*. Clarendon Press, 1988.

Ramshaw, R. S. *Power Electronics Semiconductor Switches*, 2nd ed. Chapman & Hall, 1993.

Rashid, M. H. *Power Electronics: Circuits, Devices, Applications*, 2nd ed. Prentice Hall, 1993.

Seyuier, Guy. *Power Electronic Converters: AC/DC Conversion*. McGraw-Hill, 1986.

Tarter, R. E. *Principles of Solid-State Power Conversion*. Howard Sams & Co., 1985.

Thorborg, K. *Power Electronics*. Prentice Hall, 1988.

Trzynadlowski, A. *An Introduction to Modern Power Electronics*. John Wiley, 1998.

Vithayathil, J. *Power Electronics: Principles and Applications*. McGraw-Hill, 1995.

Williams, B. W. *Power Electronics: Devices, Drivers, and Applications*. John Wiley, 1987.

Wood, Peter. *Switching Power Converters*. Krueger, 1981.

Switching-Mode Power Supplies

Billings, K. *Switch-Mode Power Supply Handbook*. McGraw-Hill, 1989.

Brown, M. *Power Supply Cookbook*, 2nd ed. Newnes, 2001.

Chryssis, George. *High-Frequency Switching Power Supplies: Theory and Design*, 2nd ed. McGraw-Hill, 1989.

Hnatek, E. *Design of Solid State Power Supplies*, 3rd ed. Van Nostrand Reinhold, 1989.

Kilgenstein, O. *Switched-Mode Power Supplies in Practice*. John Wiley & Sons, 1989.

Middlebrook, R. D., and S. Ćuk. *Advances in Switch-Mode Power Conversion*. Vols. I and II. TESLAco, 1981.

Mitchell, Daniel. *DC-DC Switching Regulator Analysis*. McGraw-Hill, 1988.

Pressman, A. I. *Switching Power Supply Design*. McGraw-Hill, 1991.

Severns, R., and G. Bloom. *Modern DC-to-DC Switchmode Power Converter Circuits*. Van Nostrand Reinhold, 1985.

Sum, Kit. *Switch Mode Power Conversion: Basic Theory and Design*. Marcel Dekker, 1984.

dc-ac Inverters and ac-ac Converters

Bedforf, B., and R. Hoft. *Principles of Inverter Circuits*. John Wiley, 1964.

Griffith, D. *Uninterruptible Power Supplies*. Marcel Dekker, 1993.

Gyugyi, L., and B. R. Pelly. *Static Power Frequency Changers*. John Wiley, 1976.

McMurray, W. *The Theory and Design of Cycloconverters*. MIT Press, 1972.

Pelly, B. R. *Thyristor Phase-Controlled Converters and Cycloconverters*. John Wiley, 1971.

Rombaut, C., Guy Seguier, and R. Bausiere. *Power Electronics Converters: AC/AC Conversion*. McGraw-Hill, 1987.

Machines and Drives

Anderson, Leonard. *Electric Machines and Transformers*. Reston, 1981.

Barton, T. H. *Rectifiers, Cycloconverters, and AC Controllers*. Clarendon Press, 1994.

Bose, B. K. *Microcomputer Control of Powerful Electronics and Drives*. IEEE Press, 1987.

Bose, B. K. *Power Electronics and AC Drives*. Prentice Hall, 1986.

Chapman, Stephen J. *Electric Machinery Fundamentals*, 3rd ed. McGraw-Hill, 1999.

Dewan, S. B., G. R. Slemon, and A. Straughen. *Power Semiconductor Drives*. John Wiley, 1984.

Dubey, G. *Power Semiconductor Controlled Drives*. Prentice Hall, 1989.

Fitzgerald, A. E., C. Kingsley, and S. Umans. *Electric Machinery*, 5th ed. McGraw-Hill, 1990.

Leonard, W. *Control of Electric Drives*. Springer-Verlag, 1985.

Mohan, Ned. *Electric Drives: An Integrative Approach*. MNPERE, 2000.

Sarma, M. *Electric Machines*, 2nd ed. West Publishing, 1994.

Sen, P. C. *Principles of Electric Machines and Power Electronics*, 2nd ed. John Wiley, 1997.

Slemon. G. *Electric Machines and Drives*. Addison-Wesley, 1992.

Vas, P. *Vector Controlled AC Machines*. Clarendon Press, 1990.

Wildi, T. *Electric Machines, Drives, and Power Systems*. John Wiley, 1989.

Magnetics

Cheng, David. *Field and Wave Electromagnetics*, 2nd ed. Addison-Wesley, 1989.

Chi-Sha, Liang, and Jin Au Kong. *Applied Electromagnetism*. PWS Engineering, 1987.

Kraus, John. *Electromagnetics*. McGraw-Hill, 1992.

Lowdon, Eric. *Practical Transformer Design Handbook*. TAB Professional and Reference Books, 1989.

McLyman, W. T. *Transformer and Inductor Design Handbook*, 2nd ed. Marcel Dekker, 1993.

MIT Staff. *Magnetic Circuits and Transformers*. MIT Press, 1965.

Ozenbaugh, R. *EMI Filter Design*. Dekker, 1996.

Schwarz, Steven E. *Electromagnetics for Engineers*. Saunders College Publishing, HRW, 1990.

Tihanyi, L. *Electromagnetic Compatibility in Power Electronics*. IEEE Press, 1995.

Power Devices

Baliga, B. J. *Modern Power Devices*. John Wiley & Sons, 1997.

Baliga, B. J., and D. Chen. *Power Transistors: Devices, Design, and Applications*. IEEE Press, 1984.

Blicher, B. *Field Effect and Bipolar Power Transistor Physics*. Academic Press, 1981.

Cobbold, Richard. *Theory and Applications of Field Effect Transistor*. John Wiley, 1970.

Dubey, G. K., S. R. Doradla, A. Joshi, and R. M. K. Sinha. *Thyristorised Power Controllers*. John Wiley, 1986.

Ghandhi, S. *Semiconductor Power Devices*. John Wiley, 1977.

Grant, D. A., and J. Gowar. *Power MOSFETS: Theory and Applications*. John Wiley, 1989.

Jaecklin, A. *Power Semiconductors Devices and Circuits*. Plenum Press, 1992.

Oxner, E. *Power FETs and Their Applications*. Prentice-Hall, 1982.

Streetman, B. *Solid-State Electrical Devices*, 3rd ed. Prentice Hall, 1990.

Tarter, R. E. *Solid-State Conversion Handbook*. John Wiley, 1993.

Taylor, B. E. *Power MOSFET Design*. John Wiley, 1993.

Warner, R. M., and B. L. Grung. *MOSFET: Theory and Design*. Oxford, 1999.

PSPICE Modeling

Basso, C. *Switch-Mode Power Supply Spice Cookbook*. McGraw-Hill, 2001.

Connelly, J., and P. Choi. *Macromodeling with SPICE*. Prentice Hall, 1992.

Gottling, James G. *Hands on PSPICE*. Houghton Mifflin, 1995.

Massobrio, G., and P. Antognetti. *Semiconductor Device Modeling with PSPICE*. McGraw-Hill, 1993.

Ramshaw, R., and D. Schuurman. *PSPICE Simulation of Power Electronics Circuits*. Chapman & Hall, 1997.

Rashid, M. *SPICE for Power Electronics and Electric Power*. Prentice Hall, 1993.

Roberts, G., and A. Sedra. *SPICE*, 2nd ed. Oxford University Press, 1997.

Tuinenga, P. *SPICE: A Guide to Circuit Simulation and Analysis Using PSPICE*, 3rd ed. Prentice Hall, 1995.

Other Related Textbooks

Bose, B. K. *Modern Power Electronics: Evaluation, Technology, and Applications*. IEEE Press, 1992.

Datta, S. *Power Electronics and Controls*. Reston, 1985.

Heydt, G. *Electric Power Quality*. Stars in a Circle Publications, 1991.

Kazimierczuk, M., and D. Czarkowski. *Resonant Power Converters*. John Wiley, 1995.

Kislovski, R. Redl, and N. Sokal, *Dynamic Analysis of Switching-Mode DC/DC Converters*. Van Nostrand Reinhold, 1991.

Tarter, R. E. *Solid State Power Conversion Handbook*. John Wiley & Sons, 1993.

PAPER REFERENCES

Overview and Applications of Power Electronics

Bose, B. K. "Power Electronics—A Technology Review." IEEE Proceedings, pp. 1303–1334, August 1992.

Bose, B. K. "Power Electronics and Motion Control—Technology Status and Recent Trends." IEEE-Power Electronics Specialists Conference (PESC '92), pp. 3–10, June 1992.

Bose, B. K. "Recent Advances in Power Electronics." IEEE-IECON '90 Conference Records, pp. 829–838, 1990.

Ehsani, M., and K. R. Ramani. "Recent Advances in Power Electronics and Applications." IEEE-SouthCon '94, pp. 8–13, March 1994.

Massie, Lowell D. "Future Trends in Space Power Technology." *IEEE Aerospace and Electronics Systems*, Vol. 6, No. 11, pp. 8–13, November 1991.

McMurry, William. "Power Electronics in the 1990's." IEEE-Applied Power Electronics Conference (APEC '90), pp. 839–843, March 1993.

Mohan, Ned. "Power Electronic Circuits: An Overview." IECON Vol. 3 '88, pp. 522–527, 1988.

Quigley, Richard. "More Electric Aircraft." IEEE-APEC '93, pp. 906–911, March 1993.

Wacks, Kenneth. "The Impact of Home Automation on Power Electronics." IEEE-APEC '93, pp. 3–9, March 1993.

Weinberg, A., and J. Schreuders. "A High Power High Voltage DC/DC Converter for Space Applications." IEEE-PESC, June 1985.

Wyk, J. D. van, H. Skudelny, and A. Muller-Hellmann. "Power Electronics: Control of the Electro-Mechanical Energy Conversion Process and Some Applications." *IEE Proceedings*, Vol. 133, Pt. B, No. 6, pp. 369–399, November 1986.

Semiconductor Devices

Adler, M. S., K. Owyang, B. J. Baliga, and R. Kokosa. "The Evaluation of Power Device Technology." *IEEE Transactions on Electronic Devices*, Vol. ED-31, No. 11, pp. 1570–1591, November 1984.

Adler, M. S., S. W. Westbrook, and A. J. Yerman. "Power Semiconductor Devices—An Assessment." IEEE Industry Applications Society Conference, pp. 723–728, 1980.

Baliga, B. J., M. S. Adler, R. P. Love, P. V. Gray, and N. D. Zammer. "The Insulated Gate Transistor—A New Three Terminal MOS-Controlled Bipolar Power Device." *IEEE Transactions on Electron Devices*, Vol. 31, No. 6, pp. 821–828, June 1984.

Blackburn, D. L. "Status and Trends in Power Semiconductor Devices." *5th European Conference on Power Electronics and Applications*, Vol. 2, pp. 619–625, 1993.

Lorenz, P. Marz, and Amann. "Rugged Power MOSFET—A Milestone on the Road to a Simplified Circuit Engineering." SIEMENS application notes on S-FET application, 2000.

Ohno, E. "The Semiconductor Evolution in Japan—A Four Decade Long Maturity Thriving to an Indispensable Social Standing." *International Power Electronics Conference*, Vol. 1, pp. 1–10, 1990.

Pelly, B. "Power Semiconductor Devices: A Status Review." IEEE International Power Semiconductor Converter Conference, 1982.

Qian, J., A. Aslam, and I. Batarseh. "Turn-off Switching Loss Model and Analysis of IGBT Under Different Switching Operation Modes." IECON '95, Vol. 1, pp. 240–245, November 1995.

Sul, S., F. Profumo, G. Cho, and T. Lipo. "MCTs and IGBTs: A Comparison of Performance in Power Electronics Circuits." IEEE Power Electronics Specialists Conference Records, pp. 163–169, June 1989.

Temple, V. "MOS-Controlled Thyristors: A New Class of Power Devices." *IEEE Transactions on Electronic Devices*, Vol. ED-33, No. 10, pp. 1609–1618, October 1986.

Temple, V., S. Arthur, D. Watrous, R. De Doncker, and H. Metha. "Megawatt MOS Controlled Thyristor for High Voltage Power Circuits." IEEE Power Electronics Specialists Conference Records, pp. 1018–1025, June 1992.

Yue, Y., J. J. Liou, and I. Batarseh. "A Steady-State and Transient IGBT Model Valid for all Free-Carrier Injunction Conditions." 12th IEEE Applied Power Electronics Conference, Atlanta, GA, pp. 168–174, February 1997.

Switch-Mode PWM dc-dc Converters

Brakus, B. "100 Amp Switched Mode Charging Rectifier for Three-Phase Mains." *IEEE/INTELEC Proceedings*, pp. 72–78, 1984.

Ćuk, S., and R. D. Middlebrook. "A General Unified Approach to Modeling Switching DC-to-DC Converters in Discontinuous Conduction Mode." IEEE Power Electronics Specialists Conference, pp. 36–57, 1977.

Chen, Q., F. C. Lee, and M. M. Jovanovic. "DC Analysis and Design of Multiple-Output Forward Converters with Weighted Voltage-Mode Control." IEEE Applied Power Electronics Conference, pp. 449–455, March 1993.

Erickson, R. W. "Synthesis of Switched-Mode Converters." IEEE Power Electronics Specialists Conference, pp. 9–22, June 1983.

Makowski, M. S. "On Topological Assumptions on PWM Converters—A Reexamination." IEEE Power Electronics Specialists Conference, pp. 141–147, June 1993.

Maksimović, D., and S. Ćuk. "General Properties and Synthesis of PWM DC-DC Converters." IEEE Power Electronics Specialists Conference, pp. 515–525, June 1989.

Maksimović, D., and S. Ćuk. "A Unified Analysis of PWM Converters with Discontinuous Modes." *IEEE Transactions on Power Electronics*, Vol. 6, No. 3, pp. 476–490, July 1991.

Middlebrook, R. D., and S. Ćuk. *Advances in Switched-Mode Power Conversion*, Vols. I and II. TESLAco, 1981.

Middlebrook, R. D and S. Ćuk "A New Optimum Topology Switching DC-to-DC Converter." IEEE Power Electronics Specialists Conference (PESC), 1977.

Moore, E. T., and T. G. Wilson. "Basic Considerations for DC-DC Conversion Networks." *IEEE Transactions on Magnetics*, p. 620, September 1966.

Mweene, L., C. Wright, and M. Schlecht. "A 1 kW, 500 kHz Front-End Converter for a Distributed

Power Supply System." IEEE Applied Power Electronics Conference, pp. 423–432, 1989.

Wester, G. W., and R. D. Middlebrook. "Low-Frequency Characterization of DC-DC converters." *IEEE Transactions on Aerospace Electronic Systems*, Vol. AES-9, No. 3, pp. 376–385, 1973.

Isolated dc-dc Converters

Maksimović, D., and S. Ćuk. "Switching Converters with Wide DC Conversion Range." *IEEE Transactions on Power Electronics*, Vol. 6, No. 1, pp. 151–157, January 1991.

Massey, R. P., and E. C. Snyder. "High-Voltage Single-Ended DC-DC Converter." IEEE Power Electronics Specialists Conference (PESC '77), pp. 156–159, June 1977.

Matsuo, H. "Comparison of Multiple-Output DC-DC Converters Using Cross Regulation." IEEE Power Electronics Specialists Conference, pp. 169–185, June 1979.

Middlebrook, R. D., and S. Ćuk. "Isolation and Multiple Outputs of a New Optimum Topology Switching DC-to-DC Converter." IEEE Power Electronics Specialists Conference, June 13–15, pp. 256–264, 1978.

Redl, R., M. Domb, and N. Sokal. "How to Predict and Limit Volt-Second Unbalance in Voltage-Fed Push-Pull Power Converters." *PCI Proceedings*, pp. 314–330, 1983.

Redl, R., and N. Sokal. "Push-Pull Current-Fed Multiple-Output DC-DC Power Converter with Only One Inductor and with 0–100% Switch Duty Ratio." IEEE Power Electronics Specialists Conference (PESC '82), pp. 341–345, June 1982.

Thottuvelil, V. J., T. G. Wilson, and H. A. Owen. "Analysis and Design of a Push-Pull Current-Fed Converter." IEEE Power Electronics Specialists Conference, pp. 192–203, June 1981.

Varga, C. "Designing Low-Power Off-Line Flyback Converters Using the Si9120 Switch-Mode Converter IC." Application Note AN90-2, Siliconix Inc., 1990.

Soft-Switching and Resonant Converters

General Papers on Resonant Converters

Divan, D. M. "The Resonant DC Link Converter—A New Concept in Static Power Conversion." 1986 IEEE-IAS Annual Meeting Record, pp. 648–656, 1986.

Freeland, S. "I. A Unified Analysis of Converters with Resonant Switches. II. Input-Current Shaping for Single-Phase AC-DC Power Converters," Ph.D. thesis, California Institute of Technology, 1988.

Freeland, S. and R. D. Middlebrook. "A Unified Analysis of Converters with Resonant Switches." IEEE Power Electronics Specialists Conference, pp. 20–30, 1986.

Johnson, S. D., A. F. Witulski, and R. W. Erickson. "A Comparison of Resonant Topologies in High Voltage Applications." *IEEE Transactions on Aerospace and Electronic Systems*, Vol. 24, No. 3, pp. 263–274, July 1988.

Kassakian, John G., and Martin F. Schlecht. "High Frequency High-Density Converters for Distributed Power Supply Systems." *IEEE Proceedings*, Vol. 76, No. 4, April 1988.

Liu, K., and F. C. Lee. "Resonant Switches—A Unified Approach to Improve Performances of Switching Converters." IEEE INTELEC Conference Records, pp. 344–351, 1984.

Liu, K., R. Oruganti, and F. C. Lee. "Resonant Switches—Topologies and Characteristics." IEEE Power Electronics Specialists Conference, pp. 106–116, January 1987.

Liu, K., R. Oruganti, and F. C. Lee. "Resonant Switches: Topologies and Characteristics." IEEE Power Electronics Specialists Conference, pp. 106–116, 1985.

Ngo, K. "Generalization of Resonant Switches and Quasi-Resonant DC-DC Converters," IEEE Power Electronics Specialists Conference, pp. 395–403, 1987.

Owen, E. L. "Origins of the Inverter." *IEEE Industry Applications*, Vol. 2, p. 64, January 1996.

Patterson, O. D., and D. M. Divan. "Pseudo-Resonant Full-Bridge DC-DC Converter." IEEE Power Electronics Specialists Conference, pp. 424–430, 1987.

Steigerwald, R. L. "A Comparison of Half-Bridge Resonant Converter Topologies." *IEEE Transactions on Industrial Electronics*, Vol. IE-31, No. 2, pp. 181-191, May 1984.

Quasi-Resonant Converters

Erickson, R., A. Hernandez, A. Witulski, and R. Xu. "A Nonlinear Resonant Switch." *IEEE Transactions on Power Electronics*, Vol. 4, No. 2, pp. 242–252, April 1989.

Maksimović, D., and S. Ćuk. "A General Approach to Synthesis and Analysis of Quasi-Resonant Converters." *IEEE Transactions on Power Electronics*, Vol. 6, No. 1, pp. 127–140, January 1991.

Ngo, K. D. T. "Generalization of Resonant Switches and Quasi-Resonant DC-DC Convert-

ers." IEEE Power Electronics Specialists Conference, pp. 395–403, 1987.

Schlecht, M. F., and L. F. Casey. "Comparison of the Square-Wave and Quasi-Resonant Topologies." IEEE Applied Power Electronics Conference, pp. 124–134, 1987.

Vorperian, V., R. Tymerski, and F. C. Lee. "Equivalent Circuit Models for Resonant and PWM Switches." *IEEE Transactions on Power Electronics*, Vol. 4, No. 2, pp. 205–214, April 1989.

Zero-Current Switching

Hopkins, D., M. Joranovic, F. Lee, and W. Stephenson. "Hybridized Off-line 2-MHz Zero-Current-Switched Quasi-Resonant Converter."

IEEE Transactions on Power Electronics, Vol. 4, No. 1, pp. 147–154, January 1989.

Vinciarelli P. "Forward Converter Switching at Zero Current." U.S. Patent 4,415,959, November 1983.

Zero-Voltage Switching

Cho, J. G., J. A. Sabate, and F. C. Lee. "Novel Full Bridge Zero-Voltage-Transition PWM DC/DC Converter for High Power Applications." IEEE Applied Electronics Conference, pp. 143–149, 1994.

Farrington, R., M. Jovanovic, and F. C. Lee. "Constant-Frequency Zero-Voltage-Switched Multi-Resonant Converters: Analysis, Design, and Experimental Results." IEEE Power Electronics Specialists Conference, pp. 197–205, 1990.

Hayes, J. G., N. Mohan, and C. P. Henze. "Zero-Voltage Switching in a Digitally Controlled Resonant DC-DC Power Converter." *IEEE Applied Power Electronics Conference Proceedings*, pp. 360–367, 1988.

Henze, C. P., H. C. Martin, and D. W. Parsley. "Zero-Voltage Switching in High Frequency Power Converters Using Pulse Width Modulation." IEEE Applied Power Electronics Conference, 1988.

Liu, K., and F. C. Lee. "Zero Voltage Switching Technique in DC-DC Converters." IEEE Power Electronics Specialists Conference, pp. 58–70, 1986.

Qian, J., I. Batarseh, and M. Ehsani. "Analysis and Design of a Zero-Voltage-Switching Isolated Boost Converter." High Frequency Power Conversion Conference (HFPC '95), San Jose, CA, pp. 1201–1207, May 6–12, 1995.

Qian, J., I. Batarseh, K. Siri, and M. Ehsani. "A Novel Zero-Voltage-Switching Boost Converter Using a Nonlinear Magnetizing Inductor of the Transformer." IEEE Applied Power Electronics Conference (APEC '95), pp. 490–495, March 1995.

Redl, R., L. Belogh, and D. Edwards. "Optimum ZVS Full-Bridge DC/DC Converter with PWM Phase-Shift Control: Analysis, Design Considerations, and Experimental Results." IEEE Applied Power Electronics Conference, pp. 159–165, 1994.

Quasi-Square PWM Converters

Barbi, I., D. Martins, and R. Prado. "Effects of Nonlinear Resonant Inductor on the Behavior of Zero-Voltage Switching Quasi-Resonant Converters." IEEE Power Electronics Specialists Conference, pp. 522–527, 1990.

Jang, Y. "New Quasi–Square Wave and Multi-Resonant Integrated Magnetic Zero-Voltage Switching Converters." IEEE Power Electronics Specialists Conference, pp. 721–727, 1993.

Maksimović, D. "Design of the Zero-Voltage-Switching Quasi-Square-Wave Resonant Switch." IEEE Power Electronics Specialists Conference, pp. 323–329, 1993.

Vorperian, Vatche. "Quasi-Square-Wave Converters: Topologies and Analysis." *IEEE Transactions on Power Electronics*, Vol. PE-3, No. 2, pp. 183–191, April 1988.

Clamped-Mode Resonant Converters

Jones, D. V. "A New Resonant-Converter Topology," *High Frequency Power Conversion Conference Proceedings*, pp. 48–52, 1987.

Soksatra, Somboon. "Generalization of Resonant Converters." Ph.D. thesis, University of Illinois at Chicago, June 1991.

Tsai, F. S., P. Materu, and F. C. Lee. "Constant Frequency, Clamped Mode Resonant Converters." IEEE Power Electronics Specialists Conference, pp. 557–566, June 1987.

Series Resonant Converters

King, R., and T. A. Stuart. "Inherent Overload Protection for the Series Resonant Converter." *IEEE Transactions on Aerospace and Electronic Systems*, Vol. 19, No. 6, pp. 820–830, November 1983.

King, R., and T. Stuart. "A Normalized Model for the Half Bridge Series Resonant Converter." *IEEE Transactions on Aerospace and Electronic Systems*, pp. 180–193, 1981.

Lee, C. Q., and K. Siri. "Analysis and Design of Series Resonant Converter by State Plane Diagram." *IEEE Transactions on Aerospace and Electronic Systems*, Vol. 22, No. 6, pp. 757–763, November 1986.

Murai, Y., and T. A. Lipo. "High Frequency Series Resonant DC Link Power Conversion." IEEE Industry Applications Society Annual Meeting, pp. 648–656, 1988.

Ngo, K. D. T. "Analysis of a Series Resonant Converter Pulse-Width-Modulated Current-Controlled for Low Switching Loss." IEEE Power Electronics Specialists Conference, pp. 527–536, June 1987.

Oruganti, R. "State-Plane Analysis of Resonant Converters." Ph.D. dissertation, Virginia Polytechnic Institute, 1987.

Oruganti, R., and F. C. Lee. "Resonant Power Processors, Part 1: State Plane Analysis." *IEEE Transactions on Industry Applications*, Vol. 21, pp. 1453–1460, November–December 1985.

Oruganti, R., and F. C. Lee. "Resonant Power Processors: Part 1–State Plane Analysis." IEEE-IAS Annual Meeting Conference Records, pp. 860–867, 1984.

Trabert, S., and R. Erickson. "Steady-State Analysis of the Duty Cycle Controlled Series Resonant Converter." IEEE Power Electronics Specialists Conference, pp. 545–556, 1987.

Vorperian, V., and S. Ćuk. "A Complete DC Analysis of the Series Resonant Converter." IEEE Power Electronics Specialists Conference, pp. 85–100, June 1982.

Witulski, A. F., and R. W. Erickson. "Design of the Series Resonant Converter for Minimum Component Stress." *IEEE Transactions on Aerospace and Electronic Systems*, Vol. 22, No. 4, pp. 356–363, July 1986.

Witulski, A., and R. Erickson. "Steady-State Analysis of the Series Resonant Converter." *IEEE Transactions on Aerospace and Electronic Systems*, Vol. 21, No. 6, pp. 791–799, November 1985.

Parallel Resonant Converters

Batarseh, I. "State-Plane Approach for the Analysis of Half-Bridge Parallel Resonant Converters." *IEE Proceedings—Circuits, Devices and Systems*, Vol. 142, No. 3, pp. 200–204, August 1995.

Batarseh, I., and C. Q. Lee. "High Frequency Link Parallel Resonant Converter." *IEE Proceedings—Electronic Circuits and Systems*, Vol. 138, No. 1, pp. 34–37, February 1991.

Batarseh, I., and C. Q. Lee. "Steady State Analysis of the Parallel Resonant Converter with LLCC-Type Commutation Network." *IEEE Transactions on Power Electronics*, Vol. 6, No. 3, pp. 525–538, July 1991.

Bhat, A., and M. Swamy. "Analysis and Design of a High-Frequency Parallel Resonant Converter Operating Above Resonance." *IEEE Transactions on Aerospace and Electronic Systems*, Vol. 25, No. 4, pp. 449–458, July 1989.

Cosby, M., and R. Nelms. "Designing a Parallel-Loaded Resonant Inverter for an Electronic Ballast Using the Fundamental Approximation." IEEE Applied Power Electronics Conference, pp. 413–423, 1993.

Johnson, S. "Steady-State Analysis and Design of the Parallel Resonant Converter." M.S. thesis, University of Colorado, Boulder, 1986.

Johnson, S., and R. Erickson. "Steady-State Analysis and Design of the Parallel Resonant Converter." *IEEE Transactions on Power Electronics*, Vol. 3, No. 4, pp. 93–104, January 1988.

Kazimierczuk, M., W. Szaraniec, and S. Wang. "Analysis and Design of Parallel Resonant Converter at High Q_L." *IEEE Transactions on Aerospace and Electronic Systems*, Vol. 28, pp. 35–50, January 1992.

Liu, R., I. Batarseh, and C. Q. Lee. "Comparison of Capacitively and Inductively Coupled Parallel Resonant Converters." *IEEE Transactions on Power Electronics*, Vol. 8, No. 4, pp. 445–454, October 1993.

Oruganti, R., and F. C. Lee. "State Plane Analysis of the Parallel Resonant Converter." IEEE Power Electronics Specialists Conference, pp. 56–73, June 1985.

Siri, K., I. Batarseh, and C. Q. Lee. "Frequency Response for the Conventional Parallel Resonant Converter Based on the State-Plane Diagram." *IEEE Transactions on Circuits and Systems*, Vol. 39, CAS-I, pp. 33–42, January 1993.

Steigerwald, R. "Analysis of a Resonant Transistor DC-DC Converter with Capacitive Output Filter." *IEEE Transactions on Industrial Electronics*, Vol. IE-32, No. 4, pp. 439–444, November 1985.

Multi-Resonant Converters

Lee, F. C., W. A. Tabisz, and M. M. Jovanovic. "Recent Developments in High-Frequency Quasi-Resonant and Multi-Resonant Converter Technologies." European Power Electronics Conference Records, Aachen, Germany, 1989.

Tabisz, W. A., M. M. Jovanovic, and F. C. Lee. "High Frequency Multi-Resonant Converter Technology and Its Applications." IEE International Conference on Power Electronics and Variable Speed Drives, pp. 1–8, 1990.

Tabisz, W. A., and F. C. Lee. "Zero-Voltage-Switching Multi-Resonant Technique—A Novel Approach to Improve Performance of High-Frequency Quasi-Resonant Converters." IEEE Power Electronics Specialists Conference, pp. 9–17, 1988.

Class E Converters

Lohn, K. "On the Overall Efficiency of the Class-E Power Converter." IEEE Power Electronics Specialists Conference, pp. 351–358, 1986.

Omori, H., T. Iwai, et al. "Comparative Studies between Regenerative and Non-Regenerative Topologies of Single-Ended Resonant Inverters." High Frequency Power Conversion Conference, 1987.

Raab, F. H. "Idealized Operation of Class-E Tuned Power Amplifier." *IEEE Transactions on Circuits and Systems*, Vol. 24, No. 12, pp. 725–735, December 1977.

Redl, R., B. Molnar, and N. Sokal. "Class E Resonant Regulated DC-DC Power Converters: Analysis of Operation and Experimental Results at 1.5 MHz." *IEEE Transactions on Power Electronics*, Vol. 1, No. 2, pp. 111–120, 1986.

Sokal, N., and A. Sokal. "Class-E, a New Class of High Efficiency Tuned Single-Ended Switching Power Amplifiers." *IEEE Journal of Solid State Circuits*, Vol. SC-10, pp. 168–176, June 1975.

Song, J., A. Greenwood, and I. Batarseh. "Analysis and Design of Zero-Voltage-Switching Class-E Converter." IEEE Southeast Conference '96, Tampa, FL, pp. 545–550, April 1996.

High-Order Resonant Converters

Batarseh, I. "Analysis and Design of High Order Parallel Resonant Converters." Ph.D. thesis, University of Illinois at Chicago, June 1990.

Batarseh, I. "Resonant Converters with Three and Four Energy Storage Elements." *IEEE Transaction on Power Electronics*, Vol. 8, No. 1, pp. 64–73, January 1994.

Batarseh, I., and C. Q. Lee. "High Frequency High Order Parallel Resonant Converter." *IEEE Transactions on Industrial Electronics*, Vol. 36, No. 4, pp. 485–495, November 1989.

Batarseh, I., and C. Q. Lee. "Steady State Analysis of the Parallel Resonant Converter with LLCC-Type Commutation Network." *IEEE Transactions on Power Electronics*, Vol. 6, No. 3, pp. 525–538, July 1991.

Batarseh, I., and R. Severns. "Resonant Converter Topologies with Three and Four Energy Storage

Elements." *High Frequency Power Conversion '92 Conference Proceedings,* pp. 374–383, May 4–8, 1992.

Batarseh, I., R. Liu, and A. Upadhyay. "Theoretical and Experimental Studies of the LCC-Type Parallel Resonant Converter." *IEEE Transactions on Power Electronics*, Vol. 5, No. 2, pp. 140–150, April 1990.

Bhat, A. K. S. "Analysis and Design of LCL-Type Series Resonant Converter," IEEE INTELEC, pp. 172–178, 1990.

Bhat, A. K. S. "Analysis and Design of a Series-Parallel Resonant Converter with Capacitive Output Filter." IEEE IAS, pp. 1308–1314, 1990.

Bhat, A. K. S. "A Unified Approach for the Steady State Analysis of Resonant Converters." *IEEE Transactions on Industrial Electronics*, pp. 251–259, August 1991.

Bhat, A. K. S., and S. B. Dewan. "Analysis and Design of a High Frequency Resonant Converter Using LCC-Type Commutation." *IEEE Conference Records*, pp. 657–663, 1986.

Bhat, S., and S. B. Dewan. "A Generalized Approach for the Steady State Analysis of Resonant Inverter." IEEE-APEC Conference Records, pp. 664–671, 1986.

Lee, C. Q., R. Liu, and I. Batarseh. "Parallel Resonant Converter with LLC-Type." *IEEE Transactions on Aerospace and Electronic Systems*, Vol. 25, No. 6, pp. 844–847, November 1989.

Liu, R., I. Batarseh, and C. Q. Lee. "The LLC-Type and the Class-E Resonant Converters." High Frequency Power Conversion Conference Records, Naples, FL, pp. 1486–1496, May 1989.

Liu, R., C. Q. Lee, and A. Upadhyay. "Experimental Study of the LLC-Type Series Resonant Converter," IEEE-APEC, pp. 31–37, 1991.

Severns, R. "Generalized Topologies for Converters with Reactive Energy Storage." IEEE-IAS Annual Conference Records, pp. 1147–1151, October 1989.

Severns, R. "Topologies for Three Element Resonant Converters." IEEE-APEC '90, Los Angeles, CA, pp. 712–722, March 12–16, 1990.

Steigerwald, R. "A Comparison of Half-Bridge Resonant Converter Topologies." IEEE PESC Conference, pp. 135–144, 1987.

Steigerwald, R. L. "High Frequency Resonant Transistor DC-DC Converters." *IEEE Transactions on Industrial Electronics*, Vol. 31, No. 2, pp. 181–191, May 1984.

PWM Inverters and Harmonics Cancellation

Boost, M., and P. D. Ziogas. "State-of-the-Art PWM Techniques: A Critical Evaluation." IEEE Power Electronics Specialists Conference, pp. 425–433, 1986.

Broek, H. W. van der, H. C. Skudelny, and G. V. Stanke. "Analysis and Realization of a Pulse Width Modulator Based on Voltage Space Vectors." *IEEE Transactions on Industry Applications*, Vol. 24, No. 1, pp. 142–150, January 1988.

Divan, D. M., T. A. Lipo, and T. G. Habetler. "PWM Techniques for Voltage Source Inverters." IEEE Applied Power Electronics Conference, 1990.

Ehsani, M. "A Tutorial on Pulse Width Modulation Techniques in Inverters and Choppers." IEEE Applied Power Electronics Conference, 1981.

Enjeti, P., P. Ziogas, and J. Lindsay. "Programmed PWM Techniques to Eliminate Harmonics: A Critical Evaluation." *IEEE Transactions on Industry Applications*, Vol. 26, No. 2, pp. 302–316, 1990.

Hingorani, N. G. "Flexible ac Transmission." *IEEE Spectrum*, pp. 40–45, 1993.

Holtz, J. "Pulsewidth—A Survey." *IEEE Transactions on Industrial Electronics*, Vol. 39, No. 5, pp. 410–420, 1992.

Kato, T. "Precise PWM Waveform Analysis of Inverter for Selected Harmonic Elimination." IEEE-IAS Annual Meeting, pp. 611–616, 1986.

Kheraluwala, M. H., and D. M. Divan. "Delta Modulation Strategies for Resonant Link Inverters." *IEEE Transactions on Power Electronics,* Vol. 5, No. 2, pp. 220–228, April 1990.

Murai, Y., T. Watanabe, and H. Iwasaki. "Waveform Distortion and Correction Circuit for PWM Inverters with Switching Lag-Times." IEEE-IAS Annual Meeting, pp. 436–441, 1985.

Patel, H., and R. G. Hoft. "Generalized Techniques of Harmonic Elimination and Voltage Control in Thyristor Inverters: Part I—Harmonic Elimination." *IEEE Transactions on Industry Applications*, Vol. IA-9, No. 3, May–June 1973.

Patel, H., and R. G. Hoft. "Generalized Techniques of Harmonic Elimination and Voltage

Control in Thyristor Inverters: Part II—Voltage Control Techniques." *IEEE Transactions on Industry Applications*, Vol. IA-10, No. 5, September–October 1974.

Pitel, I. J., S. N. Talukdar, and P. Wood. "Characterization of Programmed-Waveform Pulse-width Modulation." *IEEE Transactions on Industry Applications*, Vol. IA-16, No. 5, 1980.

Plunkett, A. B. "A Current-Controlled PWM Transistor Inverter Drive." IEEE-IAS Annual Meeting, pp. 785–792, 1979.

Salazar, L., and G. Joos. "PSPICE Simulation of Three-Phase Inverters by Means of Switching Functions." *IEEE Transactions on Power Electronics*, Vol. 9, No. 1, pp. 35–42, January 1994.

Schlecht, M. F. "Novel Topological Alternatives to the Design of a Harmonic-Free Utility/DC Interface." IEEE Power Electronics Specialists Conference (PESC), pp. 206–214, 1983.

Weischedel, H. R., and G. R. Westerman. "A Symmetry Correcting Pulsewidth Modulator for Power Conditioning Applications." *IEEE Transactions on Industry Applications*, Vol. IA-9, No. 3, pp. 318–322, 1973.

Ziogas, P. D., Y. G. Kang, and V. R. Stefanović. "Rectifier-Inverter Frequency Changers with Suppressed DC Link Components." *IEEE Transactions Industry Applications*, Vol. IA-22, No. 6, pp. 1026–1036, November 1986.

Magnetics and Magnetic Components

Carsten, B. "High Frequency Conductor Losses in Switchmode Magnetics." High Frequency Power Converter Conference, pp. 155–176, 1986.

Ćuk, Slobodan, and William Polivka. "Analysis of Integrated Magnetics to Eliminate Current Ripple in Switching Converters." PCI '83 Conference Records, April 19–21, pp. 239–264, 1983.

Dauhajre, A., and R. D. Middlebrook. "Modeling and Estimation of Leakage Phenomena in Magnetic Circuits." IEEE Power Electronics Specialists Conference, pp. 213–226, 1986.

Dowell, P. L. "Effects of Eddy Currents in Transformer Windings" *IEE Proceedings*, Vol. 113, No. 8, pp. 1387–1394, 1966.

El-Hamamsy, S., and E. Chang. "Magnetics Modeling for Computer-Aided Design of Power Electronics Circuits." IEEE Power Electronics Specialists Conference, pp. 635–645, 1989.

Goldberg, A. F., and M. F. Schlecht. "The Relationship Between Size and Power Dissipation in a 1–10MHz Transformer." IEEE Power Electronics Specialists Conference, pp. 625–634, 1989.

Jiles, D. C., and D. L. Atherton. "Ferromagnetic Hysteresis." *IEEE Transactions on Magnetics*, Vol. 19, No. 5, pp. 2183–2185, September 1983.

Kelley, A. W., and W. F. Yadusky. "Phase-Controlled Rectifier Line-Current Harmonics and Power Factor as a Function of Firing Angle and Output Filter Inductance." *IEEE Applied Power Electronics Conference*, pp. 588–597, 1990.

Kelley, A., and W. F. Yadusky. "Rectifier Design for Minimum Line-Current Harmonics and Maximum Power Factor." *IEEE Transactions*

on Power Electronics, Vol. 7, No. 2, pp. 332–341, April 1992.

Ngo, K. D. T., R. P. Alley, A. J. Yerman, R. J. Charles, and M. H. Kuo. "Evaluation of Trade-offs in Transformer Design for Very-Low-Voltage Power Supply with Very High Efficiency and Power Density." IEEE Applied Power Electronics Conference, pp. 344–353, 1990.

Redl, R. "Power Electronics and Electromagnetic Compatibility." IEEE-PESC '96, pp. 15–21, 1996.

Schwarz, F. C. "An Improved Method of Resonant Current Pulse Modulation for Power Converters." IEEE Power Electronics Specialists Conference, pp. 194–204, June 1975.

Urling, A. M., V. A. Niemela, G. R. Skutt, and T. G. Wilson. "Characterizing High-Frequency Effects in Transformer Windings—A Guide to Several Significant Articles." IEEE Applied Power Electronics Conference, pp. 373–385, 1989.

Vandelac, J. P., and P. Ziogas. "A Novel Approach for Minimizing High Frequency Transformer Copper Losses." IEEE Power Electronics Specialists Conference, pp. 355–367, 1987.

Wong, R. C., H. A. Owen, and T. G. Wilson. "Parametric Study of Minimum Converter Loss in an Energy-Storage Dc-to-Dc Converter." IEEE Power Electronics Specialists Conference, pp. 411–425, 1982.

Zhang, I. "Integrated EMI/Thermal Design for Switching Power Supplies." M.S. thesis, Virginia Polytechnic and State University, 1998.

Power Electronics Education

Web-Based Teaching

Batarseh, Issa. "Web-Based Power Electronics I Course at the University of Central Florida." reach.ucf.edu/~EEL5245.

Batarseh, Issa, T. Vining, and A. Campsi. "Web-Based Delivery of the First Power Electronics Course." National Science Foundation Workshop Proceedings, University of Central Florida, Orlando, November 11–13, 2000.

Chowdhury, B. H. "Power Education at the Crossroads." *IEEE Spectrum*, pp. 64–68, October 2000.

Elbuluk, Malik, Celal Batur, and Iqbal Husain. "A Proposal for Internet-Based Certificate Program in Motion Control System." National Science Foundation Workshop Proceedings, University of Central Florida, Orlando, November 11–13, 2000.

Hietpas, Steven M. "Using Multimedia Tools for Teaching Electric Drives." National Science Foundation Workshop Proceedings, University of Central Florida, Orlando, November 11–13, 2000.

Krein, Philip. "Experience with a Full Web-Based Power Electronics Course." National Science Foundation Workshop Proceedings, University of Central Florida, Orlando, November 11–13, 2000.

Mohan, Ned, William P. Robbins, and Christopher P. Henze. "Multi-Media Delivery of Modern Power Curriculum." National Science Foundation Workshop Proceedings, University of Central Florida, Orlando, November 11–13, 2000.

Pittet, A., and V. Rajagopalan. "Web-Based Learning Module for Power Electronics: A Case Study." National Science Foundation Workshop Proceedings, University of Central Florida, Orlando, November 11–13, 2000.

Rahman, M. F., and Karanayil. "The Laboratory and Modeling Studies Associated with the Professional Elective on Power Electronics at the University of New South Wales, Australia." National Science Foundation Workshop Proceedings, University of Central Florida, Orlando, November 11–13, 2000.

Rashid, Muhammad. "A Web-Based Self-Study Course on Fundamentals of Power Electronics." National Science Foundation Workshop Proceedings, University of Central Florida, Orlando, November 11–13, 2000.

Ubell, R. "Engineers Turn to e-Learning." *IEEE Spectrum*, pp. 59–63, October 2000.

Wu, Thomas. "Web-Based EMC Course at the University of Central Florida." National Science Foundation Workshop Proceedings, University of Central Florida, Orlando, November 11–13, 2000.

Curriculum Design

Blaabjerg, F., J. Pedersen, and B. Alvsten. "Project- and Problem-Oriented Education: An Efficient Way to Reach a High Level in Education of Power Electronic Engineers." National Science Foundation Workshop Proceedings, University of Central Florida, Orlando, March 24–26, 1996.

Duk, R., N. Mohan, and W. Robbins. "A Survey of Power Electronics Education in U.S. and Canadian Universities." National Science Foundation Workshop Proceedings, University of Central Florida, Orlando, March 24–26, 1996.

Ehsani, M., and K. R. Ramani. "Recent Advances in Power Electronics and Applications." IEEE SouthCon '94, pp. 8–13, March 1994.

Habetler, T., R. Bass, H. Puttgen, and W. Sayle. "Power Electronics Education in the Ever-Expanding EE Curriculum." National Science Foundation Workshop Proceedings, University of Central Florida, Orlando, March 24–26, 1996.

Kemnitz, Debra, A. Khan, and I. Batarseh. "Power Electronics Education: Courses and Laboratory." IEEE SouthCon '95, Ft. Lauderdale, FL, pp. 240–245, March 1995.

Krein, Philip T. "Power Electronics as an Opportunity in Multidisciplinary Design Curricula." National Science Foundation Workshop Proceedings, University of Central Florida, Orlando, March 24–26, 1996.

Mohan, Ned. "Teaching of First Course on Power Electronics." NSF-Faculty Workshop Proceedings, University of Minnesota, June 1998.

O'Connell, Robert M. "Power Electronics at Missouri: An Evolutionary Program." National Science Foundation Workshop Proceedings, University of Central Florida, Orlando, March 24–26, 1996.

Robbins, Bill, Tore Undeland, and Ned Mohan. "Workshop to Promote Power Electronics Education." National Science Foundation Report, University of Minnesota, August 22–24, 1991.

Shenai, K., C. Q. Lee, and I. Batarseh. "An Integrated Power Electronics Curriculum." National Science Foundation Workshop Proceedings (see FloridaPec.engr.ucf.edu), University of Central Florida, Orlando, March 24–26, pp. 21–26, 1996.

Stankovic, Aleksandar M. "Towards an Effective and Integrated Power Electronics Curriculum." National Science Foundation Workshop Proceedings, University of Central Florida, Orlando, March 24–26, 1996.

Velez-Reyes, M., and E. Tranter. "Center for Power Electronic Systems: Education and Outreach Programs." National Science Foundation Workshop Proceedings, University of Central Florida, Orlando, November 11–13, 2000.

Power Systems and Drive Curriculum

Alvarado, F. L., and D. W. Novotny. "Power Electronics, Power Systems and Machines: A Modern Power Curriculum." Department of Electrical and Computer Engineering, University of Wisconsin-Madison.

Batarseh, I., A. Gonzalez, Z. Qu, and A. Khan. "Proposed Power Electronics Curriculum." IEEE SouthCon '96, pp. 251–262, March 1996.

Domijan, A., and R. Shoults. "Electric Power Engineering Laboratory Resources of the U.S. and Canada." *IEEE Transactions on Power Systems*, Vol. 3, No. 3, August 1988.

Elbuluk, Malik E., and Iqbal Husain. "Towards a Comprehensive Program in Power Electronics and Motor Drives." National Science Foundation Workshop Proceedings, University of Central Florida, Orlando, March 24–26, 1996.

Hess, Herbert L. "Incorporating Electronic Motor Drives into the Existing Undergraduate Electric Energy Conversion Curriculum." National Science Foundation Workshop Proceedings, University of Central Florida, Orlando, March 24–26, 1996.

Mayher, Jeffrey. "The Development of an Electric Machinery and Drives Laboratory." National Science Foundation Workshop Proceedings, University of Central Florida, Orlando, March 24–26, 1996.

Venkata, Mani, and Chen-Ching Liu. "Role of Power Electronics in Power Engineering." National Science Foundation Workshop Proceedings, University of Central Florida, Orlando, March 24–26, 1996.

Computer Simulation

Bass, Richard M. "Simulation Laboratory for Teaching Switching Power Supplies." National Science Foundation Workshop Proceedings, University of Central Florida, Orlando, March 24–26, 1996.

Ben-Yaakov, S. "Average Simulation of PWM Converters by Direct Implementation of Behavioral Relationships." *International Journal of Electronics*, Vol. 77, pp. 731–746, 1994.

Ben-Yaakov, S., and D. Adar (Edry). "Average Models as Tools for Studying the Dynamics of Switch Mode DC-DC Converters." IEEE Power Electronics Specialists Conference (PESC '94), Taipei, Taiwan, pp. 1218–1225, 1994.

Hart, D. W. "Circuit Simulation as an Aid in Teaching the Principles of Power Electronics." IEEE *Transactions on Education*, Vol. 6, No. 1, pp. 1–10, February 1993.

Hart, Daniel W. "Using PSPICE for Undergraduate Design Projects in Power Electronics." ASEE Annual Conference Proceedings, pp. 1717–1720, June 1991.

Lauritzen, Peter. "Integration of Laboratories and Computer Simulation in an Undergraduate Power Electronics Design Course." National Science Foundation Workshop Proceedings, University of Central Florida, Orlando, March 24–26, 1996.

"PSPICE Models of Power Electronic Circuits." University of Minnesota, 1986.

Zhu, G., and I. Batarseh. "Correcting the PSPICE Large-Signal Model for PWM Converters Operating in DCM." *IEEE Transactions on Aerospace and Electronic Systems*, Vol. 36, No. 2, pp. 718–721, April 2000.

Laboratory Design

Batarseh, I. "Course and Laboratory Instructions in Power Electronics." IEEE Power Electronics Specialists Conference (PESC '94), Vol. 2, Taipei, Taiwan, pp. 1359–1368, June 1994.

Bonert, R. "A Flexible Modular Hardware for Power Electronic Laboratories and Projects." National Science Foundation Workshop Proceedings, University of Central Florida, Orlando, March 24–26, 1996.

Krein, Philip. "A Broad-Based Laboratory for Power Electronics and Electric Machines." IEEE-PESC '93 Conference Record, June 20–24, Seattle, pp. 959–964, 1993.

Krein, Philip. "Summary Guide to the Power Electronics Laboratory at the University of Illinois." National Science Foundation Workshop Proceedings, University of Central Florida, Orlando, March 24–26, 1996.

Krein, P., and P. Sauer. "An Integrated Laboratory for Electric Machines, Power Systems and Power Electronics." *IEEE Transactions on Power Systems*, Vol. 7, pp. 1060–1066, August 1992.

Sloane, Thomas. "Laboratories for an Undergraduate Course in Power Electronics." IEEE-PESC '93 Conference Record, June 20–24, Seattle, pp. 965-971, 1993.

Torrey, David A. "Considerations in the Development of Power Electronics Teaching Laboratories." National Science Foundation Workshop Proceedings, University of Central Florida, Orlando, March 24–26, 1996.

Torrey, David. "A Project Oriented Power Electronics Laboratory." IEEE-PESC '93 Conference Record, June 20–24, Seattle, WA, pp. 972–978, 1993. (Also see *IEEE Transactions on Power Electronics*, Vol. 9, pp. 250–255, 1944.)

Index